S0-ASN-384

BASIC ELECTRONICS FOR SCIENTISTS

BASIC ELECTRONICS FOR SCIENTISTS

Second Edition

James J. Brophy
Illinois Institute of Technology

McGraw-Hill Book Company

New York St. Louis San Francisco Düsseldorf Johannesburg
Kuala Lumpur London Mexico Montreal New Delhi Panama
Rio de Janeiro Singapore Sydney Toronto

BASIC ELECTRONICS FOR SCIENTISTS

Library of Congress Catalog Card Number 72-167491

07-008129-8

1 2 3 4 5 6 7 8 9 0 K P K P 7 9 8 7 6 5 4 3 2 1

This book was set in Baskerville by Black Dot, Inc., and printed and bound by Kingsport Press, Inc. The designer was Richard Paul Kluga; the drawings were done by John Cordes, J. & R. Technical Services, Inc. The editors were Bradford Bayne and David Damstra. Matt Martino supervised production.

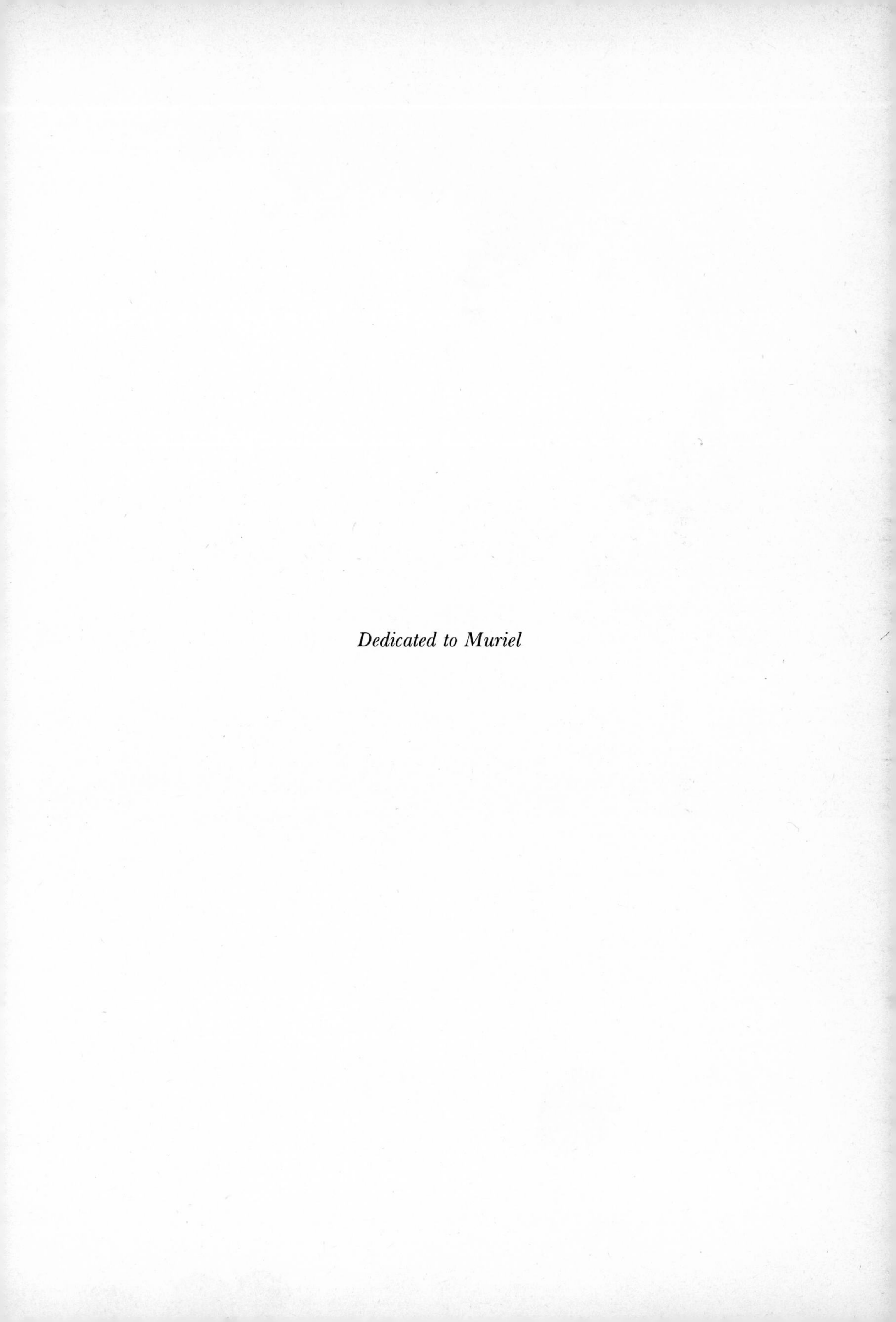

Dedicated to Muriel

CONTENTS

PREFACE

Electronic measurement and control pervade all corners of science and engineering. The great power and versatility of electronic devices and, consequently, their widespread application make it imperative that science and engineering students obtain a working familiarity with electronics. Yet this familiarity need not be as intensive as that achieved in the training of electronic engineers.

This text is written to provide the undergraduate science or engineering student with a basic understanding of electronic devices and circuits. This understanding should be sufficient to appreciate the operation and characteristics of the many electronic instruments he will use in his professional career. The analysis of circuits, rather than their design, is emphasized since such a student will design only the simplest of circuits. On the other hand, he must appreciate the operation of quite complicated instruments.

In order to display adequately the physical basis of electronic circuits, it is necessary in several instances to employ a few simple differential equations. The solutions are always immediately given in the text, however, so that a knowledge of differential equations is not necessary. The author feels that this approach results in a more satisfying exposition for the scientifically oriented student than, for example, simply assuming the validity of Ohm's law. Complex-number representation necessary in ac-circuit theory is introduced by sample manipulations of addition and multiplication. It is assumed that the student possesses a general acquaintance with electricity and properties of materials as covered in beginning physics courses. Circuit theory is treated by starting with direct currents, however, so that parts of the early chapters may be used as refresher material.

This text is written from the point of view of an experimentalist and concurrent laboratory experience is strongly recommended. Since, however, curriculum pressures often obviate laboratory class hours, an attempt has been made to provide an acceptable substitute. Two laboratory exercises per chapter are included to acquaint students with quantitative aspects of laboratory work. These laboratory exercises can be carried out entirely without actual measurements, but the approach is as though the student were engaged in a laboratory experiment. This effect may be enhanced by simple Fortran programs which simulate laboratory data collection, but this addition is not mandatory. If a laboratory course is included, the exercises are also quite suitable for actual experimental investigations.

This second edition has been modified in several areas, while retaining the emphasis and thrust of the original text. The treatment of integrated circuits and of field-effect transistors is greatly enlarged at the expense of vacuum tube circuits. The operation of active devices is described in terms of current-voltage characteristics, and the physical explanations for current-voltage characteristics of all devices are collected into a single chapter. Thus it is possible for students to develop an appreciation for the physics of active devices or, alternatively, to cope successfully with the application of nonlinear devices in circuits on the basis of empirical current-voltage characteristics alone. Increased emphasis is also placed on digital electronics in view of the ever-expanding scope of digital instrumentation and control.

For many years I have felt that a working familiarity with electronics contributes immeasurably to a professional scientific or engineering career. If this text makes it possible for others to attain such familiarity, I will be well satisfied. *I am deeply indebted to many colleagues who, through their publications, have provided much background material for this book. I must also express my sincere thanks to those who kindly read and criticized the manuscript, most particularly to Richard J. Higgins for his many valuable suggestions, and to James E. Vandendorpe for providing Fortran programs for the laboratory exercises. Lastly, the constant inspiration by Muriel, my wife, has made the completion of this second edition possible.*

James J. Brophy

BASIC ELECTRONICS FOR SCIENTISTS

chapter one

DIRECT CURRENT CIRCUITS

The operation of any electronic device, be it as complicated as a television receiver or as simple as a flashlight, can be understood by determining the magnitude and direction of electric currents in all parts of its functional unit, the circuit. In fact, it is not possible to appreciate how any given circuit functions without a detailed knowledge of the currents in its individual components.

Even the most complicated circuits can be examined in easy stages by first considering each part separately and subsequently noting how the various subcircuits fit together. Therefore, circuit analysis should start by treating elementary configurations under the simplest possible conditions. Circuits in which the currents are steady and do not vary with time are called direct current circuits. Such dc circuits, which are considered in this chapter, are important and relatively simple to understand.

INTRODUCTORY CONCEPTS

Potential Difference One of the fundamental properties of electrons is their electric charge, which has the magnitude 1.6029×10^{-19} coulomb (C). If two electric charges (such as two electrons) are near one another, they exert a force on each other given by *Coulomb's law*

$$F = \frac{1}{4\pi\epsilon_0}\frac{q_1 q_2}{r^2} \tag{1-1}$$

where F is the force in newtons, q_1 and q_2 are the electric charges, r is the distance in meters between the charges, and $\epsilon_0 = 8.85 \times 10^{-12}\ \mathrm{C^2/N \cdot m^2}$ is a constant which depends upon the system of units used to measure the force, charge, and distance. The particular system of units chosen in Eq. (1-1) is called the *rationalized meter-kilogram-second system* (*mks* system). It is selected here because the conventional electrical quantities important in circuit analysis are part of this system.

The experimental fact that an electric charge experiences a force due to another charge located some distance away may be usefully represented by saying that the first charge sets up an *electric field* in the surrounding space and that the field produces the observed force on the second charge. Thus, we may say that the force on q_2 corresponding to Eq. (1-1) is caused by an electric field $\mathcal{E}$ at the position of q_2 such that

$$F = q_2\mathcal{E} \tag{1-2}$$

This relation defines an electric field as the force per unit charge

$$\mathcal{E} = \frac{F}{q_2} \tag{1-3}$$

and the units of electric field are clearly newtons per coulomb. The direction of the electric field at any point is simply the direction of the force on a positive charge at that point.

The idea of an electric field removes the conceptual difficulty of how two objects can "push" each other even though they are separated by some distance. It also makes it possible to determine the force on an electric charge using Eq. (1-2) in situations where the electric field in a region is known, whereas the geometric arrangement of charges generating the field is not. This situation is very common in electric-circuit analysis.

Suppose an electric charge q is placed in an electric field $\mathcal{E}$. According to Eq. (1-2) a force $q\mathcal{E}$ is exerted on the charge by the field. This means that the charge has potential energy by virtue of being located in the field. The change in potential energy of the charge in moving from point a to point b in the electric field is simply the mechanical work required to move it from a to b against the force due to the field. If there is no friction or acceleration, the electric force $q\mathcal{E}$ must be balanced by an externally

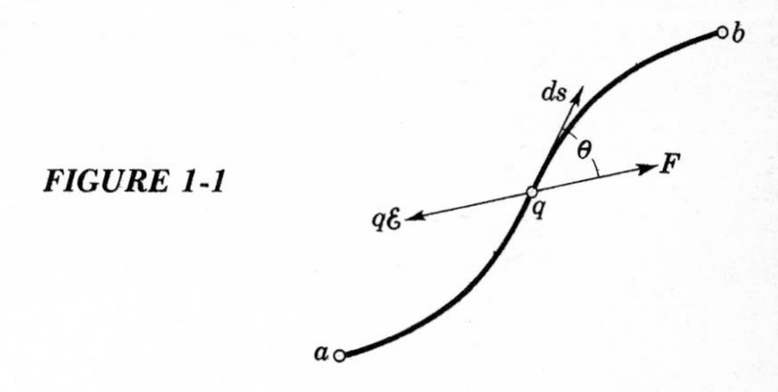

FIGURE 1-1

applied force F at all points along the path, as in Fig. 1-1. Then, from the usual definition of work

$$W = \int_a^b F \cos \theta \, ds \tag{1-4}$$

where θ is the angle between the applied force F and the direction of motion ds. Using Eq. (1-2)

$$W = -q \int_a^b \mathcal{E} \cos \theta \, ds \tag{1-5}$$

The integral in Eq. (1-5) is defined as the electric *potential difference* V between a and b, and is given by

$$V = \frac{W}{q} \tag{1-6}$$

Notice that the units of potential difference are work per unit charge, newton-meters per coulomb or joules per coulomb. Because potential difference is used so frequently in electric circuits, the unit joules per coulomb has been named a *volt* (abbreviated as V), in honor of the early worker in electricity, Alessandro Volta.

Note that V is the potential difference between the two points a and b and that it is incorrect to speak of the potential of one point without reference to another point. The point b is said to be at a higher potential than point a if work is done against electric forces when a positive charge is moved from a to b. In circuit analysis it is common to consider the potential of several points all with reference to one given place, usually taken to be at zero potential and called the *ground.* In this case it is satisfactory to speak of the potential at a point, so long as the implied reference point is clearly understood.

According to Eq. (1-5), the potential difference V is

$$V = -\int_a^b \mathcal{E} \cos \theta \, ds \tag{1-7}$$

The quantity $\mathcal{E} \cos \theta$ is simply the component of the field in the direction of the path s, which we call $\mathcal{E}_s$. Notice that if we write

$$\mathcal{E}_s = -\frac{dV}{ds} \tag{1-8}$$

and substitute Eq. (1-8) into (1-7)

$$V = -\int_a^b \mathcal{E}_s \, ds = \int_a^b \frac{dV}{ds} \, ds = \int_a^b dV = V_b - V_a = V \tag{1-9}$$

That is, (1-7) is satisfied by an expression of the form of (1-8). This means that (1-7) and (1-8) are alternative expressions relating potential difference and electric field. In particular, (1-8) states that the component of electric field in a given direction is just the negative of the space rate of change of the potential in that direction. This rate of change is called the *potential gradient* and is measured in volts per meter (V/m). According to Eq. (1-8) the same units of measurement are appropriate for electric field as well.

Current and Current Density The motion of electric charges, as, for example, in response to an electric field, constitutes an electric *current.* Specifically, current is defined as the time rate at which charge is transported past a given point, so that

$$I = \frac{dq}{dt} \tag{1-10}$$

Many materials, most notably metals such as copper and silver, contain free electrons which can move when acted upon by an electric field. Consider the section of copper wire shown in Fig. 1-2 in which there is an electric field $\mathcal{E}$ directed from right to left. Because the electronic charge is negative, the free electrons in the metal are urged from left to right and their motion constitutes an electric current in the wire.

Suppose each electron moves with an average velocity v as the net result of the accelerating force due to the field and collisions within the metal. Then in a small interval of time dt each electron advances a distance $v\,dt$. The electrons which cross the shaded plane during this interval are those contained in a section of wire of length $v\,dt$ or volume $Av\,dt$, where A is the cross section of the wire. If there are n free electrons per

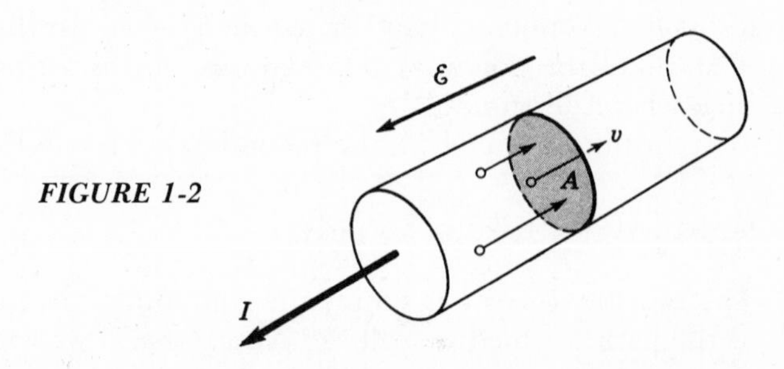

FIGURE 1-2

cubic meter, the number is $nAv\,dt$. Denoting by e the electronic charge, the total charge dq crossing the plane in time dt is therefore

$$dq = nevA\,dt \tag{1-11}$$

According to Eq. (1-10), the current resulting from the electronic motion is

$$I = \frac{dq}{dt} = nevA \tag{1-12}$$

Current is expressed in coulombs per second (C/s), which is termed an *ampere* (A) in honor of the French scientist André Marie Ampère.

Note that the direction of the current is opposite to the motion of the electrons. Although this may seem incongruous, the current in gaseous and liquid conductors, for example, is transported by charges of both signs, and these move in opposite directions. Obviously, whichever direction is assigned to the current, one or the other of the charges moves in the opposite direction. By common agreement the current direction was arbitrarily defined to be the direction of motion of positive charges. This convention was settled upon before it was known that the free charges in metals are electrons. Actually, Benjamin Franklin's original definitions of positive and negative electricity determined that the electronic charge is a negative quantity, although the existence of electrons was unknown at the time.

In many instances it is useful to consider the *current density* in a conductor, the ratio of the current to the cross-sectional area. Thus, in the case of the wire discussed above, the current density J is

$$J = \frac{I}{A} = nev \tag{1-13}$$

in which the right side is independent of the geometry of the conductor.

Ohm's Law Each free electron in a wire carrying a current is accelerated by the electric field until it loses its velocity as the result of a collision within the metal. After every collision the electron starts from rest as sketched in Fig. 1-3 and then accelerates once more so that the net result is an average velocity v. This average velocity increases linearly with the applied field $\mathcal{E}$ so that

$$v = \mu\mathcal{E} \tag{1-14}$$

where the quantity μ is called the *mobility* of the electron. The mobility is a property of the material; it is large for materials which are good conductors and small for those which are poor conductors, if n is the same in both cases.

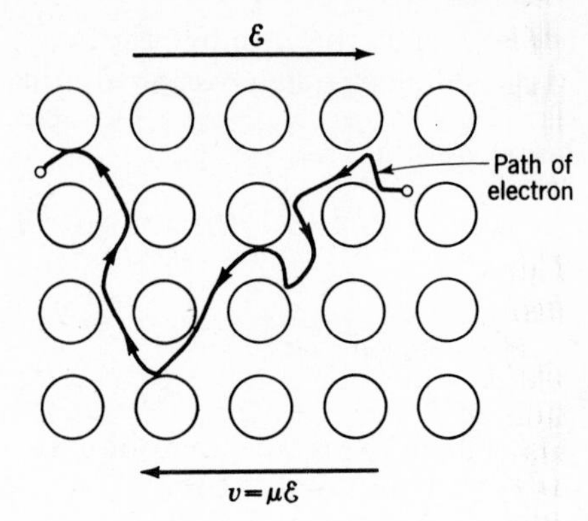

FIGURE 1-3 *Path of free electron in solid conductor is result of acceleration caused by electric field and many collisions.*

Substituting Eq. (1-14) into (1-13)

$$J = ne\mu\mathcal{E} \tag{1-15}$$

and the ratio of the current density to the electric field $J/\mathcal{E}$, which depends only upon the wire material, is called the *conductivity* σ of the metal.

$$\frac{J}{\mathcal{E}} = \sigma = ne\mu \tag{1-16}$$

The left side of Eq. (1-16) is one form of *Ohm's law,* named after the German scientist Georg Simon Ohm, who first discovered experimentally the proportionality between the current density and the electric field in a metallic conductor. The conductivities of several metals at room temperature are listed in Table 1-1, together with corresponding values of the reciprocal of the conductivity, called the *resistivity* $\rho = 1/\sigma$, which is also commonly used.

TABLE 1-1 CONDUCTIVITIES AND RESISTIVITIES OF METALS AND ALLOYS

Material	*Conductivity,* $10^6\ (\Omega \cdot m)^{-1}$	*Resistivity,* $10^{-8}\ \Omega \cdot m$
Aluminum	38	2.6
Brass	17	6
Carbon	0.029	3.5×10^2
Constantan (Cu 60, Ni 40)	2.0	50
Copper	58	1.7
Manganin (Cu 84, Mn 12, Ni 4)	2.3	44
Nichrome	1.0	100
Silver	68	1.5
Tungsten	18	5.6

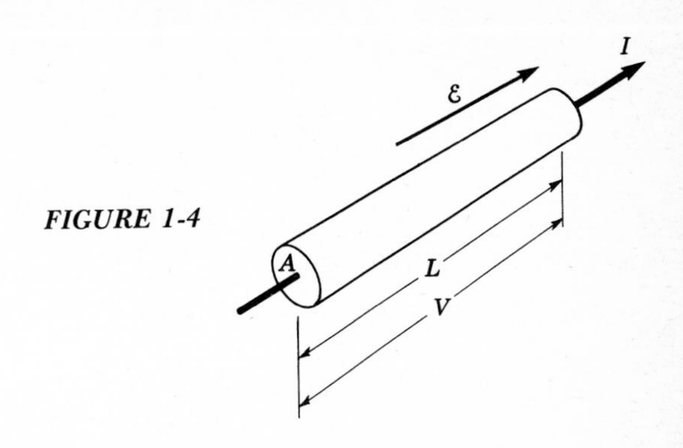

FIGURE 1-4

Consider now the metallic conductor L m long in Fig. 1-4 having a cross-sectional area of A m^2 which is carrying a current of I A. The magnitude of the electric field $\mathcal{E}$ may be written, according to Eq. (1-8), in terms of the potential difference V between the ends of the wire as

$$\mathcal{E} = \frac{V}{L} \tag{1-17}$$

Substituting this expression into Ohm's law, Eq. (1-16), results in

$$J = \frac{\sigma}{L} V \tag{1-18}$$

From the definition of current density, Eq. (1-13),

$$J = \frac{I}{A} = \frac{\sigma}{L} V \tag{1-19}$$

After rearranging and introducing resistivity as the reciprocal of the conductivity, Eq. (1-19) becomes

$$V = \frac{\rho L}{A} I \tag{1-20}$$

The quantity $\rho L/A$ is known as the *resistance* of the conductor. Specifically, the resistance R is

$$R = \frac{\rho L}{A} \tag{1-21}$$

According to Eq. (1-21) the resistance of a wire depends not only upon the material of the wire, through its resistivity, but also upon its cross-sectional area and length. That is, a long thin wire has a greater resistance than a short thick wire of the same material. The unit of resistance is called the *ohm,* and the commonly adopted symbol used to designate a resistance in ohms is the Greek letter omega, Ω. Equation (1-21) shows that resistivity is given in *ohm-meters* and, correspondingly, the units of conductivity are termed *reciprocal ohm-meters* (see also Table 1-1). The reciprocal of resistance is *conductance,* which is measured in units of reciprocal ohms, often called *mhos.*

Finally, from Eq. (1-20), Ohm's law may be written

$$V = RI \tag{1-22}$$

This more familiar form, which is fundamental to circuit analysis, states that whenever a conductor of resistance R carries a current I, a potential difference V must be present across the ends of the conductor. Note that, according to Eq. (1-22), an ohm is equivalent to a volt per ampere.

Joule's Law The kinetic energy of the electrons in a conductor, which results from acceleration by the electric field, is dissipated in inelastic collisons within the conductor and converted to heat energy. Consequently the temperature of a conductor carrying a current must increase slightly, and it is apparent that electric power is expended in forcing a current through the resistance of the conductor.

In order to calculate the rate at which energy must be supplied to the conductor of Fig. 1-4, note that a charge dq goes through a potential difference V in moving from one end of the wire to the other. According to Eq. (1-6), the energy dW required is

$$dW = V\,dq \tag{1-23}$$

which, from the definition of current, Eq. (1-10), is

$$dW = VI\,dt \tag{1-24}$$

Therefore, the rate at which energy is converted to heat, that is, the power P, is

$$P = \frac{dW}{dt} = IV \tag{1-25}$$

This expression may be written in terms of the resistance of the conductor using Ohm's law. The result,

$$P = I^2R \tag{1-26}$$

is known as *Joule's law,* after Sir James Prescott Joule, who discovered experimentally that the rate of development of heat in a resistance is proportional to the square of the current.

According to Joule's law, electric power is dissipated in a conductor whenever it carries an electric current. This effect is put to use in incandescent lamps, where a thin metal filament is heated to white heat by the current, and also in electric fuses, in which the conductor melts when the current exceeds a predetermined value. On the other hand, the size of wires, and therefore their resistance, is selected so that the power loss is small and the temperature rise negligible when the wire is carrying less than the maximum design current. The joule heat in a conductor is commonly spoken of as the "I-squared-R" loss. Note that the unit of power, according to Eq. (1-25), is a joule per second, which is called a *watt* (W), in honor of James Watt, developer of the steam engine.

CIRCUIT ELEMENTS

Resistors An electrical component very frequently used in electronic circuits is the *resistor,* which is a circuit element having a specified value of resistance. Resistance values commonly encountered range from a few ohms to thousands of ohms, or *kilohms* (abbreviated as kΩ), and even millions of ohms, or *megohms* (designated by MΩ). Lumped resistances that resistors introduce into a circuit are large compared with those of wires and contacts between wires. According to Ohm's law, a potential difference develops across the resistor as a result of current in it at the place in the circuit where the resistor is inserted. The conventional symbol for a resistor in a circuit diagram is a zigzag line, as illustrated in Fig. 1-5.

Some resistors are constructed from a long, very fine wire wound on an insulating support. Resistance values can be increased by decreasing the cross-sectional area of the wire and by increasing its length, as Eq. (1-21) shows, and by selecting wire materials having a large resistivity (see Table 1-1). Such *wire-wound* resistors commonly employ metal-alloy wires which have resistivities relatively independent of temperature. Typical materials are manganin and constantan. Wire-wound resistors are used where it may be necessary to dissipate sufficient joule heat for the temperature of the resistor to rise significantly. The resistance of wire-wound resistors can be determined quite precisely by choosing the proper wire length, so wire-wound resistors are also useful in applications where accurate resistance values are desired.

Thin-film resistors are made by depositing a thin film of a metal on a cylindrical insulating support. High-resistance values are a consequence

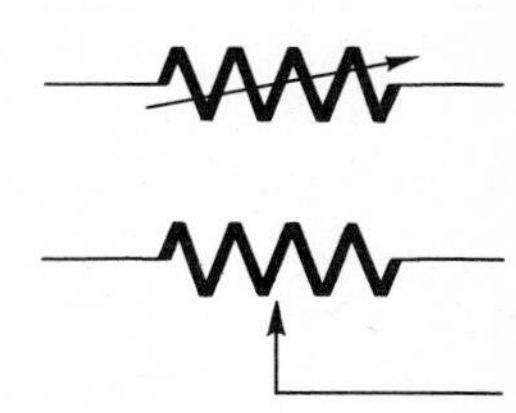

FIGURE 1-5 *Conventional circuit symbols for fixed (top) and variable resistors.*

of the thinness of the film. Because of the difficulty in producing uniform films, it is not possible to control resistance values as precisely as in the case of wire-wound resistors. However, thin-film resistors are free of troublesome inductance effects common in wire-wound units (Chap. 2), and this is important in high-frequency circuits. Thin-film resistors fabricated from nonmetallic materials, particularly finely divided carbon granules, are also common. Carbon itself has very high resistivity, as do the points of contact between the granules. In fact, it is possible to achieve such high resistance values with carbon granules that in many situations it is unnecessary to employ thin films at all and the resistance element is a simple rod of pressed carbon granules. Such units are known as *composition* resistors.

Both thin-film and composition resistors are provided with insulation and wire leads to facilitate inserting them into circuits. It is common practice to provide colored markings which denote the resistance value of each unit according to a universal *resistor color code.* In addition, the physical size of the resistor is a rough indication of the maximum permissible power the unit is capable of dissipating without appreciable increase in temperature caused by joule heating. Thus, for example, common resistor power ratings are 1 W, $\frac{1}{2}$ W, and $\frac{1}{4}$ W, although other values are used as well. Examples of typical thin-film and composition resistors are shown in Fig. 1-6.

It is often necessary to vary the resistance of a resistor while it is permanently connected in a circuit. Such *variable resistors* employ a mechanical slider or arm which rides over the resistance element, thus selecting the length of the element included in the circuit. Both wire-wound and composition resistance elements are commonly made circular so that the position of the slider may be adjusted by rotation of a shaft. The circuit symbols for variable resistors are of two types, as in Fig. 1-5, depending upon whether two or three terminals are provided for external connections. A variable resistor having two terminals is called a *rheostat,* while one with three is known as a *potentiometer.* Obviously a potentiometer, with its terminals at each end of the resistance element and a third termi-

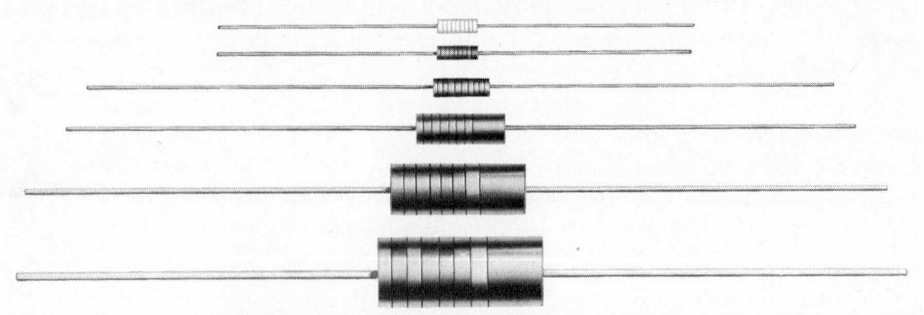

FIGURE 1-6 *Typical composition resistors.* (Allen-Bradley Co.)

nal attached to the slider, can be used as a rheostat if one of the resistance-element terminals is ignored.

Batteries According to Joule's law, electric energy is dissipated in any conductor when it carries a current. In simple dc circuits the source of this energy, which must be supplied in order to maintain the current, is often a chemical *battery*. Other sources of dc electric power will be considered in a later chapter. In a battery, chemical energy is converted into electric energy, and the chemical reactions maintain a potential difference between the battery terminals whether or not a current is present. This potential difference is commonly referred to as an *electromotive force*, abbreviated *emf*, in order to distinguish it from the potential difference which appears across a resistance in accordance with Ohm's law. As a battery continues to supply the energy necessary to maintain current in a circuit, the chemical constituents eventually become depleted and the battery is said to be *discharged*. Depending upon the particular chemical nature of the battery, it may be possible to *charge* it, that is, return it to its original chemical composition, by passing a current between its terminals in a direction opposed to the internal emf. The symbol for a battery in circuit diagrams, Fig. 1-7, consists of a short heavy line parallel to a longer thin line. It is always assumed, if not explicitly indicated, that the longer line represents the higher, or positive, terminal of the internal emf. Since the internal emf is a potential difference, its unit is the volt.

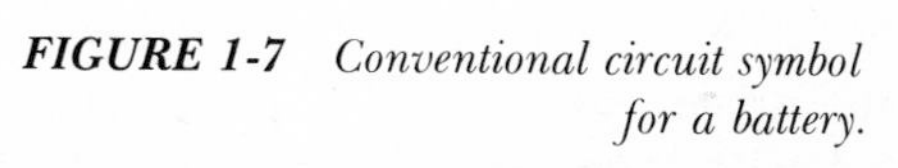
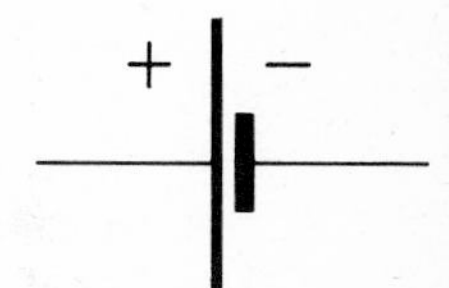

FIGURE 1-7 *Conventional circuit symbol for a battery.*

The carbon-zinc battery is by far the most common, and least expensive, source of electrical energy. Although it is conventionally referred to as a *dry cell*, it actually consists of a moist paste of zinc chloride, ammonium chloride, and manganese dioxide (called the *electrolyte*) contained between a zinc electrode and a carbon electrode. The zinc and carbon electrodes serve as the terminals of the battery. The operation of such a cell is briefly as follows. At the zinc electrode, zinc atoms are dissolved into solution as doubly charged zinc ions. The zinc electrode becomes negatively charged because each zinc atom leaves behind two electrons. At the carbon electrode ammonium ions reacting with manganese dioxide withdraw electrons from the carbon, and thus it becomes charged positively. If the negative zinc electrode is connected externally through a circuit to the positive carbon electrode, electrons can flow between them to complete the chemical reaction.

Notice that in order for the chemical reaction to continue, zinc ions must move away from the negative electrode and the reaction products near the positive terminal must likewise move away from the carbon electrode. Thus, current is carried internally to the battery by means of ions moving in the electrolyte, and this is a source of internal resistance. Current in the internal resistance has the effect of reducing the terminal voltage of the battery. The terminal voltage of the dry cell slowly decreases with use as the internal resistance increases because of depletion of the manganese dioxide. The internal resistance eventually becomes so large that the battery is useless.

If the dry cell is left idle for some time before it is completely discharged, the internal resistance gradually reduces because of internal diffusion of the ions. On the other hand, if a dry cell is allowed to age for extended periods (more than one year) internal ionic diffusion increases the internal resistance so much that the cell becomes inoperative, even though it may never have been used. The emf of a freshly prepared dry cell is 1.5 V. Higher voltages are conventionally obtained by connecting a number of individual units (Fig. 1-8); in fact the term *battery* originated from just such assemblies. Dry-cell batteries of 1.5, 9, 22.5, 45, 67.5 and 90 V are most commonly available.

The familiar lead-acid *storage battery* used in automobiles is an example of a battery that can be repeatedly recharged. The positive electrode of a fully charged storage battery is a porous coat of lead dioxide on a grid of metallic lead. The negative electrode is metallic lead, and both electrodes are immersed in a liquid sulfuric acid electrolyte at a specific gravity of about 1.3. During discharge the lead dioxide is converted to lead sulfate, which is poorly soluble and clings to the positive plate. This reaction withdraws electrons from the electrode, thus charging it positively. At the negative electrode, sulfate ions from solution produce lead sulfate and release electrons. Again the lead sulfate adheres to the electrode and at discharge both electrodes are nearly entirely converted to lead sulfate. The loss of sulfate ions from solution during discharge reduces the specific gravity to about 1.16, so that the condition of the battery may be determined by measuring the specific gravity of the electrolyte.

These chemical reactions are easily reversible, and current directed into the positive terminal acts to return the electrodes to their original chemical composition. Charging requires an external source to furnish

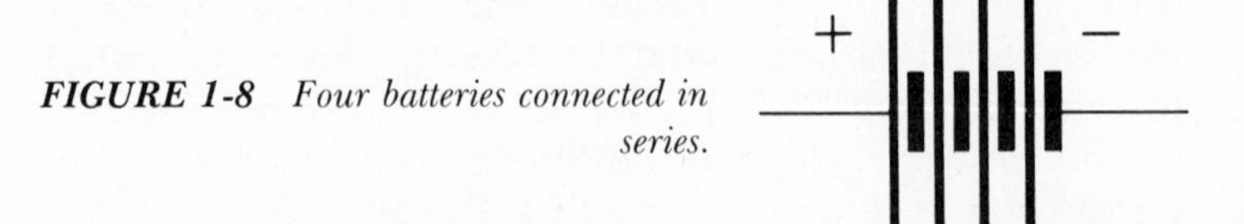

FIGURE 1-8 *Four batteries connected in series.*

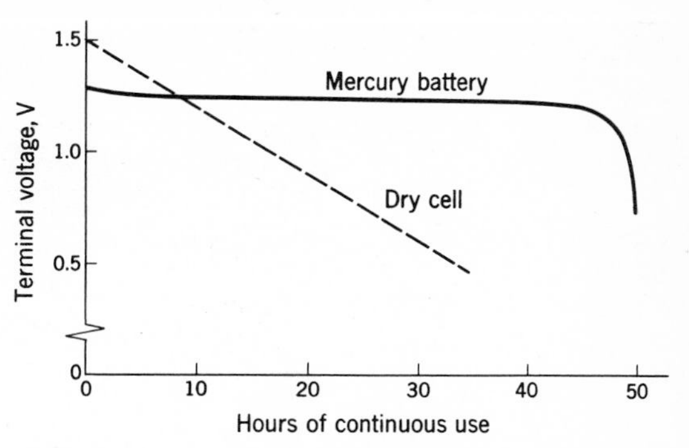

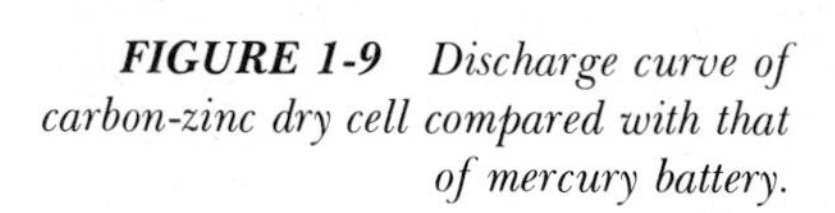
FIGURE 1-9 *Discharge curve of carbon-zinc dry cell compared with that of mercury battery.*

electric energy, after which the battery again can supply energy during discharge. Thus, the storage battery may be said to store electric energy in chemical form. In addition, the internal resistance of the lead-acid battery is very low and the battery is capable of delivering currents of several hundred amperes for short times. The fully charged cell has an emf of about 2.1 V, and commercial units are available as 6-, 12-, and 24-V batteries. It is important to maintain an idle storage battery fully charged, for otherwise the electrodes slowly become converted to a sulfate which cannot be returned to the original chemical composition by a charging current. In this condition, the electric energy capacity of the battery is reduced.

The internal resistance of the recently developed *mercury battery* does not change appreciably during discharge. This means that the terminal voltage remains essentially constant throughout the useful life. It then falls precipitously when the battery is exhausted, as illustrated in Fig. 1-9. The constant-voltage characteristic of mercury batteries is important in those electronic applications where the proper operation of a circuit depends critically upon the battery voltage. Such situations are not uncommon in vacuum-tube and transistor circuits. In addition, the constant-voltage feature means that the mercury battery is useful as a voltage standard in electrical measurement circuits. The mercury battery has a zinc amalgam for one electrode and mercuric oxide and carbon for the other. The chemical reactions at the electrodes are somewhat similar to those of the dry cell, and the potential developed is 1.35 V.

Other recent battery types include the *alkaline* battery and the *nickel-cadmium* battery. The alkaline battery is chemically quite similar to the dry cell, but has a strongly basic electrolyte between the electrodes. This, together with a modified electrode structure, lowers internal resistance, increases energy capacity, and improves shelf life. The nickel-cadmium battery can be repeatedly recharged like the lead storage battery, but is completely sealed, since gas evolution during charging acts as a self-regulating mechanism to prevent the buildup of a large gas pressure. This

FIGURE 1-10 *Typical modern batteries.* (Union Carbide Co.)

feature, and the fact that a liquid electrolyte is not required, compensates for the high cost of this battery. Typical modern batteries are illustrated in Fig. 1-10.

SIMPLE CIRCUITS

Series Circuits If several electric components, such as resistors, are connected so that the current is the same in every one, the components are said to be in a *series* circuit. Consider the simple series circuit comprising the battery and three resistors illustrated in Fig. 1-11*a*. The current I results in a potential difference between the terminals of each resistor which is given by Ohm's law. That is,

$$V_1 = R_1 I \qquad V_2 = R_2 I \qquad \text{and} \qquad V_3 = R_3 I \tag{1-27}$$

Clearly, the sum of these voltages is equal to the battery emf, or

$$V = V_1 + V_2 + V_3 \tag{1-28}$$

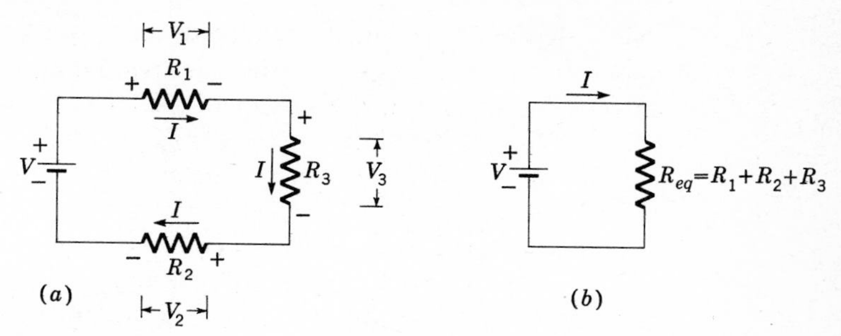

FIGURE 1-11 *(a) Simple series circuit and (b) its equivalent.*

Equation (1-28) is a simple example of a principle of electronic circuits which is considered in greater detail in the next section. The equation states that the algebraic sum of the potential differences around any complete circuit is equal to zero. Note the polarity distinction between the potential difference at the terminals of a resistor compared with that of a source of emf such as a battery: the current direction is *into* the positive terminal of a resistance while it is *out of* the positive terminal of an emf source. Since, according to Fig. 1-4, the potential decreases in the direction of the current through a resistance, the potential difference is commonly referred to as the *IR drop* across the resistor.

If the *IR* drops of Eq. (1-27) are inserted into Eq. (1-28), the result is

$$V = IR_1 + IR_2 + IR_3 = I(R_1 + R_2 + R_3) \tag{1-29}$$

Thus, the current in the series circuit is

$$I = \frac{V}{R_1 + R_2 + R_3} = \frac{V}{R_{eq}} \tag{1-30}$$

where the equivalent resistance R_{eq} is defined as

$$R_{eq} = R_1 + R_2 + R_3 \tag{1-31}$$

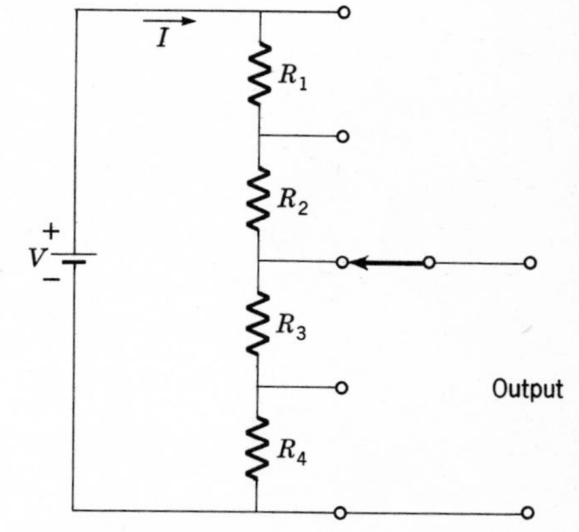

FIGURE 1-12 *Potential-divider circuit.*

Evidently the equivalent resistance of any number of resistors connected in series equals the sum of their individual resistances. Insofar as the current is concerned, the circuit of Fig. 1-11*b* containing the one single resistor R_{eq} is equivalent to that of Fig. 1-11*a*, which has three resistors.

A useful circuit based on the series connection of resistors is the *potential divider*, Fig. 1-12, in which the junction between each pair of resistors is connected to a terminal of a *multiple-tap* selector switch. By positioning the switch on each of its various taps it is possible to present a given fraction of the battery voltage V at the output terminals. The division of the potential V among the various taps depends upon the magnitudes of the resistances in the potential divider. Obviously, if the series resistors are replaced by a potentiometer, the output voltage may be set at any desired fraction of V. This is the principle of the volume control in radio and television receivers.

Parallel Circuits Another way of connecting electric components, such as resistors, is shown in Fig. 1-13. Here the potential difference across each resistor in the circuit is the same; this form of connection is called a *parallel circuit*. The current in each resistor is given by Ohm's law as

$$I_1 = \frac{V}{R_1} \qquad I_2 = \frac{V}{R_2} \qquad \text{and} \qquad I_3 = \frac{V}{R_3} \tag{1-32}$$

In this case, the sum of the currents equals the battery current,

$$I = I_1 + I_2 + I_3 \tag{1-33}$$

Substituting for the currents from Eq. (1-32), this becomes

$$I = \frac{V}{R_1} + \frac{V}{R_2} + \frac{V}{R_3} = V\left(\frac{1}{R_1} + \frac{1}{R_2} + \frac{1}{R_3}\right) \tag{1-34}$$

FIGURE 1-13 *Parallel-connected resistors.*

Now, in order to determine the equivalent resistance for parallel resistors, we define R_{eq}, using Ohm's law, as

$$V = IR_{eq} \tag{1-35}$$

Inserting Eq. (1-35) into Eq. (1-34)

$$I = \frac{V}{R_{eq}} = V\left(\frac{1}{R_1} + \frac{1}{R_2} + \frac{1}{R_3}\right) \tag{1-36}$$

So that

$$\frac{1}{R_{eq}} = \frac{1}{R_1} + \frac{1}{R_2} + \frac{1}{R_3} \tag{1-37}$$

which states that for any number of resistors in parallel the reciprocal of the equivalent resistance equals the sum of the reciprocals of the individual resistances.

Consider the special case of two resistors in parallel,

$$\frac{1}{R_{eq}} = \frac{1}{R_1} + \frac{1}{R_2} = \frac{R_2 + R_1}{R_1 R_2} \tag{1-38}$$

so that

$$R_{eq} = \frac{R_1 R_2}{R_1 + R_2} = \frac{R_1}{1 + R_1/R_2} \tag{1-39}$$

Suppose now R_1 is fixed and R_2 can take any value; according to Eq. (1-39) R_{eq} is always less than R_1. A similar argument applies if R_2 is fixed and R_1 can vary. This proves that the equivalent resistance of the combination of two resistors in parallel is smaller than that of either resistor. The same result holds for any number of parallel resistors.

Networks *Network* connections of series and parallel resistances can be analyzed by successive applications of Eqs. (1-31) and (1-37). Consider, for example, the network of Fig. 1-14*a* with the resistance values as marked on the circuit diagram. The parallel combination of R_5 and R_6, each of which is 10 Ω, may be replaced by a 5-Ω resistor since according to Eq. (1-37)

$$\frac{1}{R_{eq}} = \frac{1}{R_5} + \frac{1}{R_6} = \frac{1}{10} + \frac{1}{10} = \frac{2}{10} \tag{1-40}$$

$$R_{eq} = 5\ \Omega$$

Therefore the network is reduced to that shown in Fig. 1-14*b*. Next, the combination of R_{eq} with R_4 (= 10 Ω) is, by Eq. (1-31),

$$R'_{eq} = R_{eq} + R_4 = 5 + 10 = 15\ \Omega \tag{1-41}$$

and the network is now Fig. 1-14*c*. R'_{eq} and R_3 are in parallel, so that their equivalent is

$$R''_{eq} = \frac{R'_{eq} R_3}{R'_{eq} + R_3} = \frac{15 \times 15}{15 + 15} = \frac{225}{30} = 7.5\ \Omega \tag{1-42}$$

Lastly, the series combination of R''_{eq}, R_1, and R_2 is simply

$$R_T = R''_{eq} + R_1 + R_2 = 7.5 + 5 + 5 = 17.5\ \Omega \tag{1-43}$$

and the entire network of Fig. 1-14*a* may now be replaced by its simple

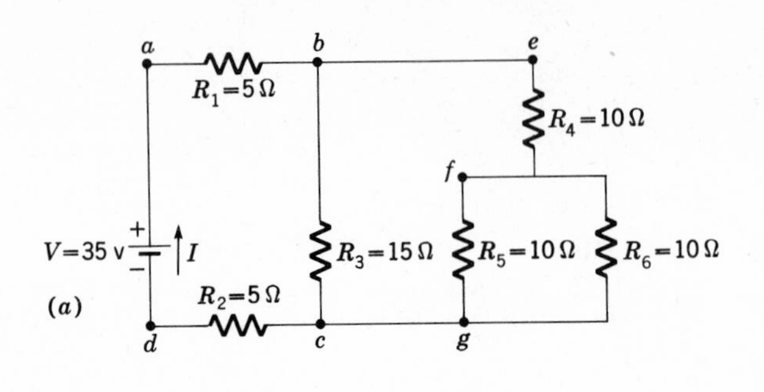

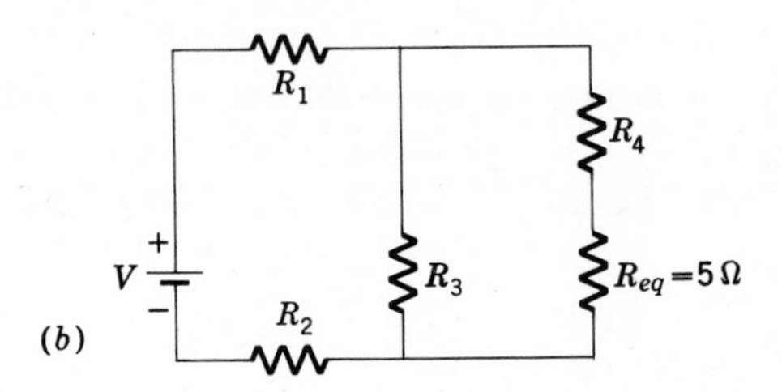

FIGURE 1-14 *Network reduction by series and parallel equivalents.*

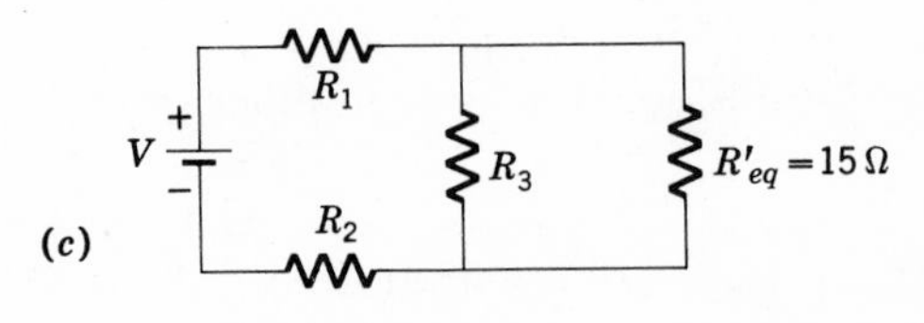

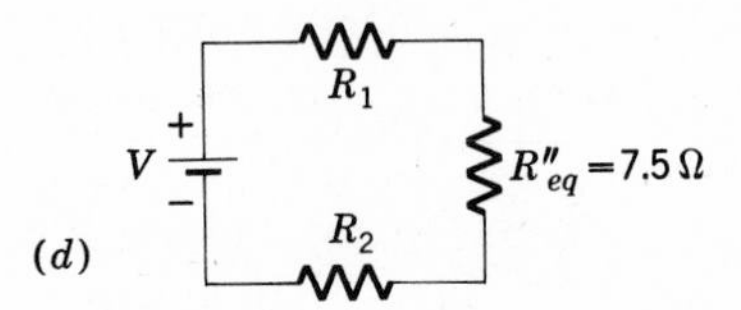

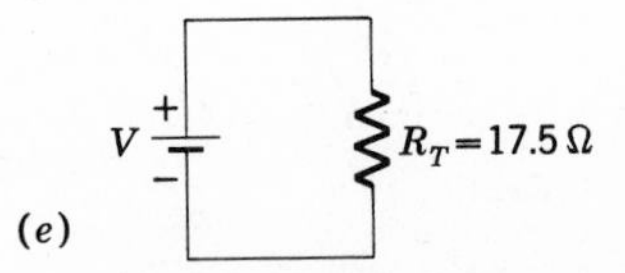

equivalent, Fig. 1-14*e*, where R_T represents the resistance of the entire network. The current in the battery is therefore

$$I = \frac{V}{R_T} = \frac{35}{17.5} = 2 \text{ A} \tag{1-44}$$

Suppose it is desired to determine the current I_3 in R_3. This is accomplished by first calculating the potential difference V_3 between points

b and c in the circuit diagram. The IR drop across R_1 is $IR_1 = 2 \times 5 = 10$ V, and a similar value applies to the IR drop across R_2. According to Eq. (1-28)

$$V = V_1 + V_3 + V_2 \tag{1-45}$$

Thus

$$V_3 = V - V_1 - V_2 = 35 - 10 - 10 = 15 \text{ V} \tag{1-46}$$

The current in R_3 is therefore

$$I_3 = \frac{V_3}{R_3} = \frac{15}{15} = 1 \text{ A} \tag{1-47}$$

By similar reasoning it is possible to determine the current in each resistor.

As a second example, note the network reduction illustrated in Fig. 1-15*a* to *d*. Resistors R_4 and R_5 are in parallel and their equivalent in

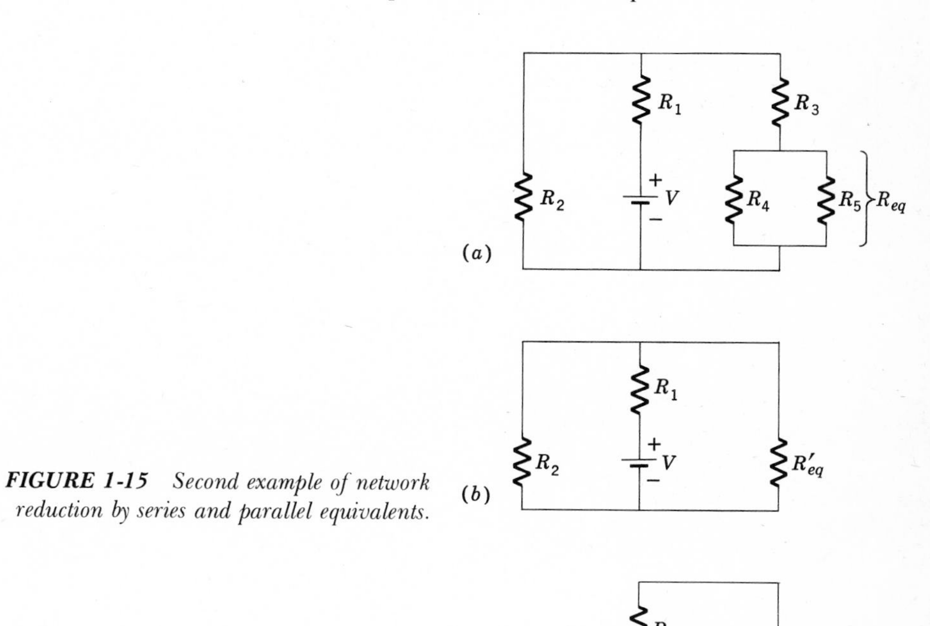

FIGURE 1-15 *Second example of network reduction by series and parallel equivalents.*

combination with R_3 results in R'_{eq}. Using the definition of parallel resistors, R_2 and R'_{eq} are seen to be in parallel and therefore may be replaced by R''_{eq}. Finally, R_T is simply the sum of R''_{eq} and R_1. Again, the current in each resistor may be determined by successive applications of Eq. (1-28).

CIRCUIT ANALYSIS

Kirchhoff's Rules It is not possible to reduce many of the networks important in electronics to simple series-parallel combinations, so that more powerful analytical methods must be used. Two simple extensions of Eqs. (1-28) and (1-33), known as *Kirchhoff's rules,* are most helpful in this connection. Consider first the simple parallel circuit, Fig. 1-13, redrawn as in Fig. 1-16 to illustrate the idea of a *branch point,* or *node,* of a circuit. A node is the point at which three (or more) conductors are joined. Kirchhoff's first rule is that the algebraic sum of the currents at any node is zero. Symbolically

$$\Sigma I = 0 \tag{1-48}$$

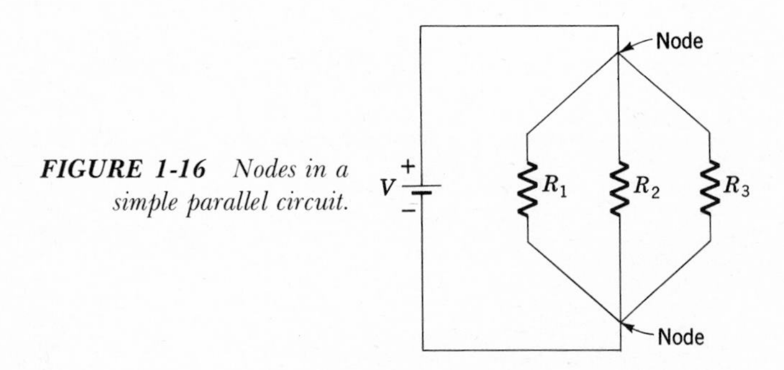

FIGURE 1-16 *Nodes in a simple parallel circuit.*

Note that Eq. (1-48) is essentially a statement of continuity of current; it may also be looked upon as a result of the conservation of electric charge.

Kirchhoff's second rule has already been applied implicitly in using Eq. (1-28) to calculate I_3 in Fig. 1-14*a*. It states that the algebraic sum of the potential differences around any complete loop of a network is zero. Symbolically

$$\Sigma V = 0 \tag{1-49}$$

A loop of a network is understood to be any closed path such as *abcda* in Fig. 1-14*a* which returns to the same point. Other examples of complete loops in the same network are *befgcb* and *daefgd*. Equation (1-49) is a consequence of the conservation of energy.

In applying Kirchhoff's rules to any network the first step is to assign a current of arbitrary direction to each of the resistances in the network. The polarity of the potential difference across each resistor is then marked on the circuit diagram using the convention already noted that the current enters the positive terminal of a resistance. The polarity of emf sources are, of course, specified in advance from the circuit diagram itself. Kirchhoff's rules are then applied to the various nodes and circuit loops to obtain a sufficient number of simultaneous equations to solve for the total number of unknown currents.

It is true that if a network contains m nodes and n unknown currents there are $m - 1$ independent equations which result from Eq. (1-48). Similarly, there are $n-(m-1)=n-m+1$ independent equations derived from Eq. (1-49). The total number of independent equations obtained from Kirchhoff's rules applied to any network is therefore $(m - 1) + n - (m - 1) = n$. This is just the number of unknown currents and the network solution is therefore completely determined. It is generally possible to write down more node and loop equations than are needed but only n of them are truly independent.

The solution of these independent equations often results in certain of the currents being negative. This means that the original arbitrarily assigned current direction is, in fact, incorrect and the current is actually in the opposite direction. Thus, it is not necessary to know the true current direction in advance. Once the various currents have been calculated, the IR drops in any portion of the circuit are determined using Ohm's law.

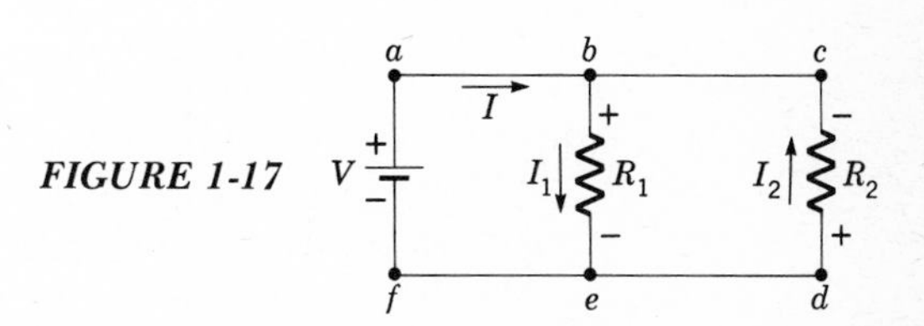

FIGURE 1-17

The technique of applying Kirchhoff's rules to a network can best be illustrated with a few examples. Consider first the simple parallel-resistor circuit of Fig. 1-17. The current direction in each resistor has been arbitrarily selected and the polarity of the IR drops marked in accordance with these assigned directions. Note that this network has only two nodes, one at b and the other at e. Therefore, there is only $2 - 1 = 1$ independent node equation. Considering the branch point at b, Eq. (1-48) yields

$$I - I_1 + I_2 = 0 \tag{1-50}$$

Notice at branch point e the current equation is

$$-I + I_1 - I_2 = 0 \tag{1-51}$$

Clearly Eq. (1-51) is simply the negative of (1-50) and the two relations are therefore not independent. Either equation may be used in the solution of the network.

Consider now the loop *abef*. According to Eq. (1-49)

$$V - I_1R_1 = 0 \tag{1-52}$$

Similarly, around the loop *abcdef*

$$V + I_2R_2 = 0 \tag{1-53}$$

Since there are three unknown currents, there must be $3 - 2 + 1 = 2$ independent loop equations and these are (1-52) and (1-53). Note, however, that around loop *bcde*

$$I_1R_1 + I_2R_2 = 0 \tag{1-54}$$

This is not an independent relation, as may be shown by subtracting Eq. (1-52) from (1-53). The result is Eq. (1-54). Thus these three loop equations are not independent, and any two may be used in solving the network.

Choose Eqs. (1-50), (1-52), and (1-54) as the three independent equations to solve for the three unknown currents. The solution is accomplished by first solving Eq. (1-52) for I_1,

$$I_1 = \frac{V}{R_1} \tag{1-55}$$

Next, I_2 is determined from (1-54),

$$I_2 = -\frac{I_1R_1}{R_2} \tag{1-56}$$

Substituting (1-56) into (1-50),

$$I - I_1 - \frac{I_1R_1}{R_2} = 0 \tag{1-57}$$

Substituting from (1-55) for I_1.

$$I - \frac{V}{R_1} - \frac{VR_1}{R_1R_2} = I - V\left(\frac{1}{R_1} + \frac{1}{R_2}\right) = 0 \tag{1-58}$$

The current I is therefore

$$I = V\left(\frac{1}{R_1} + \frac{1}{R_2}\right) \tag{1-59}$$

which is quite equivalent to the solution corresponding to Eq. (1-34) arrived at by considering parallel resistors.

Finally, I_2 is determined by substituting for I_1 in Eq. (1-56),

$$I_2 = -\frac{VR_1}{R_1R_2} \tag{1-60}$$

or

$$I_2 = -\frac{V}{R_2} \tag{1-61}$$

According to the minus sign in Eq. (1-61), this current is actually in the opposite direction to that assumed in Fig. 1-17. Correspondingly, the IR drop across R_2 has the opposite polarity from that shown on the circuit diagram.

More complicated networks require more than three equations, and it is usually desirable to employ the standard method of determinants to solve the set of simultaneous equations. This technique, illustrated in the following section, has the considerable advantage that it is possible to solve directly for only those currents that are of interest. Often, only one or two of the currents in a network are of direct concern, and in this case a complete solution for all the unknowns is superfluous.

Wheatstone Bridge In this section Kirchhoff's rules are used to analyze the *Wheatstone bridge* circuit illustrated in Fig. 1-18. This extremely useful circuit was developed in 1843 by Charles Wheatstone and is widely used in electrical measurements to determine values of unknown resistances. The manner in which it is used may be understood from an analysis of the circuit. According to Kirchhoff's rule for branch points a, b, and d,

$$\begin{aligned} I - I_1 - I_2 &= 0 \\ I_1 - I_3 + I_5 &= 0 \\ I_3 + I_4 - I &= 0 \end{aligned} \tag{1-62}$$

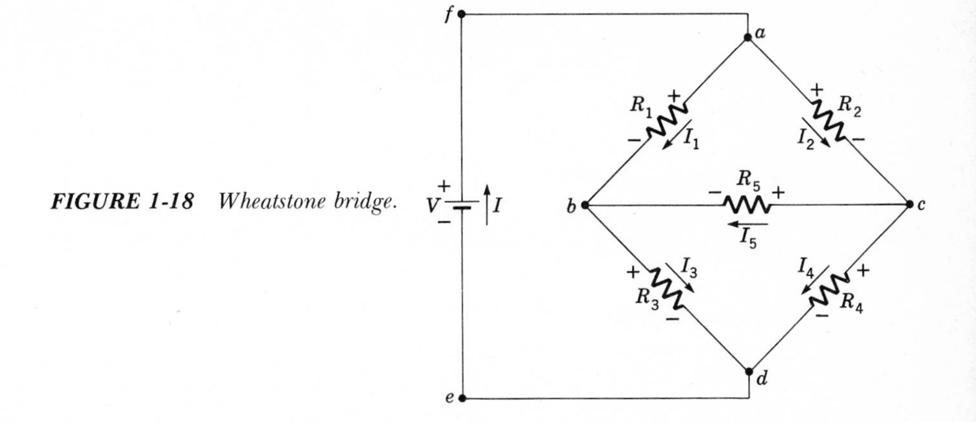

FIGURE 1-18 *Wheatstone bridge.*

Because there are four branch points in the Wheatstone bridge circuit, these three current equations are independent, so the fourth one, which could be written for branch point c, is not used.

Applying Kirchhoff's rule to loops *abdefa, acba,* and *bcdb,* the equations are

$$
\begin{aligned}
-I_1R_1 - I_3R_3 + V &= 0 \\
-I_2R_2 - I_5R_5 + I_1R_1 &= 0 \\
I_5R_5 - I_4R_4 + I_3R_3 &= 0
\end{aligned}
\qquad (1\text{-}63)
$$

Note carefully the indicated polarities of the various *IR* drops as they are encountered in traversing each loop. Since there are six unknown currents, $6 - 4 + 1 = 3$ loop equations are needed here and any others are redundant.

Equations (1-62) and (1-63) are six equations in six unknowns. Therefore, in applying the method of determinants to these simultaneous equations it is necessary to evaluate two sixth-order determinants in calculating each current. The total solution involves seven different such determinants. While the evaluation of a sixth-order determinant is straightforward and a number of standard approaches exist to reduce the order before final evaluation, the complete solution of seven sixth-order determinants is quite laborious. Therefore, although the solution to the set of Eqs. (1-62) and (1-63) is in principle accomplished, it is useful to seek alternative methods.

The analysis of complex networks can usually be simplified by the use of *loop currents.* This technique, known as *Maxwell's method,* after James Clerk Maxwell, in effect applies both Kirchhoff's rules simultaneously and thereby reduces the number of simultaneous equations needed. The loop currents are drawn around any complete loop as the three illustrated for the case of the Wheatstone bridge in Fig. 1-19. After the

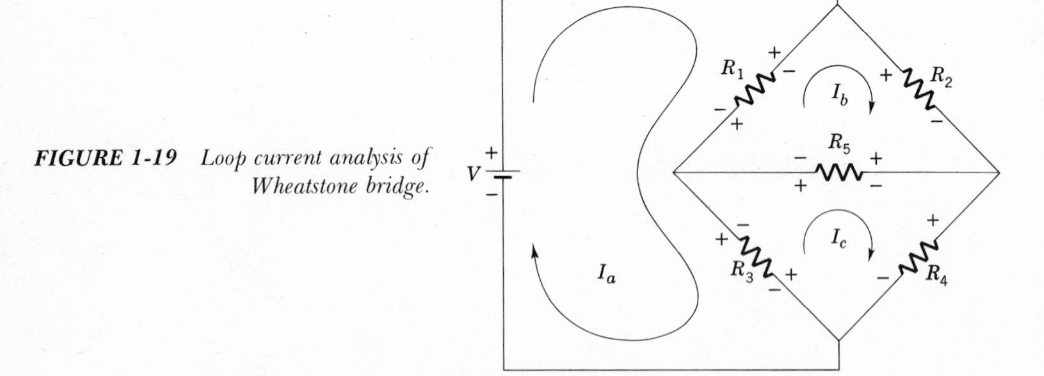

FIGURE 1-19 *Loop current analysis of Wheatstone bridge.*

polarity of the *IR* drops are indicated in accordance with the current directions, the usual voltage equations around each loop are written. Thus, with reference to Fig. 1-19

$$
\begin{aligned}
V - R_1(I_a - I_b) - R_3(I_a - I_c) &= 0 \\
-R_2 I_b - R_5(I_b - I_c) + R_1(I_a - I_b) &= 0 \\
R_3(I_c - I_a) + R_5(I_c - I_b) + R_4 I_c &= 0
\end{aligned} \tag{1-64}
$$

Here again, note the polarity of the IR drops and the current directions. Upon rearranging

$$
\begin{aligned}
-(R_1 + R_3)I_a + R_1 I_b + R_3 I_c &= -V \\
R_1 I_a - (R_1 + R_2 + R_5)I_b + R_5 I_c &= 0 \\
-R_3 I_a - R_5 I_b + (R_3 + R_4 + R_5)I_c &= 0
\end{aligned} \tag{1-65}
$$

The solution of Eqs. (1-65) for any current, say I_b, using determinants is found by forming a ratio in which the denominator is the determinant of the coefficients of the currents and the numerator is a similar determinant with the coefficients of the unknown current replaced by the right side of the equation. That is, the solution for I_b is

$$
I_b = \frac{\begin{vmatrix} -(R_1 + R_3) & -V & R_3 \\ R_1 & 0 & R_5 \\ -R_3 & 0 & R_3 + R_4 + R_5 \end{vmatrix}}{\begin{vmatrix} -(R_1 + R_3) & R_1 & R_3 \\ R_1 & -(R_1 + R_2 + R_5) & R_5 \\ -R_3 & -R_5 & R_3 + R_4 + R_5 \end{vmatrix}}
$$

$$
= \frac{VR_5R_3 + VR_1(R_3 + R_4 + R_5)}{\Delta} \tag{1-66}
$$

where Δ signifies the denominator. Similarly, I_c is

$$
I_c = \frac{\begin{vmatrix} -(R_1 + R_3) & R_1 & -V \\ R_1 & -(R_1 + R_2 + R_5) & 0 \\ -R_3 & -R_5 & 0 \end{vmatrix}}{\Delta}
$$

$$
= \frac{VR_1R_5 + VR_3(R_1 + R_2 + R_5)}{\Delta} \tag{1-67}
$$

Now, the current through R_5, which in Fig. 1-18 is labeled I_5, is

$$
\begin{aligned}
I_5 &= I_b - I_c \\
&= \frac{V}{\Delta}(R_5R_3 + R_1R_3 + R_1R_4 + R_1R_5 - R_1R_5 - R_1R_3 - R_2R_3 - R_5R_3) \\
&= \frac{V}{\Delta}(R_1R_4 - R_2R_3)
\end{aligned} \tag{1-68}
$$

Equation (1-68) is a most important relation for the Wheatstone bridge. Note that if

$$R_1R_4 = R_2R_3$$

or

$$\frac{R_1}{R_2} = \frac{R_3}{R_4} \tag{1-69}$$

then I_5 is zero, independent of the applied voltage. If the resistances in the arms of the bridge obey the ratios indicated in Eq. (1-69), the bridge is said to be *balanced*. Thus, for example, if R_1, R_2, and R_3 are known resistances and I_5 is zero, the value of R_4 may be immediately calculated from the condition for balance, Eq. (1-69).

In the common version of a Wheatstone bridge, resistances R_1 and R_2 are connected to a switch to give decade values of the ratio R_2/R_1, and R_3 is a continuously variable calibrated resistor. Once the bridge is balanced by adjusting R_3, the value of the unknown R_4 is simply $(R_2/R_1)R_3$. The decade values of the ratio (R_2/R_1) may range from 10^{-3}, 10^{-2}, and 10^{-1} to 1, 10, 10^2, and 10^3, so that a very wide range of resistance values can be measured. In practice, a current-indicating instrument is connected in the position of R_5 to indicate balance. Note that this meter need not be calibrated, since it is only used to indicate the balance condition, that is, a zero current.

Equation (1-68) is an example of a case in which useful information concerning the circuit is derived without carrying through the complete solution for all currents. It is often possible to draw the loop currents in such a way that only one current need be determined. The facility of choosing loop currents that minimizes the effort required to solve any given network is attained with experience. Note that all branches in the network must be traversed by at least one loop current in order to achieve a solution. Also, the voltage equation around each loop, as in Eq. (1-65), has the same form in all cases: the coefficient of the loop's own current is the sum of the resistances around the loop; the coefficient of all other currents is a resistance common to both loops; and, the right-hand side is the algebraic sum of all emfs encountered in traversing the loop.

Potentiometer Circuit A very accurate way to compare two potential differences uses the *potentiometer circuit*, a simple version of which is illustrated in Fig. 1-20. Here an accurately variable precision potentiometer is connected in series with a variable resistance and a battery. The slider of the precision resistor is connected to one external terminal through a current-indicating meter and one end of the resistor is connected to the other output terminal. Suppose a specific value of the current $I = V/(R + R_A)$ is selected by adjusting the variable resistor R_A. Then, the potential V' of the slider on R is simply IR', where R' is the resistance between the end of the resistor and the slider. Since I and R are known, the position

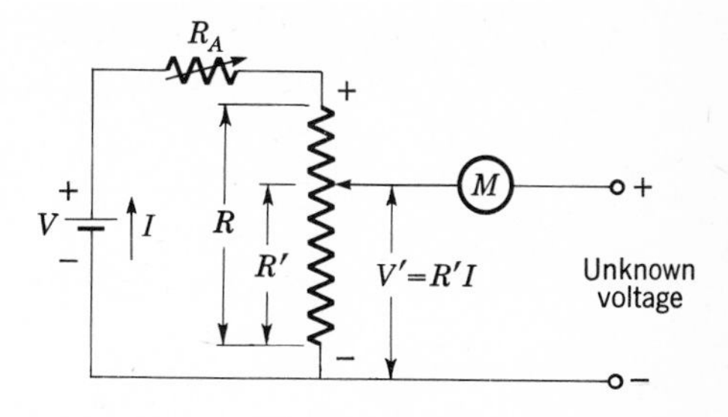

FIGURE 1-20 *Simple potentiometer circuit for measuring unknown voltage.*

of the slider can be calibrated in terms of potential difference in volts. If now an unknown voltage, as for example a battery, is connected to the output terminals, and the slider is adjusted until the current indicated by the meter M is zero, the value of the unknown voltage is equal to IR'.

The potentiometer is a comparison measuring device which determines the value of the unknown potential difference in terms of the voltage of a standard battery. To show how this is accomplished, refer to the practical circuit in Fig. 1-21. Here, V_S is the emf of a standard such as a mercury battery or, more usually, a *Weston cell,* which is a special battery with an extremely stable emf. Suppose now R_A is adjusted until the meter M_1 indicates zero current. This means that

$$IR_S = V_S \tag{1-70}$$

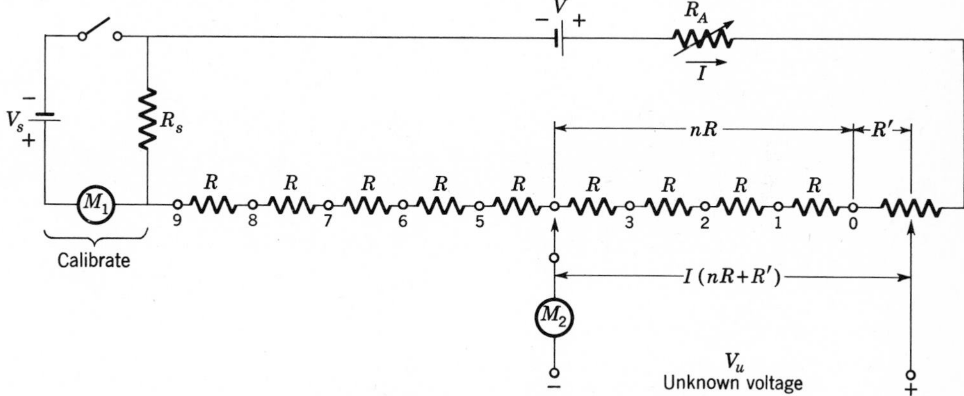

FIGURE 1-21 *Practical potentiometer circuit.*

In this circuit the precision variable resistance is composed on nine identical series resistors, each of value R, and a variable resistance R'. The current in the output circuit, indicated by M_2, is adjusted to zero with the unknown voltage source V_u connected to the output terminals. This is done by selecting the proper switch position for the switch and the slider. When the current in M_2 is zero

$$V_u = I(nR + R') \tag{1-71}$$

where n is the tap number of the selector switch. Substituting for I in Eq. (1-71) from Eq. (1-70)

$$V_u = V_S \frac{nR + R'}{R_S} \tag{1-72}$$

According to this expression, the unknown voltage is determined entirely in terms of the standard emf and the resistances of the potentiometer circuit. Rearranging (1-72)

$$\frac{V_u}{V_S} = \frac{nR + R'}{R_S} \tag{1-73}$$

which demonstrates directly how the unknown and standard voltages are compared in terms of resistances. Note, particularly, that neither the current I nor the battery voltage V need be known.

The accuracy of the potentiometer circuit depends upon the precision with which the various resistors are constructed and the mechanical stability of the slider on the variable resistance. Accuracy can be improved by combining fixed resistors selected by a tap switch together with the continuously variable resistor, as in Fig. 1-21. This is so because the potential drop across the variable resistor is only one-tenth of that, say, in the circuit of Fig. 1-20 and voltage errors caused by mechanical irregularities in the slider are reduced by the same factor. The variable resistor is ordinarily a wire of uniform resistance wound in the form of a helix. The moving slider is attached to a shaft so that the slider travels along the helix as the shaft is rotated.

A great virtue of the potentiometer circuit is that no current flows in the measuring circuit at balance. This means that the unknown potential is measured under effectively open-circuit conditions and the measurement is unperturbed by internal IR drops. Actually, a small current may be present, depending upon the sensitivity of the null indicator, and sensitive electronic amplifiers are often used in place of the meter to minimize the null current.

EQUIVALENT CIRCUITS

Thévenin's Theorem Many times the analysis of electronic circits is facilitated by replacing all or part of a network by an *equivalent circuit* which, for certain purposes, has the same characteristics as the original. An example of this possibility has already been discussed in connection with series and parallel combination of resistors. There, an entire network of resistances was replaced by a single equivalent resistance in order to calculate the current. In other situations, particularly in connection with vacuum-tube and transistor circuits, equivalent circuits may be used to represent the behavior of electronic devices. By replacing, say, a vacuum tube with its equivalent circuit it becomes possible to determine the electric characteristics of the device by circuit analysis.

One of the most useful equivalent circuits is the one that results from *Thévenin's theorem,* which states that any network of resistors and batteries having two output terminals may be replaced by the series combination of a resistor and a battery. The equivalence of the Thévenin circuit with the two-terminal network is demonstrated in the following way. Loop currents of the network are chosen such that only the current I_L is in the resistor R_L, which is connected to the output terminals of the network, Fig. 1-22*a*. The circuit equations are solved for I_L in the usual fashion. According to the previous discussion of loop currents, the equation for I_L has the form

$$\Delta I_L = V_1 D_1 + V_2 D_2 + \cdots + V_L D_L + \cdots + V_n D_n \tag{1-74}$$

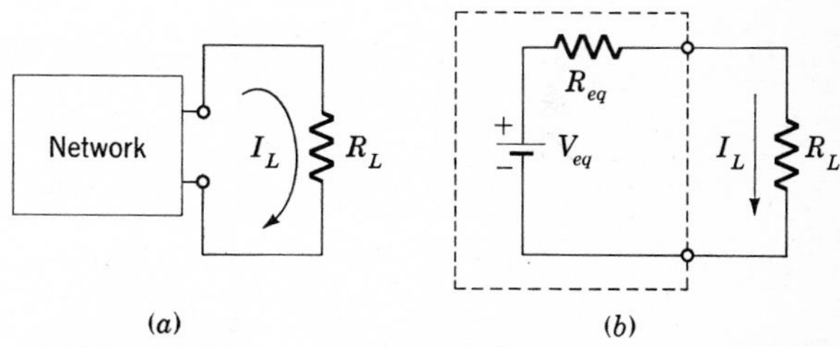

FIGURE 1-22 *(a) Two-terminal network and (b) its Thévenin equivalent circuit.*

if there are n loops in the network. In Eq. (1-74), Δ is the network determinant, the D's are coefficients made up of combinations of the resistors of the network, and each V is the algebraic sum of the emfs in a loop. In particular, V_L and D_L refer to the loop including the load resistor R_L. After dividing both sides by D_L the right side of Eq. (1-74) is just the sum of several voltages, which is identified by an equivalent emf

$$\frac{\Delta}{D_L} I_L = V_{eq}$$

The ratio Δ/D_L may be written as the load resistor R_L plus an equivalent resistance R_{eq}

$$(R_L + R_{eq})\, I_L = V_{eq} \tag{1-75}$$

But Eq. (1-75) is just the equation of the Thévenin equivalent circuit, Fig. 1-22*b*. Referring to the current through the load resistor, Fig. 1-22*b* is equivalent to the original complicated network in Fig. 1-22*a*.

The form of the Thévenin equivalent circuit, Fig. 1-22*b*, shows immediately how proper values for the quantities V_{eq} and R_{eq} are determined. The equivalent emf is just the voltage at the output terminals when the load current is zero, or the *open-circuit* voltage. The equivalent resistance is just the resistance between the terminals with the batteries of the network replaced by a zero resistance (or by their internal resistances, if these are appreciable).

Alternatively, it is often convenient to determine the equivalent resistance by measuring the *short-circuit* current, that is, I_L when $R_L = 0$. The value of R_{eq} is then just V_{eq} divided by the short-circuit current, according to Eq. (1-75). Note also that R_{eq} is also equal to the load resistance for which the voltage across the load is equal to $\frac{1}{2}V_{eq}$. This is useful in situations where the short-circuit current cannot be determined easily.

Consider the Thévenin equivalent of the simple circuit in Fig. 1-23. The equivalent emf is just

$$V_{eq} = \frac{VR_2}{R_1 + R_2} \tag{1-76}$$

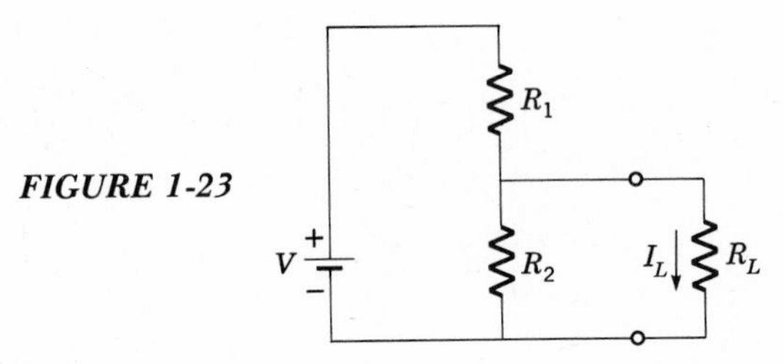

FIGURE 1-23

Replacing the battery with a zero resistance, the resistance between the terminals is just the parallel combination of R_1 and R_2. Accordingly,

$$R_{eq} = \frac{R_1R_2}{R_1 + R_2} \tag{1-77}$$

The load current is then

$$I_L = \frac{V_{eq}}{R_L + R_{eq}} = \frac{VR_2}{R_1R_2 + R_L(R_1 + R_2)} \tag{1-78}$$

The right side of Eq. (1-78) may be derived by dc analysis of Fig. 1-23. Finally, note that the output voltage, that is, the voltage across R_L, is, from Eq. (1-78),

$$V_L = I_LR_L = \frac{V_{eq}R_L}{R_L + R_{eq}} = \frac{V_{eq}}{1 + R_{eq}/R_L} \tag{1-79}$$

This shows that if R_L is large compared with R_{eq}, the output voltage is essentially equal to V_{eq}.

To illustrate the power of the equivalent-circuit method, consider the Wheatstone bridge circuit of Fig. 1-24*a*. The current through R_5 is determined by first replacing the remainder of the circuit with its Thévenin equivalent. Replacing the battery with a short circuit puts R_3 in parallel

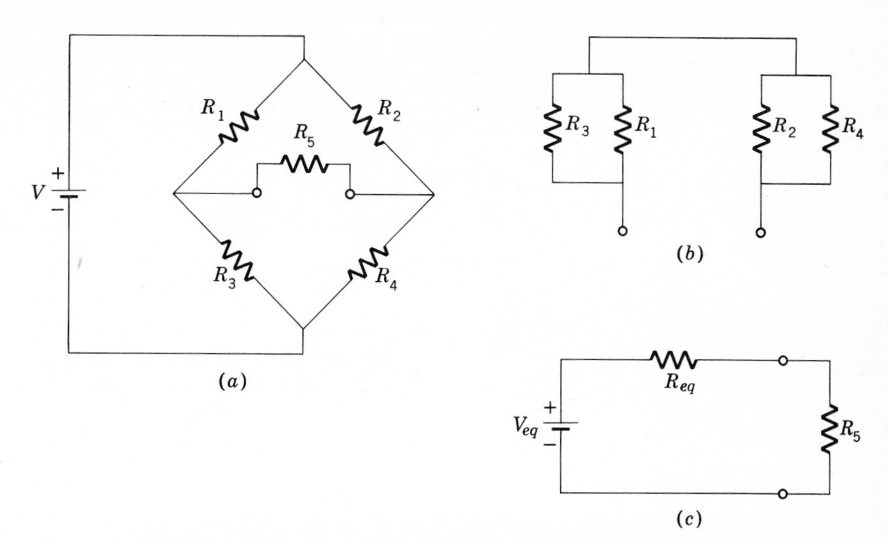

FIGURE 1-24 (a) Conventional Wheatstone bridge circuit; (b) after replacing battery with a short circuit in order to calculate R_{eq}; (c) the Thévenin equivalent.

with R_1 and this combination in series with the parallel combination of R_2 and R_4 across the output terminals, as illustrated in Fig. 1-24*b*. Therefore, R_{eq} in the equivalent circuit, Fig. 1-24*c*, is

$$R_{eq} = \frac{R_1 R_3}{R_1 + R_3} + \frac{R_2 R_4}{R_2 + R_4} \tag{1-80}$$

The open-circuit voltage at the output terminals is simply the potential difference between the junction of R_1 and R_3 and the junction of R_2 and R_4. This potential difference is found by subtracting the IR drop across R_2 from the IR drop across R_1. Therefore the equivalent battery is

$$V_{eq} = \frac{VR_1}{R_1 + R_3} - \frac{VR_2}{R_2 + R_1} \tag{1-81}$$

Finally, according to the equivalent circuit, Fig. 1-24*c*, the current through I_5 is

$$I_5 = \frac{V_{eq}}{R_{eq} + R_5} \tag{1-82}$$

The ease and rapidity with which this result is obtained should be compared with that necessary using Kirchhoff's rules. Note that the balance condition, Eq. (1-69), follows immediately from Eqs. (1-81) and (1-82), since $I_5 = 0$ at balance.

Norton's Theorem A second form of equivalent circuit useful in situations as, for example, transistor circuits where current sources, rather than emfs,

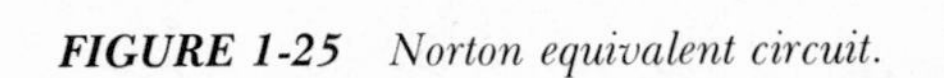

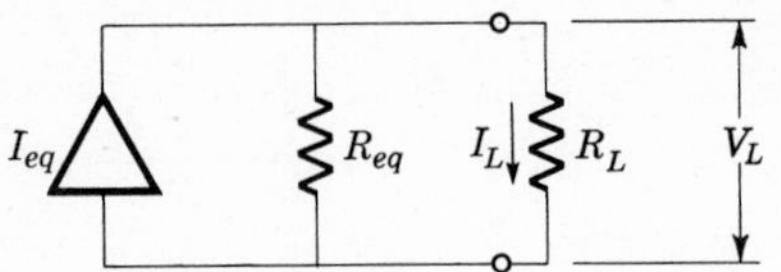

FIGURE 1-25 *Norton equivalent circuit.*

are of major interest, is one given by *Norton's theorem.* Norton's theorem states that any network of batteries and resistors having two output terminals can be replaced by a parallel combination of a current source I_{eq} and a resistance R_{eq}. The current source I_{eq} is the short-circuit current in the output terminals of a circuit, that is, the load current when the load resistance is zero. The resistance R_{eq} is the same as for Thévenin's theorem.

Norton's equivalent circuit is shown in Fig. 1-25, where the triangle represents the current source I_{eq}. No simple electric component acts as a current source the way a battery acts as a voltage source. Nevertheless, the idea of a current source is conceptually very useful in circuit analysis. Note that in Fig. 1-25 the voltage across the load resistor R_L is simply

$$V_L = I_{eq} \frac{R_{eq}R_L}{R_{eq} + R_L} \tag{1-83}$$

and the output current in R_L is

$$I_L = \frac{V_L}{R_L} = I_{eq} \frac{R_{eq}}{R_{eq} + R_L} = \frac{I_{eq}}{1 + R_L/R_{eq}} \tag{1-84}$$

According to Eq. (1-84), if R_L is small compared with R_{eq}, the output load current is approximately equal to I_{eq}. This corresponds to Eq. (1-79) for the Thévenin equivalent circuit.

Since it is possible to represent any network by either the Thévenin or the Norton equivalent circuit, it must be possible to convert from one equivalent circuit to the other. Referring to Fig. 1-26*a* and *b*, the short-circuit current through the load is V_{eq}/R_{eq} for the Thévenin equivalent circuit, and, according to Eq. (1-84), is equal to I_{eq} for the Norton equivalent circuit. For both circuits to represent the same network, it must be true that

$$I_{eq} = \frac{V_{eq}}{R_{eq}} \tag{1-85}$$

R_{eq}

$V_{eq} = I_{eq} R_{eq}$

$I_{eq} = \frac{V_{eq}}{R_{eq}}$ R_{eq}

(*a*) (*b*)

FIGURE 1-26 *Relationship between (a) Thévenin equivalent circuit and (b) Norton equivalent circuit.*

Thus, it is a simple matter to convert from one equivalent circuit to another. Which circuit is used to represent any given network is entirely a matter of choice and convenience.

Maximum Power Transfer In many electronic circuits, such as a radio transmitter or phonograph amplifier, it is important to transfer efficiently the maximum amount of electric power from the circuit to the load, which may be an antenna or loudspeaker. Therefore, it is of interest to determine circuit conditions for which it is possible to achieve the maximum power transfer. Suppose the network is represented by its Thévenin equivalent circuit shown in Fig. 1-27 and that the load connected to the output terminals is represented by the resistance R_L. The subscripts on the equivalent battery and equivalent resistance have been eliminated for convenience.

According to Joule's law, the power delivered to the load resistance is

$$P = I^2R_L = \left(\frac{V}{R + R_L}\right)^2 R_L$$

$$= \frac{V^2/R_L}{(1 + R/R_L)^2} \tag{1-86}$$

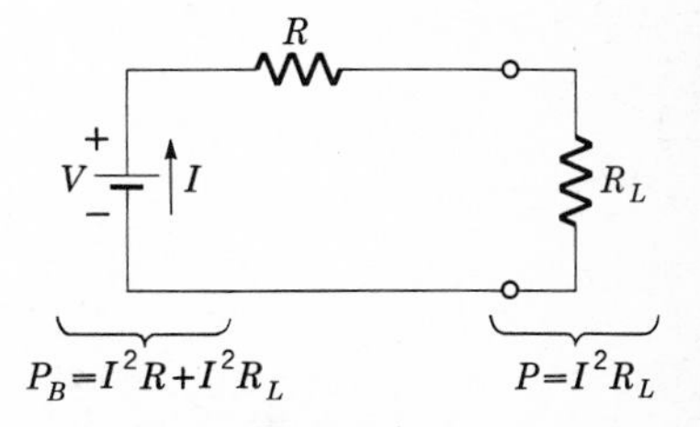

FIGURE 1-27 *Thévenin's equivalent circuit used to examine maximum power transfer to load resistance R_L.*

According to Eq. (1-86), the power in the load is zero if the load resistance is very small and is also zero when the load resistance is very large. Thus, there must be an optimum load resistance for which the power in R_L is a maximum.

To find the condition for *maximum power transfer*, differentiate Eq. (1-86) with respect to R_L and equate the result to zero,

$$\frac{dP_L}{dR_L} = \frac{V^2}{R_L} \frac{2R/R_L^2}{(1 + R/R_L)^3} + \frac{V^2}{(1 + R/R_L)^2} \frac{-1}{R_L^2} = 0 \tag{1-87}$$

$$\frac{2R}{R_L} = 1 + \frac{R}{R_L} \tag{1-88}$$

so that

$$R_L = R \tag{1-89}$$

This means that maximum power is delivered to the load when the load resistance is equal to the internal resistance of the network delivering the power. When the load resistance is equal to the internal resistance of the network, the load is said to be *matched* to the circuit. Note that with a matched load the ratio of the maximum power transferred to the power taken from the battery, $I^2R + I^2R_L$, is

$$\frac{P_{\text{max}}}{P_B} = \frac{I^2R_L}{I^2R_L + I^2R_L} = \frac{1}{2} \tag{1-90}$$

which means that half the power is lost in the internal resistance of the network and that the maximum efficiency is therefore 50 percent.

Since the equivalent circuit of Fig. 1-27 represents any network, the results represented by Eqs. (1-89) and (1-90) apply equally well to all circuits. The use of the equivalent-circuit concept in this connection has thus made it possible to prove a very general result quite easily.

ELECTRICAL MEASUREMENTS

D'Arsonval Meter By far the most common design for electric current-measuring instruments is the *d'Arsonval meter,* named after its inventor. A multiturn coil of fine wire wound on an aluminum frame is pivoted between the poles of a horseshoe permanent magnet (Fig. 1-28*a*). Two fine spiral springs serve to position the coil and to carry the current to be measured. A pointer attached to the coil indicates the current on a scale as the coil rotates in response to the interaction between the current in the coil and the magnetic field of the magnet. A soft-iron pole piece is fitted between the poles of the magnet so that the sides of the coil move in a radially directed field.

In considering the torque on a one-turn coil (Fig. 1-28*b*) note that the force on the left-hand conductor is IlB. This force is directed out of the plane of the figure. A similar force, oppositely directed, is exerted on the right-hand conductor. Therefore, the mechanical torque on the coil caused by the current I is simply $IlB \times 2r = IBA$, where A is the area of the coil. For a coil of n turns the total torque is n times this value.

The torque of restitution exerted by the two spiral springs is proportional to the angular deflection θ and may be written $K\theta$, where K is the spring constant. Equating this to the torque on the coil,

$$nIBA = K\theta \tag{1-91}$$

from which

$$\theta = \frac{nAB}{K} I \tag{1-92}$$

which means that the deflection of the pointer is directly proportional to the current. According to Eq. (1-92) the sensitivity, that is, the deflection

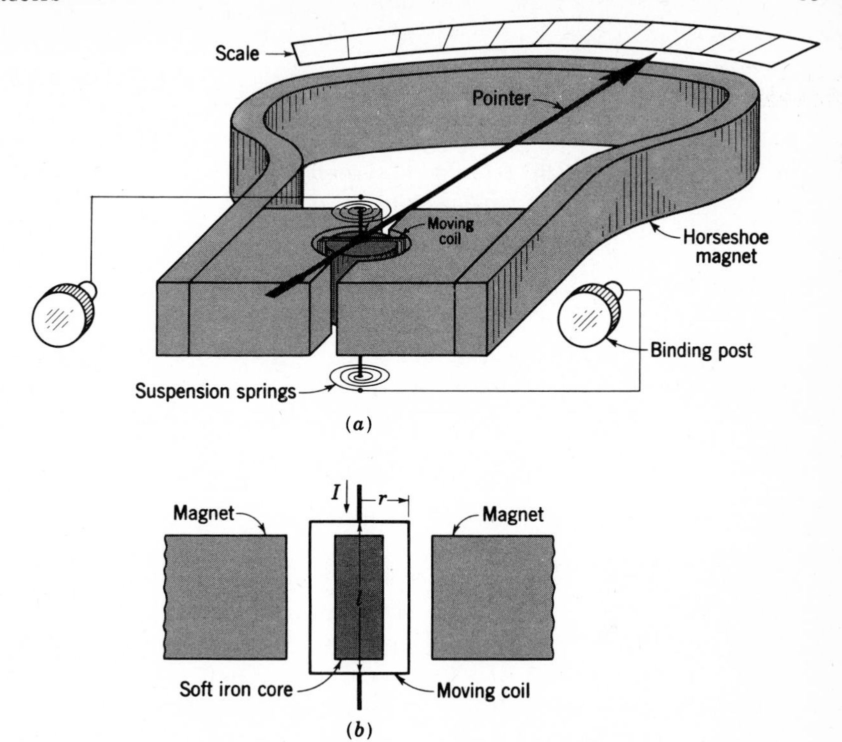

FIGURE 1-28 *(a) Sketch of the essential features of d'Arsonval meter and (b) a cross section of moving coil and magnet structure.*

for a given current, of a d'Arsonval meter is improved by increasing the magnetic field of the magnet, the coil area, and the number of turns on the coil, or by decreasing the torque constant of the springs.

The size of the coil and springs is dictated by mechanical ruggedness, since a large coil suspended by weak springs is subject to damage from mechanical shock and vibration. Also, it is not desirable to increase n inordinately in an effort to improve sensitivity, since the resistance of the coil is increased thereby. This may adversely affect the operation of the meter, as explained in a later section. The magnetic field is limited to that available from conventional permanent magnets. In spite of these practical limitations, common d'Arsonval meters have full-scale deflections for currents as small as 10^{-3} A (1 *milliampere,* abbreviated mA), or even 50×10^{-6} A (50 *microamperes,* abbreviated μA). Laboratory instruments, which may be shock-mounted and therefore designed for maximum sensitivity, are capable of measuring 10×10^{-12} A (10 *picoamperes,* abbreviated pA).

Ammeters and Voltmeters The d'Arsonval meter is a current-sensitive device, or *ammeter.* It is often convenient to change the current required for full-scale deflection in order to increase the range of currents over

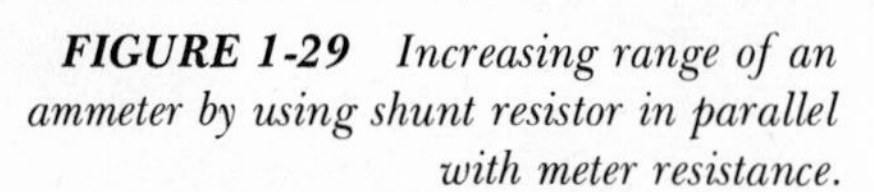

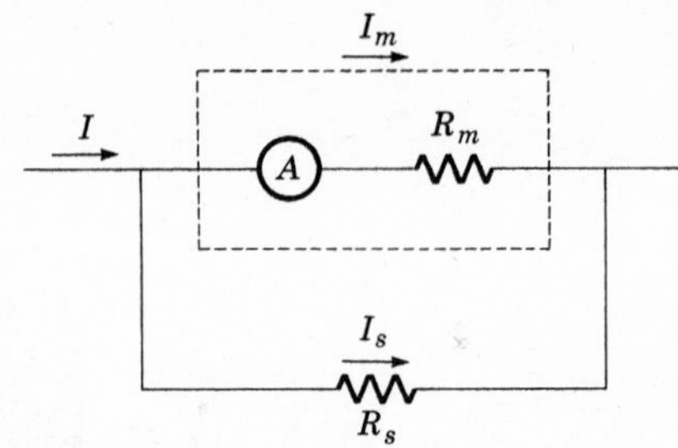

FIGURE 1-29 Increasing range of an ammeter by using shunt resistor in parallel with meter resistance.

which the meter is useful. This is accomplished by *shunting* a portion of the current around the ammeter with a parallel resistance, as diagramed in Fig. 1-29. Note that the internal resistance of the ammeter's coil R_m is indicated explicitly. Following Kirchhoff's rules, $I = I_m + I_s$ and $I_m R_m = I_s R_s$, so that the current to be determined is

$$I = I_m + \frac{I_m R_m}{R_s} = I_m \left(1 + \frac{R_m}{R_s}\right) \tag{1-93}$$

If, for example, the shunt resistance is one-ninth of the meter resistance, $1 + R_m/R_s = 10$, and the full-scale deflection is extended to ten times the inherent sensitivity of the meter.

It is always necessary to consider the effect of the meter resistance on the circuit. Suppose it is desired to measure the current in R_L of the Thévenin equivalent circuit of Fig. 1-30 using an ammeter having an internal resistance of R_m. With the ammeter connected into the circuit the current is

$$I = \frac{V}{R + R_L + R_m} \tag{1-94}$$

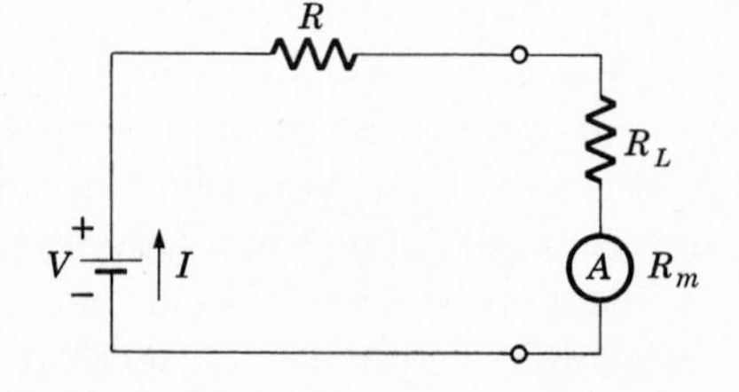

FIGURE 1-30 Effect of ammeter resistance on current in circuit.

Unless $R_m \ll R + R_L$, the current indicated by the ammeter is different from the true current. For this reason it is always desirable that the internal resistance of an ammeter be small. On the other hand, in circumstances where the internal resistance may not be small compared with circuit resistances, it is possible to correct for the disturbing influence of the meter resistance and thus determine the true current.

Since the full-scale deflection of an ammeter may be attributed to the

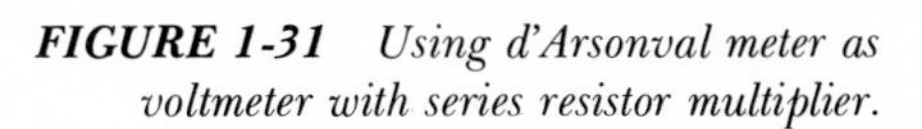

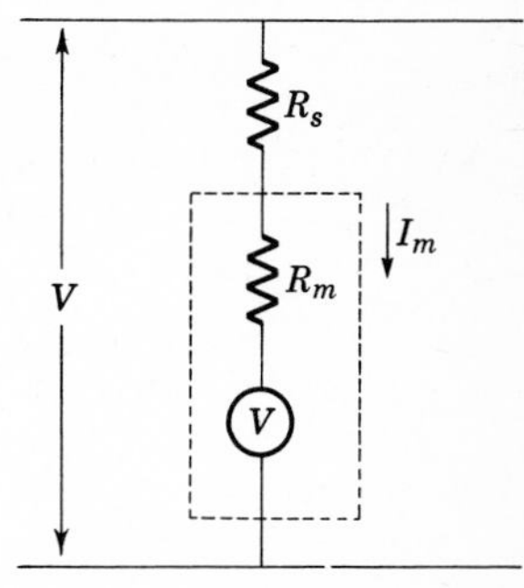

FIGURE 1-31 *Using d'Arsonval meter as voltmeter with series resistor multiplier.*

voltage $V_m = R_m I$ across the meter resistance R_m, a d'Arsonval meter also is a *voltmeter*. Here again it is useful to change the range of any given voltmeter by introducing a resistance in series with the meter. Referring to the circuit of Fig. 1-31, the voltage to be measured is

$$V = I_m(R_m + R_s) \tag{1-95}$$

and it is obvious that the series resistance R_s increases the maximum full-scale voltage of a given meter. It is common practice to provide several such series resistance *multipliers* to allow a given meter to be used over a wide range of voltages.

The effect of a voltmeter on the circuit to which it is attached must be considered, just as in the case of an ammeter. This is so because the voltmeter requires a small current to deflect the pointer and this current must be supplied by the circuit. If the meter current is not negligible compared with the normal currents in the circuit, the voltmeter is said to *load* the circuit and a correction must be applied to the indicated meter reading to determine the true voltage in the absence of the disturbing meter influence.

Sensitive d'Arsonval meters are useful as voltmeters since they require only a very small current to achieve full-scale deflection. It is common practice to specify the sensitivity of a voltmeter in terms of the ratio of its internal resistance to the full-scale voltage in units of *ohms per volt.* Note that, according to Eq. (1-95), the ratio of $R_m + R_s$ to the voltage required for full-scale deflection is simply the current sensitivity of the meter, and therefore the two specifications are quite equivalent. For example, a voltmeter which uses a d'Arsonval meter having a full-scale sensitivity of 1 mA is rated at $1/10^{-3} = 1000\ \Omega/\mathrm{V}$. This means that the voltmeter has an internal resistance of 100,000 Ω on the 100-V scale, etc. Similarly, a 20,000-Ω/V voltmeter (employing a 50-μA meter) has a resistance of 2 megohms (MΩ) on its 100-V scale.

A convenient technique for determining the resistance of a portion of a network is to measure both the current and the voltage and then apply Ohm's law. There are two distinct ways that the meters can be connected in this *voltmeter-ammeter method* (Fig. 1-32*a* and *b*). The choice between the

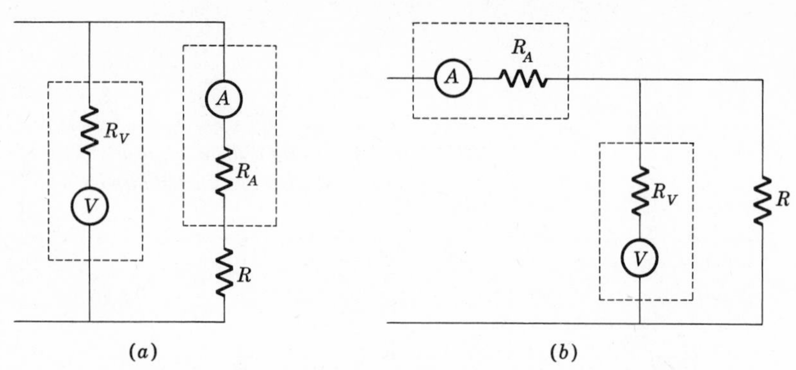

FIGURE 1-32 *Two ways of connecting voltmeter and ammeter to measure resistance R.*

two possibilities depends upon the relative values of the meter resistances and the circuit resistances, as can be shown in the following way. Consider first the circuit in Fig. 1-32*a*. From Kirchhoff's rules

$$V = AR + AR_A \tag{1-96}$$

where V and A are the meter readings. The unknown resistance is given by

$$R = \frac{V}{A} - R_A \tag{1-97}$$

which shows that the true resistance is smaller than the indicated V/A ratio.

Similarly, in the circuit of Fig. 1-32*b*, the current A divides between the parallel paths R and R_V so that

$$V = \frac{R_V R}{R_V + R} A \tag{1-98}$$

Solving for the unknown resistance, the result is, after some rearrangement,

$$R = \frac{V}{A} \frac{1}{1 - (V/A)/R_V} \tag{1-99}$$

According to Eqs. (1-97) and (1-99) the first circuit is most useful when the ammeter resistance is small compared with the unknown (or to V/A), while the second circuit applies when the voltmeter resistance is large compared with the unknown. In either case, the unknown resistance is then given simply by the ratio V/A.

Ohmmeters and Multimeters A simple extension of the voltmeter-ammeter circuit can be used as an *ohmmeter* in which the meter scale is calibrated directly in ohms. In a typical circuit (Fig. 1-33) the same meter is used suc-

FIGURE 1-33 A simple ohmmeter circuit.

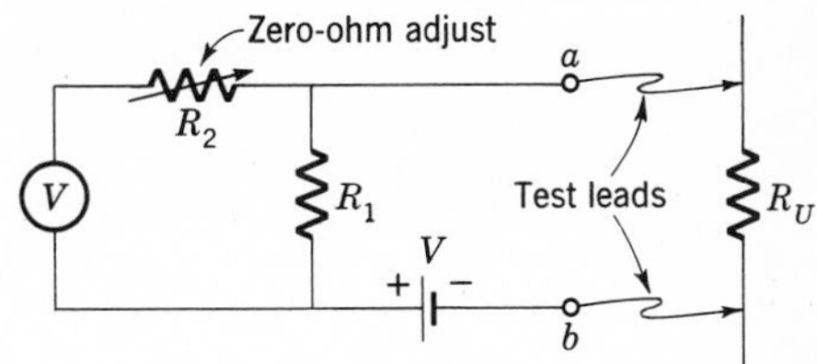

cessively to measure first the voltage across the unknown resistor and then the current in it. The way the meter scale is calibrated directly in ohms can be understood from the following analysis. First, suppose the terminals a and b in Fig. 1-33 (which are usually connected to *test leads* to facilitate connecting the ohmmeter to the unknown resistor) are shorted together. The voltmeter then measures the battery voltage V.

Next, the test leads are connected to the unknown resistance. If the voltage across R_1, as measured by the meter, is now V_R, Ohm's law gives

$$R_U + R_1 = \frac{V}{V_R/R_1} \tag{1-100}$$

Solving for R_U

$$R_U = R_1 \left(\frac{V}{V_R} - 1\right) \tag{1-101}$$

According to Eq. (1-101) the unknown resistance can be calculated directly from the two meter readings, but it is more useful to calibrate the meter scale directly in ohms in the following way.

The variable resistance R_2 is used to adjust the meter reading to full-scale when the test leads are first shorted together. This point on the scale is then "zero ohms," and is so marked. Suppose that when the test leads are connected to the unknown resistor, the meter deflects to half-scale. This means $V_R = V/2$ and, according to Eq. (1-101), $R_U = R_1$. Thus, the midpoint on the scale can be marked with the resistance corresponding to R_1. Similarly, note that the quarter-scale reading, $V_R = V/4$, corresponds to $3R_1$, while a zero reading indicates an open circuit, or infinite ohms. The scale of an ohmmeter is clearly nonlinear although it is not difficult to use since it is direct-reading in ohms.

According to Eq. (1-101) the midscale reading of an ohmmeter depends upon R_1, so that by selecting different values for R_1 it is possible to encompass a wide range of unknown resistances. Equation (1-101) assumes that the meter current is negligible, which may not be true on high-resistance ranges where R_1 is large. Therefore, practical ohmmeter circuits are slightly more complicated than the elementary one illustrated by Fig. 1-33, but the principle of operation is identical. Note that an ohmmeter cannot be used to determine resistance values in a circuit which is in operation, for erroneous readings result due to IR drops in the circuit itself.

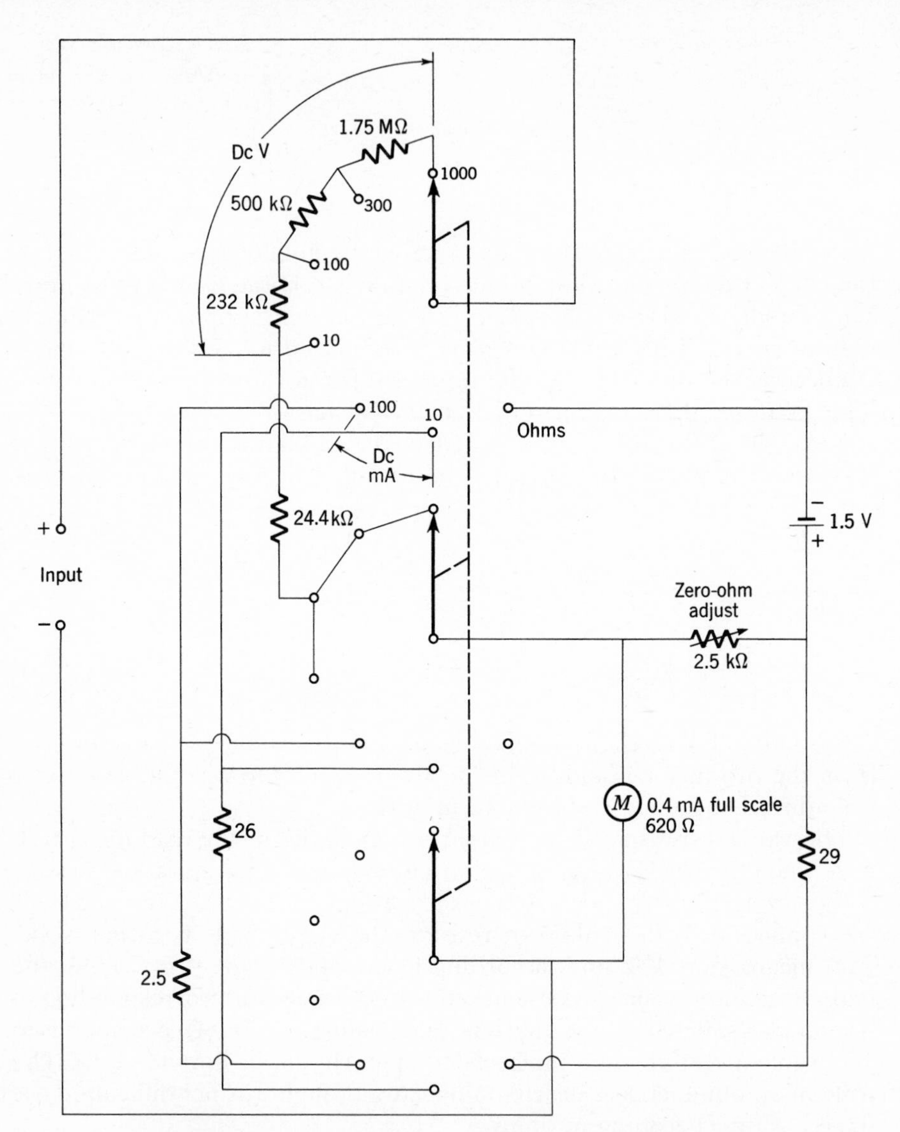

FIGURE 1-34 *Circuit diagram of a simple volt-ohm-milliammeter (VOM).*

It is convenient to include the functions of a voltmeter, ammeter, and ohmmeter within one instrument since all three employ the same basic d'Arsonval meter. In such an instrument, commonly termed a *multimeter* or *VOM* (volt-ohm-milliammeter), switches or a number of terminals select the function and range to be used. The circuit in Fig. 1-34 is an example of an elementary instrument which has four voltage ranges, two current ranges, and a single ohms scale. By carefully tracing through the connections on this circuit diagram it is possible to draw the simple functional circuits for each use and in this way subdivide the analysis of the circuit into easy stages.

The VOM circuit of Fig. 1-34 employs a multitap switch with three wipers which are mechanically linked to move together. More elaborate multimeters have considerably more complicated switching arrangements in order to accommodate additional ranges and other functions.

SUGGESTIONS FOR FURTHER READING

A. M. P. Brookes: "Basic Electric Circuits," The Macmillan Company, New York, 1963.

Leigh Page and Norman Ilsley Adams: "Principles of Electricity," D. Van Nostrand Company, Inc., Princeton, N.J., 1931.

M. E. Van Valkenburg: "Network Analysis," Prentice-Hall, Inc., Englewood Cliffs, N.J., 1955.

EXERCISES

1-1 Calculate the resistance of a copper wire 1 m long and 0.5 mm in diameter. Repeat for a similar-sized nichrome wire.

Ans.: $8.65 \times 10^{-2}\ \Omega$; $5.1\ \Omega$

1-2 What is the maximum current that can be in a 1-W 1-MΩ resistor? In a $\frac{1}{2}$-W 10,000-Ω resistor?

Ans.: 10^{-3} A; 7.07×10^{-3} A

1-3 Given that the maximum-current capability of a flashlight dry cell is 0.5 A, what is the internal resistance of the cell? Compare with the internal resistance of a storage battery, if the maximum current in this case is 500 A.

Ans.: 3 Ω; $4.2 \times 10^{-3}\ \Omega$

1-4 Assume that Fig. 1-11*a* represents a battery power source connected to a load R_3 by means of copper wires of resistance R_1 and R_2. If $V = 10$ V, the load is 5 Ω, and the wires are 0.5 mm in diameter and 100 m long, calculate the current, the power delivered to the load, the power lost in the wires, and the voltage across the load resistor.

Ans.: 1.07 A; 5.7 W; 4.9 W; 5.4 V

1-5 Design a voltage divider (Fig. 1-12) in which the output voltages possible are 1.0, 2.0, 5.0, and 10.0 V, if the battery voltage is 10 V and no current is taken by the output terminals.

1-6 How many identical 1-W resistors, and of what resistance value, are needed to yield an equivalent 1000-Ω 10-W resistor? Obtain two different solutions.

Ans.: Ten 10-kΩ resistors in parallel; ten 2.5-kΩ resistors in series-parallel

1-7 Determine the current through each resistor of Fig. 1-14*a*. Verify that the total I^2R loss in the resistors equals the power delivered by the battery.

Ans.: 2 amp in R_1 and R_2; 1 amp in R_3 and R_4; 0.5 amp in R_5 and R_6

1-8 In Fig. 1-15*a*, if $R_1 = 2\ \Omega$, $R_2 = 5\ \Omega$, $R_3 = 2\ \Omega$, $R_4 = 5\ \Omega$, $R_5 = 10\ \Omega$, and $V = 10$ V, find the total current supplied by the battery.

Ans.: 2.18 A

1-9 Find the resistance of the network in Fig. 1-35 between the input terminals. What voltage applied to the input results in a 1-A current in the 4-Ω resistor?

Ans.: 8 Ω; 72 V

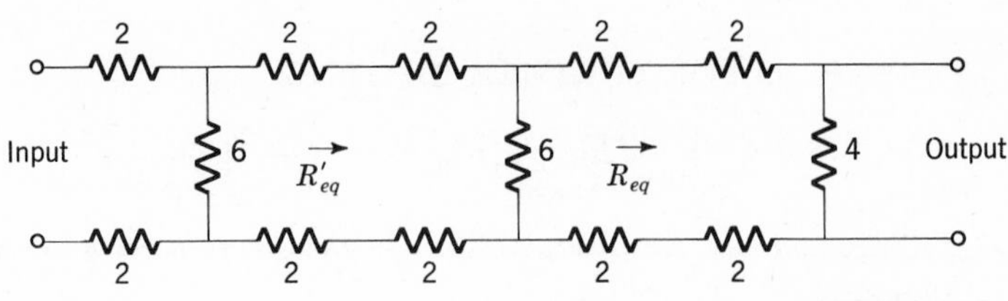

FIGURE 1-35

1-10 Using Kirchhoff's rules find the current in the 4-Ω resistor in the network in Fig. 1-36.

Ans.: 0.12 A

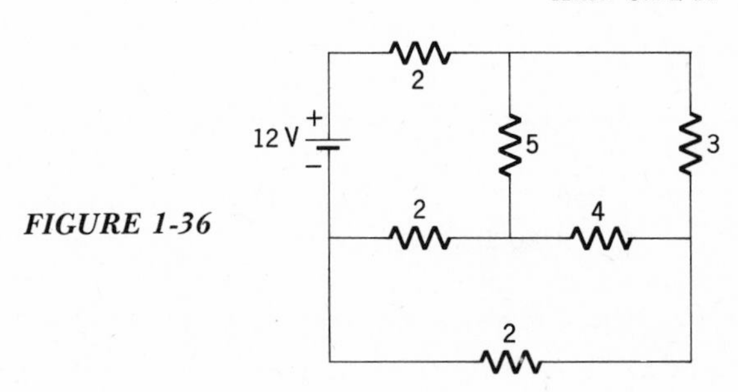

FIGURE 1-36

1-11 Find the current in each resistor of the circuit in Fig. 1-37.

Ans.: 1.15 A; 0.883 A; 0.267 A

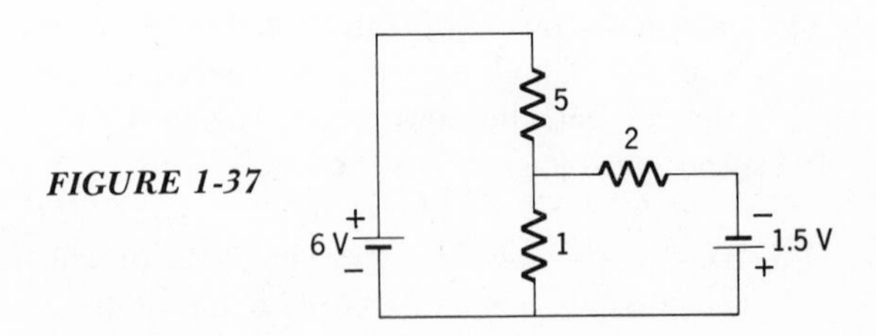

FIGURE 1-37

1-12 Note that in the Wheatstone bridge (Fig. 1-18) at balance $I_5 = 0$. Using this condition, compare the voltage drops in corresponding arms of the bridge and thus derive the balance condition, Eq. (1-69).

1-13 Solve the Wheatstone-bridge network of Fig. 1-19 by drawing loop currents such that there is only one current in R_5. Use the expression for this current to derive the balance condition.

1-14 Suppose that three arms of a Wheatstone bridge are 10-Ω resistors while the fourth is 10.1 Ω. If the battery voltage is 1.5 V, what is the current in a 100-Ω detector? Repeat if two of the three arms are 100-Ω resistors, while the third is 10 Ω.

Ans.: 4.1×10^{-4} A; 1.05×10^{-5} A

1-15 The detector in a potentiometer has a minimum sensitivity of 0.005 μA and an internal resistance of 25 Ω. What is the minimum error possible in measuring the voltage of an unknown?

Ans.: 1.25×10^{-7} V

1-16 Design an *Ayrton shunt* as in Fig. 1-38 for a 50-μA meter which has an internal resistance of 1000 Ω, if the desired current ranges are 10 mA, 100 mA, 1 A, and 10 A.

Ans.: 4.518 Ω; 0.4523 Ω; 4.523×10^{-2} Ω; 5.025×10^{-3} Ω

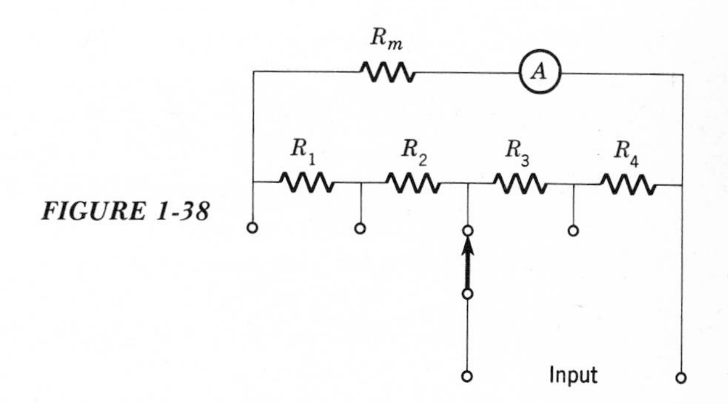

FIGURE 1-38

1-17 Suppose, in the circuit of Fig. 1-30, $R = 1000$ Ω, $R_L = 5000$ Ω, and $R_m = 1000$ Ω. If the indicated current is 1.5 mA, what is the true current when the ammeter is not present?

Ans.: 1.75 mA

1-18 The voltage in a circuit as measured with a 20,000-Ω/V meter on the 500-V scale is 200 V. On the 100-V scale the reading is 95 V. What is the true voltage?

Ans.: 278 V

1-19 What are the meter readings in the two versions of the voltmeter-ammeter method of Fig. 1-32 if the voltmeter is 1000 Ω/V, the internal resistance of the ammeter is 100 Ω, the "unknown" resistance is 1000 Ω, and the applied voltage is 10 V? Which of the two versions is more satisfactory?

Ans.: 10 V, 9.1×10^{-3} A; 9.008 V, 9.92×10^{-3} A; circuit *b* is better

1-20 Draw the scale of an ohmmeter, assuming $R_1 = 10{,}000$ Ω in Eq. (1-101).

LABORATORY EXERCISES

1-A The Wheatstone Bridge The Wheatstone bridge circuit is a most useful method to measure values of unknown resistors. This exercise examines several factors that determine the precision of measurement possible with this circuit. The technique used is to measure the values of several unknown resistors and to observe how well the balance condition, Eq. (1-69), can be obtained in practice.

In the Wheatstone bridge circuit of of Fig. 1-24 consider R_4 to be an "unknown" resistor of 110 Ω and R_3 to be a calibrated, variable resistor which is adjusted to achieve balance. Take $R_1 = R_2 = 100\ \Omega$ and $V = 12$ V. Prepare a plot of the voltage across the detector, R_5, as a function of R_3 for the region near balance and assuming that R_5 is large enough that the current in R_5 is negligible compared to current in the bridge arms. It is simplest to do this using the Thévenin equivalent circuit of the bridge. If you wish, write a simple Fortran program to carry out the multiple calculations.

Although high-resistance detectors are used in Wheatstone bridges, often the detector is a simple d'Arsonval meter which has low internal resistance. Plot the detector voltage versus R_3 if $R_5 = 100\Omega$, that is, if the detector resistance is matched to the internal resistance of the bridge at balance. Note the difference between the slopes of the two curves and decide which configuration is more sensitive.

Suppose the unknown resistor $R_4 = 1100\ \Omega$. Set up the bridge with $R_2 = 1000\ \Omega$ and $R_1 = 100\ \Omega$ so that $R_2/R_1 = 10$. Again, prepare a plot of the voltage across R_5 as a function of R_3 in the region near balance. Do this for both types of detectors in the previous paragraph. Now compare the slopes of the curves at balance and decide which method is better.

It is useful to determine a quantitative value for the precision of measurement in these several cases. Assume that the properties of both detectors are such that the balance condition can be determined to ± 0.01 V. From the slopes of the curves at balance calculate the expected precision in each case. Is a high-resistance or a low-resistance detector more suitable, assuming both have the same precision? Where possible, is an equal-arm bridge more desirable than an unequal-arm bridge? Discuss how you would decide between a precise low-resistance detector and a less precise high-resistance detector if you must use an unequal-arm bridge.

1-B Thévenin's Theorem Very frequently it is useful to determine experimentally the Thévenin equivalent circuit of a network. This is accomplished by measuring the current-voltage relationship at the terminals of the network and interpreting the experimental data in terms of the series combination of a voltage

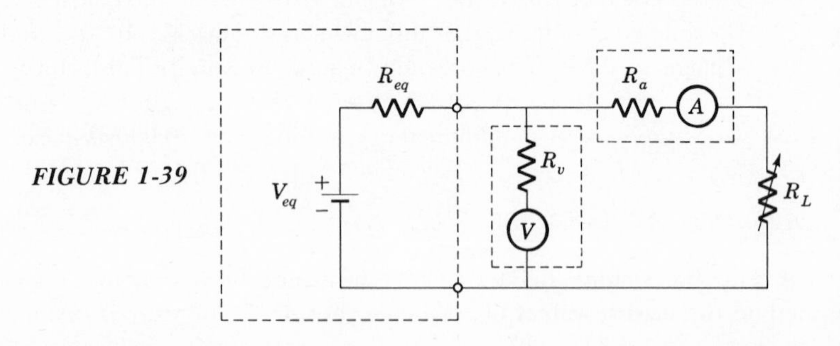

FIGURE 1-39

source and a resistor. Thus, for example, V_{eq} in Fig. 1-39 is the terminal voltage when the load current is zero, and R_{eq} is the ratio of V_{eq} divided by the short-circuit current.

The internal resistances of practical meters do not, however, permit true open-circuit or short-circuit conditions, and experimental results must be interpreted to account for the loading effects of the measuring instruments. This is illustrated in this experiment by examining the current-voltage data of an unknown *black box* having two terminals. The data is obtained by measuring the current in and the voltage across a variable-load resistance connected to the terminals. After the parameters of the Thévenin equivalent circuit are obtained the data is also used to investigate conditions for maximum power transfer to the load.

The current-voltage data pairs pertaining to Fig. 1-39 are given in Table 1-2. Knowing that the voltmeter has a sensitivity of 1000 Ω/V and that it is used on the 1-V scale, use this data to calculate V_{eq}, R_{eq}, and R_a. Probably the clearest way to do this is to plot the current as a function of voltage and interpret the curve in terms of an expression between current and voltage developed from your analysis of the circuit.

Plot the product IV, that is, the measured load power, as a function of V and note the point at which maximum power is delivered. Also plot the measured resistance V/I and determine the resistance for maximum power transfer. Compare this to R_{eq} and comment on the comparison.

The Thévenin equivalent does not specify uniquely the actual circuit in the black box. See if you can devise at least one practical circuit containing two voltage sources and two resistors, which is represented by the data in Table 1-2.

TABLE 1-2

I, − mA	V, volts
0 ($R_L = \infty$)	0.468
1.0	0.405
2.0	0.342
3.0	0.279
4.0	0.216
5.0	0.153
6.0	0.090
6.4 ($R_L = 0$)	0.064

chapter two

ALTERNATING CURRENTS

The currents and voltages in most practical electronic circuits are not steady but vary with time. For example, when a circuit is used to measure some physical quantity, such as the temperature of a chemical reaction, the voltage or current in the circuit representing the temperature may vary arbitrarily. Similarly, detection of nuclear disintegrations results in a series of rapid voltage pulses of very short duration. In order to understand such effects it is necessary to study the properties of time-varying currents.

The simplest time-varying current alternates direction periodically and accordingly is called an alternating current, abbreviated ac. Obviously an ac circuit is one in which alternating currents are active, but direct currents may be present as well. Most of the concepts developed for dc circuits in the previous chapter carry over to ac circuits. Two new elements in addition to resistance are important in ac circuits and these are treated in this chapter.

SINUSOIDAL SIGNALS

Frequency, Amplitude, and Phase The simplest alternating waveform is *sine-wave* voltage or current, which varies sinusoidally with time. A sinusoidal waveform is generated by the variation of the vertical component of a vector rotating counterclockwise with a uniform angular velocity ω as in Fig. 2-1. One complete revolution is termed a *cycle* and the time interval required for one cycle is called the *period* T. The number of cycles per second is the *frequency* f, and therefore

$$f = \frac{1}{T} \tag{2-1}$$

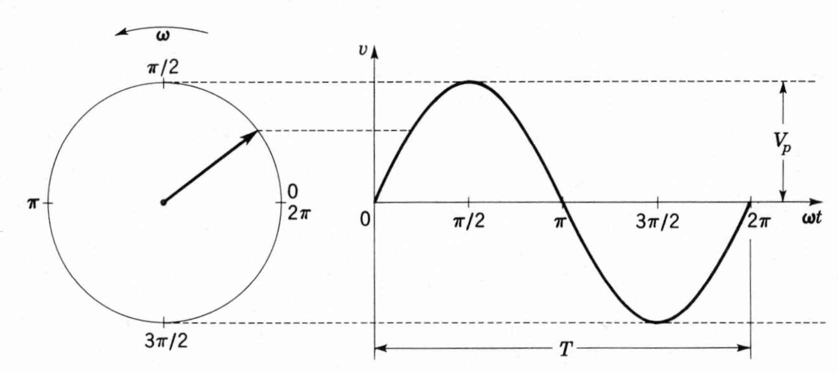

FIGURE 2-1 *Generation of sine wave by vertical component of rotating vector.*

The scope of frequencies encountered in electronic circuits is often as small as a few cycles per second, which is called a *hertz* (abbreviated Hz) in honor of Heinrich Hertz, discoverer of radio waves. The range of frequency extends to *kilohertz* (kHz, 10^3 Hz) and megahertz (MHz, 10^6 Hz), on up to the gigahertz (GHz, 10^9 Hz) range.

Since there are 2π radians in one complete revolution and this requires T s,

$$\omega = \frac{2\pi}{T} = 2\pi f \tag{2-2}$$

If the length of the vector is V_p, the instantaneous value at any time t is just

$$v = V_p \sin \omega t \tag{2-3}$$

The value V_p is the *maximum* or *peak amplitude* of the sine wave.

If two sinusoidal waveforms have the same frequency but pass through zero at different times, they are said to be out of *phase*, and the angle between the two rotating vectors is called the *phase angle*. In Fig. 2-2 the

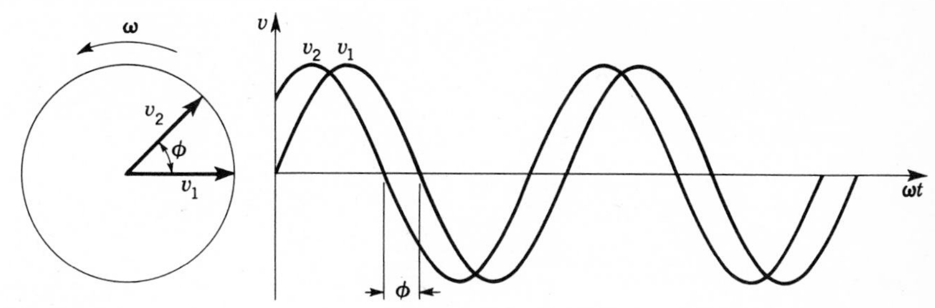

FIGURE 2-2 *Illustrating phase angle between two sinusoidal voltages.*

voltage v_2 is *leading* the sine-wave voltage v_1 because it passes through zero first and the phase difference is the angle ϕ. Note that it is only possible to specify the phase angle between two sine waves if they have the same frequency. A voltage sine wave is completely described by its frequency and amplitude unless it is compared with another signal of the same frequency. In this case the most general equation for the voltage must include the phase angle

$$v = V_p \sin(\omega t + \phi) \tag{2-4}$$

Note that lowercase symbols are used to indicate time-varying voltages (and currents), while capital letters refer to constant values or to dc quantities.

Rms Value It is often necessary to compare the magnitude of a sine-wave current with a direct current. This is accomplished by comparing the joule heat produced in a resistor by the two currents. That is, the *effective* value of a sinusoidal current is equal to the direct current which produces the same joule heating as the alternating current. To determine this value, the heating effect of an alternating current is calculated by averaging the I^2R losses over a complete cycle. Therefore, the average power is given by

$$P = \frac{1}{T}\int_0^T i^2R\,dt = \frac{I_p{}^2R}{T}\int_0^T \sin^2 \omega t\,dt \tag{2-5}$$

The integral of $\sin^2 \omega t$ is a standard form evaluated in integral tables, so that

$$P = \frac{I_p{}^2R}{T}\left[\frac{t}{2} - \frac{\sin 2\omega t}{4\omega}\right]_0^T = \frac{I_p{}^2RT}{2T} \tag{2-6}$$

Since the joule heating in a resistor caused by a direct current is equal to I^2R, the effective value of an alternating current I_e is simply

$$I_e{}^2R = \frac{I_p{}^2R}{2} \tag{2-7}$$

or

$$I_e = \frac{I_p}{\sqrt{2}} \tag{2-8}$$

According to Eq. (2-8) the effective value of a sine wave is simply its peak value divided by the square root of two. The effective value is commonly referred to as the *root-mean-square* or *rms* value. Voltmeters and ammeters capable of measuring ac signals are almost universally calibrated in terms of rms readings to facilitate comparison with dc meter readings. It is generally understood that ac currents and voltages are characterized by their rms values, unless stated otherwise.

Power Factor Suppose that the current and voltage in some portion of a circuit are given by

$$i = I_p \sin \omega t$$
$$v = V_p \sin (\omega t + \phi) \tag{2-9}$$

where the phase angle ϕ is introduced to account for the possibility that the current and voltage are not in phase. The instantaneous power p is then

$$p = vi = V_p I_p \sin \omega t \sin (\omega t + \phi) \tag{2-10}$$

According to Eq. (2-10) the instantaneous power in this part of the circuit varies with time and may even become negative, as illustrated by the waveforms in Fig. 2-3. The interpretation of negative power in Eq. (2-10) is that during some portion of a cycle this part of the circuit gives up electrical power to the rest of the circuit. For the remainder of the cycle the circuit delivers power to the part under investigation.

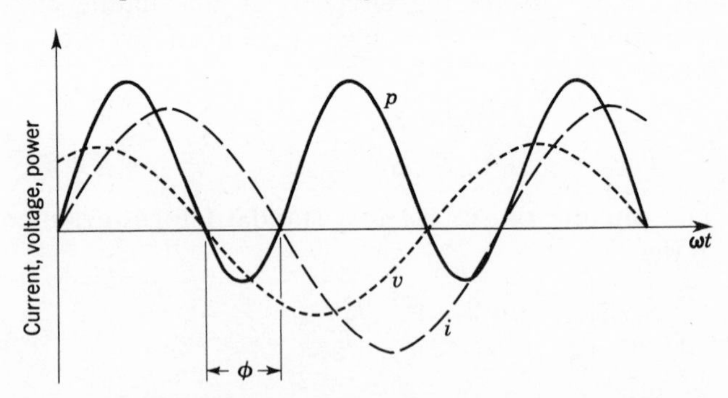

FIGURE 2-3 *Instantaneous power in an ac circuit.*

The average power is found by averaging Eq. (2-10) over a complete cycle,

$$P = \frac{1}{T}\int_0^T vi\,dt = \frac{V_pI_p}{T}\int_0^T \sin\omega t \sin(\omega t + \phi)\,dt \tag{2-11}$$

The second factor under the integral is expanded using a standard trigonometric identity

$$P = \frac{V_pI_p}{T}\left(\cos\phi \int_0^T \sin^2\omega t\,dt + \sin\phi \int_0^T \sin\omega t\,dt\right) \tag{2-12}$$

Both integrals are standard forms and may be evaluated directly to give

$$P = \frac{V_pI_p}{2}\cos\phi \tag{2-13}$$

$$P = VI\cos\phi \tag{2-14}$$

where V and I are rms values.

The meaning of Eq. (2-14) is that the useful power in ac circuits depends not only upon the current and voltage in the circuit but also upon the phase difference between them. The term $\cos\phi$ is called the *power factor* of the circuit. Note that when the phase angle is 90° the power factor is zero and no useful electrical power is delivered. Therefore, it is possible for the current and voltage to be very large and, consequently, the instantaneous power large, yet the average power to be zero. On the other hand, when the current and voltage are in phase, the power factor is unity and the power is equal to the current times the voltage, as in a dc circuit.

CAPACITANCE

Capacitors According to Coulomb's law, Eq. (1-1), and the definition of electric field, Eq. (1-2), the electric field due to a point charge is

$$\mathcal{E} = \frac{1}{4\pi\epsilon_0}\frac{q}{r^2} \tag{2-15}$$

Using the relation between field and potential gradient, Eq. (1-8),

$$\frac{dV}{ds} = -\frac{1}{4\pi\epsilon_0 r^2}q \tag{2-16}$$

which may be written in the form

$$dV = \frac{-ds}{4\pi\epsilon_0 r^2}q \tag{2-17}$$

Note that the potential difference in Eq. (2-17) increases linearly with q. Insofar as the relation between potential difference and charge is concerned, the value of the fraction depends only upon the geometry of the situation and is a constant for a given shape and arrangement of conductors and charges. This geometric constant, the ratio of charge to potential difference, is called the *capacitance C*. Therefore, the relation between charge and potential difference is written as

$$Q = CV \tag{2-18}$$

Capacitance is important in ac circuits because a voltage which changes with time gives rise to a time-varying charge according to Eq. (2-18), and this is equivalent to a current. For example, differentiating both sides of Eq. (2-18) with respect to time, and using the definition of current, Eq. (1-10), the result is

$$\frac{dv}{dt} = \frac{1}{C}\frac{dq}{dt} = \frac{i}{C} \tag{2-19}$$

Integrating Eq. (2-19)

$$v = \frac{1}{C}\int_0^t i\, dt \tag{2-20}$$

According to Eq. (2-20), a varying voltage appears across a capacitance whenever a current which varies with time is present.

The unit of capacitance is the *farad*, named in honor of Michael Faraday. It is equal to a coulomb per volt, according to (2-18). Actually the magnitude of the farad is too large to be convenient so that practical values are 10^{-6} farad (1 *microfarad*, abbreviated μF) or even 10^{-12} farad (1 *picofarad*, abbreviated pF).

Circuit elements having specific values of capacitance are known as *capacitors*. Most of the capacitors used in electronic circuits consist of two conducting plates separated by a small air gap or thin insulator. The capacitance of such a parallel-plate capacitor is given by

$$C = \frac{\epsilon_0 A}{d} \tag{2-21}$$

where A is the area of the plates and d is their separation. Note that the capacitance depends entirely upon geometrical factors, in accordance with the preceding discussion. According to Eq. (2-21), the capacitance is increased by making the area of the plates large and the separation between plates small.

It turns out that a solid insulating material between the parallel plates also increases the capacitance of a capacitor. The insulator, in effect, permits a greater charge on the plates at a given voltage. The increased capacitance is accounted for by multiplying the right side of Eq. (2-21) by the *dielectric constant* of the insulator. For example, the dielectric constant of mica is about 6 and of paper is about 2, so that the capacitances of capacitors made from these materials are greater by factors of 6 and 2, respectively, than a parallel-plate capacitor with air between the plates.

Conventional capacitors are made of two thin metal foils separated by a thin insulator or *dielectric*, such as paper or mica. This sandwich is then rolled or folded into a compact size and covered with an insulating coating. One axial wire lead is attached to each plate. In order to increase the capacitance, it is desirable for the insulator to be as thin as possible. This can only be done at the expense of reducing the maximum voltage that can be applied before the insulator ruptures because of the intense electric field. Another important factor is the resistivity of the insulator. Thin, large-area shapes increase the *leakage* resistance between the plates and thus degrade the capacitor. Mica and paper dielectric capacitors are available in capacitances ranging from 0.001 to 1 μF and can be used in circuits where the maximum voltage is of the order of hundreds of volts.

Ceramic and plastic-film capacitors are also used, generally with metal-film plates deposited directly on the dielectric. Plastic dielectrics have very high resistivity, which means that the leakage resistance is extremely small. The large dielectric constant of many ceramic materials provides large capacitance values in a small package.

In several applications, most notably transistor circuits, very large capacitance values are desirable and leakage resistances are of secondary concern. *Electrolytic* capacitors made of an oxidized metal foil in a conducting paste or solution are used to achieve high capacitance values.

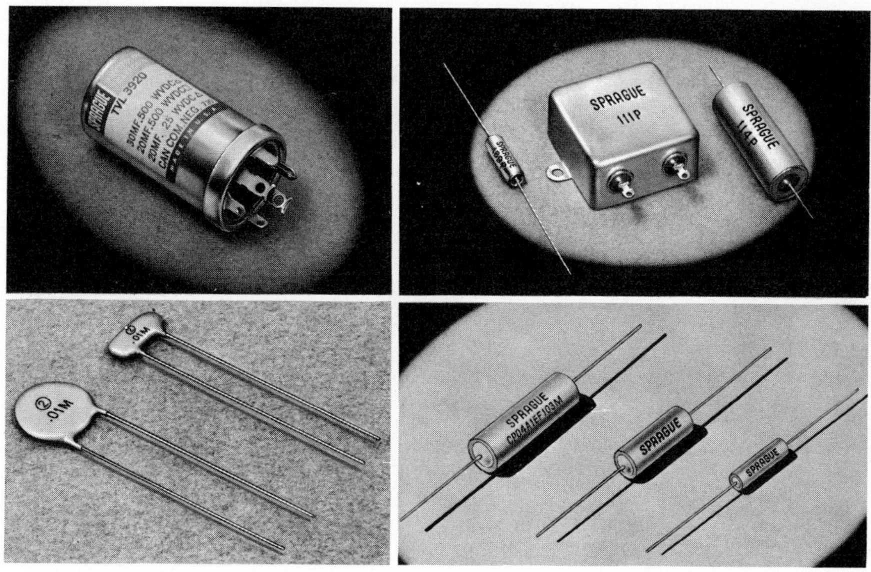

FIGURE 2-4 *Typical capacitors.* (Sprague Electric Company).

The thin oxide film is the dielectric between the metal foil and the solution. Because the film is extremely thin the capacitance is quite large. Several metals, such as tantalum and aluminum, can be used in electrolytic capacitors, and capacitance values range from 1 to 10^4 μF. The largest capacitances are useful in circuits where the applied voltage does not exceed a few volts, because the oxide dielectric is so thin. Electrolytic capacitors can only be used in circuit situations where the metal foil never becomes negative with respect to the solution. If the foil becomes negative, electrolytic action destroys the film and the capacitor becomes useless. Some typical capacitors are shown in Fig. 2-4.

It is often convenient to vary the capacitance in a circuit without removing the capacitor. The components described above are *fixed* capacitors, since it is clearly not easy to vary their capacitance. The common *variable* capacitor is made of two sets of interleaved plates, one immobile and the other set attached to a shaft. Rotating the shaft effectively changes the area in Eq. (2-21), thus changing the capacitance. Because the dielectric is air and it is necessary to make the separation plates relatively large to assure that they do not touch, maximum capacitance values are limited to about 500 pF. In the fully unmeshed state a 500-pF variable capacitor may have a minimum capacitance of 10 pF or so. Mica *trimmer* capacitors use a mica dielectric. The separation between plates is adjusted with a screwdriver. They are commonly used where variation in capacitance is only occasionally necessary. The range of capacitance values is about the same as for air dielectric capacitors.

Conventional symbols for capacitors in circuit diagrams are reminiscent of the parallel-plate construction, as illustrated in Fig. 2-5. The slightly curved plate is the negative terminal in the case of electrolytic capacitors, and the outside plate on foil capacitors. This is sometimes significant in electronic circuits in order to connect the capacitor properly, even though foil capacitors are not sensitive to voltage polarity.

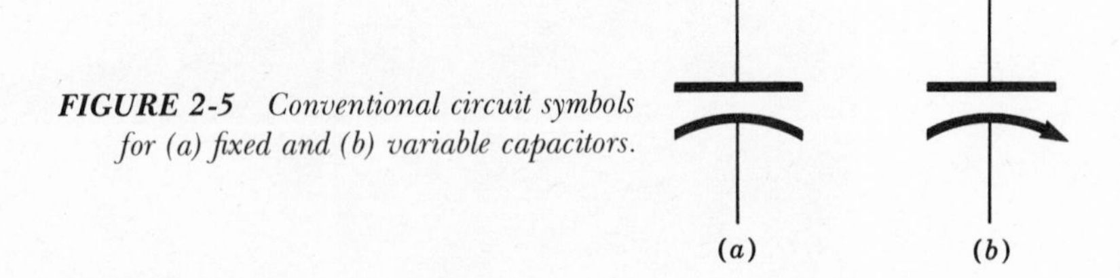

FIGURE 2-5 *Conventional circuit symbols for (a) fixed and (b) variable capacitors.*

Series and Parallel Capacitors Consider the equivalent capacitance C_{eq} of three separate capacitors connected in series as shown in Fig. 2-6. The sum of the voltages across each capacitor is the total potential V, so that using Eq. (2-18)

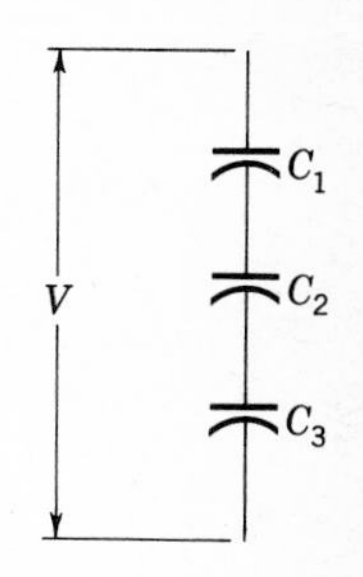

FIGURE 2-6 *Capacitors in series.*

$$V = \frac{Q_1}{C_1} + \frac{Q_2}{C_1} + \frac{Q_3}{C_3} \tag{2-22}$$

Since the capacitors are connected in series, the same electrical charge $Q = C_{eq}V$ resides on each one. This is so because, for example, the charge on the left plate of C_2 must have come from the right plate of C_1, etc. Therefore, $Q = Q_1 = Q_2 = Q_3$, and

$$\frac{Q}{C_{eq}} = \frac{Q}{C_1} + \frac{Q}{C_2} + \frac{Q}{C_3} \tag{2-23}$$

so that

$$\frac{1}{C_{eq}} = \frac{1}{C_1} + \frac{1}{C_2} + \frac{1}{C_3} \tag{2-24}$$

Evidently the reciprocal of the equivalent capacitance of any number of capacitors connected in series equals the sum of the reciprocals of the individual capacitances.

The equivalent of parallel-connected capacitors, Fig. 2-7, is found using

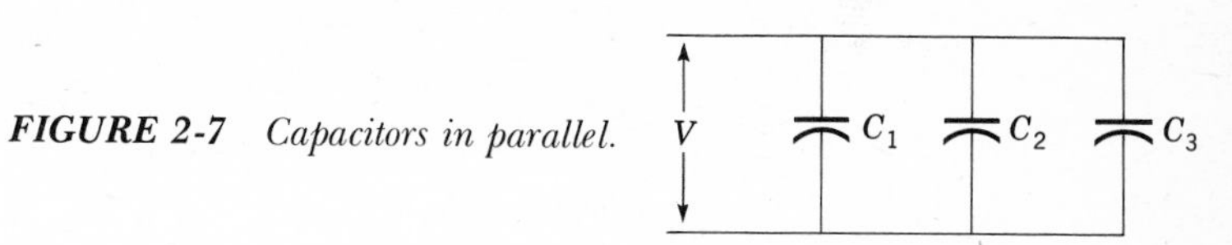

FIGURE 2-7 *Capacitors in parallel.*

similar reasoning. Obviously the voltage across all the capacitors is the same, while the total charge is the sum of the individual charges on each capacitor,

$$Q = Q_1 + Q_2 + Q_3 \tag{2-25}$$

Therefore

$$\frac{Q}{V} = \frac{Q_1}{V} + \frac{Q_2}{V} + \frac{Q_3}{V}$$

and

$$C_{eq} = C_1 + C_2 + C_3 \tag{2-26}$$

This means that the total equivalent capacitance of any number of capacitors connected in parallel is equal to the sum of the individual capacitances.

Differentiating and Integrating Circuits An elementary but very useful circuit employs a capacitor and resistor connected in series, as shown in Fig. 2-8. This *RC filter* is connected to a source of a sinusoidal voltage

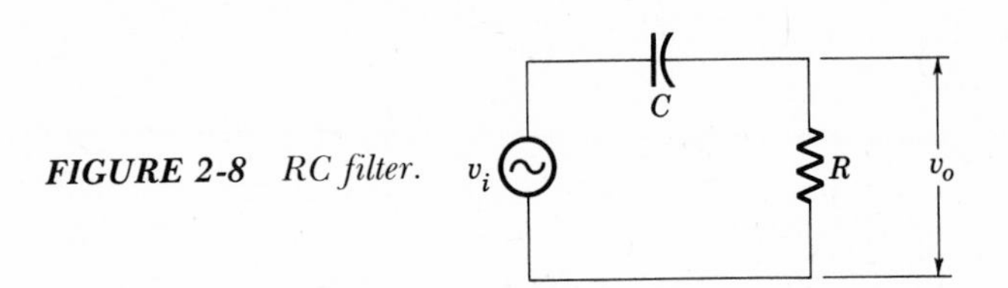

FIGURE 2-8 *RC filter.*

symbolized on the circuit diagram by a circle containing a one-cycle sine wave. If this voltage is represented by

$$v_i = V_p \sin \omega t \tag{2-27}$$

the current in the circuit is determined in the following way. According to Kirchhoff's rules the sum of the voltages around the loop must be equal to zero at every instant. This means that the source voltage must equal the voltage across the capacitor plus the IR drop across the resistor, or

$$vi = \frac{Q}{C} + Ri \tag{2-28}$$

where i is the instantaneous current. After differentiating each term with respect to time and putting $i = dQ/dt$, Eq. (2-28) becomes

$$\frac{i}{C} + R\frac{di}{dt} = \frac{dv_i}{dt} = \omega V_p \cos \omega t \tag{2-29}$$

Rearranging

$$R\frac{di}{dt} + \frac{1}{C}i = \omega V_p \cos \omega t \tag{2-30}$$

The solution of this differential equation is the circuit current i.

To solve this circuit differential equation, note that, because the applied signal has the form of Eq. (2-27), it is likely that the current is also sinusoidal of frequency ω. Therefore, we assume that the current is given by

$$i = I_p \sin (\omega t + \phi) \tag{2-31}$$

where I_p and ϕ are to be determined. Differentiating Eq. (2-31)

$$\frac{di}{dt} = \omega I_p \cos(\omega t + \phi) \tag{2-32}$$

Now Eqs. (2-31) and (2-32) are substituted into the circuit differential equation, Eq. (2-30). The result is

$$R\omega I_p \cos(\omega t + \phi) + \frac{I_p}{C} \sin(\omega t + \phi) = V_p \omega \cos \omega t \tag{2-33}$$

and the equation is solved by choosing values for I_p and ϕ that make Eq. (2-33) true for all values of t. This substitution has changed the differential equation into a trigonometric equation.

The values for I_p and ϕ which satisfy Eq. (2-33) are found by first expanding each term in Eq. (2-33) using a trigonometric identity. This gives, after dividing through by ωI_p,

$$R(\cos \omega t \cos \phi - \sin \omega t \sin \phi) + \frac{1}{\omega C}(\sin \omega t \cos \phi + \cos \omega t \sin \phi) = \frac{V_p}{I_p} \cos \omega t \tag{2-34}$$

Collecting terms in sin ωt and cos ωt,

$$\left(R \cos \phi + \frac{1}{\omega C} \sin \phi - \frac{V_p}{I_p}\right) \cos \omega t + \left(\frac{1}{\omega C} \cos \phi - R \sin \phi\right) \sin \omega t = 0 \tag{2-35}$$

Now consider that $t = 0$; then $\sin \omega t = 0$ and

$$R \cos \phi + \frac{1}{\omega C} \sin \phi - \frac{V_p}{I_p} = 0 \tag{2-36}$$

Similarly, suppose $\omega t = \pi/2$, so that $\cos \omega t = 0$. Then

$$\frac{1}{\omega C} \cos \phi - R \sin \phi = 0 \tag{2-37}$$

In order that Eq. (2-35) be satisfied for all values of t, both Eqs. (2-36) and (2-37) must be true. Equation (2-37) may be solved immediately for ϕ,

$$\frac{1}{\omega C} \cos \phi = R \sin \phi$$

$$\tan \phi = \frac{\sin \phi}{\cos \phi} = \frac{1}{R\omega C}$$

so that

$$\phi = \arctan \frac{1}{R\omega C} \tag{2-38}$$

The solution for ϕ is used in Eq. (2-36) to solve for I_p. This is done with the aid of the diagram in Fig. 2-9, which can be used to evaluate sin ϕ

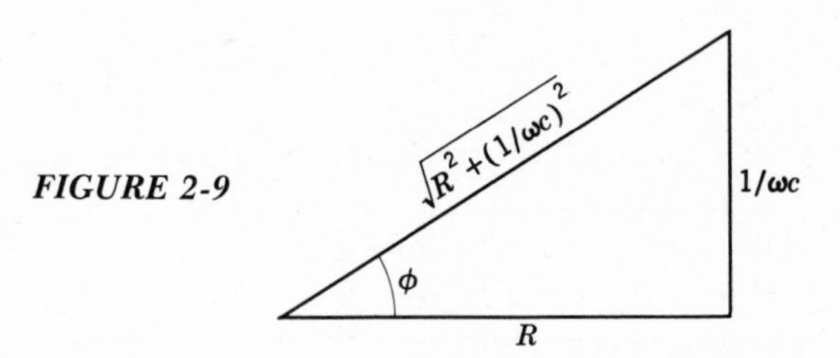

FIGURE 2-9

and cos ϕ. The sides of the triangle in Fig. 2-9 follow directly from Eq. (2-38). Substituting into Eq. (2-36)

$$R\frac{R}{\sqrt{R^2+(1/\omega C)^2}} + \frac{1}{\omega C}\frac{1/\omega C}{\sqrt{R^2+(1/\omega C)^2}} - \frac{V_p}{I_p} = 0 \tag{2-39}$$

So finally

$$I_p = \frac{V_p}{\sqrt{R^2+(1/\omega C)^2}} \tag{2-40}$$

Thus the current in the circuit is

$$i = \frac{V_p}{\sqrt{R^2+(1/\omega C)^2}} \sin(\omega t + \phi) \tag{2-41}$$

where

$$\phi = \arctan \frac{1}{\omega RC}$$

That Eq. (2-41) is indeed a solution of the circuit differential equation may be verified by directly substituting into (2-30).

Note that according to Eq. (2-38), the phase angle is positive. This means that the current leads the voltage, which is characteristic of a capacitive circuit. When $1/\omega RC \to 0$ at high frequencies, the phase angle is zero and the current is in phase with the voltage. At very low frequencies, $1/\omega RC \to \infty$, and the phase angle approaches $\pi/2$.

Suppose the voltage across the resistance is considered as an output voltage while v_i is the input voltage. The output voltage is, from Ohm's law,

$$v_o = Ri = RI_p \sin(\omega t + \phi) \tag{2-42}$$

Consider only the peak value of v_0, which is, using (2-42) and (2-40),

$$V_o = \frac{RV_p}{\sqrt{R^2 + (1/\omega C)^2}}$$

The ratio of the output voltage to the input voltage is then

$$\frac{V_o}{V_p} = \frac{1}{\sqrt{1 + (1/\omega RC)^2}} \tag{2-43}$$

A plot of Eq. (2-43), Fig. 2-10, shows that the output voltage V_o is very

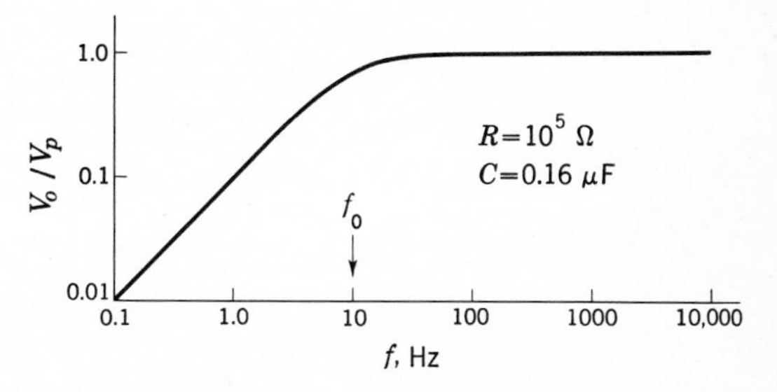

FIGURE 2-10 *Frequency characteristic of an RC high-pass filter.*

small at low frequencies and is equal to the input voltage at high frequencies. Since low frequencies are attenuated while high frequencies are not, this circuit is called an *RC high-pass filter*. Consider the frequency f_0 where

$$2\pi f_0 RC = 1 \tag{2-44}$$

According to Eq. (2-43) this is where

$$\frac{V_o^2}{V_p^2} = \frac{1}{2}$$

Since the output power in R is proportional to the voltage squared, f_0 is called the *half-power* frequency.

Suppose that the series resistance and capacitance in the simple RC filter are small enough to make $\omega RC \ll 1$ over a given frequency range. Under this condition the output voltage is, from Eqs. (2-42) and (2-43),

$$v_o = V_p \omega RC \sin(\omega t + \pi/2)$$
$$v_o = V_p \omega RC \cos \omega t \tag{2-45}$$

But, note that the time derivative of the input signal is

$$\frac{dv_i}{dt} = \omega V_p \cos \omega t \tag{2-46}$$

Therefore, combining Eqs. (2-45) and (2-46),

$$v_o = RC \frac{dv_i}{dt} \tag{2-47}$$

The interpretation of Eq. (2-47) is that when $\omega RC \ll 1$ the RC filter circuit performs the operation of differentiation. That is, the output voltage signal is the time derivative of the input voltage. This useful property is applied extensively in electronic circuits, most notably in electronic computers.

Correspondingly, the voltage across the capacitor is, by Eq. (2-20), the integral of the current. Therefore, the voltage across the capacitor v_c is, ignoring the limits of integration in Eq. (2-20) for the moment,

$$v_c = \frac{1}{C} \int I_p \sin(\omega t + \phi)\, dt = -\frac{I_p}{\omega C} \cos(\omega t + \phi) \tag{2-48}$$

Suppose now $\omega RC \gg 1$, which may be accomplished by making R and C very large. Then, $I_p = V_p/R$, $\phi = 0$, and Eq. (2-48) becomes

$$v_c = -\frac{V_p}{RC\omega} \cos \omega t \tag{2-49}$$

Introducing the integral of the input voltage results in

$$v_c = \frac{1}{RC} \int v_i\, dt \tag{2-50}$$

which means that under these conditions the RC circuit performs the operation of integration.

It is permissible to ignore the limits of integration in deriving Eq. (2-50), because we are only concerned with steady-state conditions after any initial transient voltages, which may accompany turning on the input voltage, have died away. The possibility of transient effects in RC circuits is considered in a later section. The RC circuit is a *low-pass* filter when the output voltage signal is taken from the capacitor (see Exercise 2-4). That is, the circuit as drawn in Fig. 2-11 severely attenuates high-frequency signals

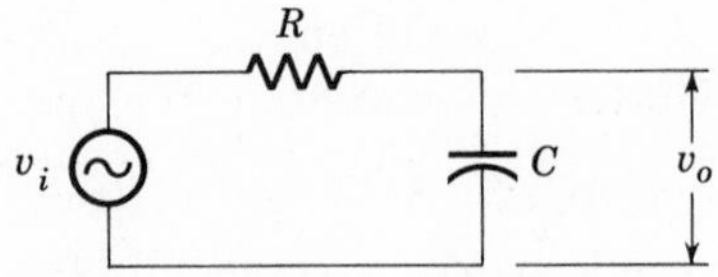

FIGURE 2-11 RC low-pass filter.

and transmits low-frequency signals undiminished. The transition frequency from one domain to the other is marked approximately by the half-power frequency, where $\omega_0 RC = 1$, as in the case of the high-pass filter.

The output voltage is only a small fraction of the input voltage when the circuit is used to differentiate or integrate, according to Eqs. (2-47) and (2-50). This is not a serious disadvantage in practical applications, however, since the output signal may be subsequently increased in magnitude using vacuum-tube or transistor amplifier circuits described in later chapters.

INDUCTANCE

Induced Voltage All the properties of both dc and ac circuits examined so far stem primarily from the force between two electric charges at rest, as expressed by Coulomb's law. It is an experimental fact that two charges in motion, that is, two electric currents, exert forces on each other in addition to the Coulomb force. As in the case of stationary charges, the force between currents exists even though the two currents are separated by a distance. This effect is interpreted by introducing the idea of a *magnetic field* set up in the space surrounding an electric current. The interaction of this field with the second current is said to produce the force. The analogy between the magnetic field of currents and the electric field of charges in the respective situations of charges in motion and at rest is quite obvious. It should be emphasized that both the magnetic field and the electric field are manifestations of the fundamental properties of electrons but that the idea of magnetic and electric fields in no way really explains these fundamental properties. The fields are a great conceptual aid in dealing with the behavior of electrons, however.

The experimental relation giving the magnetic field B at a point which is r meters away from a current I in a conductor of length l is called *Ampère's law*,

$$B = \frac{\mu_0}{4\pi} \frac{Il \sin \theta}{r^2} \tag{2-51}$$

In Eq. (2-51) μ_0 is a constant equal to 12.57×10^{-7} W/A · m (weber per ampere meters) in mks units. The angle θ is between the current direction and the radius vector to the point. The magnetic field is perpendicular to the plane containing the current direction and the radius vector and is measured in webers per square meter in the mks system.

According to Ampère's law, the magnetic field caused by a current I is directly proportional to the current. All the other factors in Eq. (2-51) depend only upon the geometry of the circuit. The total magnetic flux N encompassed by a circuit is given by the magnetic field times the area of the circuit, A, so that using Eq. (2-51) the magnetic flux is written

$$\begin{aligned} N = BA &= \frac{\mu_0 Al \sin \theta}{4\pi r^2} I \\ &= LI \end{aligned} \tag{2-52}$$

Here N is the magnetic flux intercepted by the circuit loop caused by the current I in the same circuit. The constant L, called the *inductance*, depends upon the geometric shape of the circuit and the magnetic properties of nearby materials.

A changing magnetic flux in a circuit induces a voltage in the circuit given by

$$v = -\frac{dN}{dt}$$

which is known as *Faraday's law* after Michael Faraday who first investigated induced emfs. Introducing Eq. (2-52) into Faraday's law, the result is

$$v = -L\frac{di}{dt} \tag{2-53}$$

According to Eq. (2-53), whenever the current in any circuit changes with time an emf is induced in the circuit. The magnitude of the emf depends on the rate of change of current and also upon the geometry as determined by the inductance of the circuit. The minus sign in Eq. (2-53) means that the polarity of the emf is such that it opposes the change of current. That is, if the current is increasing with time the induced emf acts to reduce the current.

The inductance of simple circuits is small enough so that the induced emf may usually be ignored. This is true except at the very highest frequencies of interest in electronic circuits, where the rate of change of current is great. Circuits used at such frequencies are kept as small as possible to minimize inductive effects. By contrast, it is possible to produce electric components which have appreciable inductance, and these are very useful in ac circuits.

Inductors Electric components with appreciable inductance are called *inductors,* or *inductances,* and in some applications, *chokes.* They consist of many turns of wire wound adjacent to one another on the same support. In this way every single coil of wire links the magnetic-flux density produced by the current in all other coils, and the total flux intercepted by all the coils together can be made large. The unit of inductance, defined using either Eq. (2-52) or Eq. (2-53), is called the *henry* after Joseph Henry, an early American investigator of inductive effects.

High-frequency electronic circuits frequently employ inductances of the order of 10^{-6} henry (*microhenry,* or μH) which may be a helical coil of a few turns on, say, a 1-cm-diameter support. A few tens of turns produce inductance values in the 10^{-3} henry (*millihenry,* or mH) range. Large inductances for use at low frequencies are obtained by winding many hundreds of turns of wire on a core of a ferromagnetic material such as iron. The magnetic properties of these materials are such that the magnetic flux density is increased appreciably. In this fashion inductances of several hundred henrys are attained. Several typical inductors are pictured in Fig. 2-12.

The cores of iron-core inductors are laminated in order to interrupt currents induced in the metal core by the changing magnetic flux. This

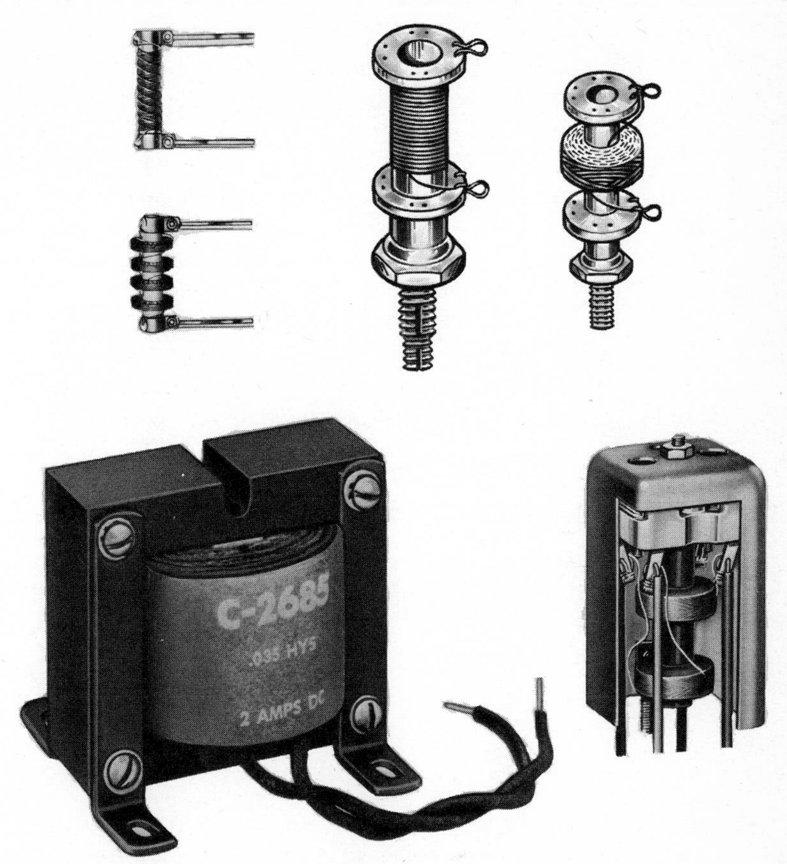

FIGURE 2-12 *Typical inductors used in electronic circuits.* (J. W. Miller Co. and Essex International, Inc.)

reduces the I^2R losses of these so-called *eddy currents* in the core. Individual laminations are stacked on top of each other and separated with insulating varnish to make a core of desired size. The circuit symbol for an inductor is a helical coil, as shown in Fig. 2-13. Parallel lines along the helix, also illustrated in Fig. 2-13, signify a magnetic core.

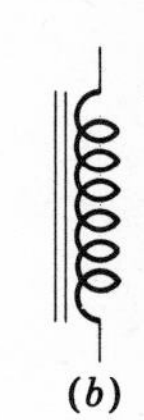

FIGURE 2-13 *Circuit symbols for (a) inductor and (b) iron-core inductor.*

(*a*) (*b*)

Variable inductances can be achieved by moving one portion of the windings relative to the other, but such components are not widely used and most inductors are fixed. In many applications it is necessary to take account of the resistance of the wire in the windings and the capacitance between layers of the winding in determining the effect of a choke in a given ac circuit.

The total flux of inductors in series is equal to the inductance of each inductor times the common current, so according to Eq. (2-52) the total inductance is the sum of the individual inductances. This is the same as found for resistors in series in Chap. 1. Although it is not immediately obvious, inductors in parallel also combine as do resistors in parallel. This is discussed in greater detail in the next chapter.

It is usually advantageous to employ a ferromagnetic core in an inductor because the inductance is appreciably increased thereby. This means that fewer turns of wire are needed and consequently that the inevitable resistance associated with the coil is minimized. The benefits of this may be illustrated in the following way. Consider the simple series circuit of an inductor and a resistor, Fig. 2-14, in which the resistor represents the

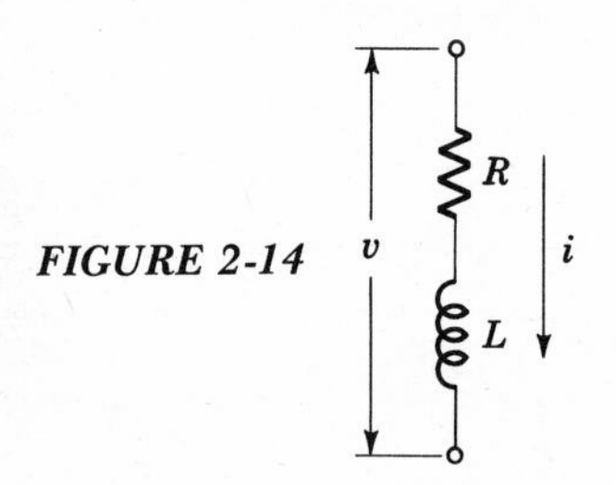

FIGURE 2-14

resistance of the wire in the coil of the inductor. The voltage across the series combination is given by Ohm's law and Eq. (2-53)

$$v = Ri + L\frac{dI}{dt} \tag{2-54}$$

The second term on the right-hand side of Eq. (2-54) is positive because of the meaning of the minus sign in Faraday's law. That is, the induced voltage in an inductor opposes the current. Therefore, the polarity of the voltage is the same as the IR drop across the resistor.

The circuit differential equation, Eq. (2-54), may be solved by a method identical to that used in connection with (2-30), but it is simpler to proceed as follows. Assume the current is sinusoidal:

$$i = I_p \sin \omega t \tag{2-55}$$

Differentiating,

$$\frac{di}{dt} = \omega I_p \cos \omega t \tag{2-56}$$

Introducing Eqs. (2-55) and (2-56) into (2-54) the voltage is immediately given by

$$v = RI_p \sin \omega t + \omega L I_p \cos \omega t \qquad (2\text{-}57)$$

This expression may be put into a more illustrative form by introducing the phase angle between the voltage and the current with the aid of Fig.

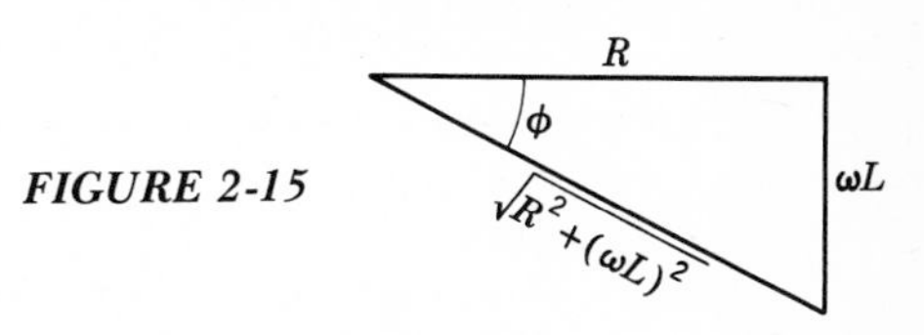

FIGURE 2-15

2-15. Expressing the coefficients of $\sin \omega t$ and $\cos \omega t$ in terms of $\cos \phi$ and $\sin \phi$, Eq. (2-57) becomes

$$v = I_p \sqrt{R^2 + (\omega L)^2} \, (\cos \phi \sin \omega t + \sin \phi \cos \omega t) \qquad (2\text{-}58)$$

The expression in parentheses is just the trigonometric identity for $\sin(\omega t + \phi)$, so that

$$v = I_p \sqrt{R^2 + (\omega L)^2} \sin (\omega t + \phi) \qquad (2\text{-}59)$$

which should be compared to the analogous expression, Eq. (2-41), for the series combination of a resistor and a capacitor.

The phase angle between the voltage and the current is, using Fig. 2-15,

$$\phi = \arctan \frac{\omega L}{R} \qquad (2\text{-}60)$$

According to Eq. (2-60) the phase angle in a highly inductive circuit where R is small approaches $\pi/2$ and the current lags the voltage, which which is the reverse of the situation in the capacitive circuit.

According to Fig. 2-15, when R is appreciable the phase angle departs from $\pi/2$ and the inductor cannot be considered to be a true inductance. Thus, any opportunity to reduce R by reducing the number of turns in the coil of the inductor through the use of a magnetic core is important. The ratio of inductive reactance to resistance, as in Eq. (2-60), is called the *quality factor* or, more usually, the Q of the inductor,

$$Q = \frac{\omega L}{R} \qquad (2\text{-}61)$$

Values of Q in the range from ten to several hundred are common in electronic circuits.

The resistive component in a practical inductor may also include the power lost in the magnetic material as a result of the rapidly changing magnetic field in the core. As mentioned earlier, metallic cores must be laminated to reduce eddy-current losses and the thickness of the lamination determines the maximum useful frequency of the inductor. This technique reaches its practical limit in powdered iron cores in which the iron is present as fine particles. *Ferrite* cores, made of high-resistivity ferromagnetic materials are used at high frequencies because their high resistivity makes eddy-current losses negligible. Such materials are not as useful as iron at low frequencies because magnetic-saturation effects limit the maximum power levels of the inductor.

The inductance of a coil wound on a ferromagnetic core is increased over that of an air core by a factor called the *permeability* of the magnetic material of the core. The permeability of ferromagnetic materials is rather analogous to the dielectric constant of insulators discussed previously. Typical values of permeability for several materials are shown in Fig. 2-16. Note that there is a considerable degradation in the permeability

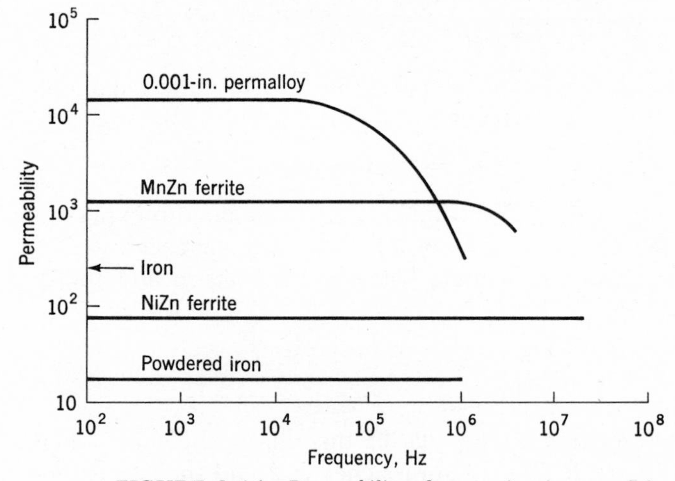

FIGURE 2-16 *Permeability of magnetic-core materials.*

of iron when it is powdered to reduce eddy-current losses and that extremely thin laminations are required in the case of metallic ferromagnetic materials such as permalloy to achieve even modest high-frequency performance. By contrast, the high-frequency properties of ferrites are very good.

The permeability and inductance both tend to decrease at high frequencies because of additional magnetic loss effects in the core. Although the origin of such losses is not clear, the decrease in effective inductance can be immediately understood from Eq. (2-59). Considering the rms values, (2-59) is written

$$V = IR\sqrt{1 + (\omega L/R)^2} \tag{2-62}$$

Squaring both sides and rearranging,

$$\left(\frac{\omega L}{R}\right)^2 = \left(\frac{V}{IR}\right)^2 - 1 \tag{2-63}$$

Finally, solving for the inductance,

$$L = \frac{1}{\omega}\left[\left(\frac{V}{I}\right)^2 - R^2\right]^{1/2} \tag{2-64}$$

this expression illustrates directly why the effective inductance decreases as the resistive component increases due to magnetic losses in the core at high frequencies.

RL ***Filter*** The series combination of a resistor and an inductor is a useful ac circuit. When connected to a sinusoidal voltage source, as in Fig. 2-17, the circuit performance may be described by the results of the previous analysis. Considering the voltage drop across R to be the output voltage of the circuit, the peak value of the output voltage is, from Eq. (2-55),

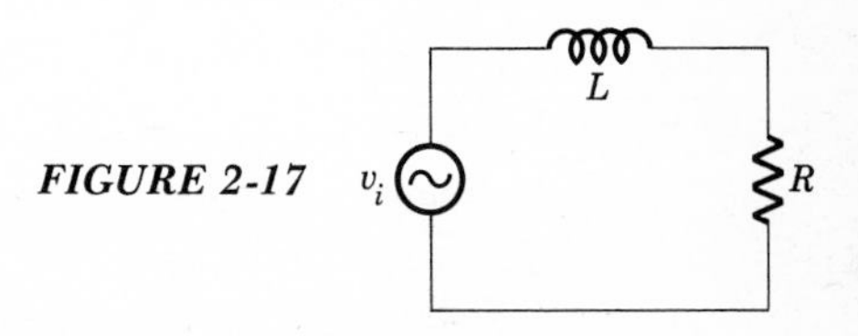

FIGURE 2-17

$$V_0 = RI_p \tag{2-65}$$

The peak value of the input voltage is, comparing Figs. 2-14 and 2-17, and using Eq. (2-59),

$$V_p = I_p\sqrt{R^2 + (\omega L)^2} \tag{2-66}$$

The ratio of output to input is therefore

$$\frac{V_o}{V_p} = \frac{1}{\sqrt{1 + (\omega L/R)^2}} \tag{2-67}$$

According to Eq. (2-67) at low frequencies, where $\omega L/R \to 0$, the output voltage is equal to the input voltage. At high frequencies the output voltage is smaller than the input and the circuit is therefore an inductive low-pass filter. The characteristic half-power frequency is given by

$$\frac{\omega_0 L}{R} = 1 \tag{2-68}$$

The frequency response characteristic for an *RL* low-pass filter, Eq. (2-67), is identical in shape to the *RC* low-pass filter (Exercise 2-5), and the circuit used depends upon the particular application and circuit conditions. For example, the *RL* circuit is common in situations where appreciable direct current is transferred from input to output while ac signals are attenuated. The *RL* circuit is more efficient because the I^2R losses are a minimum, since there is no appreciable series resistance (compare Figs. 2-11 and 2-17). In this application the inductance, in effect, chokes off the alternating current; this is the origin of the term choke introduced earlier.

TRANSIENT CURRENTS

Inductive Time Constant The study of ac circuits to this point has implicitly assumed that the rms values of the sinusoidal emfs are constant. Transient effects are associated with sudden changes in the voltages applied to ac networks, as, for example, when an emf is first applied. These transient currents dissipate rapidly, leaving steady-state currents in the network which persist as long as the emf is applied. Although most often the steady-state currents are of primary concern, in many cases transient effects are important as well. This is particularly so in determining the response of a network to isolated voltage pulses.

Transient currents in ac networks are found by solving the circuit differential equation, taking into account the magnitude of the current at the time the applied voltage is changed. Actually, the sum of the transient current and the steady-state current is the complete solution of the circuit differential equation. It is generally satisfactory to consider the two aspects of ac-circuit analysis separately because the response of circuits is most often of interest in connection either with steady emfs alone or with transient effects alone.

It is convenient to investigate first the transient currents in a simple *LR* series circuit incorporating a battery and a two-position switch, as shown in Fig. 2-18. Suppose the switch is suddenly connected to terminal 1.

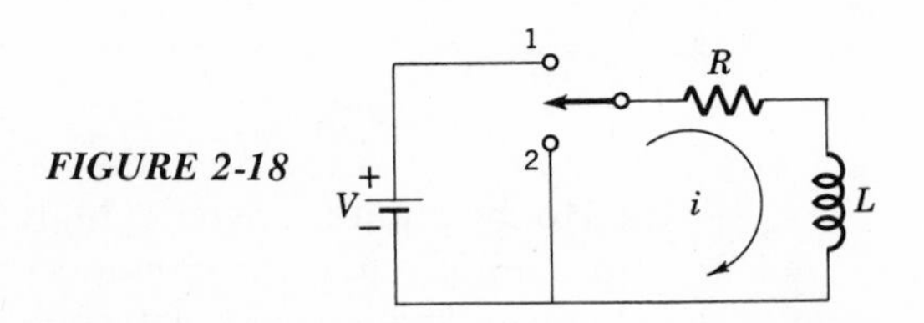

FIGURE 2-18

The circuit differential equation is, from Kirchhoff's rule,

$$V = Ri + L\frac{di}{dt} \tag{2-69}$$

This equation is solved by first rewriting it as

$$\frac{L}{R}\frac{di}{dt} = \frac{V}{R} - i$$

and then

$$\frac{di}{V/R - i} = \frac{R}{L}\,dt \tag{2-70}$$

Both the left side and the right side of Eq. (2-70) are standard integrals, so that upon integrating,

$$-\ln\left(\frac{V}{R} - i\right) = \frac{R}{L}t + K \tag{2-71}$$

where K is a constant to be evaluated later. Another way of writing Eq. (2-71) is

$$\frac{V}{R} - i = \exp\left(\frac{-t}{L/R} - K\right) = Ae^{-t/(L/R)} \tag{2-72}$$

where the constant A replaces $\exp -K$ for simplicity. In order to evaluate A, note that the current is zero at the time the switch is closed at $t = 0$. Therefore

$$\frac{V}{R} - 0 = Ae^0 = A$$

and the final solution for the current is, after substituting for A in (2-72),

$$i = \frac{V}{R}(1 - e^{-t/(L/R)}) \tag{2-73}$$

The quantity L/R, which has the dimensions of seconds, is called the *time constant*

$$\tau = \frac{L}{R} \tag{2-74}$$

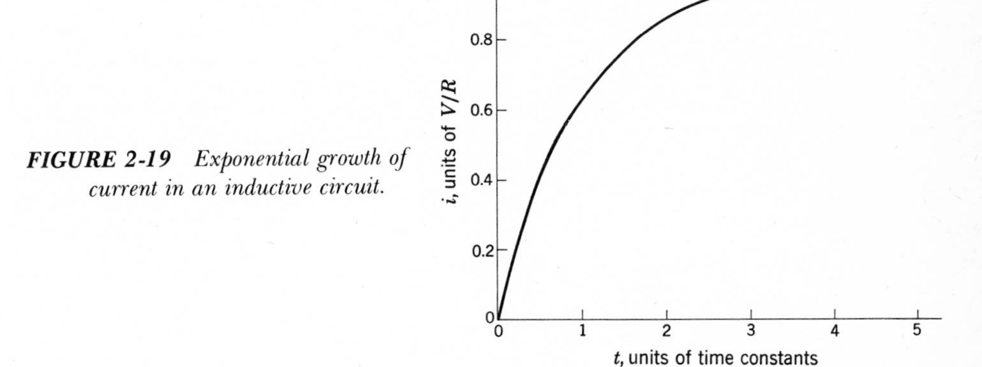

FIGURE 2-19 *Exponential growth of current in an inductive circuit.*

The growth of current in an inductive circuit according to Eq. (2-73) is shown in Fig. 2-19. Note that the current starts rapidly from zero and approaches a steady-state value given by the dc current V/R. After a time equal to one time constant the current is equal to $1-e^{-1}=1-0.368=63\%$ of its final value. The actual time at which this current is attained depends upon the magnitude of the inductance and resistance. A long time is achieved for large values of L and small values of R, and vice versa, according to Eq. (2-74).

Suppose now the switch in Fig. 2-18 is suddenly connected to terminal 2. This removes the emf from the circuit, and the circuit differential equation becomes

$$L\frac{di}{dt}+Ri=0 \tag{2-75}$$

This may be rewritten as

$$\frac{di}{i}=-\frac{R}{L}\,dt \tag{2-76}$$

Integrating both sides

$$\ln i=-\frac{R}{L}t+K'$$

where K' is a constant. This result may be put in the form

$$i=Be^{-t/\tau}$$

where $B=\exp K'$ and $\tau=L/R$. Since $i=V/R$ at $t=0$, the final solution is

$$i=\frac{V}{R}e^{-t/\tau} \tag{2-77}$$

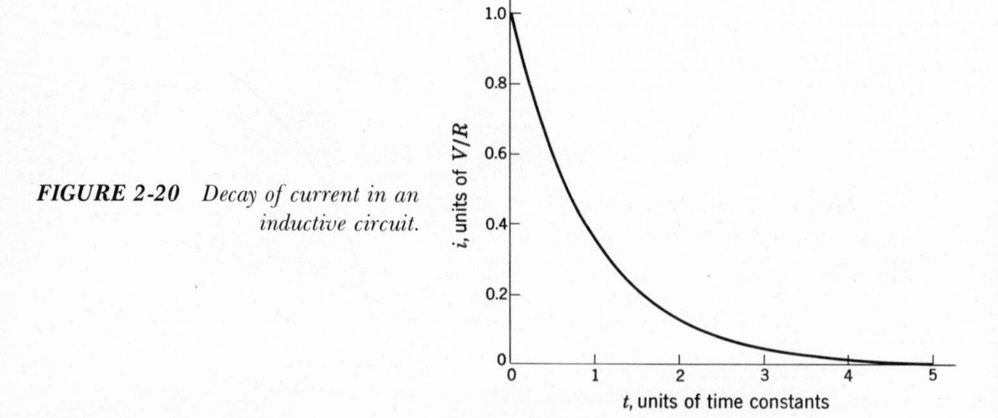

FIGURE 2-20 Decay of current in an inductive circuit.

Evidently, the current decays exponentially to zero from the dc value at a rate governed by the circuit time constant. The growth and decay of current in an inductive circuit are symmetrical, as can be seen by comparing Figs. 2-19 and 2-20.

Capacitive Time Constant The simple RC circuit is analogous to the inductive case, as may be illustrated by considering the circuit in Fig. 2-21.

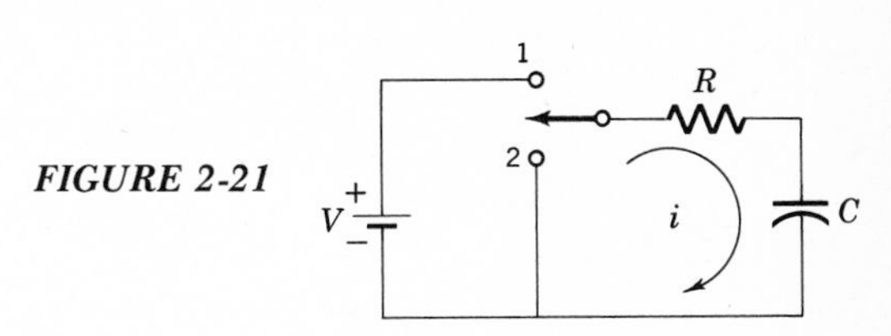

FIGURE 2-21

The circuit equation with the switch in position 1 is

$$V = Ri + \frac{q}{C} \tag{2-78}$$

Differentiating with respect to t,

$$0 = R\frac{di}{dt} + \frac{i}{C} \tag{2-79}$$

or

$$\frac{di}{i} = -\frac{1}{RC}\,dt$$

By analogy with Eq. (2-76), the solution is

$$i = Ae^{-t/RC} \tag{2-80}$$

Evidently the time constant of the RC circuit is given by

$$\tau = RC \tag{2-81}$$

The constant A is evaluated by noting that at $t = 0$ the charge on the capacitor is zero, so that the voltage across C is also zero. The initial current is simply V/R, according to Eq. (2-78). Therefore, the current is

$$i = \frac{V}{R}\,e^{-t/RC} \tag{2-82}$$

The capacitive charging current starts at a high value and falls exponentially to zero as the capacitor becomes charged, Fig. 2-22. According to Eq. (2-81) large values of resistance and capacitance result in long characteristic time constants.

The decay of current upon closing the switch to position 2 is identical

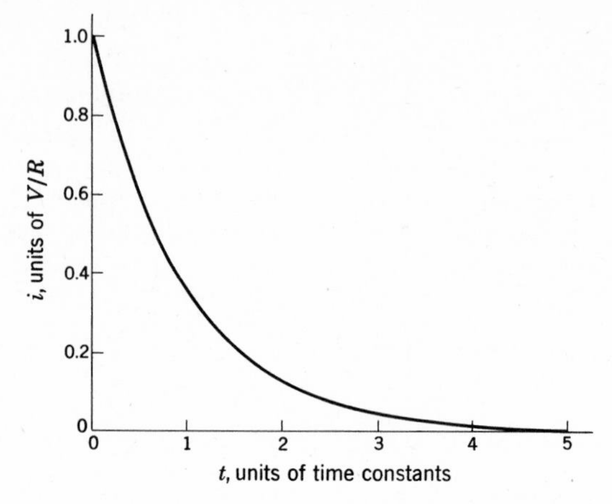

FIGURE 2-22 Charging current in a capacitive circuit.

with that given by Eq. (2-82). This can be seen by writing the circuit voltage equation for this case

$$\frac{q}{C} + Ri = 0$$

Differentiating with respect to t, the circuit differential equation is

$$\frac{i}{C} + R\frac{di}{dt} = 0 \tag{2-83}$$

which is identical to Eq. (2-79). Furthermore, the potential across the capacitor is equal to V when the capacitor is fully charged. This means that the initial discharge current is V/R, just as in the charging-current case.

The voltage across the capacitor at any time may be determined by calculating the charge on the capacitor using the solution (2-82). The variation of the capacitor voltage with time is analogous to Figs. 2-19 and 2-20, except that the steady-state asymptote upon charging is equal to V.

Transient currents also accompany the turning on of ac voltage sources.

FIGURE 2-23

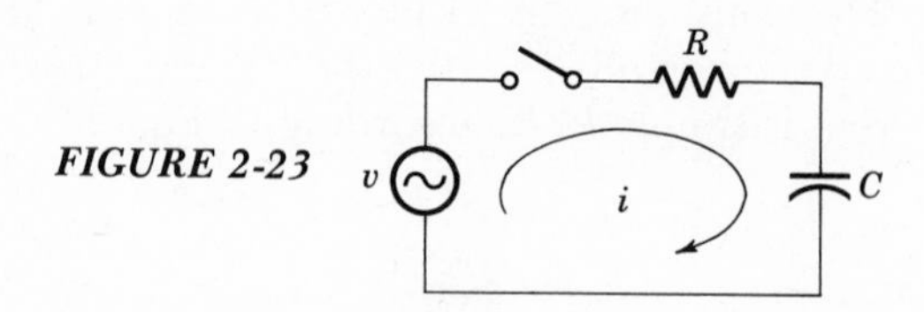

For example, consider the circuit in Fig. 2-23, for which the circuit differential equation is

$$R\frac{di}{dt} + \frac{1}{C}i = \frac{dv}{dt} \tag{2-84}$$

It is convenient to express the applied voltage as

$$v = V_p \sin(\omega t + \theta) \tag{2-85}$$

where the phase angle θ is introduced in order to choose the instantaneous input voltage amplitude at $t = 0$. Since the full solution of Eq. (2-84) involves a transient current and a steady-state current of frequency ω, the total current is

$$i = i_1 + i_2 \tag{2-86}$$

where i_1 is the transient current and i_2 is the steady-state current.

Substituting Eq. (2-86) into the differential equation (2-84) and rearranging, the result is

$$\left(R\frac{di_1}{dt} + \frac{1}{C}i_1\right) + \left(R\frac{di_2}{dt} + \frac{1}{C}i_2 - \frac{dv}{dt}\right) = 0 \tag{2-87}$$

This equation is solved if both parentheses are equal to zero. The first one is just Eq. (2-79), for which the solution previously found is Eq. (2-80). Similarly, the second parentheses is just Eq. (2-29), and the solution is Eq. (2-41). Using these expressions, the total current is written

$$i = Ae^{-t/\tau} + \frac{V_p}{\sqrt{R^2 + (1/\omega C)^2}} \sin(\omega t + \theta + \phi) \tag{2-88}$$

where

$$\tau = RC$$

$$\phi = \arctan 1/R\omega C$$

and the constant A is determined from the voltage of the source at the instant the switch is closed. Note that Eq. (2-88) consists of a transient part together with a steady-state term, as discussed above.

Suppose the applied voltage is at its peak, that is, $v = V_p$, when the switch is closed at $t = 0$. This means we choose $\theta = \pi/2$, according to Eq. (2-85), and the initial current is V_p/R. Thus, Eq. (2-88) for $t = 0$ becomes

$$\frac{V_p}{R} = A + \frac{V_p}{\sqrt{R^2 + (1/\omega C)^2}} \sin\left(\frac{\pi}{2} + \phi\right) \tag{2-89}$$

This is solved for the constant A, which is then substituted into Eq. (2-88). The result is the final solution

$$i = \frac{V_p}{R}\frac{1}{1 + (\omega RC)^2} e^{-t/\tau} + \frac{V_p}{\sqrt{R^2 + (1/\omega C)^2}} \sin\left(\omega t + \frac{\pi}{2} + \phi\right) \tag{2-90}$$

In order for the transient term of Eq. (2-90) to have an appreciable magnitude compared with the steady-state term, the denominator of its coefficient must not be too large. This means that

$$(\omega RC)^2 < 1$$

or

$$\tau < \frac{1}{\omega} \tag{2-91}$$

According to this expression, the time constant of the transient current is smaller than one period of the sine-wave current. Therefore, any transient current is negligible after only a fraction of a cycle of the ac wave, and the transient may therefore be ignored. If, on the other hand, the time constant is much greater than one period, the transient persists over many cycles, but, according to Eq. (2-91), the amplitude is small. This reasoning demonstrates why the steady-state solution alone is sufficiently accurate for most purposes.

Ringing Sudden changes of voltage in a *RLC* circuit may produce *ringing,* in which a decaying ac current is generated at a frequency characteristic of the circuit, called the *resonant* frequency. The amplitude of the ac current dies away exponentially as in nonresonant circuits. For example, after the switch is closed in the circuit of Fig. 2-24, an ac current may be

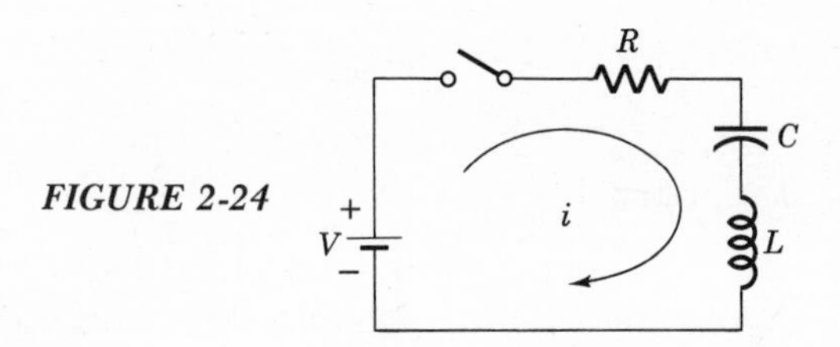

FIGURE 2-24

generated which decreases relatively slowly. The phenomenon is analogous to striking a bell a sharp mechanical blow with a hammer, whence the origin of the term ringing.

Resonant circuits are examined in greater detail in the next chapter. To examine ringing in the circuit of Fig. 2-24 start by applying Kirchhoff's rules,

$$V = Ri + \frac{q}{c} + L\frac{di}{dt} \tag{2-92}$$

which becomes, after differentiating,

$$\frac{d^2i}{dt^2} + \frac{R}{L}\frac{di}{dt} + \frac{1}{LC}i = 0 \tag{2-93}$$

To solve this circuit differential equation it is assumed, on the basis of previous analyses, that an ac current is present which has an amplitude that decays exponentially with time. Therefore, the current is written as

$$i = Ae^{-at} \sin(\omega t + \phi) \tag{2-94}$$

where A, a, ω, and ϕ are constants. Equation (2-94) is differentiated once, twice, and all three expressions introduced into the circuit differential equation. After some algebraic manipulation, (2-93) reduces to

$$\left[\left(a^2 - \omega^2 - \frac{Ra}{L} + \frac{1}{LC}\right) \sin(\omega t + \phi) + \left(\frac{R\omega}{L} - 2a\omega\right) \cos(\omega t + \phi)\right] Ae^{-at} = 0 \tag{2-95}$$

In order for (2-94) to be a solution of (2-93) the coefficients of both terms in (2-95) must separately be equal to zero. From the coefficient of the cosine term,

$$a = \frac{R}{2L} \tag{2-96}$$

while the coefficient of the sine term gives an expression for the frequency

$$\omega^2 = \frac{1}{LC} - \frac{R^2}{4L^2} \tag{2-97}$$

Therefore, the current in this circuit is sinusoidal at a frequency given by (2-97) and with an amplitude that decreases with a time constant determined from (2-96). A sketch of the current behavior is illustrated in Fig. 2-25.

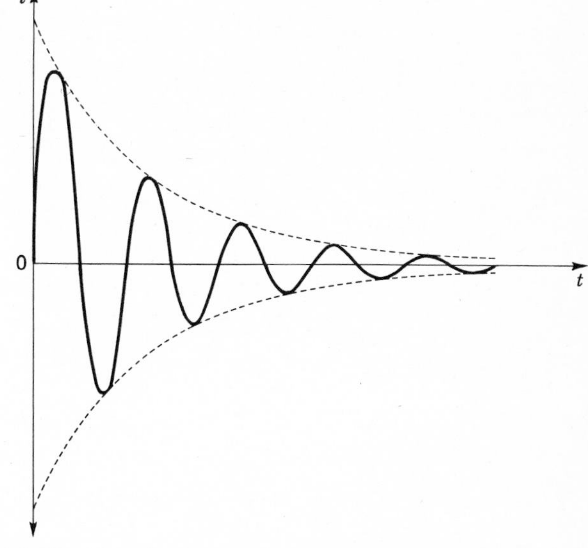

FIGURE 2-25 *Transient current in an RLC circuit can be decaying oscillation.*

It is conceivable that values of the circuit elements are such that

$$\frac{R^2}{4L^2} > \frac{1}{LC} \tag{2-98}$$

in which case the ac frequency calculated from (2-97) is imaginary. The interpretation of this mathematical result may be developed from the mathematical expression for the sine of an imaginary angle.

$$\sin j\omega t = \frac{j}{2}(e^{\omega t} - e^{-\omega t}) \tag{2-99}$$

Therefore, Eq. (2-94) reduces to a simple exponential decay with a time constant determined by the R, C, and L values in the circuit. In effect, according to (2-98), when the circuit resistance is very large the ac current is damped out by the joule losses in the resistance, and the circuit is said to be *overdamped*. The point of *critical damping* is when the frequency calculated from (2-97) vanishes and the exponential decay is governed solely by (2-96).

COMPLEX WAVEFORMS

Fourier Series Most steady ac signals of practical interest are not simple sine waves, but often have more complex waveforms. Nevertheless, the analysis based on sine waves can apply because complex waveforms may be represented by summing sine waves of various amplitudes and frequencies. Consider, for example, the sum of the two sine waves shown in Fig. 2-26, one of which is twice the frequency of the other. The resulting waveform is more complex than either constituent. If this periodic vol-

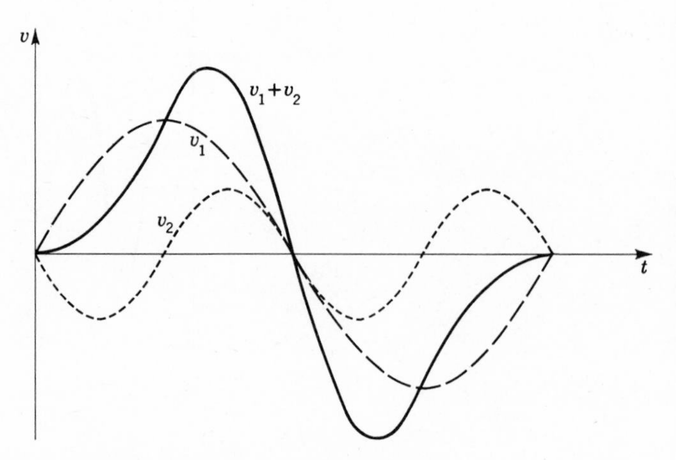

FIGURE 2-26 *Sum of two sine waves is complex waveform.*

tage is applied to a circuit, the resulting current may be determined by calculating the current caused by each sinusoidal voltage component separately. In this way the effect of any complex waveshape can be investigated.

Any single-valued periodic waveform may be represented by a *Fourier series* which is a summation of a *fundamental* (or lowest) frequency together with its *harmonics* (integral multiples of the fundamental frequency). The mathematical proof of this representation and the procedure by which the proper series is constructed to represent any given waveshape are not of direct concern here. It is interesting, however, to note the harmonic composition of several waves of practical importance to demonstrate the main features of the series representation.

The Fourier series for a square wave of peak voltage V_p and frequency ω is given by

$$v = \frac{4V_p}{\pi}\left(\sin \omega t + \tfrac{1}{3}\sin 3\omega t + \tfrac{1}{5}\sin 5\omega t + \tfrac{1}{7}\sin 7\omega t + \cdots\right) \tag{2-100}$$

The faithfulness of the series representation improves as the number of terms included in Eq. (2-100) is increased, as illustrated graphically in Fig. 2-27. This means that the higher frequencies are necessary in order

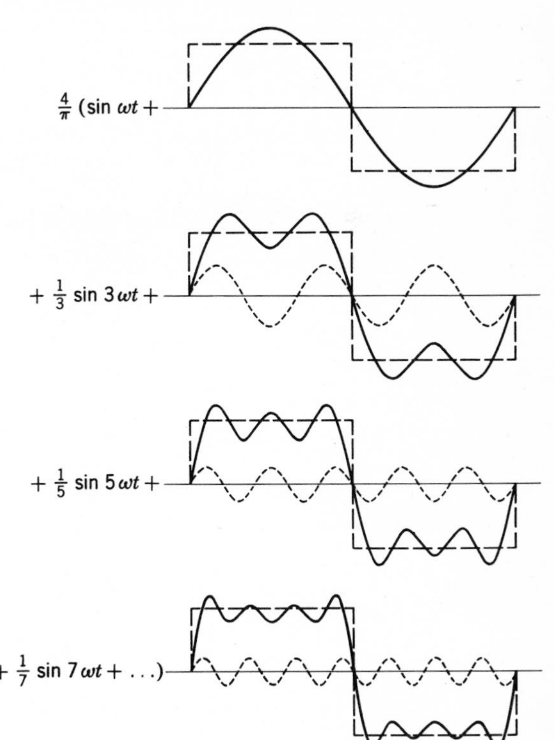

FIGURE 2-27 *Many frequency components are present in a square wave. Faithfulness of representation improves as number of harmonics is increased.*

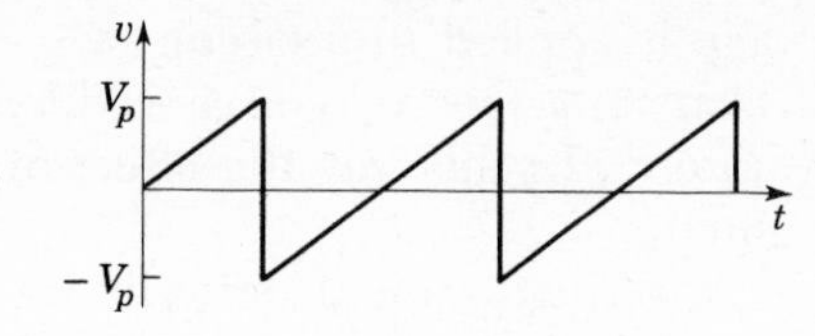

FIGURE 2-28 *Sawtooth wave.*

to reproduce the sharp corners of the square wave. A second example is the sawtooth wave (Fig. 2-28) which is represented by

$$v = \frac{2V_p}{\pi}\left(\sin \omega t - \tfrac{1}{2} \sin 2\omega t + \tfrac{1}{3} \sin 3\omega t - \tfrac{1}{4} \sin 4\omega t + \cdots\right) \tag{2-101}$$

Here, again, the high-frequency terms must be included in order to obtain the sharp corner of the sawtooth waveform (Exercise 2-6).

According to Eqs. (2-100) and (2-101), complex waveforms, such as the square wave and the sawtooth wave, are composed of a wide spectrum of frequencies. If the waveform of a given input signal applied to any network is to be preserved at the output, the attenuation of the network must be independent of frequency. For example, a square wave applied to an *RC* high-pass filter can result in a square-wave output signal only if the fundamental frequency of the square wave is greater than the cutoff frequency of the filter. It is equally important that the relative phases of the harmonics remain constant in traversing the network if the waveform is to be preserved.

On the other hand, the harmonic composition of a wave is often purposefully altered by a network in order to change one waveform into another. Suppose a square wave is applied to an *RC* differentiating circuit. The output waveform is a series of alternating positive and negative sharp pulses, as sketched in Fig. 2-29*b*. In order for the circuit to differentiate a square wave adequately, it is necessary for the highest frequency

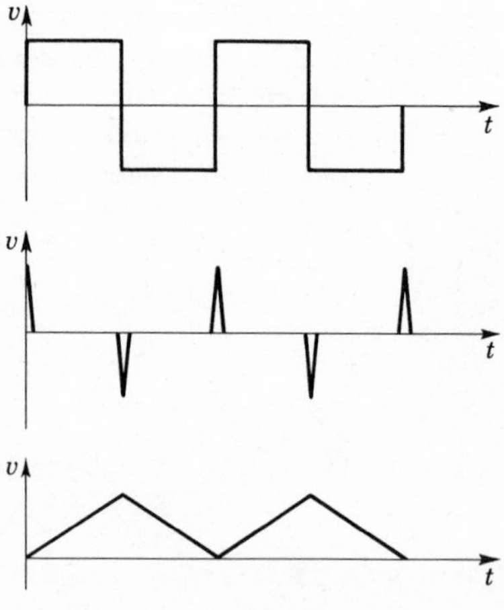

FIGURE 2-29 *(a) Square wave, (b) differentiated square wave, (c) integrated square wave.*

term in the Fourier series representation to satisfy the approximation $\omega RC \ll 1$ leading to Eq. (2-47). This is usually easily accomplished in practice, and the simple differentiating circuit is widely used to generate pulses from a square wave.

A second example is the integration of a square wave using a simple RC circuit. The result is a triangular wave, Fig. 2-29*c*. Here the fundamental frequency of the square wave must be large enough that the approximation $\omega RC \gg 1$ leading to Eq. (2-50) is satisfied.

Effective Value The effective value of a complex waveform is defined in exactly the same way as in the case of the sinusoidal wave discussed earlier.

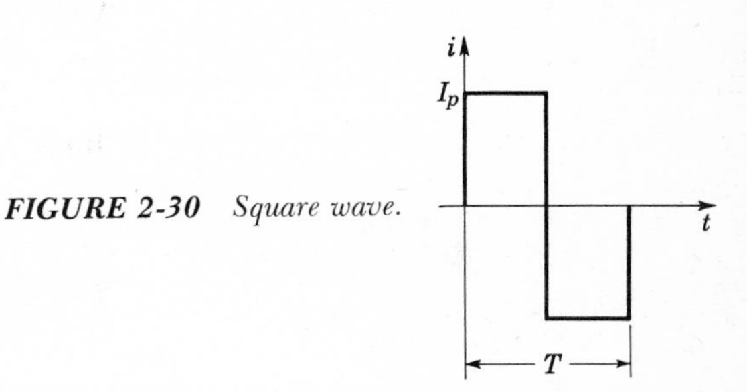

FIGURE 2-30 *Square wave.*

Consider, for example, the square wave of period T and maximum value I_p illustrated in Fig. 2-30. The effective current is just

$$I_e^2 = \frac{1}{T}\int_0^T I_p^2\,dt = \frac{I_p^2}{T}\int_0^T dt = I_p^2 \tag{2-102}$$

or

$$I_e = I_p \tag{2-103}$$

This means that the effective value of a square wave is equal to the maximum value. Since the joule heat in a resistor depends upon the square of the current, the current direction is immaterial and Eq. (2-103) is just what is expected.

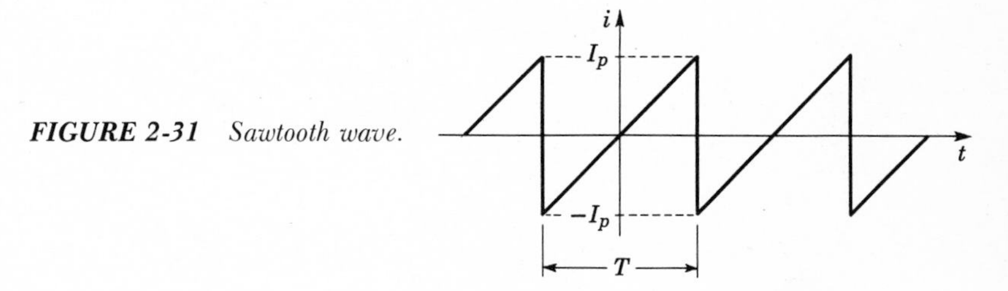

FIGURE 2-31 *Sawtooth wave.*

Consider next the simple triangular sawtooth waveform illustrated in Fig. 2-31. The current varies as

$$i = \frac{2I_p}{T} t \tag{2-104}$$

over the interval from $-T/2$ to $T/2$. Therefore, the rms value is

$$\begin{aligned} I_e^2 &= \frac{1}{T} \int_{-T/2}^{T/2} \left(\frac{2T_p}{T} t \right)^2 dt = \frac{4I_p^2}{T^3} \int_{-T/2}^{T/2} t^2\, dt \\ &= \frac{4I_p^2}{T^3} \left[\frac{t^3}{3} \right]_{-T/2}^{T/2} = \frac{4I_p^2}{3T^3} \left(\frac{T^3}{8} + \frac{T^3}{8} \right) = \frac{I_p^2}{3} \end{aligned} \tag{2-105}$$

$$I_e = \frac{I_p}{\sqrt{3}} \tag{2-106}$$

Another waveform of interest is a series of half sine waves, Fig. 2-32.

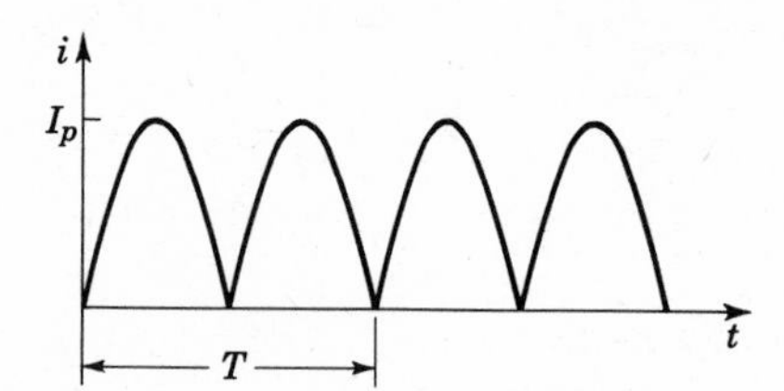

FIGURE 2-32 *Waveform of half-sinusoids.*

The effective value can be found by the procedure used above. This waveform has, however, a dc component, as can be shown by computing the average value,

$$\begin{aligned} I_{\text{dc}} &= \frac{2}{T} \int_0^{T/2} I_p \sin \omega t\, dt = \frac{2I_p}{\omega T} \left[-\cos \omega t \right]_0^{T/2} \\ &= \frac{4I_p}{\omega T} = \frac{4I_p}{2\pi} \end{aligned} \tag{2-107}$$

$$I_{\text{dc}} = \frac{2}{\pi} I_p \tag{2-108}$$

This dc component contributes to the heating effect of the complete waveform.

For most purposes the rms value of the ac components alone is of interest. It is convenient to consider the Fourier series of complex waveforms. Since the heating effect of each frequency component is independent of the others, the effective value of any complex waveform is given by

$$I_e^2 = I_{\text{dc}}^2 + I_1^2 + I_2^2 + I_3^2 + \cdots \tag{2-109}$$

where I_{dc} is the average value and $I_1, I_2, I_3, \ldots$ are the rms values of each frequency component. It is often sufficient to consider only the fundamental frequency, since it predominates.

The Fourier series of the waveform in Fig. 2-32 is

$$i = \frac{2}{\pi} I_p - \frac{4I_p}{3\pi} \cos 2\omega t - \frac{4I_p}{15\pi} \cos 4\omega t + \cdots \tag{2-110}$$

Note that the first term corresponds to Eq. (2-108), in conformity with the definition of the dc component. The effective value of the ac components alone is, using Eqs. (2-8) and (2-109),

$$I_{\text{rms}}^2 = \frac{I_p^2}{2} \left[\left(\frac{4}{3\pi} \right)^2 + \left(\frac{4}{15\pi} \right)^2 + \cdots \right] \cong \frac{1}{2} \left(\frac{4I_p}{3\pi} \right)^2 \tag{2-111}$$

$$I_{\text{rms}} = \frac{2\sqrt{2} I_p}{3\pi} \tag{2-112}$$

Thus, the ratio of the rms value of the ac components to the dc component is

$$\frac{I_{\text{rms}}}{I_{\text{dc}}} = \frac{2\sqrt{2}\ I_p/3\pi}{2I_p/\pi}$$

$$= \frac{\sqrt{2}}{3} \tag{2-113}$$

The preceding results are used in Chap. 4 in connection with further applications of inductance and capacitance filters.

Square-wave Response It is useful to investigate the response of circuits to square-wave voltages by the transient methods of the previous section. This is possible since a square wave can be considered to be a dc voltage which is swtiched on and off at the frequency of the wave. Transient-current analysis complements the Fourier-series method and is often simpler. The transient technique yields the output waveform directly and makes it unnecessary to determine the response of the circuit to each of the harmonics.

Consider first the high-pass RC filter of Fig. 2-33 to which is applied

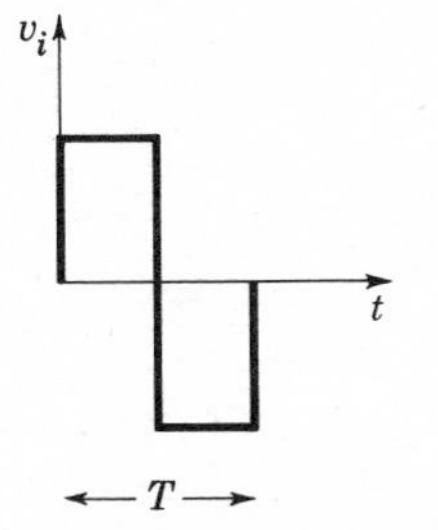

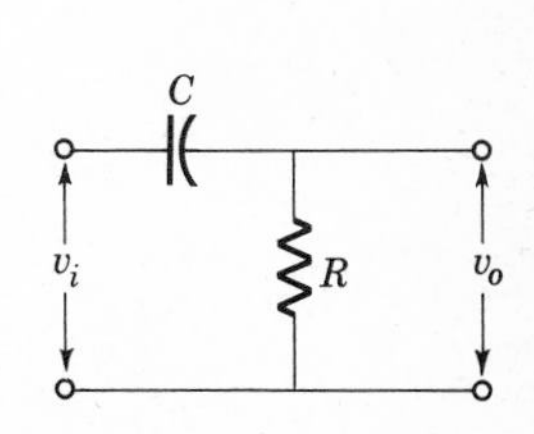

FIGURE 2-33 *Square-wave input signal applied to high-pass filter.*

the square wave of period T as indicated on the diagram. According to Eq. (2-82) the transient current accompanying each half-cycle of the square wave has the form

$$i = Ae^{-t/\tau} \tag{2-114}$$

where the constant A depends upon the current in the circuit resulting from the previous transient. For example, if $\tau = T/2$, the current has decayed 63 percent of the way at the time the square-wave voltage reverses. This means that the voltage across the capacitor causes a transient current in the reverse direction, and so forth. The output voltage ($= Ri$) is a succession of exponential transients, as shown in Fig. 2-34. These waveforms

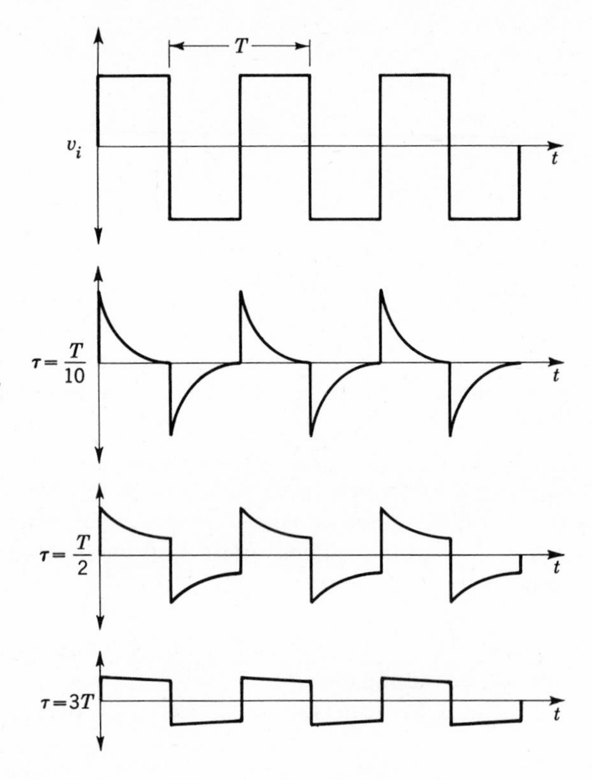

FIGURE 2-34 *Square wave (top) input to a high-pass filter produces different output waveforms depending upon ratio of period to circuit time constant.*

correspond to different values of the RC time constant compared with the period of the square wave.

Note that when the circuit time constant is very small compared with the period of the square wave, the output waveform approaches the differential of the square wave. This is an alternative approach to the study of differentiating circuits discussed in the previous section.

The low-pass filter circuit, Fig. 2-35, may be analyzed in the same fashion. The voltage across the capacitor is an exponential, as discussed earlier, and the output waveform, Fig. 2-36, again depends upon the relative magnitude of the circuit time constant compared with the period. In particular, note that when the circuit time constant is large the capaci-

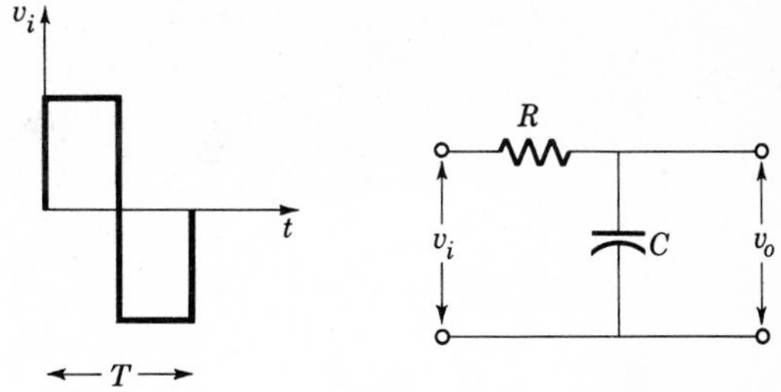

FIGURE 2-35 *Square-wave input signal applied to a low-pass filter.*

tor voltage rise is essentially linear during the pulse time. The output voltage is thus just the integral of the input signal, again in conformity with harmonic-circuit analysis.

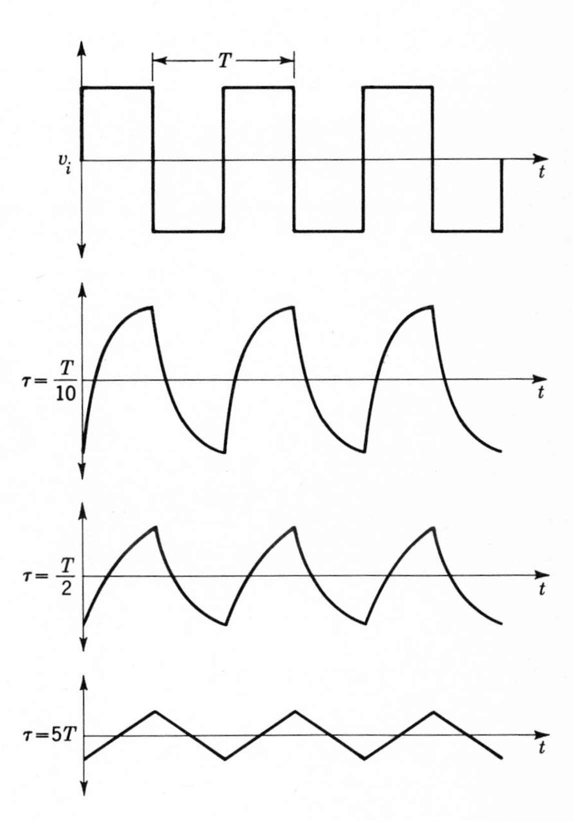

FIGURE 2-36 *Square-wave input to low-pass filter produces different output waveforms depending upon ratio of period to circuit time constant.*

Oscilloscope Undoubtedly the most useful instrument for studying complex waveforms is the *cathode-ray oscilloscope,* which is capable of displaying voltage waveforms visually. The heart of the oscilloscope is the cathode-ray tube, in which the position of a thin beam of electrons is electrically controlled to "paint" the waveform on a fluorescent screen. In the cathode-ray tube (CRT) an *electron gun* directs a high-velocity focused beam of electrons onto a glass face-plate covered with fluorescent material that emits light when bombarded by electrons. The beam passes between a pair of horizontal deflecting plates and a pair of vertical deflecting plates on its way to the screen, as sketched in Fig. 2-37. The beam produces a

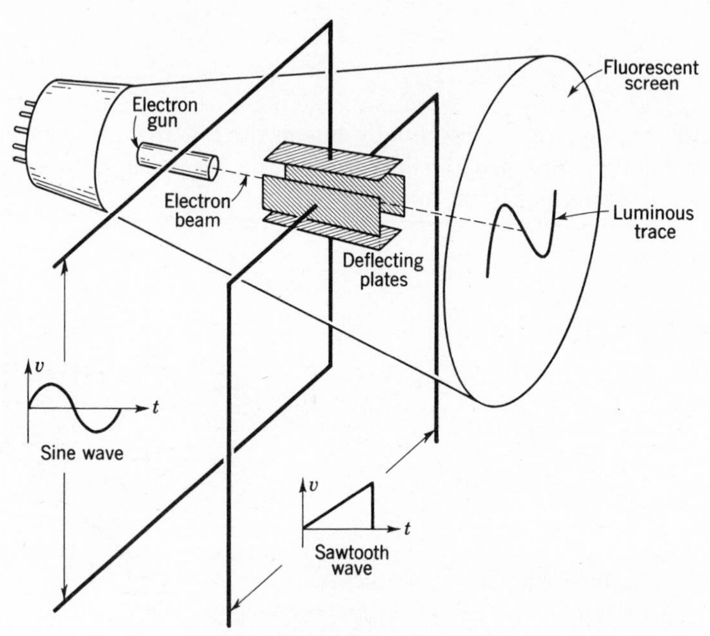

FIGURE 2-37 *Sketch of cathode-ray tube.*

spot of light in the center of the screen when the voltage on the deflection plates is equal to zero. The position of the spot on the screen is easily changed by deflecting the electron beam with voltages applied to either pair of deflecting plates. The beam is deflected as a result of the sidewise force on the electrons caused by the electric field between each pair of deflecting plates. Practical oscilloscopes include a number of auxiliary circuits to increase the versatility of the instrument, as discussed in a subsequent chapter.

The way in which the oscilloscope displays a waveshape can be understood by referring to Fig. 2-38. Suppose a sine-wave voltage of sufficient amplitude to cause a peak-to-peak deflection of about one-half the screen diameter is applied to the vertical pair of deflection plates. A sawtooth voltage waveform of the same frequency is applied to the horizontal deflecting plates with sufficient amplitude to cause the beam to sweep one-half the screen diameter in this direction. If both waveforms start from zero at the same instant, the spot begins in the center of the screen. The sawtooth wave on the horizontal plates causes the spot to move uniformly

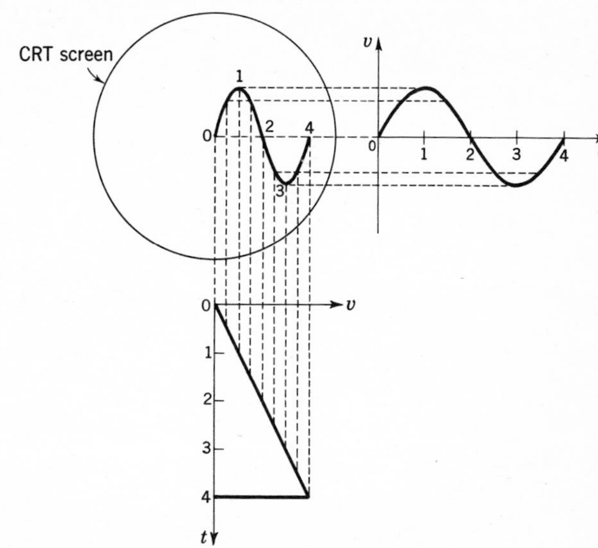

FIGURE 2-38 *Oscilloscope having a sawtooth horizontal-deflection voltage displays waveform of voltage applied to vertical plates.*

to the right while vertical motion depends upon the sine-wave voltage. The result is that the spot traces out a sine-wave pattern on the screen, as the point-by-point plot in Fig. 2-38 illustrates.

When the sawtooth voltage on the horizontal plates drops to zero at the end of the cycle, the spot rapidly returns to the starting point. The same pattern is traced out on subsequent cycles. At most frequencies of interest the spot moves too rapidly to be seen and the pattern on the screen appears to be a stationary sine wave because of the persistence of vision. The sawtooth voltage and the sine wave must start from the beginning at exactly the same instant on each cycle, for otherwise the spot does not trace out the same pattern every time. Therefore, the sawtooth *sweep* voltage is *synchronized* with the signal on the vertical plates by auxiliary circuits associated with the oscilloscope. It is also common practice to turn off the beam during the fast return of the spot to the beginning, in order to eliminate the *retrace* line from the pattern. The sweep voltage may conveniently be made a submultiple of the signal frequency in order to display more than one cycle of the waveform (Exercise 2-9).

The oscilloscope is also useful in measuring the phase angle between sine waves. This is done by applying one sine wave to the vertical plates and the other to the horizontal plates. The pattern which results can be a straight line, a circle, or an ellipse, depending upon the phase angle, as illustrated in Fig. 2-39 for the corresponding phase angles of 0, 45, and 90°. The manner in which the phase angle is calculated is as follows. Suppose the horizontal voltage is

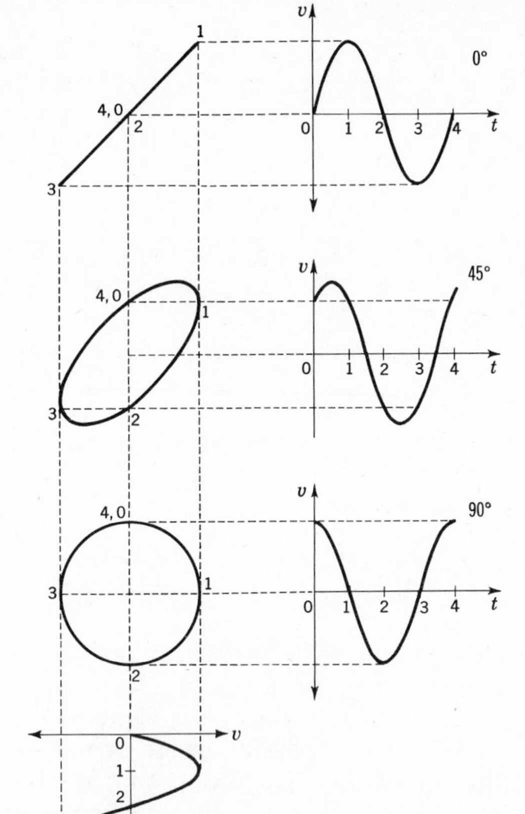

FIGURE 2-39 *Lissajous figures for phase differences of 0, 45, and 90° between horizontal- and vertical-deflection voltages.*

$$v_H = V_p \sin \omega t \tag{2-115}$$

and the vertical sine wave is given by

$$v_V = b \sin (\omega t + \phi) \tag{2-116}$$

When $t = 0$, $v_H = 0$, which means the horizontal deflection is zero. The vertical deflection at this point may be labeled a,

$$v_V = b \sin \phi = a \tag{2-117}$$

Solving for ϕ,

$$\phi = \arcsin \frac{a}{b} \tag{2-118}$$

The ratio a/b can be determined directly from the dimensions of the

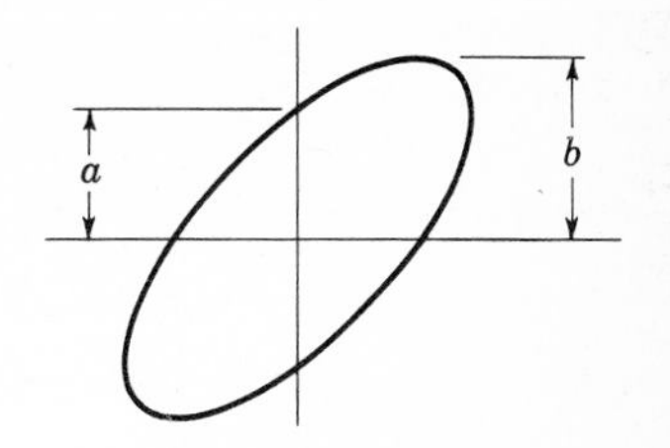

FIGURE 2-40 *Phase angle between sine-wave voltages applied to horizontal and vertical plates is determined from ratio a/b.*

pattern on the screen, as shown in Fig. 2-40. Note that the pattern must be centered on the horizontal and vertical reference lines in order for Eq. (2-118) to be valid.

The patterns in Fig. 2-39 are specific examples of so-called *Lissajous figures* which are used to determine the ratio between the frequencies of two sine waves. One sine wave is applied to the vertical deflection plates and the other is applied to the horizontal deflection plates. If the ratio of their frequencies is some integral fraction such as $\frac{1}{2}$, $\frac{1}{4}$, $\frac{2}{3}$, etc., the pattern is stationary. The frequency ratio is determined by the number of loops of the pattern touching a vertical line at the edge of the pattern compared with the number of loops touching a horizontal line at the edge of the pattern. The reason for this is that an integral number of sine waves on the horizontal deflection plates are completed in the same time that an integral number of sine waves are completed on the vertical plates. Some examples of common Lissajous figures are given in Fig. 2-41 for different

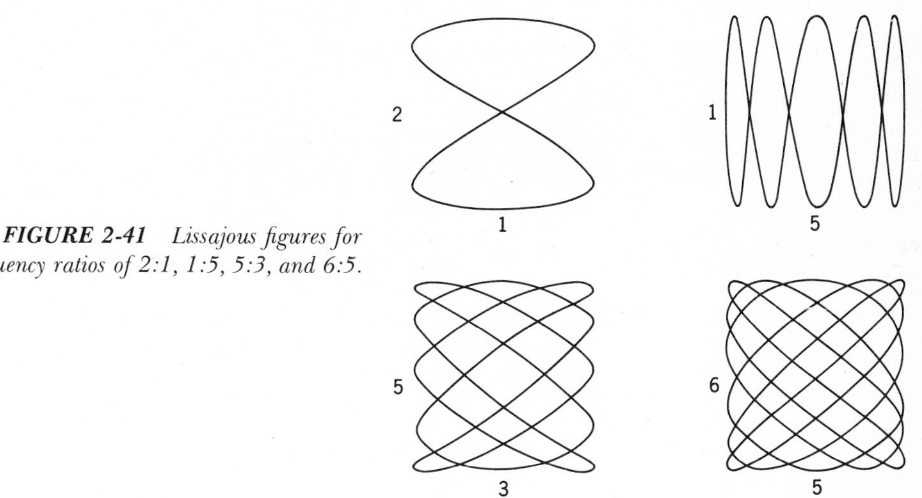

FIGURE 2-41 *Lissajous figures for frequency ratios of 2:1, 1:5, 5:3, and 6:5.*

frequency ratios. These patterns are derived in the same fashion used in connection with Figs. 2-38 and 2-39.

Several different patterns can represent the same frequency ratio, depending upon the interval between the times the two sine waves pass through zero. Figure 2-39 is an example of this for a 1:1 frequency ratio. Although in principle Lissajous figures can be used to determine the frequency ratio between nonsinusoidal waveforms, in practice the resulting patterns are confusing and difficult to interpret.

SUGGESTIONS FOR FURTHER READING

R. M. Kirchner and G. F. Corcoran: "Alternating Current Circuits," 4th ed., John Wiley & Sons, Inc., New York, 1960.

Leigh Page and Norman Isley Adams: "Principles of Electricity," D. Van Nostrand Company, Inc., Princeton, N.J., 1931.

"The Radio Amateur's Handbook" (published annually by the American Radio Relay League, West Hartford, Conn.).

EXERCISES

2-1 Compute the capacitance of a parallel-plate capacitor that has a plate area of 4 cm^2 and a separation between plates of 10^{-3} cm. What separation is needed to achieve a capacitance of 10 μF, as in a typical electrolytic capacitor?
Ans.: 3.54×10^{-10} F; 3.5×10^{-8} cm

2-2 Three capacitors are connected in series across a 100-V battery. If the capacitances are 1.0, 0.1, and 0.01 μF, respectively, calculate the potential difference across each capacitor.
Ans.: 0.90 V, 9.0 V, 90.1 V

2-3 Plot the frequency-response characteristic of an *RC* high-pass filter using Eq. (2-43) for $R = 10^6\ \Omega$ and $C = 0.01\ \mu$F. Cover the frequency interval from two decades below the half-power frequency to two decades above it.

2-4 Develop an expression analogous to Eq. (2-43) for an *RC* low-pass filter (Fig. 2-11). Plot the frequency-response characteristic for $R = 10^6\ \Omega$ and $C = 100$ pF.

2-5 Plot the frequency-response characteristic of an *RL* low-pass filter, Eq. (2-67), if $L = 10$ H and $R = 100\ \Omega$. Design an *RC* circuit having the same frequency characteristic.

2-6 Show how the Fourier series representation of a sawtooth wave, Eq. (2-101), approximates the sawtooth more accurately as additional harmonic terms are included by summing two, three, and four terms, as in Fig. 2-27.

2-7 Using the Fourier series representation of a square wave, determine the result of integrating a square wave. Do this by integrating Eq. (2-100) term by term and plotting the resulting series. Compare the waveform with the sketch in Fig. 2-29.

2-8 Using the Fourier series representation of a sawtooth wave, determine the result of differentiating a sawtooth wave. Do this by differentiating Eq. (2-101) term by term and plotting the result.

2-9 By a procedure analogous to that used in preparing Fig. 2-38, determine the observed waveform on an oscilloscope screen if the frequency of the horizontal sawtooth is one-half of the vertical sine-wave frequency.

2-10 By a procedure analogous to that used in preparing Fig. 2-39, determine the observed waveform on an oscilloscope screen if a sine wave is applied to both the horizontal and vertical deflection plates and the phase angle between them is 30°. Use the resulting pattern to confirm Eq. (2-118).

LABORATORY EXERCISES

2-A *RC* Filter Networks The simple *RC* filter network is extensively used in electronic circuits. It is very common for a number of *RC* filters to be part of a single network and so it is useful to examine the overall effect of several filters. Most often the input-output characteristics as a function of frequency are of major interest. That is, the ratio of the output voltage divided by the input voltage and the phase angle between input and output are determined.

Start by plotting the ratio of the output voltage divided by the input voltage as a function of frequency for a high-pass *RC* filter, Fig. 2-8, with $R = 10^4\ \Omega$ and $C = 0.16\ \mu F$. The frequency characteristic is similar to Fig. 2-10 except that the half-power frequency is higher. Can you state how much higher just by inspection? Note that it is most convenient to plot the frequency characteristic on log-log paper. Now plot the phase angle between the input and output voltage, taking the input signal as the reference. This time semilog paper is appropriate.

Connect a second identical high-pass filter to the output terminals. Now plot the overall frequency and phase characteristic of the two *RC* filters in cascade. To simplify the calculations, assume that the second filter does not load the first one so that the input voltage of the second filter is just the output voltage of the isolated first filter. This is not strictly true in practice (can you estimate the frequency range where the error is greatest?), but illustrates the important features. Now repeat all this for three identical filter stages. For the three cases examined, comment on the half-power frequency, the slope of the frequency response curve below the half-power frequency, and the phase shift at the half-power frequency.

The properties of a low-pass *RC* filter, Fig. 2-11, with $R = 10^5\ \Omega$ and $C = 160$ pF are examined in the same fashion. What is the half-power frequency? Cascade two identical low-pass filters and note the mirror-symmetry of the response characteristic compared to that for two high-pass filters.

For a final experiment plot the frequency and phase characteristic of two high-pass filters followed by two low-pass filter sections using the resistance and capacitance values listed above. The frequency interval between the upper and lower half-power points is known as the passband of the network. What is the phase shift between input and output over most of the frequency interval in the passband?

2-B Square Waves According to the Fourier series representation of nonsinusoidal signals, many frequencies are present simultaneously in complex waveforms. This makes it possible to examine the properties of a network without a laborious measurement of the frequency-response characteristic. To accomplish this, a known waveform is applied to the input terminals of the network and the output waveform is observed. Differences between the input and output waveforms are interpreted in terms of the frequency and phase characteristics of the circuit.

In this experiment, 1-kHz square wave signals are applied to single high-pass and low-pass *RC* filters and the output waveforms are observed. The form of the output waveform is directly related to the value of the 1-kHz fundamental frequency of the square wave compared to the half-power frequency of the filter. It is useful to examine filters having half-power frequencies of 10 kHz, 1 kHz, and 100 Hz in both the low-pass and high-pass networks.

Plot the six output waveforms specified in the previous paragraphs and compare the results qualitatively with Figs. 2-34 and 2-36. It is easiest to do this by starting with the Fourier series representation of a square wave, Eq. (2-100), and applying Eq. (2-43), and Eq. (2-67) in the case of a low-pass filter, to each harmonic component. The resulting series is summed to illustrate the output waveform. Note that it is necessary to determine the phase shift as well as the amplitude of each frequency component. To reduce the labor of these calculations, a simple Fortran computer program should be developed which can carry out the process. Use at least 10 terms (preferably 20) in the Fourier series representation of the input square wave. Perhaps you can include a graphic output format in your program to plot the output waveforms directly.

What is the relation between the fundamental frequency and the filter half-power frequency for the waveform to pass nearly unchanged? For the waveform to be integrated? Differentiated? Note particularly the output waveform in both cases when the filter half-power frequency is equal to the fundamental frequency.

chapter three

AC-CIRCUIT ANALYSIS

The principles of ac circuits treated in Chap. 2 can be used to find the currents in any network. However, solving the differential equation pertaining to each network of most practical circuits is cumbersome. More powerful techniques of ac-circuit analysis permit solutions for the circuit currents with much less labor. These techniques are, of course, based on the same differential equation of the circuit. The procedures are only slightly more complicated than for dc-circuit analysis, and both Ohm's law and Kirchhoff's rules are used in modified form. In fact, all the techniques of dc-network analysis treated in Chap. 1 are applicable.

IMPEDANCE

Reactance The rms current in a simple RC circuit is, from Eq. (2-41),

$$I = \frac{V}{\sqrt{R^2 + (1/\omega C)^2}} \tag{3-1}$$

where V is the applied voltage. Similarly, the rms current in a simple RL circuit is, from Eq. (2-59),

$$I = \frac{V}{\sqrt{R^2 + (\omega L)^2}} \tag{3-2}$$

In both cases the second term under the radical appears as important as the resistance in the circuit. As a matter of fact, in the absence of circuit resistance, the current in a capacitance is

$$I = \frac{V}{1/\omega C} \tag{3-3}$$

and the current in an inductance is

$$I = \frac{V}{\omega L} \tag{3-4}$$

which means that the quantity $1/\omega C$ determines the magnitude of the current in purely capacitive circuits, while ωL does so in purely inductive circuits. Clearly both act much the way a resistance does.

The quantity $1/\omega C$ is called the *capacitive reactance*

$$X_C = \frac{1}{\omega C} \tag{3-5}$$

and ωL is called the *inductive reactance*

$$X_L = \omega L \tag{3-6}$$

In ac circuits, the reactance of capacitors and inductors can be treated much like the resistance of resistors. Note, however, that according to Eqs. (3-5) and (3-6) capacitive reactance decreases with frequency, while inductive reactance increases with frequency. In contrast, resistance is a constant. According to Eqs. (3-3) and (3-4), the unit of reactance is a volt per ampere, or an ohm.

The second major difference between reactance and resistance is that the current and voltage in a reactance are not in phase. As found in Chap. 2, the current leads the voltage by a phase angle of 90° in a purely capacitive circuit and lags by 90° in a purely inductive circuit. The phase angle of reactances must always be taken into account in determining their effect upon the current.

Complex Impedance It is convenient to represent the two elements of reactance, the magnitude and phase angle, in such a way that the result of combining several resistances and reactances can be determined easily. This is done by representing reactance as a complex number. The real part of the complex number is associated with resistance while the imaginary part stands for reactance. Thus, for example, the series combination of a resistance R and an inductive reactance X_L is written as the complex *impedance* $\mathbf{Z}$, so that

$$\mathbf{Z} = Ze^{j\phi} = R + jX_L \tag{3-7}$$

where $j = \sqrt{-1}$. The symbol j, rather than i, is commonly used to signify $\sqrt{-1}$ in electronic-circuit analysis in order to avoid confusion with the conventional symbol for current.

Note that impedance is a complex number; hence it has both a magnitude and an angle associated with it. According to the standard graphical representation of a complex number, Fig. 3-1, the real part is plotted on

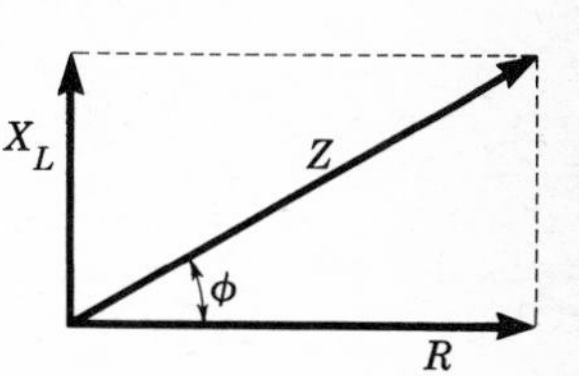

FIGURE 3-1 *Complex impedance is represented graphically by resistive component and reactive component.*

the horizontal axis and the imaginary part on the vertical axis. Therefore, the impedance angle is given by

$$\phi = \arctan \frac{X_L}{R} \tag{3-8}$$

The term impedance comes from the fact that the effect of both reactances and resistances is to impede the current in a circuit. According to Eq. (3-7) the unit of impedance is the ohm.

It is useful to recall some of the properties of complex numbers in order to facilitate their use in circuit analysis. Two complex numbers are equal only if their real and imaginary parts are equal. Therefore, if it is true that

$$R_1 + jX_1 = R_2 + jX_2 \tag{3-9}$$

then it must be that

$$R_1 = R_2$$

and (3-10)

$$X_1 = X_2$$

From this definition of equality it follows that the addition of two complex numbers is accomplished by separately adding the real and imaginary parts. That is,

$$(R_1 + jX_1) + (R_2 + jX_2) = (R_1 + R_2) + j(X_1 + X_2) \tag{3-11}$$

Clearly, the same rule applies to the subtraction of two complex numbers.

Complex numbers are multiplied as follows. According to Fig. 3-1 the complex number $\mathbf{Z}$ can be expressed in the form

$$\mathbf{Z} = Ze^{j\phi} = Z(\cos\phi + j\sin\phi) \tag{3-12}$$

The product of the numbers $\mathbf{Z}_1$ and $\mathbf{Z}_2$ is therefore

$$\begin{aligned}\mathbf{Z}_1\mathbf{Z}_2 = Z_1e^{j\phi_1}Z_1e^{j\phi_2} &= [Z_1(\cos\phi_1 + j\sin\phi_1)]\,[Z_2(\cos\phi_2 + j\sin\phi_2)] \\ &= Z_1Z_2\,[(\cos\phi_1\cos\phi_2 - \sin\phi_1\sin\phi_2) \\ &\quad + j(\sin\phi_1\cos\phi_2 + \cos\phi_1\sin\phi_2)]\end{aligned} \tag{3-13}$$

Using the trigonometric identity for the sine and cosine of the sum of angles, Eq. (3-13) becomes

$$Z_1e^{j\phi_1}Z_2e^{j\phi_2} = Z_1Z_2[\cos(\phi_1 + \phi_2) + j\sin(\phi_1 + \phi_2)] \tag{3-14}$$

Therefore, the result may be written

$$\mathbf{Z}_1\mathbf{Z}_2 = Z_1e^{j\phi_1}Z_2e^{j\phi_2} = Z_1Z_2e^{j(\phi_1+\phi_2)} \tag{3-15}$$

According to (3-15), the product of two complex numbers is a complex number of magnitude equal to the product of the individual magnitudes and with an angle which is the sum of the individual angles.

Actually, of course, (3-15) can be written down immediately from the usual rules for the multiplication of exponentials. Division of complex numbers also follows immediately,

$$\frac{\mathbf{Z}_1}{\mathbf{Z}_2} = \frac{Z_1e^{j\phi_1}}{Z_2e^{j\phi_2}} = \frac{Z_1}{Z_2}e^{j(\phi_1-\phi_2)} \tag{3-16}$$

This relation states that the result of dividing two complex numbers is a complex number whose magnitude is the ratio of the individual magnitudes and whose angle is the difference between the individual angles.

It is often useful to *rationalize* the reciprocal of a complex number. Thus

$$\frac{1}{\mathbf{Z}} = \frac{1}{Z(\cos\phi + j\sin\phi)} \tag{3-17}$$

This fraction is rationalized by multiplying numerator and denominator by the *complex conjugate*, obtained by replacing the imaginary part of the complex number by its negative. Therefore, Eq. (3-17) becomes

$$\frac{1}{\mathbf{Z}} = \frac{1}{Z(\cos\phi + j\sin\phi)}\,\frac{\cos\phi - j\sin\phi}{\cos\phi - j\sin\phi} = \frac{1}{Z}\,\frac{\cos\phi - j\sin\phi}{\cos^2\phi + \sin^2\phi}$$

$$= \frac{1}{Z}(\cos\phi - j\sin\phi) \tag{3-18}$$

Evidently the reciprocal of a complex number is simply a complex number with a magnitude given by the reciprocal of the magnitude of the complex number and an angle which is its negative. Using the exponential form

$$\frac{1}{Ze^{j\phi}} = \frac{1}{Z}\,e^{-j\phi} \tag{3-19}$$

which, again, can be written immediately.

The equivalent forms of representing a complex number

$$\mathbf{Z} = Ze^{j\phi} = Z(\cos\phi + j\sin\phi) = R + jX \tag{3-20}$$

should be noted. Evidently addition and subtraction are easiest in the component form, $R + jX$, while multiplication and division are easiest in the *polar* form, $Ze^{j\phi}$. Equation (3-20) is used to pass directly from one representation to another as needed. These operations, together with that of rationalization, are useful in computing the equivalent complex impedance of circuits comprising a number of individual impedances, as is illustrated in the next section. Note also that a complex number may be written in terms of its real and imaginary parts as

$$\mathbf{Z} = Ze^{j\phi} = \sqrt{R^2 + X^2}\,e^{j\arctan(X/R)} \tag{3-21}$$

In particular, the magnitude is equal to the square root of the sum of the squares of the real and imaginary components.

Ohm's Law for Alternating Current Since currents and voltages have both amplitudes and phase angles, it is not surprising that the complex-number notation proves equally useful to represent these quantities as in the case of impedance. When sinusoidal signals are represented by means of complex numbers, the circuit differential equations may be solved quite directly, as can be illustrated in the following way. Sinusoidal currents and voltages are written as

$$i = \mathbf{I} = I_p e^{j\omega t} = I_p(\cos\omega t + j\sin\omega t) \tag{3-22}$$

and

$$v = \mathbf{V} = V_p e^{j\omega t} = V_p(\cos\omega t + j\sin\omega t) \tag{3-23}$$

Consider now the differential equation for a simple RL series circuit, Eq. (2-54)

$$v = Ri + L\frac{di}{dt} \tag{3-24}$$

Insert (3-22) and (3-23) into (3-24), allowing for a phase angle between the current and the voltage,

$$V_p e^{j\omega t} = RI_p e^{j(\omega t+\phi)} + Lj\omega I_p e^{j(\omega t+\phi)} \tag{3-25}$$

$$= (R + j\omega L)I_p e^{j(\omega t+\phi)} \tag{3-26}$$

The quantity in parentheses is just the complex impedance of the circuit. Therefore, the circuit differential equation reduces to

$$\mathbf{V} = \mathbf{ZI} \tag{3-27}$$

This equation is the *ac form* of Ohm's law. It relates the current and the voltage in terms of the complex impedance of the circuit. Thus the solution of ac networks is reduced to simply determining the complex impedance. The validity of Eq. (3-27) rests directly upon the properties of the circuit differential equation. The complex representation of impedances is simply a very convenient computational device which greatly simplifies the solution of circuit differential equations.

Note that in the absence of reactance, Eq. (3-27) reduces to the standard dc form of Ohm's law. This means that series and parallel impedances must combine in the same way as do series and parallel resistances. Thus the equivalent impedance of a series of individual impedances is

$$\mathbf{Z}_{eq} = \mathbf{Z}_1 + \mathbf{Z}_2 + \mathbf{Z}_3 + \cdots \tag{3-28}$$

Similarly, the equivalent of parallel impedances is

$$\frac{1}{\mathbf{Z}_{eq}} = \frac{1}{\mathbf{Z}_1} + \frac{1}{\mathbf{Z}_2} + \frac{1}{\mathbf{Z}_3} + \cdots \tag{3-29}$$

In applying Eqs. (3-28) and (3-29) due considerstion must be given to the complex nature of impedances and the rules pertaining to how complex numbers combine. Several illustrative examples are considered in subsequent sections.

RLC CIRCUITS

Series Resonance As the first example of ac-circuit analysis using the complex-impedance method, consider the series *RLC* circuit of Fig. 3-2. According to Kirchhoff's law

$$v = Ri + \frac{q}{C} + L\frac{di}{dt} \tag{3-30}$$

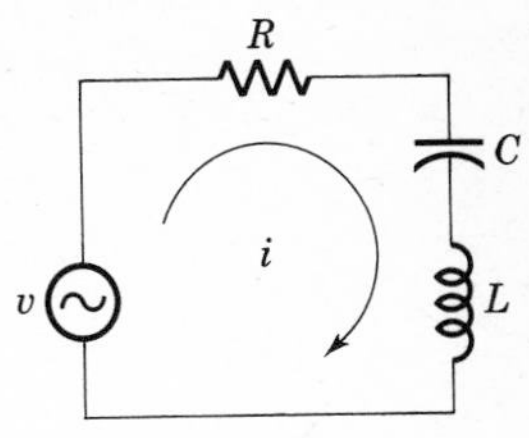

FIGURE 3-2 Series RLC circuit

The differential equation of the circuit is, after differentiating with respect to t,

$$L\frac{d^2i}{dt^2}+R\frac{di}{dt}+\frac{1}{C}i=\frac{dv}{dt} \tag{3-31}$$

Note that this is a second-order differential equation. It can be solved for the steady-state current i by the procedure used in Chap. 2. The result is

$$i=\frac{V_p}{\sqrt{R^2+(1/\omega C-\omega L)^2}}\sin(\omega t+\phi) \tag{3-32}$$

where

$$\phi=\arctan\frac{-(\omega L-1/\omega C)}{R} \tag{3-33}$$

The complex-impedance technique for the solution of the same circuit proceeds as follows. The total impedance of the series combination is, from Eq. (3-28),

$$\mathbf{Z}=R+j\omega L-j\frac{1}{\omega C}=R+j\left(\omega L-\frac{1}{\omega C}\right) \tag{3-34}$$

Using Ohm's law to calculate the current,

$$\mathbf{I}=\frac{\mathbf{V}}{\mathbf{Z}}=\frac{\mathbf{V}}{R+j(\omega L-1/\omega C)} \tag{3-35}$$

Equation (3-35) is the final solution, obtained directly in only two simple steps. The voltage **V** usually is given in terms of its rms value, so that the rms value of **I** is obtained.

To illustrate more clearly the correspondence between this solution and that given by (3-32) and (3-33), (3-35) is rationalized,

$$\mathbf{I}=\frac{V}{R+j(\omega L-1/\omega C)}\,\frac{R-j(\omega L-1/\omega C)}{R-j(\omega L-1/\omega C)}$$

$$=V\frac{R-j(\omega L-1/\omega C)}{R^2+(\omega L-1/\omega C)^2}$$

$$= \frac{V}{\sqrt{R^2 + (\omega L - 1/\omega C)^2}} \left[\frac{R}{\sqrt{R^2 + (\omega L - 1/\omega C)^2}} - j \frac{\omega L - 1/\omega C}{\sqrt{R^2 + (\omega L - 1/\omega C)^2}} \right] \tag{3-36}$$

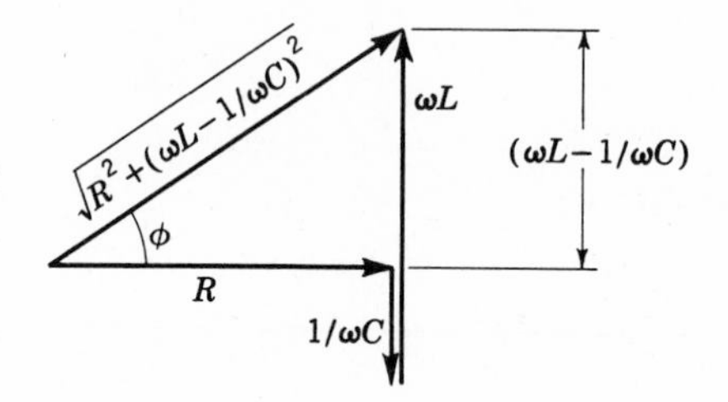

FIGURE 3-3 *Complex impedance diagram of an RLC circuit.*

Referring to the complex-impedance diagram of the circuit, Fig. 3-3, Eq. (3-36) may be written

$$\mathbf{I} = \frac{V}{\sqrt{R^2 + (\omega L - 1/\omega C)^2}} (\cos \phi - j \sin \phi)$$

$$= \frac{V}{\sqrt{R^2 + (\omega L - 1/\omega C)^2}} e^{-j\phi} \tag{3-37}$$

where

$$\phi = \arctan \frac{\omega L - 1/\omega C}{R} \tag{3-38}$$

The exact correspondence between the differential-equation solution, Eq. (3-32), and the result of the complex-impedance approach, Eq. (3-37), is evident. In subsequent sections the latter technique is used exclusively because of its simplicity.

Consider the amplitude of the current in the *RLC* circuit as a function of the frequency of the applied voltage. According to Eq. (3-37), the current is very small at low frequencies ($\omega \to 0$), because the capacitive reactance is great. Similarly, the current is small at high frequencies ($\omega \to \infty$), because the inductive reactance becomes large. Between these two extremes, the phase angle is zero when

$$\omega L - \frac{1}{\omega C} = 0 \tag{3-39}$$

At this frequency the circuit is said to be in *resonance*, and the current is a maximum,

$$I = \frac{V}{R} \tag{3-40}$$

The circuit appears as a pure resistance, and the current is in phase with the applied voltage. This is so because the capacitive reactance cancels the inductive reactance at resonance. The *resonant frequency* is, from Eq. (3-39),

$$\omega_0 = 2\pi f_0 = \frac{1}{\sqrt{LC}} \qquad (3\text{-}41)$$

FIGURE 3-4

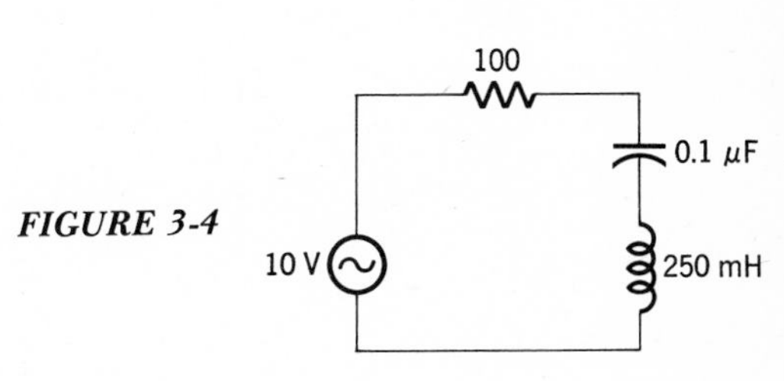

To illustrate the properties of a series resonant circuit, consider the current variation in the specific circuit of Fig. 3-4. After substituting the component values on the diagram into Eq. (3-37) it is found that the current in the circuit changes with frequency as illustrated in Fig. 3-5. The current maximum is $I = \frac{10}{100} = 0.1$ A at the resonant frequency,

$$f_0 = \frac{1}{2\pi\sqrt{LC}} = \frac{1}{6.28\sqrt{250 \times 10^{-3} \times 0.1 \times 10^{-6}}} = 1000 \text{ Hz} \qquad (3\text{-}42)$$

and the decrease on either side of resonance is clearly evident.

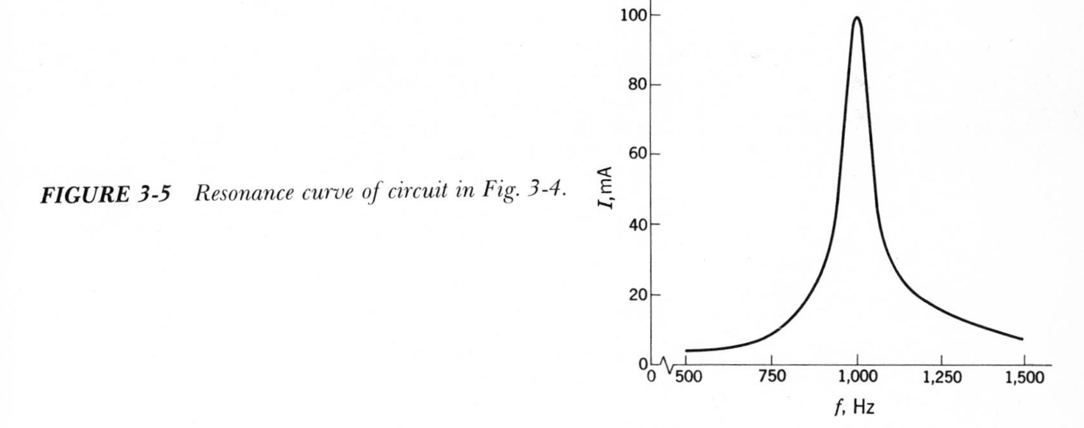

FIGURE 3-5 *Resonance curve of circuit in Fig. 3-4.*

The voltage drops around the circuit at resonance further illustrate an important feature of alternating currents. The drop across the resistor is

$$V_R = RI = 100 \times 0.1 = 10 \text{ V} \qquad (3\text{-}43)$$

The corresponding voltage drop across the capacitor is the current times the capacitive reactance

$$V_C = I\frac{1}{\omega C} = \frac{0.1}{6.28 \times 10^3 \times 10^{-7}} = 158\ \text{V} \tag{3-44}$$

and, similarly, the voltage across the inductor is

$$V_L = I\omega L = 0.1 \times 6.28 \times 10^3 \times 0.25 = 158\ \text{V} \tag{3-45}$$

It is evident that the rms voltage drops around the circuit do not sum to zero. In fact, the voltages across both reactances rise to large values at resonance. Note, however, that the phase angle of the capacitor voltage is $+90°$ with respect to the current, while the phase angle of the voltage across the inductor is $-90°$. This means that the instantaneous voltages across the two reactances cancel each other and that the drop across the resistance equals the applied emf, as confirmed by Eq. (3-43). Kirchhoff's voltage rule is valid when the phases of the currents and voltages in ac circuits are taken into account. This matter is considered further in the Exercises.

A voltmeter connected across, say, the capacitor in Fig. 3-4 indicates a voltage corresponding to Eq. (3-44) at resonance. Therefore, the capacitor (and the inductor) must be capable of withstanding high voltages without breakdown, even though the applied voltage is well below the ratings of the components. Resonance is widely used in electronic circuits to increase the current at the resonant frequency.

Parallel Resonance Resonance also occurs in a parallel circuit, such as the one shown in Fig. 3-6. The current in the circuit is obtained by first

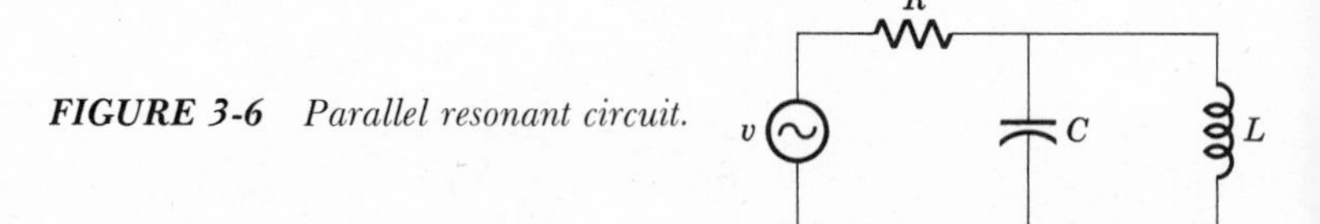

FIGURE 3-6 *Parallel resonant circuit.*

computing the total complex impedance. Since the inductance and capacitance are connected in parallel, their equivalent impedance is found with the aid of Eq. (3-29),

$$\begin{aligned}\frac{1}{\mathbf{Z}_1} &= \frac{1}{-j(1/\omega C)} + \frac{1}{j\omega L} = j\omega C + \frac{1}{j\omega L} \\ &= \frac{-\omega^2 LC + 1}{j\omega L}\end{aligned} \tag{3-46}$$

Therefore, the impedance of the LC combination is

$$\mathbf{Z}_1 = j\frac{\omega L}{1 - \omega^2 LC} \tag{3-47}$$

According to Eq. (3-47) the impedance is very large, actually infinite, when

$$\omega_0^2 LC = 1$$

or

$$\omega_0 = \frac{1}{\sqrt{LC}} \tag{3-48}$$

This is the same relation found in the case of series resonance, Eq. (3-41). Note, however, that in series resonance the impedance is a minimum, according to Eq. (3-34), while in parallel resonance Eq. (3-47) shows that the impedance is a maximum at the resonant frequency.

The total impedance of the circuit includes the series combination of R with $\mathbf{Z}_1$

$$\mathbf{Z} = R + j\frac{\omega L}{1 - \omega^2 LC} \tag{3-49}$$

so that the current in the circuit is

$$\mathbf{I} = \frac{V}{R + j\omega L/(1 - \omega^2 LC)} \tag{3-50}$$

According to Eq. (3-50) the current is zero at resonance, when the impedance becomes infinite. The voltage V then appears across the LC combination, independent of the value of R. This feature of the parallel resonant circuit is extensively used in practical circuits. The output voltage V_o is simply the current times the impedance $\mathbf{Z}_1$ or

$$\mathbf{V}_o = \mathbf{I}\mathbf{Z}_1 = \frac{V}{R + j\omega L/(1 - \omega^2 LC)} \frac{j\omega L}{1 - \omega^2 LC}$$

$$= V \frac{j\omega L}{R(1 - \omega^2 LC) + j\omega L}$$

Rationalizing

$$\mathbf{V}_o = V \frac{j\omega L}{R(1 - \omega^2 LC) + j\omega L} \frac{R(1 - \omega^2 LC) - j\omega L}{R(1 - \omega^2 LC) - j\omega L}$$

$$= V \frac{(\omega L)^2 + j\omega LR(1 - \omega^2 LC)}{(\omega L)^2 + R^2(1 - \omega^2 LC)^2}$$

$$= V \frac{1 + j(R/\omega L)(1 - \omega^2 LC)}{1 + (R/\omega L)^2(1 - \omega^2 LC)^2} \tag{3-51}$$

The magnitude of the output voltage is found by taking the square root of the sum of the squares of the real and imaginary parts. Therefore, the ratio of the output voltage to the input voltage becomes

$$\frac{V_o}{V} = \left\{\frac{1 + (R/\omega L)^2(1 - \omega^2 LC)^2}{[1 + (R/\omega L)^2(1 - \omega^2 LC)^2]^2}\right\}^{1/2}$$

$$= \frac{1}{[1 + (R/\omega L)^2(1 - \omega^2 LC)^2]^{1/2}} \tag{3-52}$$

The behavior of Eq. (3-52) is illustrated with the specific component

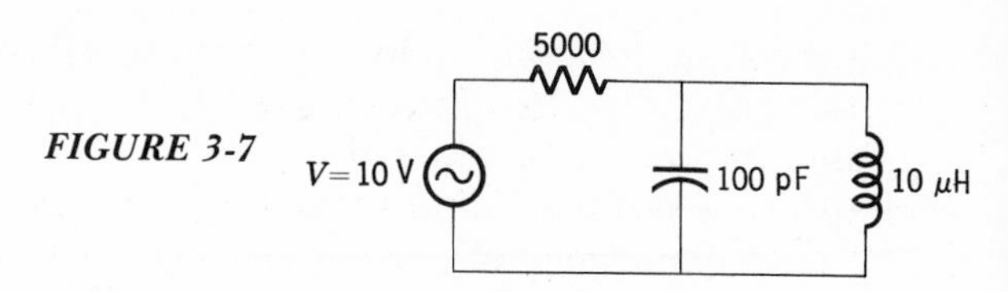

FIGURE 3-7

values given on the parallel-circuit diagram of Fig. 3-7. The output voltage rises to equal the input voltage at the resonant frequency, as shown in Fig. 3-8. Parallel resonance is commonly used in electronic circuits

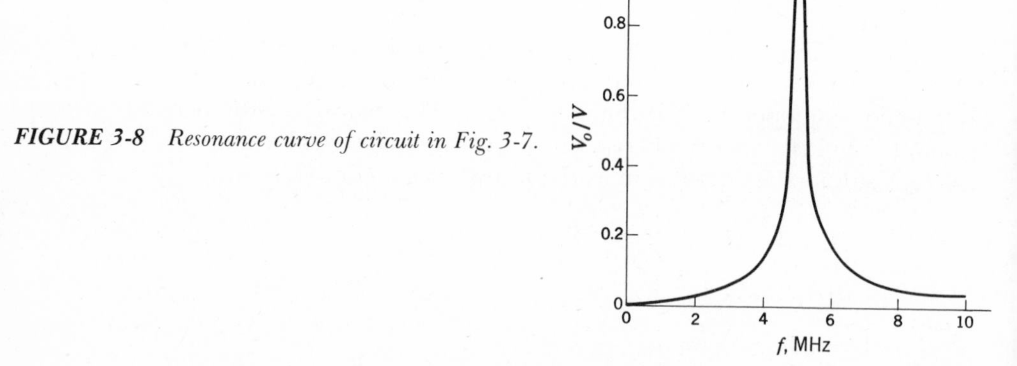

FIGURE 3-8 *Resonance curve of circuit in Fig. 3-7.*

to achieve a high impedance which develops an appreciable signal voltage at resonance. Resonance is also important in circuits designed to emphasize one single frequency over all others. By adjusting the value of, say, the capacitance, the circuit may be *tuned* to different frequencies. This is the principle used to select different channels in radio and TV receivers.

It is interesting to compare the currents in various components of a parallel resonant circuit at resonance. According to Eq. (3-52) the full input voltage is applied across both the capacitor and inductance at resonance, so that the currents are, respectively

$$I_C = \frac{V}{1/\omega C} \tag{3-53}$$

and

$$I_L = \frac{V}{\omega L} \tag{3-54}$$

Yet the total current in the circuit is zero, according to Eq. (3-50). This is another situation where the phase angle of the currents must be considered. Since the capacitive reactance equals the inductive reactance at resonance, the magnitudes of the currents in the inductance and capacitance are equal. Their phase angles are such that at, say, the upper branch point in Fig. 3-6 these two currents cancel at every instant and no current is present in the resistor. In this sense a circulating current exists in the parallel *LC* combination at resonance.

Q factor The resistance in a resonant circuit is significant in determining the current at frequencies removed from the resonant frequency, as well as at resonance. This is an important consideration since, for example, the resistance of the turns of wire is present in all inductors. The effect is similar in both series and parallel resonance cases, but the former is simpler mathematically and can be used to illustrate all the important features.

Consider the magnitude of the current in a series circuit Fig. 3-2, given by Eq. (3-37),

$$I = \frac{V}{[R^2 + (\omega L - 1/\omega C)^2]^{1/2}} \tag{3-55}$$

This can be rearranged as

$$I = \frac{V}{R} \frac{1}{\sqrt{1 + (\omega L/R)^2(1 - 1/\omega^2 LC)^2}} \tag{3-56}$$

The ratio of inductive reactance to resistance is defined in Chap. 2 and is called the Q of the circuit. For present purposes, it is convenient to consider the Q at resonance,

$$Q_0 = \frac{\omega_0 L}{R} \tag{3-57}$$

Introducing (3-57) and the resonant frequency from (3-41) into (3-56),

$$\frac{I}{I_M} = \frac{1}{\{1 + Q_0^2(\omega/\omega_0)^2[1 - (\omega_0/\omega)^2]^2\}^{1/2}} \tag{3-58}$$

where I_M is the maximum current at resonance. Finally, an alternative form to Eq. (3-58) is, after multiplying the last two quantities under the radical,

$$\frac{I}{I_M} = \left[1 + Q_0^2\left(\frac{\omega}{\omega_0} - \frac{\omega_0}{\omega}\right)^2\right]^{-1/2} \tag{3-59}$$

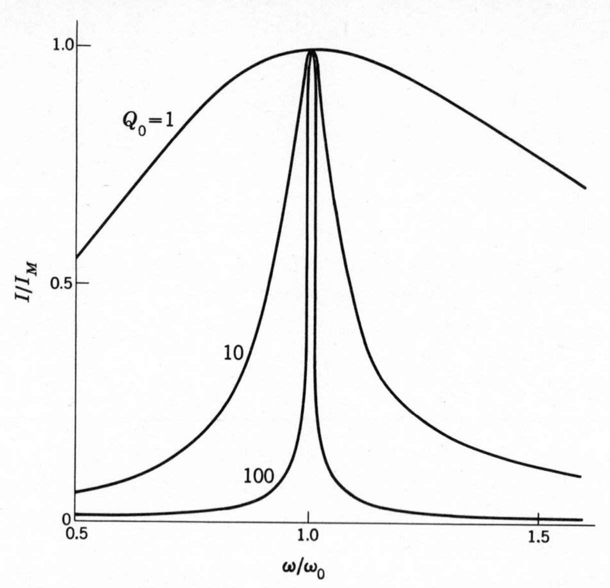

FIGURE 3-9 *Sharpness of resonance curve is greatest for high-Q circuits.*

This expression, plotted in Fig. 3-9 for representative values of Q_0, shows that a large Q_0 (that is, a small value of R) leads to a very sharply resonant circuit. Under this condition the bandwidth of the resonant circuit, that is, the frequency interval between half-power points, is simply related to Q_0 as shown in the following way. From Eq. (3-59) at the half-power frequencies,

$$2 = 1 + Q_0^2\left(\frac{\omega'}{\omega_0} - \frac{\omega_0}{\omega'}\right)^2 \qquad (3\text{-}60)$$

$$\frac{1}{Q_0} = \frac{\omega'}{\omega_0} - \frac{\omega_0}{\omega'} = \frac{\omega_0}{\omega'}\left[\left(\frac{\omega'}{\omega_0}\right)^2 - 1\right] \qquad (3\text{-}61)$$

Introducing the bandwidth $\Delta\omega$,

$$\omega' = \omega_0 + \frac{\Delta\omega}{2} \qquad (3\text{-}62)$$

$$\frac{\omega_0 + \Delta\omega/2}{\omega_0 Q_0} = \left(\frac{\omega_0 + \Delta\omega/2}{\omega_0}\right)^2 - 1 \qquad (3\text{-}63)$$

Since $\Delta\omega$ is small compared to ω_0

$$\frac{1}{Q_0} = \frac{\Delta\omega}{\omega_0} \qquad (3\text{-}64)$$

According to Eq. (3-64) the percentage bandwidth of a sharply resonant circuit is just the reciprocal of the Q at resonance.

Values of Q_0 in the range from 10 to 100 are common in electronic circuits and are useful in very selective tuned circuits, as illustrated by the two lower curves of Fig. 3-9. Special resonant components can have Q's as high as several thousand, and these result in very high-frequency selectivity indeed. On the other hand, it is sometimes desirable to broaden the frequency response of a resonant circuit so that it is useful over a wide band of frequencies on either side of the resonant frequency. In this case a resistor is purposely included to yield a lower Q than that of the inductor alone.

BRIDGE CIRCUITS

Figure 3-10 shows the ac analogy of the dc Wheatstone bridge. It has a complex impedance in each arm and uses a sine-wave generator and an

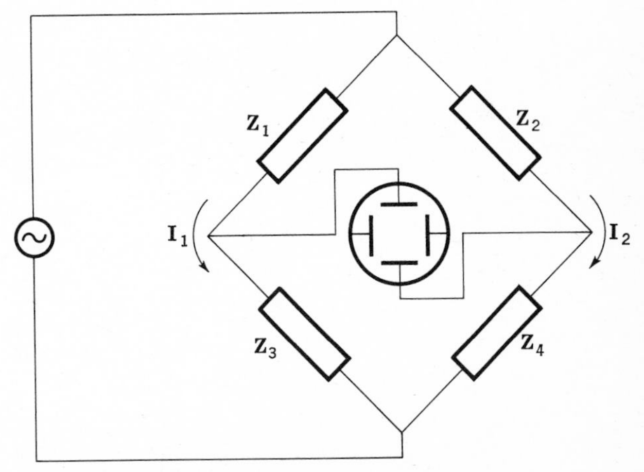

FIGURE 3-10 *The ac Wheatstone bridge uses sine-wave generator and ac detector such as an oscilloscope.*

ac detector, such as an oscilloscope. Analysis of this circuit proceeds as in the dc case considered in Chap. 1, except that complex impedances and currents are used. Only the balance condition is of interest here, and this is obtained in the following way. At balance, the voltage across the detector is zero, which means that the current in this branch is zero and therefore that the current in $\mathbf{Z}_1$ is equal to that in $\mathbf{Z}_3$; also, the current in $\mathbf{Z}_2$ is equal to that in $\mathbf{Z}_4$. Furthermore, since the voltage across the detector is zero, the voltage drops across $\mathbf{Z}_1$ and $\mathbf{Z}_2$ are equal and so are the voltage drops across $\mathbf{Z}_3$ and $\mathbf{Z}_4$.

Equating the voltage drops across corresponding arms of the bridge

$$\mathbf{Z}_1\mathbf{I}_1 = \mathbf{Z}_2\mathbf{I}_2 \tag{3-65}$$

and

$$\mathbf{Z}_3\mathbf{I}_1 = \mathbf{Z}_4\mathbf{I}_2 \tag{3-66}$$

Dividing Eq. (3-65) by Eq. (3-66), the condition for balance is found to be

$$\frac{\mathbf{Z}_1}{\mathbf{Z}_3} = \frac{\mathbf{Z}_2}{\mathbf{Z}_4} \tag{3-67}$$

which should be compared with the condition for balance of the dc Wheatstone bridge, Eq. (1-69).

Using Eq. (3-16) to carry out the ratios indicated, the result is

$$\frac{Z_1}{Z_2} e^{j(\phi_1 - \phi_2)} = \frac{Z_3}{Z_4} e^{j(\phi_3 - \phi_4)} \tag{3-68}$$

According to Eq. (3-68) the balance condition involves the equality of two complex numbers. This means that both the real and imaginary parts must be equal and implies that two independent balance adjustments are necessary in ac bridge circuits. Specific illustrations of this situation are taken up in the following sections.

Inductance and Capacitance Bridge Bridge circuits in Figs. 3-11 and 3-12 can be used to measure inductance and capacitance in the same way

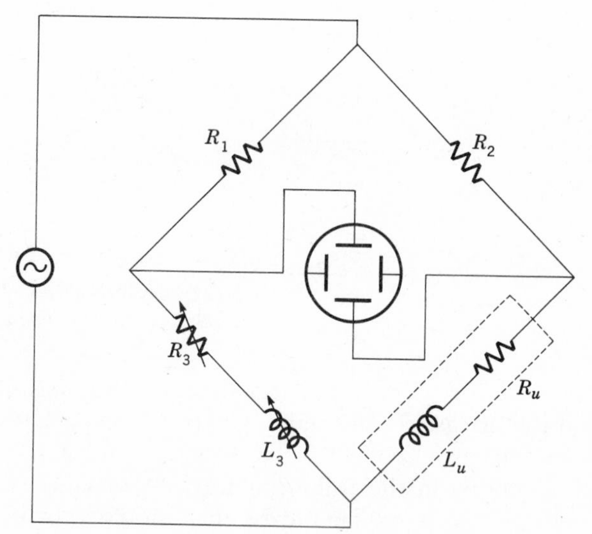

FIGURE 3-11 *Inductance bridge used to measure resistance and inductance of unknown coil.*

a Wheatstone bridge is used to measure resistance. Consider the inductance bridge, Fig. 3-11, and note the comparison with Fig. 3-10. The impedance $R_u + j\omega L_u$ represents the inductance and resistance of an unknown inductor. The balance condition, Eq. (3-67), means

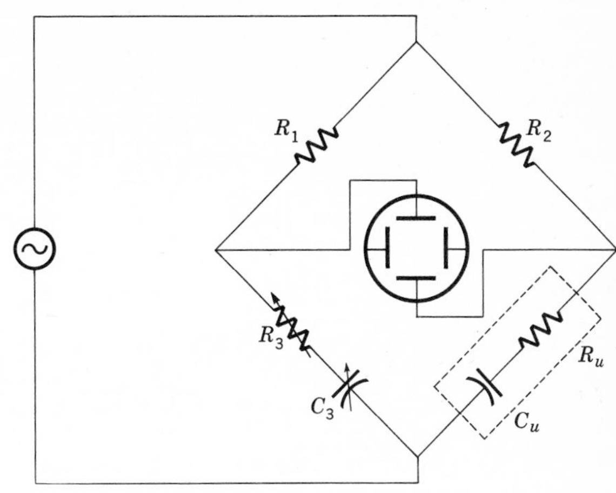

FIGURE 3-12 *Capacitance bridge.*

$$\frac{R_1}{R_2} = \frac{R_3 + j\omega L_3}{R_u + j\omega L_u} \tag{3-69}$$

Cross-multiplying,

$$R_1 R_u + j\omega R_1 L_u = R_2 R_3 + j\omega R_2 L_3 \tag{3-70}$$

After equating real and imaginary parts, the two balance conditions are written as

$$R_1 R_u = R_2 R_3 \qquad \text{and} \qquad R_1 L_u = R_2 L_3 \tag{3-71}$$

The first condition is the same as the Wheatstone bridge condition and may be carried out with dc instruments. The second equation is obtained using ac excitation of the bridge subsequent to obtaining the dc balance.

Rewriting Eqs. (3-71),

$$R_u = \frac{R_2}{R_1} R_3 \qquad \text{and} \qquad L_u = \frac{R_2}{R_1} L_3 \tag{3-72}$$

which illustrates how the ratio R_2/R_1 acts as a multiplying factor on the variable components R_3 and L_3 to yield the unknown resistance and inductance. This suggests that both R_3 and L_3 should be variable, but since variable inductors are difficult to make, the following procedure is more satisfactory. Either R_2 or R_1 is adjusted until balance is obtained with a given value of L_3. Then R_3 is adjusted to satisfy the dc balance condition. In this way the unknown inductance is compared with the standard inductance L_3, and the only variable components needed are resistances.

An interesting feature of the balance conditions, Eqs. (3-72), is that they are independent of frequency, so that the generator frequency may have any convenient value. It is usually best to choose a frequency such that the inductive reactance is approximately equal to the resistances. A similar situation exists in the case of the capacitance bridge, Fig. 3-12, which is analyzed in Exercise 3-10. Since there is no dc path in the lower arms of the capacitance bridge, both balance adjustments are made with ac excitation (this can be done in the inductance bridge case as well). Although calibrated variable capacitors are usually used in this circuit, it is equally possible to employ a fixed standard capacitor together with variable resistors. This is particularly convenient when the same instrument is used as an inductance bridge as well as a capacitance bridge by replacing the standard capacitor C_3 with a standard inductance L_3. Since it is also possible to employ a standard resistor in this position, such an instrument is quite versatile in that it can measure resistance, capacitance, and inductance.

Wien Bridge A bridge which has a parallel combination in one arm and a series combination in an adjacent arm is known as a *Wien* bridge. A useful example employing only resistors and capacitors, shown in Fig. 3-13,

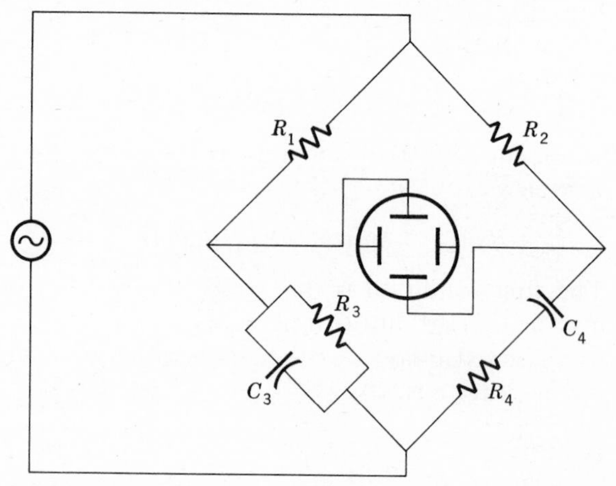

FIGURE 3-13 *Wien bridge.*

is analyzed by first calculating the impedances $\mathbf{Z}_3$ and $\mathbf{Z}_4$. Considering the parallel combination first,

$$\frac{1}{\mathbf{Z}_3} = j\omega C_3 + \frac{1}{R_3} = \frac{1 + j\omega R_3 C_3}{R_3} \tag{3-73}$$

which is rationalized to give

$$\mathbf{Z}_3 = \frac{R_3}{1 + (\omega R_3 C_3)^2}(1 - j\omega R_3 C_3) \tag{3-74}$$

The series combination is

$$\mathbf{Z}_4 = R_4 - j\frac{1}{\omega C_4} \tag{3-75}$$

Inserting Eqs. (3-74) and (3-75) into the balance equation (3-67) results in

$$\frac{R_1}{R_2} = \frac{R_3/(1 + j\omega R_3 C_3)}{R_4 - j(1/\omega C_4)} \tag{3-76}$$

Cross-multiplying,

$$(1 + j\omega R_3 C_3)\left(R_4 - \frac{j}{\omega C_4}\right) = \frac{R_2}{R_1} R_3 \tag{3-77}$$

$$R_4 + \frac{R_3 C_3}{C_4} + j\left(\omega R_3 R_4 C_3 - \frac{1}{\omega C_4}\right) = \frac{R_2}{R_1} R_3 \tag{3-78}$$

Upon equating the real and imaginary parts, the balance conditions are found to be

$$\frac{C_3}{C_4} + \frac{R_4}{R_3} = \frac{R_2}{R_1}$$

and

$$\omega^2 R_3 C_3 R_4 C_4 = 1 \tag{3-79}$$

This result differs from those for the inductance and capacitance bridges in that the frequency ω appears in the balance equations. Therefore, it is possible to achieve balance by adjusting the frequency and only one component, say R_1, rather than using two variable impedances. Alternatively, by adjusting two of the components for balance, the bridge is capable of determining the frequency of a sine-wave source.

The Wien bridge is also useful as a frequency-selective network which has properties similar to those of a resonant circuit. Since inductors suitable for use over a wide frequency interval are expensive and difficult to construct, the Wien bridge has considerable advantages in many applications. This is particularly true for low-frequency circuits where inconveniently large inductance values are necessary. Often, the resistors and capacitors in the series and parallel branches are equal, as in Fig. 3-14. This means that the characteristic frequency of the network is, from Eq. (3-79),

$$\omega_0 = \frac{1}{RC} \tag{3-80}$$

The frequency-selective properties of the Wien bridge are illustrated by calculating the output voltage of the network as a function of the frequency of the input voltage. The output voltage is obtained by subtracting

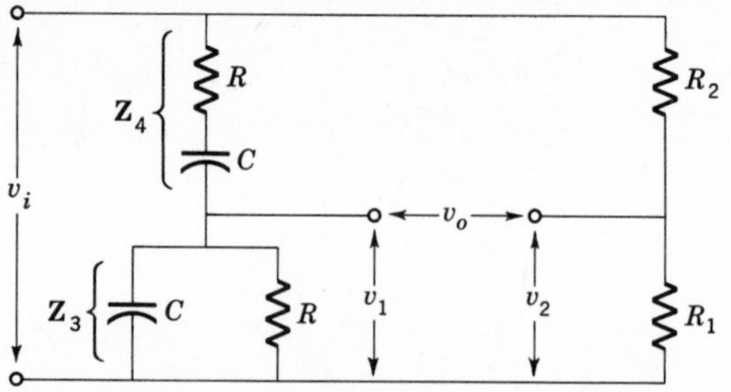

FIGURE 3-14 *Wien bridge used as frequency-selective network.*

the voltages across $\mathbf{Z}_3$ and R_1. The drop across $\mathbf{Z}_3$ is simply the current in this arm times the impedance,

$$v_1 = \frac{v_i}{\mathbf{Z}_3 + \mathbf{Z}_4}\mathbf{Z}_3 = \frac{v_i}{1 + \mathbf{Z}_4/\mathbf{Z}_3} \tag{3-81}$$

where it has been assumed that the current in the output circuit is negligible, even when the bridge is not balanced. This condition is satisfied in practice by connecting a high-impedance load to the output terminals. The impedance ratio in Eq. (3-81) is evaluated from Eqs. (3-74) and (3-75), after making all the R's and C's equal. Also, introducing ω_0,

$$\frac{\mathbf{Z}_4}{\mathbf{Z}_3} = \frac{R(1 - j\omega_0/\omega)}{R(1 - j\omega/\omega_0)/[1 + (\omega/\omega_0)^2]} = 2 + j\left(\frac{\omega}{\omega_0} - \frac{\omega_0}{\omega}\right) \tag{3-82}$$

Note that according to Eq. (3-82) the reactive term vanishes at the characteristic frequency when $\omega = \omega_0$. This means that the network is resistive

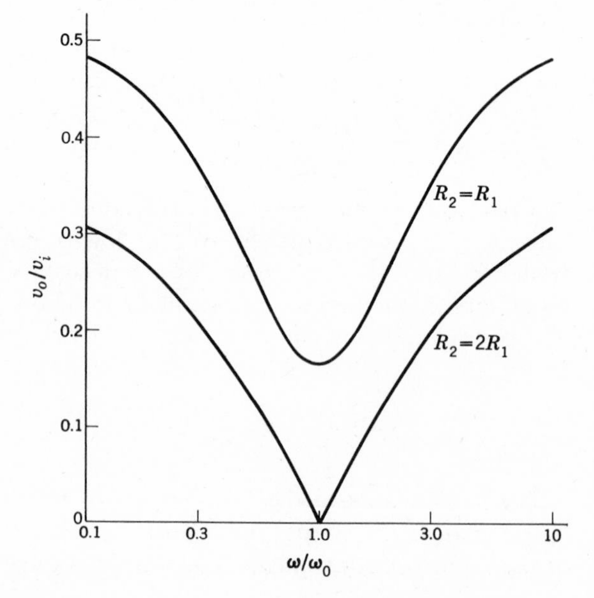

FIGURE 3-15 *Frequency characteristic of Wien bridge network.*

at this frequency (and at this frequency only) and that the output voltage is in phase with the input voltage, just as for a resonant circuit.

The drop across R_1 is

$$v_2 = \frac{v_i}{R_1 + R_2} R_1 = \frac{v_i}{1 + R_2/R_1} \tag{3-83}$$

Finally, the output voltage is

$$v_0 = v_1 - v_2 = v_i \left[\frac{1}{3 + j(\omega/\omega_0 - \omega_0/\omega)} - \frac{1}{1 + R_2/R_1} \right] \tag{3-84}$$

The ratio of the output to the input v_0/v_i can be found after rationalizing. Two plots of Eq. (3-84) are presented in Fig. 3-15 for different values of the ratio R_2/R_1. The output voltage clearly goes through a minimum at $\omega = \omega_0$, very analogous to the situation in series resonance. The minimum is quite sharp and mathematically discontinuous in the case of true balance, $R_2 = 2R_1$. Therefore, the frequency selectivity of the circuit is quite good.

Bridged-T and Twin-T Networks Although the so-called *bridged-T* filter is not a true bridge circuit, it has properties similar to those of the Wien bridge. In contrast to the Wien bridge, the bridged T, shown in Fig. 3-16, generally does not employ equal values of resistance and capacitance in the two positions. To solve for the output voltage, the current in the upper resistor is determined and the output is obtained by subtracting the voltage drop across R_2 from the input voltage.

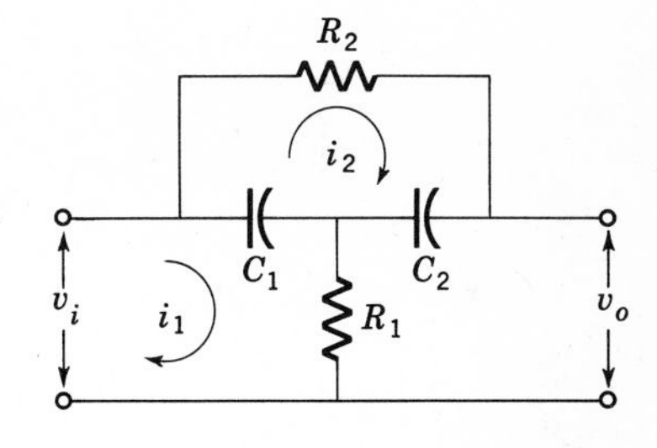

FIGURE 3-16 *Bridged-T filter.*

Kirchhoff's law applied to the i_1 loop in Fig. 3-16 is

$$v_i = i_1 \frac{-j}{\omega C_1} - i_2 \frac{-j}{\omega C_1} + i_1 R_1 \tag{3-85}$$

Adding up the voltages around the second loop,

$$0 = -i_1 \frac{-j}{\omega C_1} + i_2 \frac{-j}{\omega C_1} + i_2 R_2 + i_2 \frac{-j}{\omega C_2} \tag{3-86}$$

After simplification, i_1 is determined from (3-86),

$$i_1 = \left(1 + \frac{C_1}{C_2} + j\omega R_2 C_1\right) i_2 \tag{3-87}$$

and substituted into (3-85). The result is solved for i_2 and the output voltage is then given by

$$v_0 = v_i - R_2 i_2 = v_i - \frac{v_i \omega R_2 C_2}{\omega C_2[R_1(1 + C_1/C_2) + R_2] + j(\omega^2 R_1 C_1 R_2 C_2 - 1)} \tag{3-88}$$

The imaginary part vanishes at the critical frequency

$$\omega_0 = \frac{1}{\sqrt{R_1 C_1 R_2 C_2}} \tag{3-89}$$

which should be compared with the corresponding expression for the Wien bridge, Eq. (3-80). After inserting (3-89) into (3-88), the complex ratio of output to input voltage may be written

$$\frac{v_0}{v_i} = 1 - \left\{\left[1 + \frac{R_1}{R_2}\left(1 + \frac{C_1}{C_2}\right)\right] + j\sqrt{\frac{R_1 C_1}{R_2 C_2}}\left(\frac{\omega}{\omega_0} - \frac{\omega_0}{\omega}\right)\right\}^{-1} \tag{3-90}$$

The frequency-response characteristic of the bridged-T filter, Eq. (3-90), is similar to that of the Wien bridge, except that the minimum is not as sharp at the characteristic frequency. Nevertheless, its relative simplicity, and the fact that both input and output terminals have a common connection, make the bridged T a useful frequency-selective network. Specific applications of this circuit are considered in a later chapter.

Somewhat more complicated than the bridged T is the *twin-T* filter shown in Fig. 3-17, which may be analyzed by the technique used in the

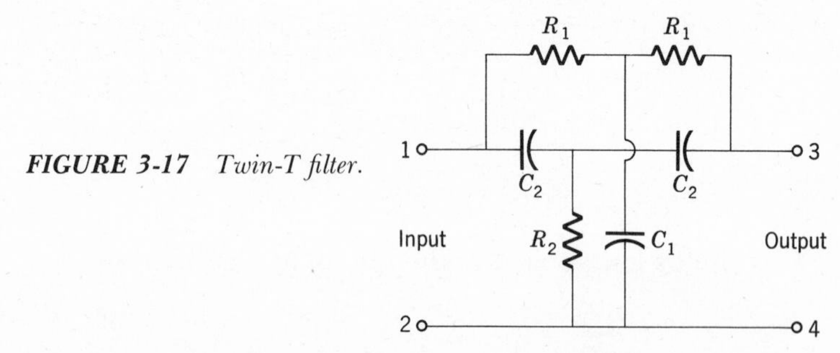

FIGURE 3-17 *Twin-T filter.*

previous section. The twin T is equivalent in response characteristic to the Wien bridge, but the minimum is much sharper. It also has the advantage of a common input-output terminal. The choice between the Wien bridge, bridged-T, and twin-T network for any given application depends on factors such as frequency range, desired performance, and complexity; all three circuits are commonly used.

TRANSFORMERS

Mutual Inductance Suppose a changing magnetic flux resulting from current in one circuit is intercepted by another circuit. According to Ampère's law, an emf is induced in the second circuit. The *mutual inductance* between circuit 1 and circuit 2 is defined by analogy with Eq. (2-52) as

$$N_2 = M_{1,2} I_1 \tag{3-91}$$

where N_2 is the magnetic flux in circuit 2 produced by the current I_1 in circuit 1. A very important application of mutual inductance is the *transformer,* which has two multiturn coils wound on the same iron core. This makes the mutual inductance between the two coils as large as possible. Schematically, a transformer appears as in Fig. 3-18*a*, with a *primary* winding which is part of one circuit and a *secondary* winding which is part of the second circuit. The symbol for a transformer is shown in Fig. 3-18*b*.

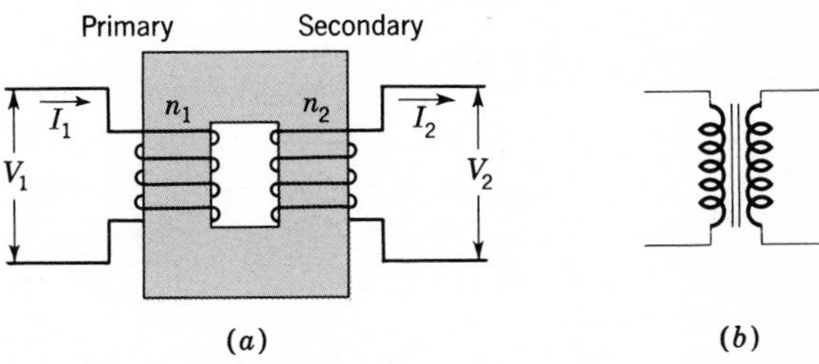

FIGURE 3-18 *(a) Sketch of a transformer and (b) circuit symbol.*

Consider an ideal transformer, in which all of the magnetic flux from the primary winding is intercepted by the secondary winding. Suppose the secondary winding is open-circuited and the primary is connected to a sinusoidal voltage source. Current in the primary winding is determined by the inductance of the primary. The voltage induced in the primary winding V_1 is proportional to the primary inductance according to Faraday's law, and the inductance is proportional to the number of turns on the primary winding. Since all the flux is also intercepted by the secondary winding, the voltage V_2 induced in the secondary is proportional to the number of turns on the secondary winding. That is,

$$\frac{V_1}{V_2} = \frac{n_1}{n_2} \tag{3-92}$$

Note that it is possible to achieve either a *step-up* transformer or a *step-down* transformer, depending upon whether $n_1 < n_2$ or $n_1 > n_2$, so that the secondary voltage is respectively greater or less than the primary voltage.

Suppose now the secondary is connected to a load such as a resistor. Current in the secondary circuit results in I^2R losses in the resistor, and this power must come from the primary winding. The way this comes about is as follows. Both the primary current and the secondary current set up magnetic flux in the core. The magnetic flux caused by the secondary current acts to weaken the magnetic flux set up by the primary current in conformity with the minus sign in Faraday's law, Eq. (2-53). The weaker magnetic flux means that the induced primary voltage is correspondingly smaller. Therefore the voltage source connected to the primary winding increases the primary current until the voltages around the primary circuit are zero again as required by Kirchhoff's rules. Thus current in the secondary winding requires current in the primary winding and the peak magnetic flux remains constant at its no-load value. The near cancellation of fluxes also means that the induced voltages are essentially independent of frequency. Since the rate of change of current increases with frequency, Eq. (2-53) suggests that the secondary voltage increases with frequency. The induced voltage in the primary winding rises correspondingly, however, so that the magnitude of the primary current is reduced, and the secondary voltage remains independent of frequency.

Note that the total magnetic flux in the core does not change with current, because the magnetic flux from the primary and secondary currents are equal and opposite. This means that, from Eq. (2-52),

$$n_1 I_1 = n_2 I_2 \tag{3-93}$$

where the fact that the inductance is proportional to the number of turns on a winding has again been used. The primary current in Eq. (3-93) really refers only to the additional current accompanying a load on the secondary. The current under no-load conditions is usually so small, however, that it may be neglected.

Transformer Ratio The ratio of the number of turns on the secondary to that on the primary is called the *transformer ratio*

$$a = \frac{n_2}{n_1} \tag{3-94}$$

The ratio is particularly significant when the impedance connected to the secondary is compared with the apparent impedance in the primary

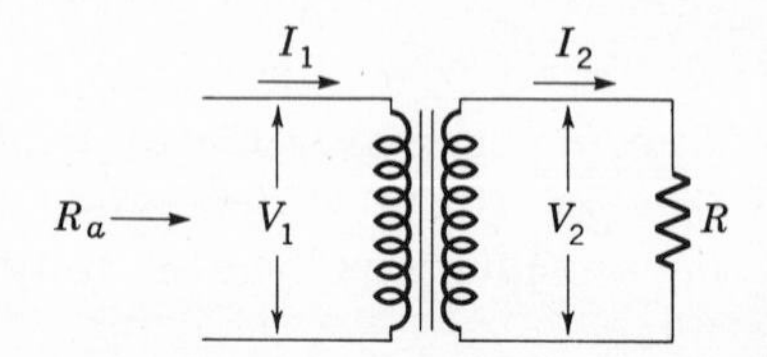

FIGURE 3-19 Apparent resistance at primary terminals of transformer differs from resistance connected to secondary.

circuit. Consider the situation depicted in Fig. 3-19, where the secondary load is a simple resistance R.

The magnitude of R determines the secondary current

$$I_2 = \frac{V_2}{R} \tag{3-95}$$

Substituting for I_2 from Eq. (3-93) and for V_2 from Eq. (3-92),

$$\frac{n_1}{n_2} I_1 = \frac{n_2 V_1}{n_1 R}$$

or

$$I_1 = \frac{a^2 V_1}{R} \tag{3-96}$$

Viewed from the primary side, the current I_1 is in an equivalent resistance R_{eq} such that

$$I_1 = \frac{V_1}{R_{eq}} \tag{3-97}$$

Comparing (3-96) and (3-97), the secondary load resistance R appears to be a resistance on the primary side given by

$$R_{eq} = \frac{R}{a^2} = \left(\frac{n_1}{n_2}\right)^2 R \tag{3-98}$$

This means that a transformer may be used to match the resistances in a circuit to obtain maximum power transfer between a given power source and a fixed-load resistance. This very useful property is applied, for example, in coupling the large internal resistance of a vacuum-tube amplifier to the low resistance of a loudspeaker load. Complex impedances are transformed by the turns ratio in exactly the same way as the resistance in Eq. (3-98).

According to Eq. (3-98) the transformer appears from the primary side as a pure resistance. In particular, the self-inductance of the primary winding is not evident. The reason for this is that the fluxes caused by current in the primary and secondary windings cancel each other. This means that a transformer acts as a device which changes the effective resistance from one circuit to another but which has no inductance of itself. Note also that the primary and secondary circuits in a transformer are not connected electrically and that the two circuits are isolated for direct currents.

Practical Transformers Most transformers are constructed in a fashion similar to iron-core chokes described in an earlier section, except, of course,

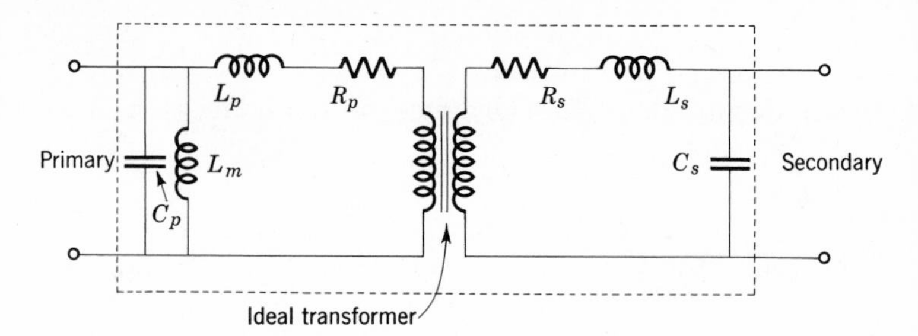

FIGURE 3-20 *Equivalent circuit of practical transformer.*

that more than one winding is present on the core. For large transformer ratios it is common practice to make the low-voltage high-current winding of heavy-gauge wire in order to reduce I^2R losses in this winding. The other winding is a large number of turns of fine wire since only a small current is present. Laminated cores are ordinarily used to reduce eddy-current losses, but transformers for frequencies in excess of 100,000 Hz generally employ high-resistivity ferrite cores.

Transformers with one primary winding and several separate secondary windings are used to supply different voltages to vacuum-tube and transistor devices. Such a *power* transformer may, for example, have a primary winding suitable for connection to a 115-V 60-Hz source with secondaries providing 700, 6.3, and 5.0 V. Applications of such power transformers are studied in a later chapter.

Although for many purposes the inherent inductive effects of the transformer windings may be neglected, in careful work it is necessary to account for the properties of the transformer more exactly. An equivalent circuit of a practical transformer is illustrated in Fig. 3-20. The inductances in the primary and secondary circuits are caused by leakage magnetic flux which does not link both windings, so that the opposing fluxes do not quite cancel. The resistances are included to account for the resistance of the wire in the windings. The inductance L_m accounts for the small magnetizing current corresponding to the no-load primary current. The capacitors on the primary and secondary sides result from the layer-to-layer capacitances between windings.

According to this equivalent circuit, a transformer is ineffective at low frequencies where the reactance of L_m becomes so small that current is shunted from the primary winding of the ideal transformer. High-frequency performance is impaired by the leakage flux inductances and winding capacitances. In spite of these limitations, transformers can be designed for effective performance over useful frequency intervals. It is common practice to indicate the approximate impedance level at which the primary and secondary windings are designed to be used, even though Eq. (3-98) suggests that only the transformer ratio is significant. The specification indicates the conditions under which the transformer inductances and resistances in Fig. 3-20 are negligible in comparison with primary and secondary load impedances, and, correspondingly, the transformer operates very nearly like an ideal transformer.

SUGGESTIONS FOR FURTHER READING

A. M. P. Brookes: "Basic Electric Circuits," The Macmillan Company, New York, 1963.

S. Fich and J. L. Potter: "Theory of AC Circuits," Prentice-Hall, Inc., Englewood Cliffs, N.J., 1959.

"The Radio Amateur's Handbook" (published annually by the American Radio Relay League, West Hartford, Conn.).

EXERCISES

3-1 Determine the current in a series *RC* circuit using the ac form of Ohm's law and show that the result is identical with the differential-equation solution in Chap. 2.

3-2 Calculate the equivalent impedance of the network shown in Fig. 3-21. Is the reactive term capacitive or inductive?

Ans.: Capacitive

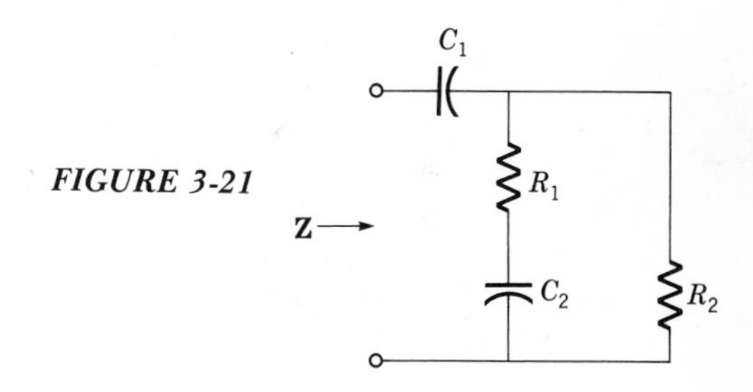

FIGURE 3-21

3-3 Determine the rms current in the 1000-Ω resistor of the circuit in Fig. 3-22. Is the current inductive or capacitive?

Ans.: 6.5 mA; capacitive

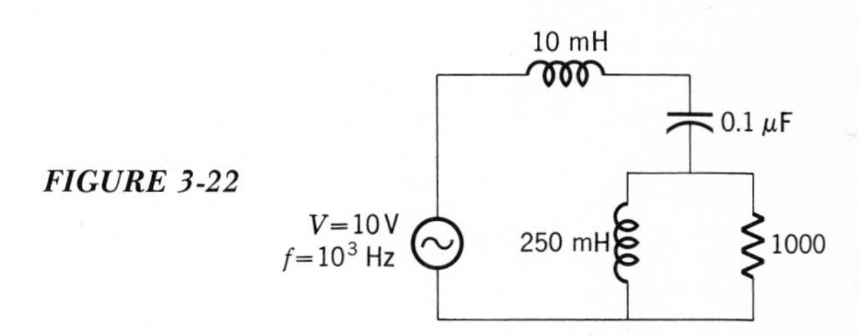

FIGURE 3-22

3-4 Calculate the equivalent impedance of the circuit in Fig. 3-23 at a frequency of 100 Hz. Repeat for 1000 Hz.

Ans.: $0.198 + j2.63 \times 10^6\ \Omega$; $384 + j2.47 \times 10^9\ \Omega$

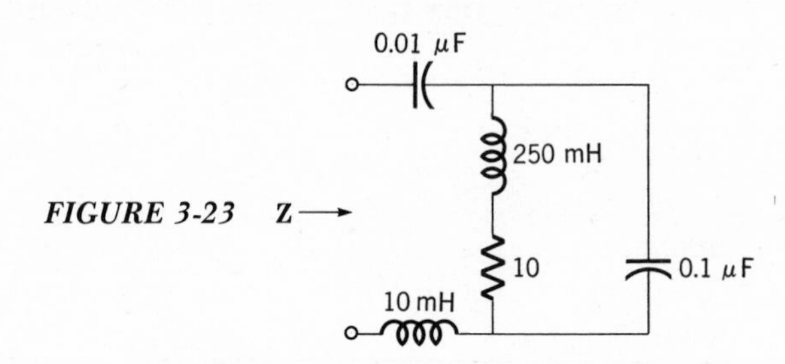

FIGURE 3-23

3-5 Determine the current and the rms voltages across each component in the circuit of Fig. 3-4 at a frequency of 900 Hz if the resistor is 100 Ω. Show that, considering the phase angles of the voltages, Kirchhoff's rule concerning the sum of the voltages around a loop is valid.

Ans.: $2.86 \times 10^{-2}e^{j1.54}$ A; $0.286e^{j1.54}$ V; $50.6e^{-j0.03}$ V; $40.6e^{j3.11}$ V

3-6 Repeat Exercise 3-5 at a frequency of 1100 Hz.

Ans.: $3.57 \times 10^{-2}e^{-j1.54}$ A; $0.357e^{-j1.54}$ V; $51.7e^{-j3.11}$ V; $61.8e^{j0.03}$ V

3-7 Is the equivalent impedance of a parallel resonant circuit inductive or reactive below resonance? Above resonance? *Hint:* Use Eq. (3-47). Compare with a series resonant circuit.

Ans.: Inductive; capacitive

3-8 The impedance at resonance of a parallel resonant circuit is limited by the resistance of the windings of the inductor. Derive an expression for the impedance of the circuit of Fig. 3-24 and calculate the value at resonance appropriate to the components given on the circuit diagram.

Ans.: $4 \times 10^5\ \Omega$

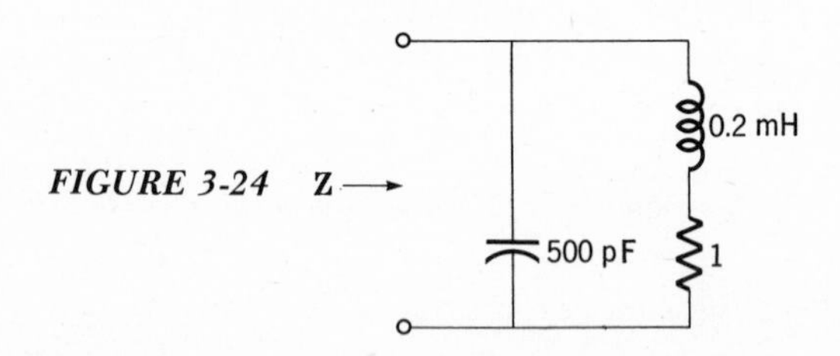

FIGURE 3-24

3-9 Compute the currents in the components of the circuit of Fig. 3-7 at resonance. Repeat for a frequency $\omega = \omega_0/2$. In both cases show that, considering the phase angles of the currents, Kirchhoff's current rule is valid.

Ans.: 0 A; $3.16 \times 10^2e^{-j1.57}$ A; $3.16 \times 10^{-2}e^{j1.57}$ A; $2 \times 10^{-3}e^{-j0.042}$ A; $2.67 \times 10^{-3}e^{-j0.042}$ A; $6.67 \times 10^{-4}e^{j3.10}$ A

3-10 Determine the balance conditions for the capacitance bridge, Fig. 3-12.

Ans.: $R_2R_3 = R_1R_u$, $R_1/R_2 = C_u/C_3$

LABORATORY EXERCISES

3-A Series and Parallel Resonance Series and parallel resonant circuits are widely used to enhance selected frequency components and suppress others. It is important to understand the apparently anomalous relationships between currents and voltages in *RLC* circuits. For example, in the case of series resonance, the rms voltage across both the inductor and the capacitor may exceed the source voltage. Similarly, the current in both reactances may be very large in a parallel resonant circuit, while the source current is small. The key to understanding these useful properties is the phase relationships between currents and voltages in the circuit.

In this experiment the phase relationships in simple *RLC* series and parallel resonant circuits are examined. To start, determine the current in the simple series circuit of Fig. 3-4 as a function of frequency to confirm the resonance curve in Fig. 3-5. Next, plot the rms voltage and the phase angle of the voltage across each of the components as a function of frequency, taking the phase angle of the source voltage as the reference.

Note the frequency for which the current is in phase with the voltage source. Over what frequency interval does the inductor and capacitor voltage exceed the source voltage? By means of simple vector diagrams corresponding to Fig. 2-2 show that the sum of the voltages across the three series components always equals the source voltage. Do this at resonance, at both half-power frequencies, and at a frequency far removed from resonance.

What is the bandwidth (frequency interval between half-power points) of this circuit? Now plot the resonance curve of the circuit if the internal resistance of the voltage source is 900 Ω. What is the bandwidth of this circuit? Compute the Q of the circuit in both cases and see if the bandwidth is equal to the resonant frequency divided by Q.

Now connect the capacitor and inductor in parallel and increase R to 10^5 Ω. Plot the resonance curve. Also plot the rms current and the phase angle of the currents in each component as a function of frequency. This time use vector diagrams to confirm that the currents always sum properly when the phase angle is taken into account. Compute the Q of the circuit.

3-B Bridged-T and Twin-T Filters Resistance-capacitance circuits having frequency-selective properties are useful complements to *RLC* resonant circuits. They are often employed in circumstances where inductors are not convenient (for example, at low frequencies) or are too bulky. Although the frequency response characteristics of several *RC* circuits resemble those of resonant circuits, they are not really the same, as illustrated in this experiment.

Plot the frequency and phase characteristic of a bridged-T filter using Eq. (3-90). Select component values such that the characteristic frequency is 1 kHz while $R_2 = 1000\ R_1$, $C_1 = C_2$, and $R_1 = 10^4$ Ω. It is probably easiest to determine the frequency and phase characteristics by using a Fortran computer program to evaluate Eq. (3-90) at each frequency. Repeat for $R_2 = 100R_1$ and $R_2 = 50R_1$, each time changing the capacitors so that the characteristic frequency stays at 1 kHz.

Contrast the frequency and phase characteristics seen here with those of *RLC* circuits such as the ones studied in Laboratory Exercise 3-A. Try to relate the filter bandwidth qualitatively to circuit values or ratios, particularly R_2/R_1.

Repeat the above investigation for the so-called twin-T filter, Fig. 3-17, for which the response characteristic is

$$\frac{v_o}{v_i} = \left[1 + j\omega_0 R_1 \frac{C_1 + 2C_2}{\omega/\omega_0 - \omega_0/\omega}\right]^{-1}$$

where the characteristic frequency is achieved when $R_1C_1 = 4R_2C_2$ so that

$$\omega_0{}^2 = \frac{2}{R_1{}^2 C_1 C_2} \tag{3-99}$$

Choose appropriate component values to produce a characteristic frequency of 1 kHz. Contrast the performance of this circuit with that of the bridged-T filter. Which has the better frequency selectivity? Which has the greater rejection at the characteristic frequency?

chapter four

RECTIFIER CIRCUITS

Resistors, capacitors, and inductors are called linear components, because the current increases in direct proportion to the applied voltage, in accordance with Ohm's law. Components for which this proportionality does not hold are termed nonlinear devices; they are widely used in practical electronic circuits. This chapter examines the properties of an important nonlinear device, the diode rectifier. The term "diode" comes from the fact that rectifiers have two active terminals, or electrodes.

A rectifier is nonlinear in that it passes a greater current for one polarity of applied voltage than for the other. If a rectifier is included in an ac circuit, the current is negligible whenever the polarity of the voltage across the rectifier is in the reverse direction. Therefore, only a unidirectional current exists and the alternating current is said to have been rectified.

A major application of rectifiers is in power-supply circuits which convert conventional 115-V 60-Hz ac line voltages to dc potentials suitable for use with vacuum tubes and transistors. Two types of rectifiers, the vacuum diode and the semiconductor junction diode, are commonly used in power supplies. Because of its superior properties, the junction diode is rapidly supplanting the vacuum diode.

NONLINEAR COMPONENTS

Current-Voltage Characteristics The most useful description of the electrical properties of a nonlinear component is the relationship between the current in the unit and the applied voltage. Indeed, this is analogous to the importance of Ohm's law for linear components since Ohm's law forms the basis for all linear-circuit operation and analysis. The relationship between current and voltage is conventionally displayed graphically in the *current-voltage* (or *IV*) *characteristic* of the device. The performance of a nonlinear circuit element in any circuit may be analyzed with the aid of the appropriate current-voltage characteristic.

There exists a rich variety of current-voltage characteristics associated with various practical electrical components, rectifiers included, which are discussed in this and subsequent chapters. In many cases it is possible to understand the origin of these nonlinear relationships in terms of basic physical principles. For example, the motions of electrons in the device may lead directly to the variation of current with applied voltage. This is the case with Ohm's law discussed in Chap. 1 where the drift of electrons in a conductor results in a linear *IV* characteristic. While an understanding of the physical principles involved is often important in the development of new device designs having favorable current-voltage properties, such an understanding is not necessary for the analysis of device performance and circuit operation.

Actually, the *IV* characteristics of practical nonlinear devices are universally determined by empirical measurement, rather than by theoretical analysis. This is so because of the complexities inherent in practical units. Characteristic current-voltage curves of commercial devices are presented in manufacturer's compilations and are widely available. All the principles of circuit analysis, in particular Kirchhoff's rules, are applicable to circuits with nonlinear components. Most often it is convenient to employ graphical techniques in the analysis because the *IV* characteristic is not expressible in a manageable mathematical form.

An elementary but very important nonlinear current-voltage character-

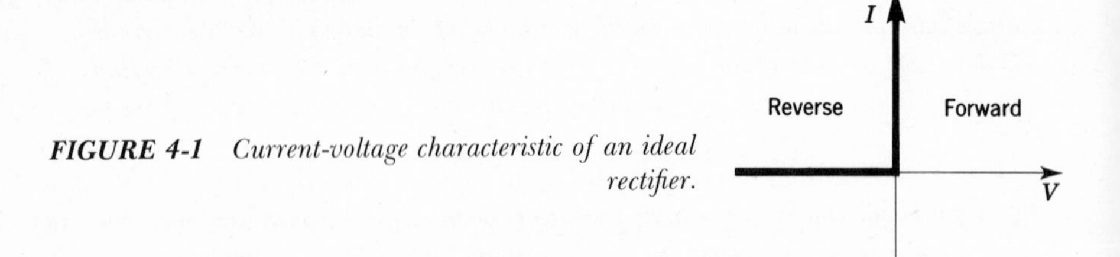

FIGURE 4-1 *Current-voltage characteristic of an ideal rectifier.*

istic is that of the *ideal rectifier,* Fig. 4-1. This shows that an ideal rectifier has zero resistance for one polarity of applied voltage and infinite resistance for the reversed polarity. It is, in effect, a voltage-sensitive switch which is closed when the voltage polarity is in the *forward* direction and open for *reverse* polarity. This ideal characteristic is approximated by several practical types of diodes and is the basis for many important electronic circuits.

The Vacuum Diode The vacuum diode comprises a hot *cathode* surrounded by a metal *anode* inside an evacuated enclosure, usually glass, as sketched in Fig. 4-2. At sufficiently high temperatures electrons are emitted from

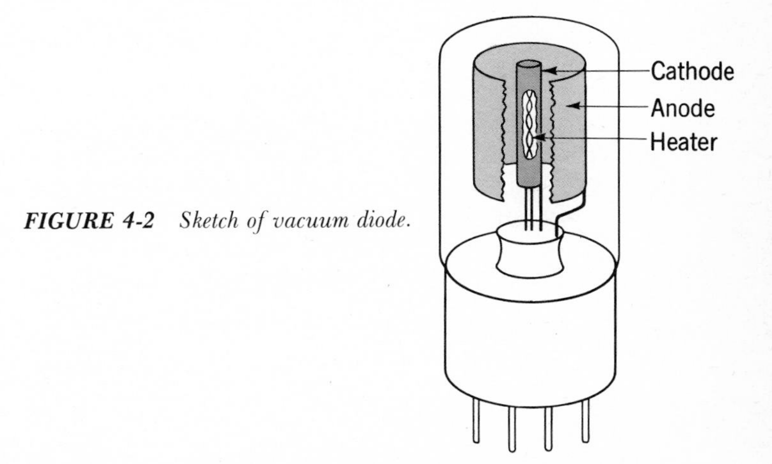

FIGURE 4-2 *Sketch of vacuum diode.*

the cathode and are attracted to the positive anode. Electrons moving from the cathode to the anode constitute a current; they do so when the anode is positive with respect to the cathode. When the anode is negative with respect to the cathode, electrons are repelled by the anode and the reverse current is zero. The space between the anode and cathode is evacuated, so that electrons may move between the electrodes unimpeded by collisions with gas molecules.

Cathodes in vacuum diodes take several different forms. The free electrons in any conductor, given sufficient energy by heating, will escape from the solid. Some materials are much more satisfactory in this respect than others, however, either because it is relatively easy for electrons to escape, or because the material can safely withstand high temperatures. Tungsten, for example, is a useful cathode material, because it retains its mechanical strength at extreme temperatures. A thin layer of thorium on the surface of a tungsten cathode filament increases electron emission, and appreciable current is attained at temperatures of about 1900 K. This form of cathode is directly heated by an electric current the way the filament in an incan-

FIGURE 4-3 *Circuit symbols for vacuum diodes of (a) filamentary-cathode and (b) heated-cathode types.*

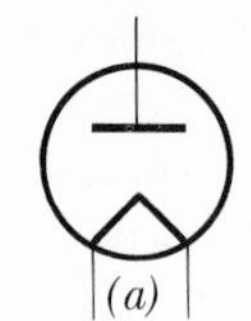

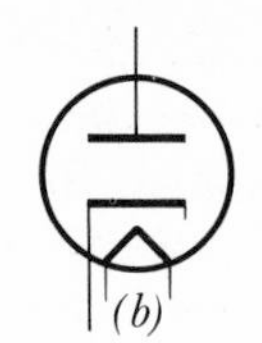

descent lamp is. It is used in vacuum diodes suitable for high-voltage applications. The conventional symbol for a vacuum diode of this type is shown in Fig. 4-3*a*.

The modern oxide-coated cathode, which consists of a metal sleeve coated with a mixture of barium and strontium oxides, is the most efficient electron emitter developed to date. Copious electron emission is obtained at temperatures near 1000 K, which means that the power required is much less than for tungsten. Usually the oxide cathode is indirectly heated by a separate heater located inside the metal sleeve, as in Fig. 4-2. This isolates the heater current electrically from the cathode connection, which is a considerable advantage in electronic circuits. Also, an ac heater current may be used without introducing undesirable temperature variations in the cathode at the frequency of the heater current. Oxide cathodes are employed in the great majority of vacuum tubes. The heater is often omitted in the conventional circuit symbol, Fig. 4-3*b*, since it is not directly an active part of the rectifier.

When the anode is negative with respect to the cathode, the electron current is zero and the reverse characteristic is essentially that of an ideal diode. The forward characteristic is determined by the motion of electrons in the space between the cathode and the anode when the anode has a positive potential. The current depends upon the separation between anode and cathode, upon the anode voltage, and upon the area of the cathode. Thus it is possible to obtain different current-voltage character-

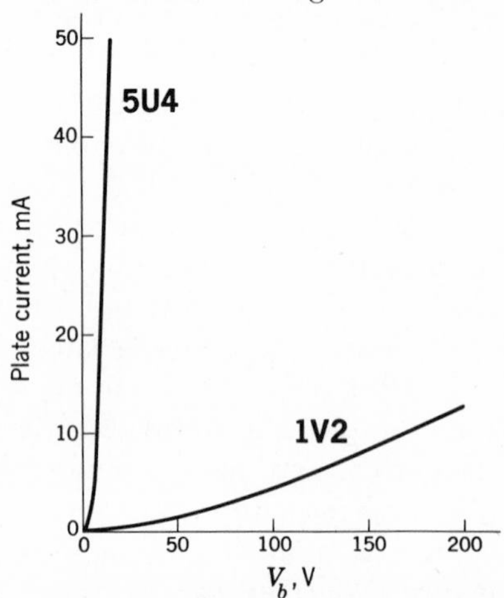

FIGURE 4-4 *Experimental forward characteristics of two practical vacuum diodes are much different because of different geometric construction.*

istics by altering the geometric shape of the anode and cathode. Two examples of practical diode characteristics are given in Fig. 4-4. The type 5U4 diode is a medium-power rectifier; the type 1V2 diode is designed for high-voltage low-current power supplies. The anode-cathode separation is considerable in the 1V2 in order to minimize the possibility of a discharge passing between anode and cathode on the reverse portion of the voltage cycle. Consequently, the current in the forward direction is smaller than in the case for the 5U4. The 5U4 is designed for use at lower voltages, so the separation between anode and cathode is smaller. The forward current is correspondingly greater at the same forward voltage.

In any practical vacuum diode the reverse current is not truly zero, because of leakage currents on the surfaces of the glass insulators and similar secondary effects. Typically, the reverse resistance is of the order of 10 MΩ. Since the forward resistance may be of the order of 100 Ω at a suitable operating potential, the ratio of the reverse resistance to the forward resistance, or the *rectification ratio,* is appreciable. The interelectrode capacitance between the cathode and anode limits the maximum frequency at which a vacuum diode is useful. The cathode-anode capacitive reactance is effectively in parallel with the electron current and tends to short out the reverse resistance at high frequencies (see Exercise 4-2).

The Junction Diode As will be shown in Chap. 5, the addition of certain foreign atoms to otherwise pure semiconductor materials such as germanium or silicon produces free electrons which carry electric current. A semiconductor crystal containing such foreign atoms is called an *n-type* crystal because of the negative charge of the current carrying electrons. Similarly, by incorporating certain other kinds of foreign atoms, the semiconductor appears to conduct current by positive current carriers. Such a crystal is called a *p-type* conductor because the current carriers have a positive charge. It is possible to change from *n*-type to *p*-type conductivity

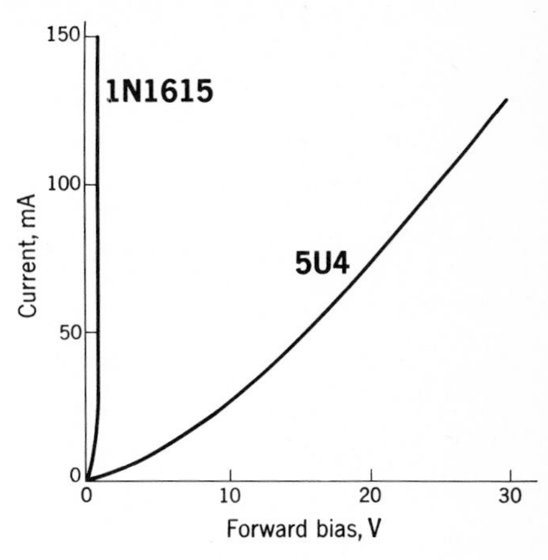

FIGURE 4-5 *Forward characteristics of silicon pn-junction rectifier, type 1N1615, and vacuum diode, type 5U4.*

in the same crystal by introducing an abrupt change from one impurity type to the other across a given region of the semiconductor. The junction between an *n*-type region and a *p*-type region in a semiconductor crystal, called a *pn junction*, exhibits distinctive nonlinear electrical characteristics.

In fact, the *junction diode* is very nearly an ideal rectifier for the voltages commonly encountered in practical applications. This can be illustrated by comparing the current-voltage characteristics of a typical silicon junction diode with the type 5U4 vacuum diode, Fig. 4-5. Note that the voltage drop in the forward direction is much less for the junction diode. The size of this junction is only about 2 mm^2, so the junction diode is physically very much smaller than the vacuum diode. A third major advantage of the semiconductor device is that a hot cathode is not required. This means that heater power is not wasted and also that no warmup time is necessary after applying power to the circuit.

The reverse current of a *pn* junction is somewhat greater than that of the vacuum diode, although it is still small enough to be negligible. Of greater concern is the fact that the reverse current increases exponentially with temperature. Even though the reverse current is of the order of only 10 μA at room temperature for a typical silicon diode, the rapid increase means that only a modest temperature rise can be tolerated. Silicon junction diodes, for example, are inoperative above about 200°C. Junction diodes used in circuits at power levels above a few watts or so are cooled in order to dissipate joule heat caused by the current in the device. It is common practice to attach junction diodes firmly to metal heat sinks having radiator fins to conduct the heat away from the diode.

In spite of their temperature sensitivity, a characteristic common to most semiconductor devices, junction diodes are extremely good rectifiers and enjoy wide application. Commercial junction diodes are made of either silicon or germanium by processes similar to those described in Chap. 5. Practical devices are completely encapsulated to protect the semiconductor surface from contamination. This is accomplished by placing the diode in a small metal can filled with an inert atmosphere or by encasing the diode in plastic. Because of their superior electrical characteristics, silicon and germanium junction diodes have largely supplanted older forms of semiconductor rectifiers made of selenium or copper oxide. It is interesting to note that the latter rectifiers were the first commercially useful semiconductor devices. The circuit symbol for all semiconductor diodes, given in Fig. 4-6, has an arrow to indicate the direction of conventional forward current in the device.

FIGURE 4-6 *Circuit symbol for semiconductor diode.*

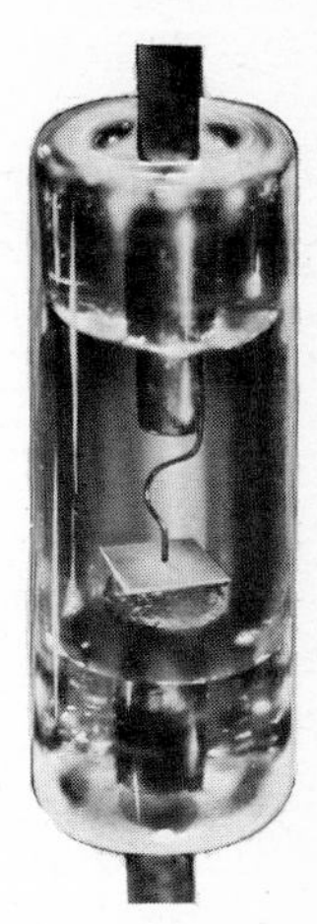

FIGURE 4-7 *Point-contact gold-bonded diode.* (Ohmite Manufacturing Company.)

The excellent forward conductivity of a *pn* junction means that practical diodes can be quite small. Therefore the stray capacitive reactances are correspondingly small and junction diodes are useful at high frequencies. This feature is enhanced in *point-contact* diodes, in which a metal probe is placed in contact with a semi-conductor crystal. During processing a minute *pn* junction is formed immediately under the point. Such tiny devices can operate at frequencies corresponding to millimeter wavelengths and are used, for example, in radar and high-speed computer circuits. A typical point-contact diode is shown in Fig. 4-7.

RECTIFIER CIRCUITS

Half-wave Rectifier An elementary rectifier circuit, Fig. 4-8, has a vacuum diode in series with an ac source and a resistive load. When the polarity

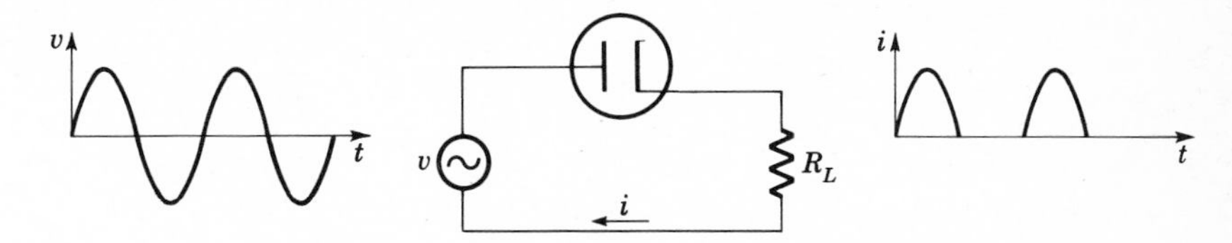

FIGURE 4-8 *Elementary half-wave rectifier circuit.*

of the source makes the anode positive with respect to the cathode, the diode conducts and produces a current in the load. On the reverse half-cycle the diode does not conduct and the current is zero. The output current is therefore a succession of half sine waves, as indicated in Fig. 4-8, and the circuit is called a *half-wave* rectifier. The average value of the half sine waves is clearly not zero, so that the output current has a dc component. That is, the input sine wave has been rectified.

The current in the circuit is determined from the voltage equation

$$v = iR_L + v_b \tag{4-1}$$

where v_b is the voltage across the diode. This equation together with the relationship between v_b and i given by the diode current-voltage characteristic are solved simultaneously. The solution is carried out graphically in the following manner. Equation (4-1) is solved for i,

$$i = \frac{v}{R_L} - \frac{1}{R_L}v_b \tag{4-2}$$

This is the equation of a straight line with slope $-1/R_L$ and intercept v/R_L, as shown in Fig. 4-9 for three different instantaneous values of v. Starting

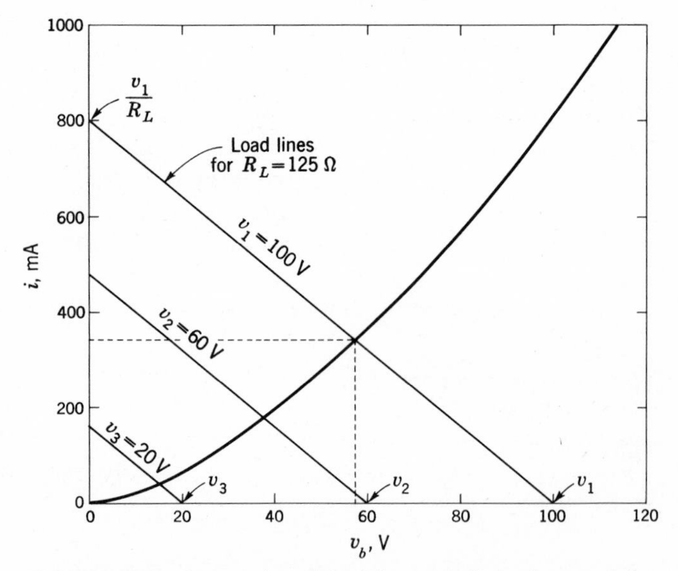

FIGURE 4-9 *Intersection of load line with diode characteristic gives current in circuit.*

with a given instantaneous value of the input voltage v_1, the intersection of this so called *load line* with the diode characteristic gives the current at the time the instantaneous input voltage is v_1. As the input voltage swings through the positive half of the cycle the current at every instant can be determined from a similar load line having a voltage intercept corresponding to the instantaneous voltage of the source.

A plot of the current as a function of applied voltage, Fig. 4-10, is called the *dynamic characteristic* of the circuit. The *static characteristic,* which is simply the current-voltage curve of the diode, and the dynamic characteristic differ because of the voltage drop across the load resistance. The voltage drop reduces the anode-cathode potential for a given input voltage.

The dynamic characteristic can be used to plot the current waveform

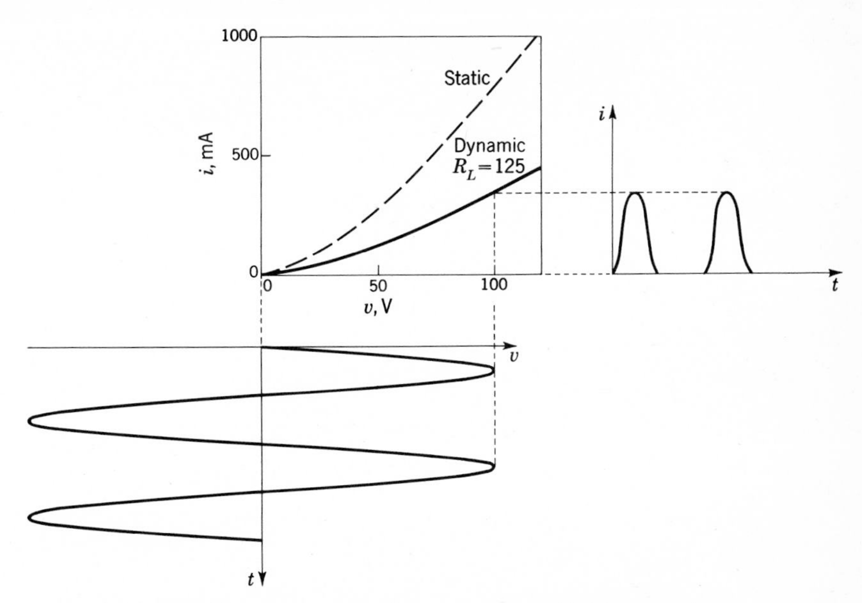

FIGURE 4-10 *Current waveform is nonsinusoidal because of curvature of rectifier dynamic characteristic.*

resulting from any input voltage waveform, as illustrated in Fig. 4-10. In particular, if the input voltage is sinusoidal, the output current is not truly a half sine wave because of curvature of the diode characteristic. For many purposes, however, this curvature may be ignored and the forward characteristic replaced by a straight line approximating the true curve. In this case the diode is represented by a fixed resistance R_p in the forward direction. The approximate current in the circuit may then be found immediately from Eq. (4-1) after replacing v_b by iR_p.

Full-wave Rectifier The half-wave rectifier is inactive during one-half of the input cycle and is therefore less efficient than is possible. By arranging two diodes as in Fig. 4-11 so that each diode conducts on alternate half-cycles, *full-wave* rectification results. This is accomplished by using a center-

FIGURE 4-11 *Full-wave rectifier.*

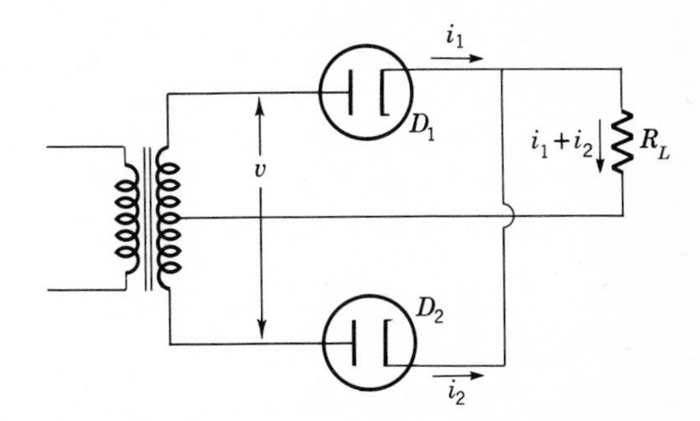

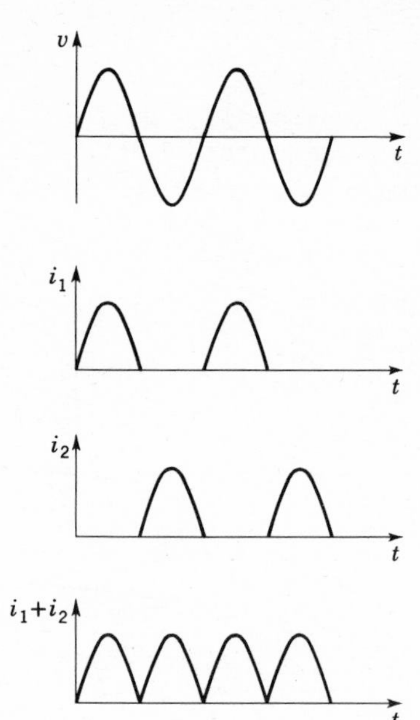

FIGURE 4-12 *Waveforms in full-wave rectifier.*

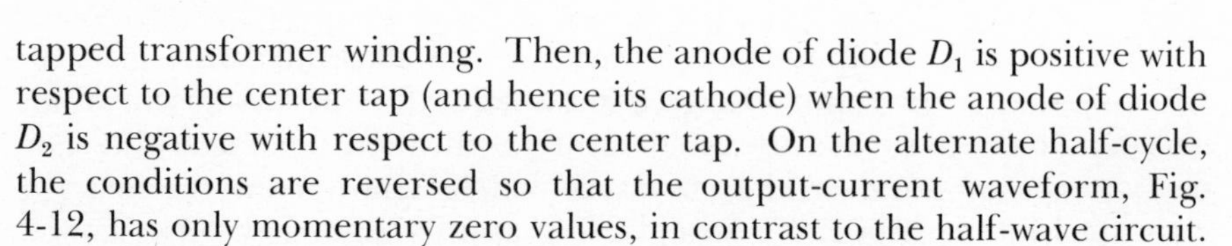

tapped transformer winding. Then, the anode of diode D_1 is positive with respect to the center tap (and hence its cathode) when the anode of diode D_2 is negative with respect to the center tap. On the alternate half-cycle, the conditions are reversed so that the output-current waveform, Fig. 4-12, has only momentary zero values, in contrast to the half-wave circuit.

The full-wave rectifier, a widely used circuit, can be analyzed exactly as in the case of the half-wave rectifier. Note that each diode in the diagram of Fig. 4-11 must withstand the full end-to-end voltage of the transformer winding. Therefore the *peak inverse voltage* rating of the diodes must be at least twice the peak output voltage. This is usually not a serious drawback except for circuits designed to operate at highest voltages. Comparing the output waveforms of half-wave and full-wave rectifiers, Figs. 4-8 and 4-12, reveals that the fundamental frequency equals the supply voltage in the half-wave rectifier but is twice the supply frequency in the case of the full-wave circuit. This is an important consideration in power-supply circuits, as will be demonstrated in a later section. The center-tapped transformer in the full-wave circuit supplies current on both half-cycles of the input voltage, which permits a more efficient transformer design than is possible in the case of the half-wave circuit. Note, however, that the output voltage is only one-half the total voltage of the transformer secondary.

Bridge Rectifier Full-wave rectification without a center-tapped transformer is possible with the *bridge rectifier*, Fig. 4-13. The operation of this circuit may be described by tracing the current on alternate half-cycles of the input voltage. Suppose, for example, the upper terminal of the transformer is positive. This means that diode D_2 conducts, as does diode D_3, and current is present in the load resistor. On the alternate half-cycle diodes D_4 and D_1 conduct and the current direction in the load resistance is the same as before. The voltage across R_L corresponds to full-wave rectification and the peak voltage is equal to the transformer voltage less the potential drops across the diodes.

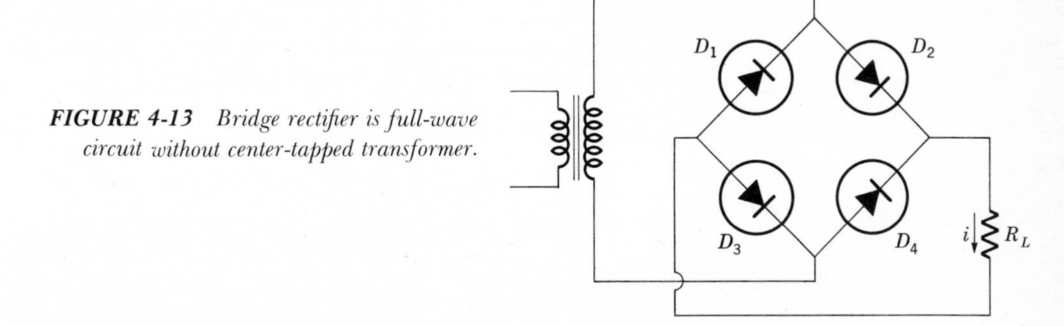

FIGURE 4-13 *Bridge rectifier is full-wave circuit without center-tapped transformer.*

Note that if vacuum diodes are used in a bridge rectifier circuit, the heater currents must be supplied from separate sources, since the diode cathodes are not at the same potential. For this reason it is common to employ junction diodes in this circuit. Since two diodes are in series with the load, the output voltage is reduced by twice the diode drop. On the other hand, the peak inverse voltage rating of the diodes need only be equal to the transformer voltage, in contrast to the previous full-wave circuit.

Voltage Doubler Consider the circuit of Fig. 4-14, in which two diodes are connected to the same voltage source but in the opposite sense. On the half-cycle during which the upper terminal of the source is positive, diode D_1 conducts and capacitor C_1 charges to the peak value of the input voltage. On the reverse half-cycle D_2 conducts and C_2 also charges to the

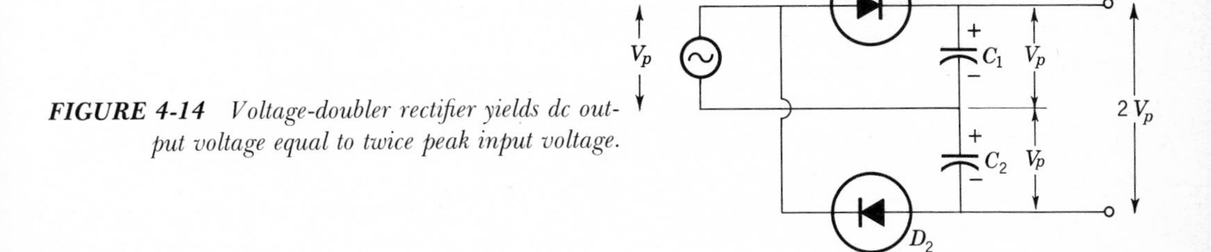

FIGURE 4-14 *Voltage-doubler rectifier yields dc output voltage equal to twice peak input voltage.*

full input voltage. Meanwhile, the charge on C_1 is retained, since the potential across D_1 is in the reverse direction. Thus both C_1 and C_2 charge to the peak supply voltage and the dc output voltage is equal to twice the peak input voltage. Accordingly, this circuit is called a *voltage doubler*.

The above analysis applies to the case when no current is delivered to the load. When a load resistance is connected, current is supplied by the discharge of the capacitors. On alternate half-cycles the capacitors are subsequently recharged. This implies, however, that the output voltage under load is no longer dc, but has an ac component. It is necessary to make the capacitance of C_1 and C_2 large enough to minimize this variation in output voltage, taking into account the current drain and the supply frequency. The filtering action of the capacitors in this circuit is treated in greater detail in the next section.

FILTERS

It is usually desirable to reduce the alternating component of the rectified waveform so that the output is primarily a dc voltage. This is accomplished by means of *filters* which are composed of suitably connected capacitors and inductances. A power-supply filter is a low-pass filter which reduces the amplitudes of all alternating components in the rectified waveform and passes the dc component. A measure of the effectiveness of a filter is given by the *ripple factor* r, which is defined as the ratio of the rms value of the ac component to the dc or average value. That is,

$$r = \frac{V_{\text{rms}}}{V_{\text{dc}}} \tag{4-3}$$

According to this definition it is desirable to make the ripple factor as small as possible. Although rectified waveforms contain many harmonics of the input voltage frequency, it is generally satisfactory to determine the ripple factor at the fundamental frequency only. This is so because the fundamental frequency is predominant and also because higher harmonics are attenuated more than the fundamental by the low-pass filter characteristic.

Capacitor Filter The simplest filter circuit consists of a capacitor connected in parallel with the load resistance, as in the typical low-voltage bridge-rectifier power supply shown in Fig. 4-15. If the reactance of the capacitor at the power-line frequency is small compared with the load resistance R_L, the ac component is shorted out and only dc current remains in the load resistor.

Characteristics of the *capacitor filter* are determined by examining waveforms in the circuit, Fig. 4-16. The capacitor is charged to the peak

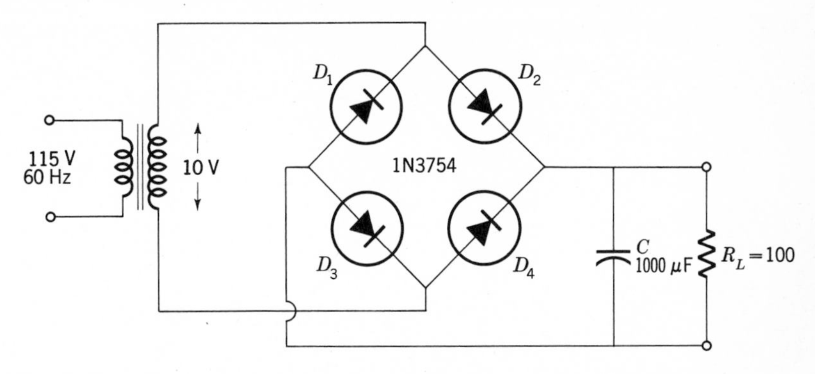

FIGURE 4-15 Practical low-voltage power supply using capacitor filter.

value of the rectified voltage V_p and begins to discharge through R_L after the rectified voltage decreases from the peak value. The decrease in capacitor voltage between charging pulses depends upon the relative values of the RC time constant and the period of the input voltage. A small time constant means the decrease is large and the ripple voltage is also large. On the other hand, a large time constant results in a small ripple component. The diodes conduct only during the portion of the cycle that the capacitor is charging, because only during this interval is the sum of the supply voltage and the capacitor voltage such that the potential across the diodes is in the forward direction.

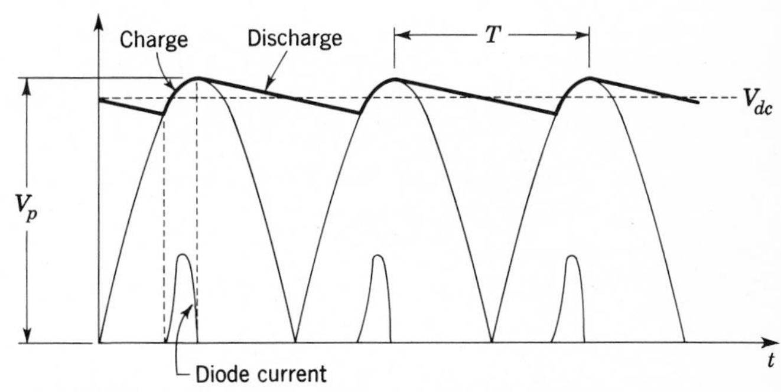

FIGURE 4-16 Output voltage of capacitor filter is dc voltage and small triangular ripple voltage.

The ripple voltage is approximately a triangular wave if the RC time constant is long compared with the period, $R_LC \gg T$. In this case the exponential decrease of capacitor voltage during one period is given approximately by $V_p \times T/R_LC$. The dc output voltage is the peak capacitor voltage minus the average ripple component, or

$$V_{\text{dc}} = V_p - \frac{V_pT}{2R_LC} = V_p\left(1 - \frac{1}{2fR_LC}\right) \tag{4-4}$$

where f is the main frequency of the rectified waveform. The effective voltage of a triangular wave is calculated in Chap. 2. In present notation

$$V_{\mathrm{rms}} = \frac{1}{2\sqrt{3}} \frac{V_p T}{R_L C} \tag{4-5}$$

Therefore, using Eq. (4-3), the ripple factor becomes

$$r = \frac{1}{2\sqrt{3}} \frac{1}{f R_L C} \tag{4-6}$$

where the approximation $R_L C \gg T$ has been used again.

This result shows that ripple is reduced by increasing the value of the filter capacitor. When the load current is equal to zero ($R_L \to \infty$), the ripple factor becomes zero, which means that the output voltage is pure dc. As the load current is increased (smaller values of R_L), the ripple factor increases. Inserting component values given on the circuit diagram of Fig. 4-15, the ripple factor is found to be 0.05. Thus the ripple voltage is about 5 percent of the dc output. Note that the ripple voltage of a full-wave rectifier is approximately one-half that of the half-wave circuit, because the frequency of the rectified component is twice as great.

The dc output voltage may be written, using Eq. (4-4), as

$$V_{\mathrm{dc}} = V_p - \frac{1}{2fC} I_{\mathrm{dc}} \tag{4-7}$$

where the approximation $I_{\mathrm{dc}} \approx V_p/R_L$ has been used. According to Eq. (4-7), the dc output voltage decreases linearly as the dc current drawn by the load increases. The constancy in output voltage with current is called the *regulation* of the power supply; Eq. (4-7) shows that a large value of filter capacitance improves the regulation. It should be noted that the decrease in output voltage given by this equation refers only to the change accompanying the increase in the ac ripple component. The IR drops associated with diode resistances and the resistance of the transformer winding further reduce the output voltage as the load current increases.

The simple capacitor filter provides very good filtering action at low currents and is often used in high-voltage low-current power supplies. Because of its simplicity, the circuit also is found in those higher current supplies where ripple is relatively less important. The dc output voltage is high, equal to the peak value of the supply voltage. The disadvantages of the capacitor filter are poor regulation and increased ripple at large loads.

***L*-section Filter** It is useful to add a series inductor to the capacitance filter, as in the circuit of Fig. 4-17. The series inductance in this *L-section* or *choke-input* filter opposes rapid variations in the current and so con-

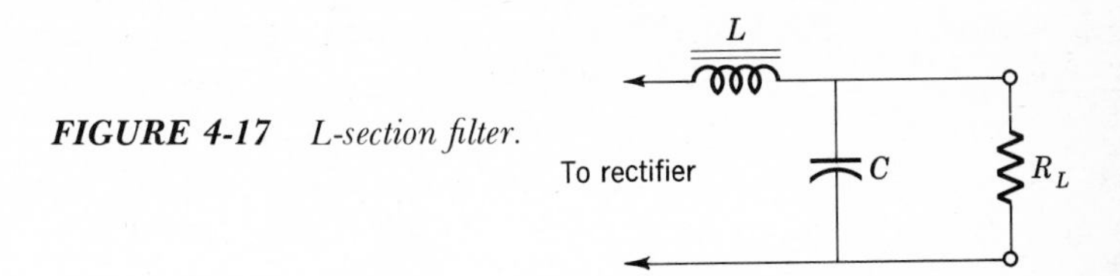

FIGURE 4-17 *L-section filter.*

tributes to the filtering action. The ripple factor may be determined by noting that the ac voltage components of the rectified waveform divide between the inductance and the impedance **Z** of the resistor-capacitor combination. Therefore

$$r = \left(\frac{V_{\text{rms}}}{V_{\text{dc}}}\right)_f = \frac{(V_{\text{rms}})_r[\mathbf{Z}/(\mathbf{Z} + X_L)]}{(V_{\text{dc}})_r} \tag{4-8}$$

where the subscript f refers to the voltage ratio at the output of the filter and the subscript r refers to the rectified waveform. Equation (4-8) may be put in a more instructive form by calculating the magnitude of the impedance and rearranging,

$$r = \left(\frac{V_{\text{rms}}}{V_{\text{dc}}}\right)_r \frac{1}{1 + X_L/\mathbf{Z}} = \left(\frac{V_{\text{rms}}}{V_{\text{dc}}}\right)_r \frac{1}{1 + \omega L\sqrt{(1/R_L)^2 + (\omega C)^2}} \tag{4-9}$$

$$r = \left(\frac{V_{\text{rms}}}{V_{\text{dc}}}\right)_r \frac{1}{1 + \omega^2 LC\sqrt{1 + (1/\omega R_L C)^2}} \tag{4-10}$$

Note that the dependence upon R_L, that is, upon the load current, is much smaller than is the case for the capacitor filter. In fact, if $\omega R_L C \gg 1$, as is done in practice, the ripple factor is independent of the load. Inserting the rms/dc ratio appropriate for a full-wave rectified waveform, Eq. (4-10) becomes

$$r = \frac{\sqrt{2}}{3} \frac{1}{\omega^2 LC} \tag{4-11}$$

According to Eq. (4-11) large values of L and large values of C improve the filtering action.

The inductance chokes off alternating components of the rectified waveform and the dc output voltage is simply the average, or dc, value of the rectified wave. For a series of half-sinusoids this means that the dc output voltage is from Eq. (2-110),

$$V_{\text{dc}} = \frac{2}{\pi} V_p \cong 0.9V \tag{4-12}$$

where V_p is the peak and V is the rms value of the transformer voltage. Therefore, the output voltage of the choke-input filter is considerably less than that of the capacitor filter.

According to Eq. (4-11) the ripple is independent of the load current. This means that there is no decrease in output voltage due to a decrease in filtering action at high currents, such as in Eq. (4-7). Therefore, the output voltage is independent of the load, except for IR drops in the diodes and transformer windings. For this reason the L-section filter is used in applications where wide variations in the load current are expected. The advantage of good regulation must be balanced against the comparatively low output voltage in considering any given application.

When the load current is zero the capacitor charges to the peak value of the input waveform as in the capacitor filter. The output voltage is then larger than that given by Eq. (4-12) and regulation is poor. As discussed in connection with the capacitor filter, in this circumstance the diodes conduct for only a part of the applied voltage cycle. The transition between this performance and that of the true L-section filter occurs at the load current for which the diodes just conduct over the entire cycle, and the voltage across the capacitor remains at the dc component of the rectified waveform, Eq. (4-12).

The transition point is determined by first calculating the ac component of the rectified current waveform in the following way. The input impedance of the filter is

$$\mathbf{Z}_f = j\omega L + \frac{R_L}{1 + j\omega C R_L} \cong j\omega L \tag{4-13}$$

where the approximation is valid since $\omega R_L C \gg 1$, as assumed previously. The peak value of the ac current to the filter (that is, the diode current) is computed from Eq. (4-13) and the leading term in the Fourier analysis of the rectified voltage waveform using Eq. (2-110),

$$I_f = \frac{1}{Z_f}\frac{4V_p}{3\pi} = \frac{2V_p}{3\pi\omega L} = \frac{V_{dc}}{3\omega L} \tag{4-14}$$

where the second harmonic 2ω of the applied voltage frequency has been inserted in conformity with the Fourier analysis, and Eq. (4-12) has also been used. For current to be present in the diodes over the entire input voltage cycle, the peak value of the ac component, I_f, must be smaller than the dc component. Accordingly, the minimum value of load current, I_{dc}, is just

$$I_{dc} = \frac{V_{dc}}{3\omega L} \tag{4-15}$$

It is common practice to include a resistor across the capacitor so that the output current never drops below the value given by Eq. (4-15) and the superior regulation properties of the L-section filter are achieved for all loads. Such a *bleeder* resistor is also useful in draining charge from the filter capacitor after the power supply is turned off, which reduces injury hazard. The minimum value of bleeder resistor required is given by

$$R_L = \frac{V_{dc}}{I_{dc}} = 3\omega L \tag{4-16}$$

A comparison of the voltage-regulation curves of a capacitor filter and a choke-input filter, Fig. 4-18, shows the poor regulation but high output

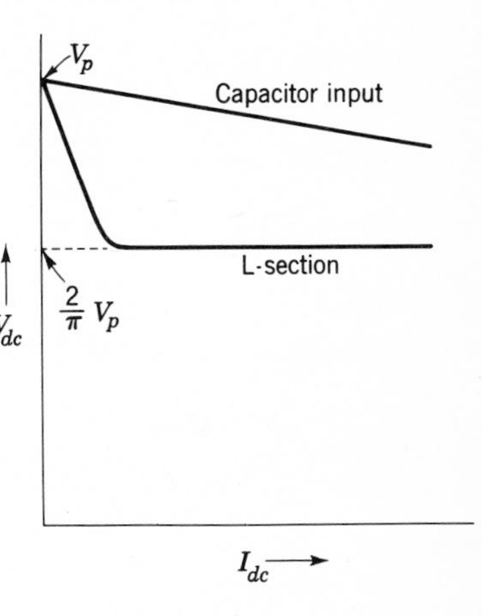

FIGURE 4-18 *Voltage-regulation characteristics of capacitor-input filter and L-section filter.*

of the former and the good regulation but lower voltage of the latter. The minimum current necessary to ensure good regulation in the case of the L-section filter is clearly evident.

π-section Filter The combination of a capacitor-input filter with an L-section filter, as shown in the power-supply diagram of Fig. 4-19, is a very popular circuit. The output voltage of this *π-section filter* is nearly that of the capacitor-input filter, and the regulation characteristics are about the same. The ripple is very much reduced by the double filtering action, however. In fact, the overall ripple factor is essentially the product of the ripple factor of the capacitor filter times the impedance ratio of the L-section filter. Therefore, using Eq. (4-6) and the impedance ratio developed for (4-11),

$$r = \frac{\pi}{\sqrt{3}} \frac{1}{\omega R_L C_1} \frac{1}{\omega^2 L C_2}$$

$$= \frac{\pi}{\sqrt{3}} \frac{X_{C_1} X_{C_2}}{X_L R_L} \tag{4-17}$$

According to Eq. (4-17), if the reactances of both capacitors are small at the ripple frequency, and that of the inductance is large, the ripple factor is small. Note that the ripple increases with load as R_L is made smaller.

In spite of the inferior regulation properties of the π-section filter, it is widely used, because of its excellent filtering action.

The practical power-supply diagram of Fig. 4-19 has several other

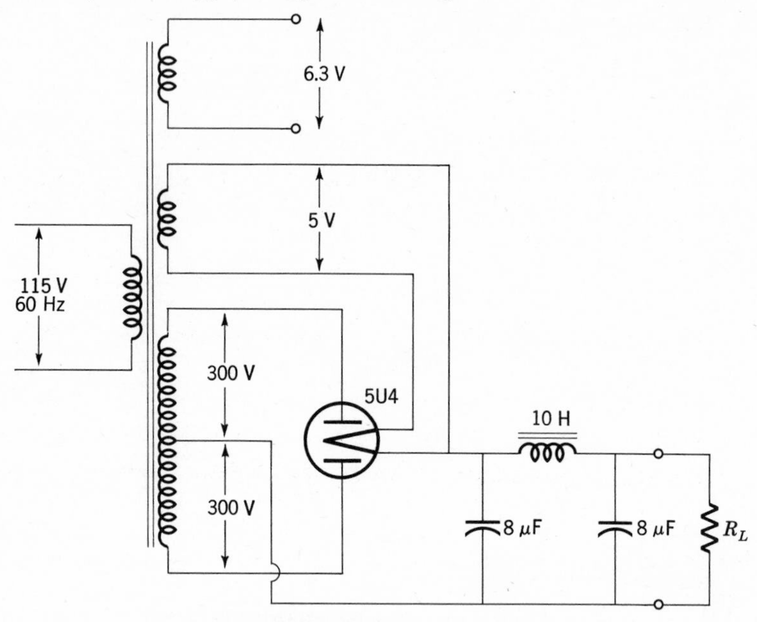

FIGURE 4-19 *A practical 300-V power supply.*

features of note. Both rectifiers are enclosed in the same glass envelope of the type 5U4 diode to simplify full-wave rectification. The 5-V heater-cathode (the oxide emitter is coated directly on the heater) is heated by a separate secondary winding on the power transformer. Finally, a 6.3-V winding is also provided to supply the heaters of other vacuum tubes in the associated circuitry. A circuit similar to Fig. 4-19 is very common in electronic devices.

Other combinations of filter types are also used. Two L-section filters, Fig. 4-20*a* for example, provide the good regulation of the choke-input filter together with very low ripple. The ripple factor of any filter design is calculated by considering each simple filter separately, as in the case of the π-section filter described above.

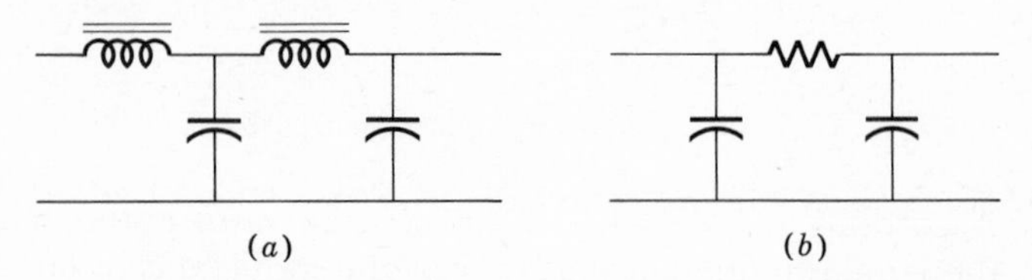

FIGURE 4-20 *(a) Double L-section filter and (b) π-section RC filter.*

A useful variant of the π-section filter particularly suitable for low-current circuits replaces the inductance with a resistor, Fig. 4-20*b;* the ripple factor is simply Eq. (4-17) with X_L replaced by the value of the series resistor R. This circuit is useful only if the current is low enough so that the IR drop across the resistor is not excessive and if regulation is of secondary concern. Within these restrictions, the circuit provides better filtering action than a simple capacitance filter.

VOLTAGE REGULATORS

It is often desirable that the voltage of a power supply remain fixed independent of the load current. Although the regulation of the choke-input filter is very good, the output voltage decreases with increasing current even in this circuit, because of the IR drops across the transformer winding resistance and the rectifier diodes. Other variables, such as changes in the line voltage and component aging, may also contribute to variations in the power-supply voltage. In order to maintain the output voltage constant, *voltage regulator* devices are made part of the power supply. These are fairly elaborate electronic circuits if very precise regulation is necessary, as discussed in a later chapter. In other cases, special components that maintain modest voltage stability are used.

Zener Diodes At some particular reverse-bias voltage the reverse current in a *pn* junction increases very rapidly. This happens when electrons are accelerated to high velocities by the electric field at the junction and

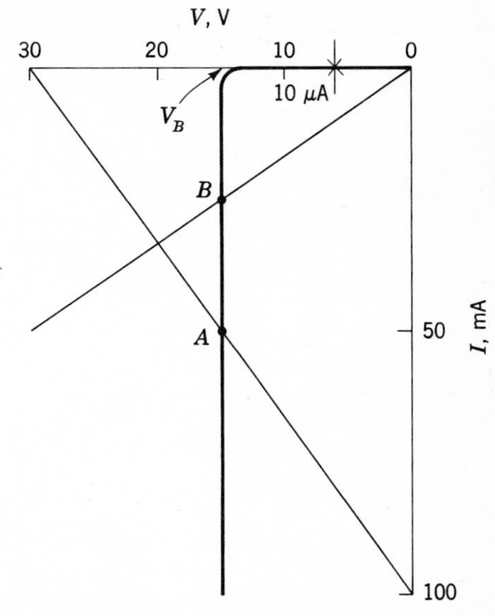

FIGURE 4-21 *Current-voltage characteristic of zener diode.*

produce other free electrons by ionization collisions with atoms. These electrons are similarly accelerated by the field and in turn cause other ionizations. This avalanche process leads to very large current and the junction is said to have suffered breakdown. The breakdown is not destructive, however, unless the power dissipation is allowed to increase the temperature to the point where local melting destroys the semiconductor. The voltage across the junction remains quite constant over a wide current range in the breakdown region. This effect can be used to maintain the output of a power supply at the breakdown voltage.

These *pn* junctions are called *zener diodes,* because Clarence Zener first suggested an explanation for the rapid current increase at breakdown. The current-voltage curve of a zener diode, Fig. 4-21, has a sharp transition potential and a flat current plateau above breakdown. Zener diodes may be obtained with breakdown voltages ranging from about two to several hundred volts and with current ratings of a few milliamperes to many amperes.

The way in which a zener diode is used to regulate the output voltage of a power supply is illustrated in Fig. 4-22. The unregulated output volt-

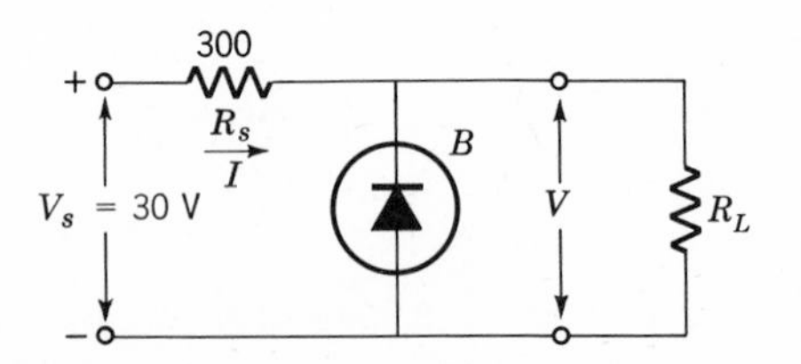

FIGURE 4-22 *Zener diode used as voltage regulator.*

age of the supply V must be greater than the zener breakdown voltage of the diode. Then the voltage drop across R_S caused by the diode current plus the diode breakdown potential adds up to the power-supply potential. As the load current increases, the diode current decreases, so the drop across R_s always is the difference between the breakdown potential and the power-supply voltage. Even if the power-supply voltage varies under load, the regulated output voltage remains constant.

The zener diode voltage regulator is analyzed by plotting the current-voltage characteristic pertaining to R_s on the same axes as the current-voltage characteristic of the diode. Kirchhoff's rule applied to the circuit of Fig. 4-22 is

$$V_s - IR_s - V = 0 \tag{4-18}$$

Solving for I,

$$I = \frac{V_s}{R_s} - \frac{1}{R_s} V \tag{4-19}$$

which is the equation of a straight line of slope $-1/R_s$ and intercept $V = V_s$.

Using numerical values given on the circuit diagram, Eq. (4-19) intersects the current-voltage characteristic of the diode at point A on Fig. 4-21. This intersection gives the current in R_s. The load current is just V_B/R_L, where V_B is the breakdown voltage of the diode.

It is useful to plot the IV characteristic of the load,

$$I_L = \frac{1}{R_L} V \tag{4-20}$$

on Fig. 4-21. The load current is given by the intersection of Eq. (4-20) with the diode characteristic, point B. Note that the load current must be less than the current in R_s, for otherwise the voltage drop across R_s is too large to maintain V_B across the diode and regulating action ceases. So long as intercept B is at a smaller current than point A, the circuit performs properly. Therefore, the maximum load current (minimum value of R_L) is just equal to the current in R_s, or, using Eqs. (4-19) and (4-20),

$$\frac{V_B}{R_{L,\,\mathrm{min}}} = \frac{V_s - V_B}{R_s} \tag{4-21}$$

$$R_{L,\,\mathrm{min}} = \frac{R_s}{V_s/V_B - 1} \tag{4-22}$$

Zener diodes also provide filtering action since they maintain the output voltage constant against changes in the power supply voltage, including ripple. For this reason it is usually possible to employ only rudimentary capacitance filtering in conjunction with voltage regulators.

Controlled Rectifiers It is often necessary to control the power delivered to some load, such as an electric motor or the heating element of a furnace. Series resistances or potentiometers waste power, a serious drawback in high-power circuits. *Controlled rectifiers* have been developed which are capable of adjusting the transmitted power with little waste.

The most satisfactory unit of this kind is the *silicon controlled rectifier*, or *SCR*. This semiconductor device contains four parallel *pn* junctions; it is described in detail in Chap. 5. For present purposes it is sufficient to note that it is similar to a junction rectifier in which forward conduction is controlled by the current in a control electrode, called the *gate*. The inclusion of the gate electrode is shown on the SCR symbol (Fig. 4-23).

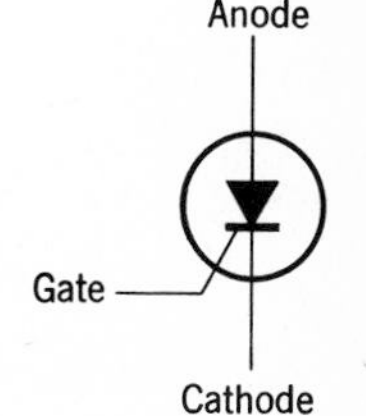

FIGURE 4-23 *Circuit symbol for silicon controlled rectifier.*

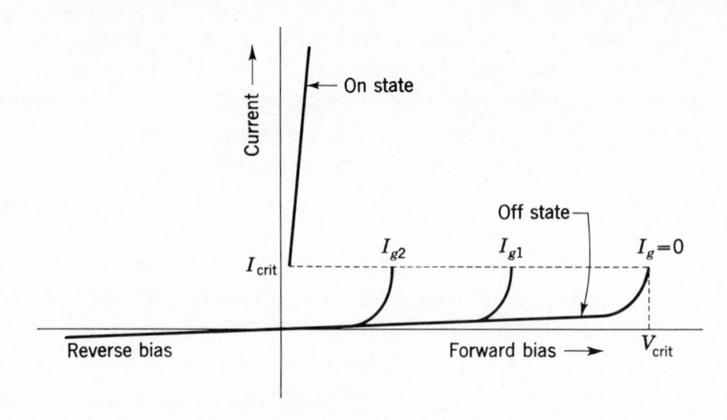

FIGURE 4-24 *Current-voltage characteristic of SCR. Unit can be triggered into on state by gate current. Here $I_{g2} > I_{g1} > 0$.*

The current-voltage characteristic of a typical SCR, Fig. 4-24, is identical to a junction rectifier in the reverse direction. The forward direction has both an "on" state, which is equivalent to normal forward conduction in a junction rectifier, and a low-current "off" state. So long as the forward anode-cathode potential remains below a certain critical value (actually it is the corresponding current which is critical), the forward current is small. Above this critical value the SCR is in the high-current low-voltage on condition. The critical current may be supplied by the gate electrode, which means that the SCR may be triggered into the on position by a small current (as low as 100 μA) supplied to the gate terminal. It is not necessary to maintain the gate current, for once the SCR is triggered into the high-conduction state it remains in this condition until the anode potential is reduced to zero.

Consider the simple SCR circuit of Fig. 4-25, in which the motor speed

FIGURE 4-25 *Simple SCR circuit to control speed of dc motor.*

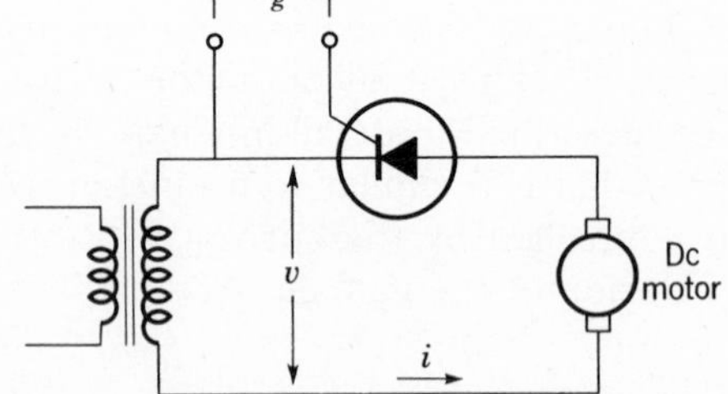

is controlled by the power delivered to it. The peak value of the transformer voltage V_p is less than the critical value, so that, unless the SCR is triggered by a gate current, the load current is zero. Suppose the gate is supplied with a current pulse each cycle that lags the transformer voltage

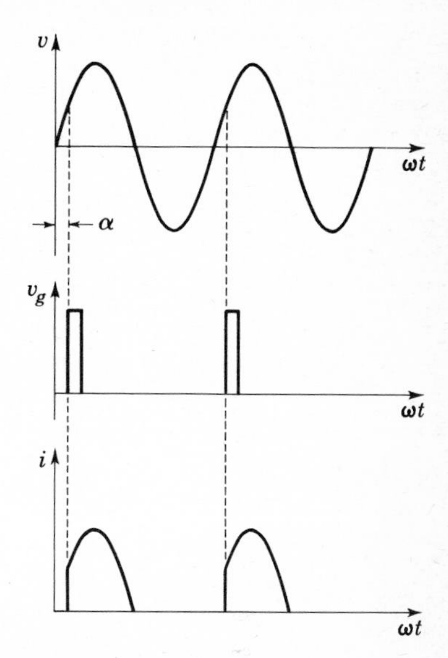

FIGURE 4-26 *Output current waveform depends upon phase angle between transformer voltage and gate-current pulse.*

by a phase angle α, as illustrated in Fig. 4-26. Forward conduction is delayed until this point in each cycle and the rectified output waveform is similar to that of a half-wave rectifier, except that the first part of each cycle is missing. The average, or dc, value of the current in the motor is found by integrating the output current over a complete cycle, or

$$I_{\mathrm{dc}} = \frac{I_p}{2\pi} \int_{\alpha}^{\pi} \sin \omega t \, d\omega t \tag{4-23}$$

where I_p is the peak value of the current. Integrating,

$$I_{\mathrm{dc}} = \frac{I_p}{2\pi} (1 + \cos \alpha) \tag{4-24}$$

According to Eq. (4-24) the motor current can be adjusted from a maximum ($\alpha = 0$) to zero ($\alpha = \pi$) simply by changing the phase angle of the gate pulses. Simple circuits are available to generate such pulses, as described in a later chapter.

It is not necessary to use pulses in the gate circuit, so long as care is taken to limit the power dissipated by the control electrode below that which might damage the SCR. Consider the *phase-shift control* circuit of Fig. 4-27, in which the gate current is shifted in phase with respect to the anode voltage. Since the gate voltage is the sum of the secondary voltage v_1 and the drop across R, the gate voltage is

$$v_g e^{j\phi_g} = v_1 - Rie^{j\phi} \tag{4-25}$$

This is easily solved for the phase angle between the gate voltage and the anode voltage, which has the same phase as v_1. The result

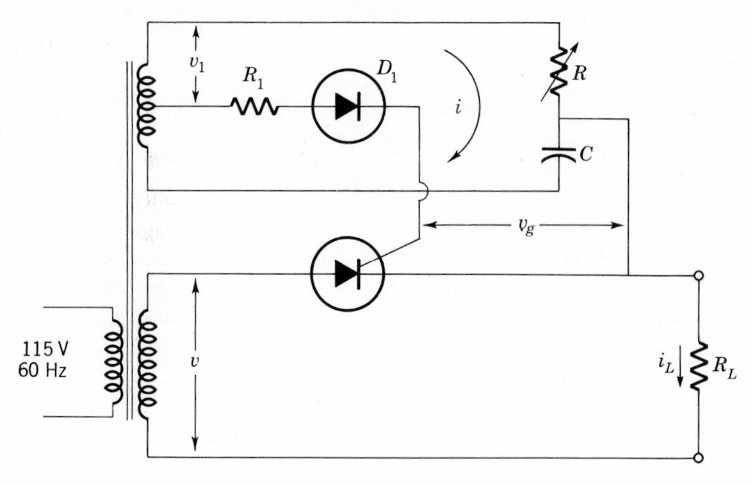

FIGURE 4-27 *Phase-shift control of SCR.*

$$\phi_g = \arctan \frac{2R\omega C}{(R\omega C)^2 - 1} \tag{4-26}$$

shows that the gate voltage may be put in phase with the anode potential ($R = 0$) or nearly 180° out of phase ($R = \infty$).

The waveforms of Fig. 4-28 show how variation of ϕ_g alters the time at which the diode conducts on each forward cycle as the gate voltage becomes larger than the critical value. Note that the gate signal remains on during part of the reverse half-cycle as well, but this has no effect upon the reverse

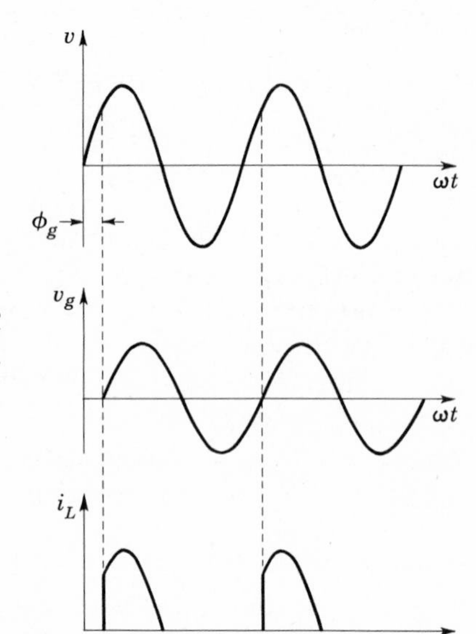

FIGURE 4-28 *Waveforms in phase-shift control SCR circuit.*

characteristic. The purpose of resistor R_1 and diode D_1 in the circuit (Fig. 4-27) is to limit the gate current to a safe value and to prevent reverse gate current, both of which are deleterious to the SCR.

DIODE CIRCUITS

Diodes prove to be useful in circuits other than power supplies. A major application is in circuits designed to operate with square-wave pulses, as will be discussed in a later chapter. The rectification characteristics of diodes are also put to work in circuits dealing with sine-wave signals. In these applications the nonlinear diode modifies the sine-wave signals in a specified manner.

Clippers Consider the diode *clipper* circuit, Fig. 4-29, in which two diodes are connected in parallel with an ac voltage source. Batteries V_1 and V_2

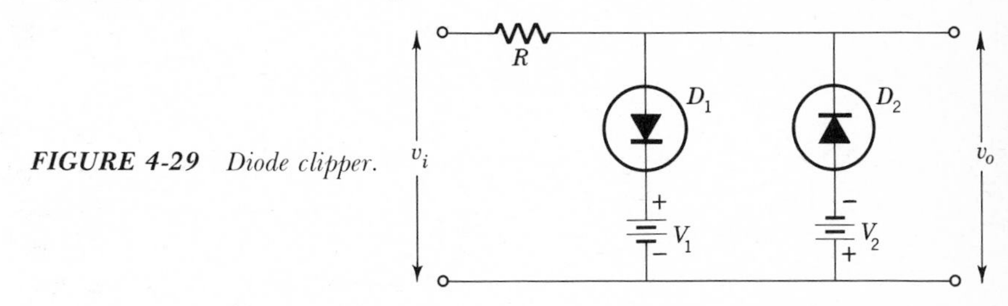

FIGURE 4-29 *Diode clipper.*

bias each diode in the reverse direction. Whenever the input voltage signal is greater than V_1, diode D_1 conducts and causes a voltage drop across the series resistor R. Similarly, every time the input voltage becomes more negative than V_2, diode D_2 conducts. Thus, the output waveform is *clipped* or *limited* to the voltages set by the reverse biases V_1 and V_2. The clipping action is most efficient when the series resistance is much greater than the load impedance.

Clipper circuits produce square waves from a sine-wave generator, as illustrated by the waveforms in Fig. 4-30. If $V_1 = V_2$ and the amplitude of the input signal is considerably greater than the bias voltages, the output

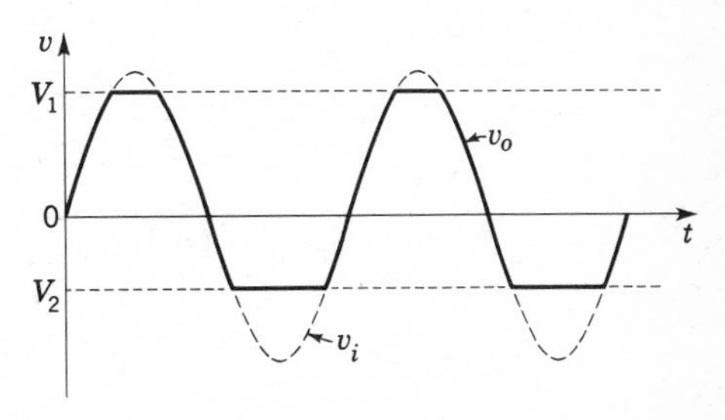

FIGURE 4-30 *Maximum amplitudes in output waveform of diode clipper are limited to values of bias voltages.*

waveform is very nearly a square wave. Note that if, say, $V_2 = 0$, the output can never swing negative, so the waveform is a train of positive-going square pulses. The clipping action operates on any input waveform; it can be adjusted by altering the reverse-bias voltages V_1 and V_2.

The clipper circuit is also a useful protective device which limits input voltages to a safe value set by the bias voltage. In radio-receiver circuits limiters are often used to reduce the effect of strong noise pulses by limiting their amplitude to that of the desired signal. The signal waveform is transmitted undistorted so long as the instantaneous amplitude remains smaller than the bias voltages.

Clamps The diode *clamp* circuit is shown in Fig. 4-31. Consider first the situation in which the bias voltage V is set equal to zero. The diode con-

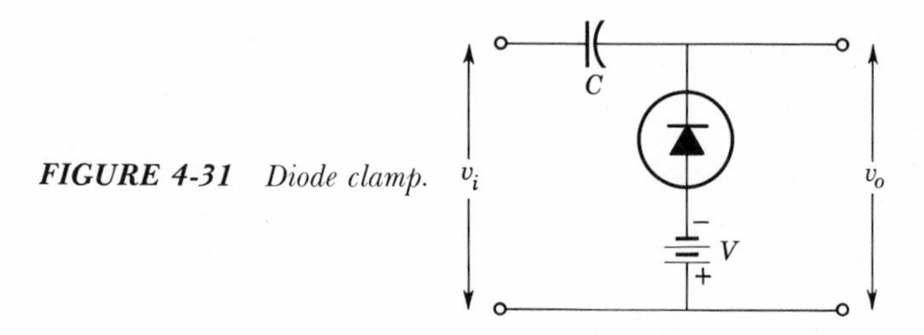

FIGURE 4-31 *Diode clamp.*

ducts on each negative cycle of input voltage and charges the capacitor to a voltage equal to the negative peak value of the input signal. If the load current is zero, the capacitor retains its charge on the positive half-cycle, since the diode voltage is in the reverse direction. Thus, the output voltage is

$$v_o = v_i + V_p \tag{4-27}$$

where V_p is the negative peak value of the input voltage. According to Eq. (4-27) the output voltage waveform replicates the input signal except that it is shifted by an amount equal to the dc voltage across the capacitor.

The output waveform corresponding to a sinusoidal input, Fig. 4-32,

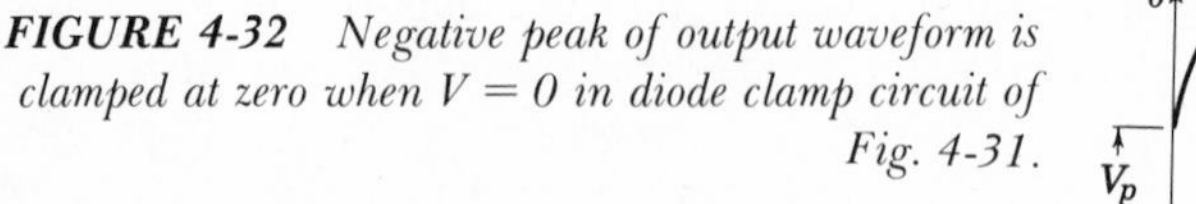

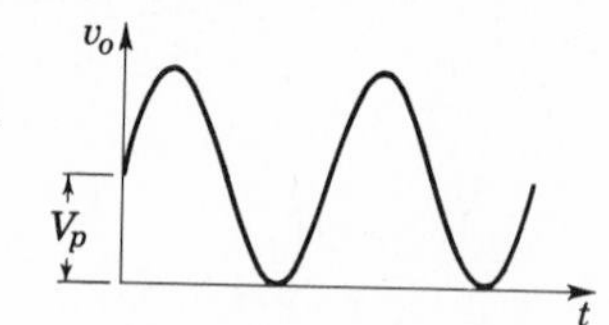

FIGURE 4-32 *Negative peak of output waveform is clamped at zero when $V = 0$ in diode clamp circuit of Fig. 4-31.*

has the negative peaks of the sine wave *clamped* at zero voltage. This is always the case, independent of the amplitude of the input voltage. Furthermore, the negative peaks are always clamped at zero volts no matter what form the input waveshape takes. When the terminals of the diode are interchanged, the same circuit analysis applies; the positive peaks of the output wave are clamped at zero.

If the bias voltage V in Fig. 4-31 is set at a potential other than zero, the capacitor charges to a voltage equal to $V_p + V$. Therefore, the negative peaks are clamped at the voltage V. Similarly, when the bias is $-V$, the negative peaks are clamped at this potential. Reversing the diode polarity makes it possible to clamp the positive peaks of the input wave at the voltage equal to the bias potential.

The diode clamp is used in circuits which require that the voltages at certain points have fixed peak values. This is important, for example, if the waveform is subsequently clipped at a given voltage level. In most diode clamp circuits it is useful to connect a large resistance across the output terminals so that the charge on the capacitor can eventually drain away. This makes it possible for the circuit to adjust to changes in amplitude of the input voltage.

Ac Voltmeters The rectifying action of diodes makes it possible to measure ac voltages with a dc voltmeter, such as the d'Arsonval meter discussed in the first chapter. Two ways in which this is accomplished in VOM meters are illustrated in Fig. 4-33. In Fig. 4-33a diode D_1 is a half-wave rectifier

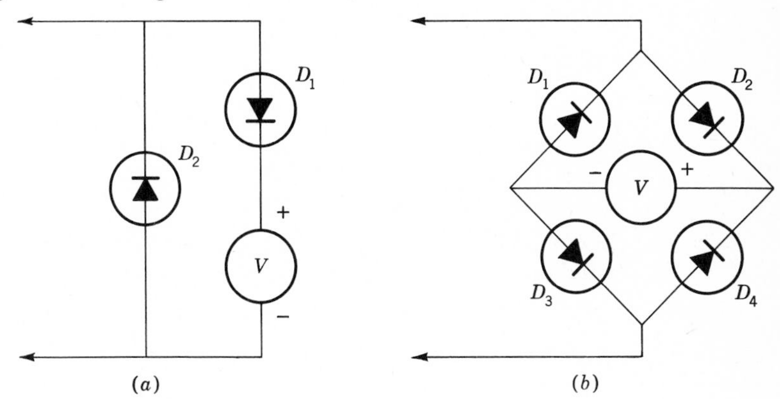

FIGURE 4-33 *(a) Half-wave and (b) full-wave ac voltmeters.*

and the dc component of the rectified waveform registers on the dc voltmeter. Diode D_2 is included to rectify the negative cycle of the input waveform. Although no meter current results from the action of diode D_2, it is included so that both halves of the ac voltage are rectified. This ensures that the voltmeter loads the circuit equally on both half-cycles and avoids possible waveform distortion. The bridge circuit, Fig. 4-33b, has greater sensitivity than the half-wave circuit because it is a full-wave rectifier.

The deflection of a d'Arsonval meter is proportional to the average current, so that the voltmeters of Fig. 4-33 measure the average value of the ac voltage. The meter scale is commonly calibrated in terms of rms readings, however, in order to facilitate comparison with dc readings. This calibration assumes that the ac waveform is sinusoidal, and, if this is not the case, the meter readings must be interpreted in terms of the actual waveform measured. Series resistor multipliers are used with ac voltmeters to increase the range, just as in the case of dc meters.

The half-wave rectifier, Fig. 4-34, makes a useful *peak-reading* voltmeter

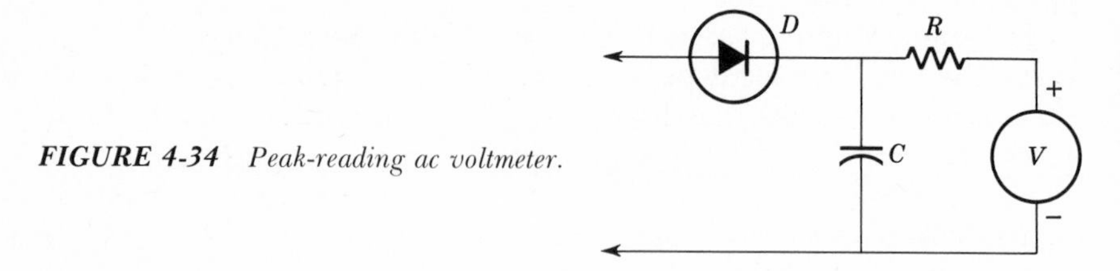

FIGURE 4-34 *Peak-reading ac voltmeter.*

if the current drain through the meter circuit is small. In this case the capacitor charges to the positive peak value of the input waveform on each half-cycle, and the dc meter deflection corresponds to this value. Here again, it is common practice to calibrate the meter scale in terms of rms readings, assuming that the unknown voltage is sinusoidal. This is accomplished by designing the scale to indicate the peak value divided by $\sqrt{2}$. In case the unknown voltage is not sinusoidal, the peak value is obtained by multiplying the scale reading by $\sqrt{2}$. A difficulty with this circuit is that the system being measured must contain a dc path. If this is not so and a series capacitor is present, the voltage across C depends upon the relative values of the two capacitors, since they are effectively in series during the forward cycle. This results in erroneous voltage readings.

This diode clamp is a peak-reading circuit which does not have this difficulty. It is customary to include an RC filter, Fig. 4-35, to remove the

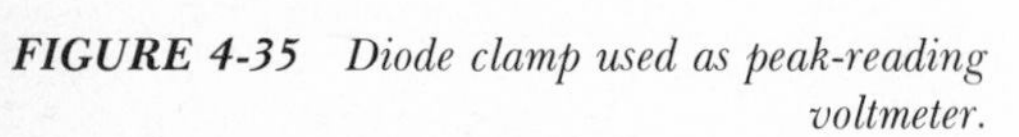

FIGURE 4-35 *Diode clamp used as peak-reading voltmeter.*

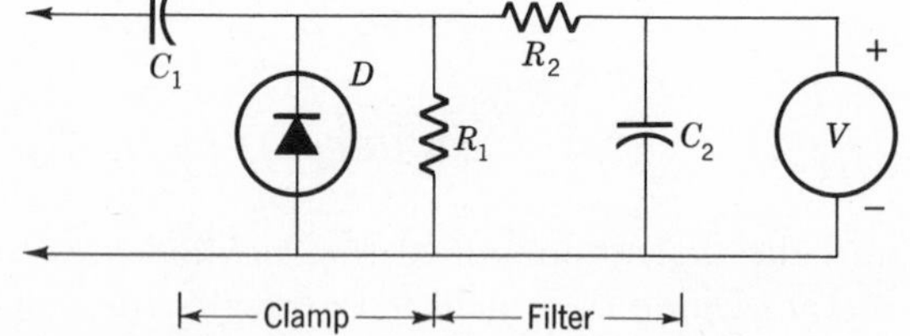

ac component of the clamped wave. Referring to the waveform in Fig. 4-32, the average value is equal to the negative peak of the input voltage, so the dc meter reading corresponds to the negative peak voltage. Note that a series capacitor in the circuit to be measured in effect becomes part of the clamping action and does not result in a false meter indication. If the filter in Fig. 4-35 is replaced with a diode peak rectifier, Fig. 4-34, the combination reads the peak-to-peak value of the input wave (Exercise 4-10). Although usually calibrated in terms of rms readings for a sine wave, a peak-to-peak voltmeter is very useful in measuring complex waveforms.

Detectors Nonlinear diode properties are useful in other ways than those corresponding to actual rectification. Suppose that the amplitude of the ac voltages applied is very small, so the diode current-voltage characteristic can be represented by the expression

$$i = a_1 v + a_2 v^2 \tag{4-28}$$

where a_1 and a_2 are constants. An expression of this form is a satisfactory approximation of the diode characteristic for sufficiently small values of applied voltage. According to Eq. (4-28), one part of the current is proportional to the square of the input voltage.

Consider the circuit shown in Fig. 4-36, in which two sinusoidal signals

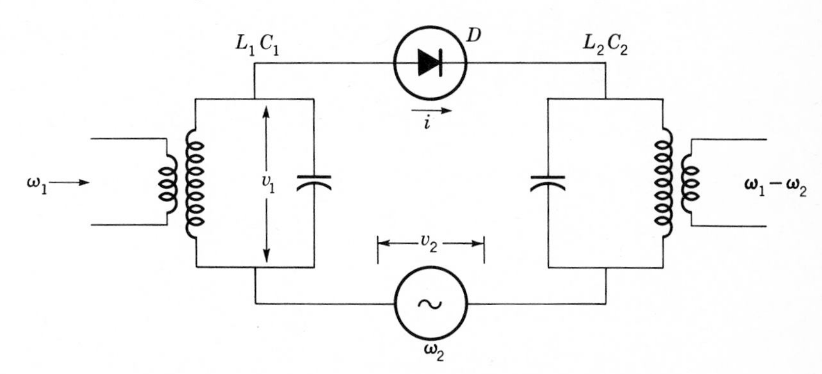

FIGURE 4-36 *Diode mixer uses nonlinear properties of diode.*

of somewhat different frequencies, ω_1 and ω_2, are supplied to a diode. The total diode voltage is the sum of the individual waves,

$$v = V_1 \sin \omega_1 t + V_2 \sin \omega_2 t \tag{4-29}$$

where V_1 and V_2 are the peak amplitudes. The current is found by substituting Eq. (4-29) into Eq. (4-28), assuming for the moment that the diode impedance is greater than other impedances in the circuit.

$$i = a_1 V_1 \sin \omega_1 t + a_1 V_2 \sin \omega_2 t + a_2 V_1^2 \sin^2 \omega_1 t + a_2 V_2^2 \sin^2 \omega_2 t + 2a_2 V_1 V_2 \sin \omega_1 t \sin \omega_2 t \tag{4-30}$$

It is instructive to express Eq. (4-30) in the form of a Fourier series type expression in which each term represents a given frequency in the total waveform. This is accomplished by introducing trigonometric substitutions for $\sin^2$ and $\sin \omega_1 t \sin \omega_2 t$ and rearranging,

$$i = \frac{a_2}{2}(V_1^2 + V_2^2) + a_1(V_1 \sin \omega_1 t + V_2 \sin \omega_2 t)$$
$$- \frac{a_2}{2}(V_1^2 \cos 2\omega_1 t + V_2^2 \cos 2\omega_2 t)$$
$$+ a_2 V_1 V_2 \cos(\omega_1 - \omega_2)t - a_2 V_1 V_2 \cos(\omega_1 + \omega_2)t \quad (4\text{-}31)$$

The first term in Eq. (4-31) is a direct current, while the second corresponds to the input voltages. The last two terms are at frequencies not present in the original inputs. Suppose ω_1 and ω_2 are not too different; then the term involving $\omega_1 - \omega_2$ is a comparatively low frequency.

In the *diode mixer* circuit of Fig. 4-36 the output circuit L_2C_2 is tuned to the frequency $\omega_1 - \omega_2$, so only this component has an appreciable amplitude at the output terminals. In effect, an input signal of frequency ω_1 is mixed with a constant-amplitude sine wave at ω_2 and is converted to an output frequency $\omega_1 - \omega_2$. This is called *heterodyning* and is of considerable importance in, for example, radio receivers. The incoming high-frequency signal is converted into a lower-frequency signal which is easier to amplify. Furthermore, changing the frequency ω_2 means that the receiver is tuned to a different ω_1, such that $\omega_1 - \omega_2$ is a constant. This is important because the amplification of the $\omega_1 - \omega_2$ signal is accomplished by fixed tuned circuits that are inexpensive and easy to keep in adjustment, yet the receiver may be tuned to a different input frequencies. In this application Fig. 4-36 is called a *first detector* because the signal frequency is modified in the first stages of the receiver.

This circuit has another important use, which may be illustrated by an alternate rearrangement of Eq. (4-31). Using the trigonometric substitution,

$$2 \sin a \sin b = \cos(a - b) - \cos(a + b) \quad (4\text{-}32)$$

Equation (4-31) may be put in the form

$$i = \frac{a_2}{2}(V_1^2 + V_2^2) + a_1 V_2 \sin \omega_2 t - \frac{1}{2}(V_1^2 \cos 2\omega_1 t + V_2^2 \cos 2\omega_2 t)$$
$$+ a_1 V_1 \sin \omega_1 t + 2a_2 V_1(V_2 \sin \omega_2 t) \sin \omega_1 t \quad (4\text{-}33)$$

Suppose ω_2 is much smaller than ω_1 and that the output circuit L_2C_2 is tuned to the frequency ω_1. Then only the last two terms in Eq. (4-33) have an appreciable output amplitude. The first of these is simply the current resulting from the input voltage at the frequency ω_1. Careful inspection of the last term in Eq. (4-33) shows that it is a sinusoidal waveform at a frequency ω_1 but with an amplitude that varies sinusoidally at a frequency corresponding to ω_2.

The waveform of the last term in Eq. (4-33), illustrated in Fig. 4-37,

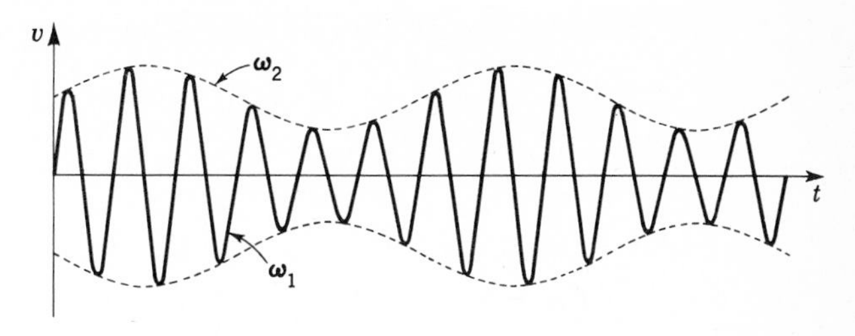

FIGURE 4-37 *Modulated waveform.*

shows how the amplitude varies. The frequency ω_1 is said to be modulated at the frequency of ω_2. This is the basis for radio and television communication wherein a high-frequency *carrier* ω_1 is modulated in accordance with a low-frequency signal ω_2 corresponding to the sound or picture to be transmitted. The high-frequency voltage is radiated from the sending station to a receiver where the waveform of the signal is recovered.

The signal is recovered by *demodulating* the modulated waveform with

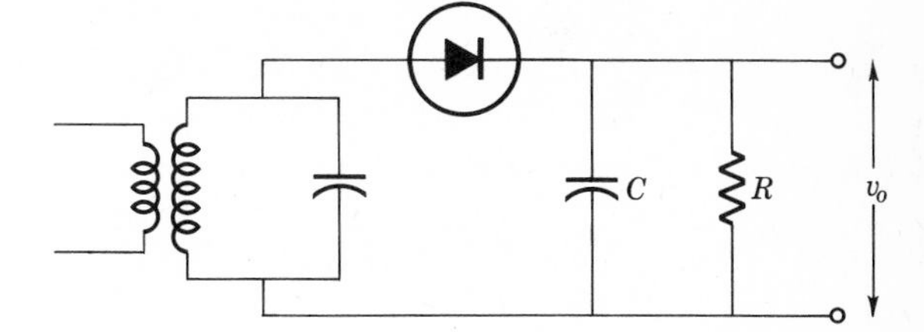

FIGURE 4-38 *Diode peak rectifier used to demodulate modulated waveform.*

a diode peak rectifier, Fig. 4-38. Since the output of the peak rectifier is equal to the peak value of the input voltage, the waveform across the output capacitor corresponds to that of modulating signal, as in Fig. 4-39.

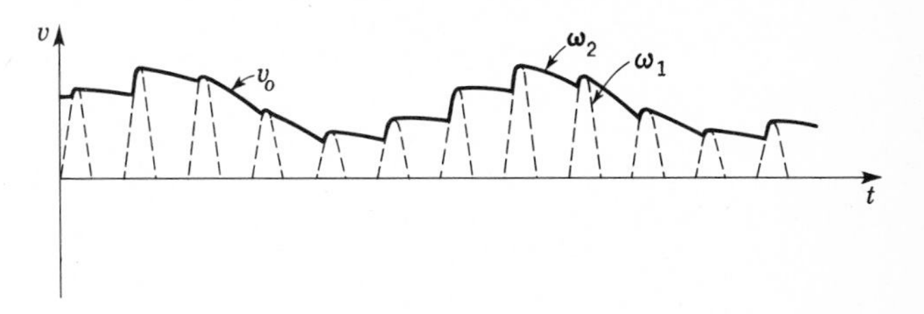

FIGURE 4-39 *Modulating waveform recovered at output of second detector.*

In this application the peak rectifier is called a second detector, because the waveform is changed for the second time in the receiver circuit. The second detector circuit, Fig. 4-38, is widely used in radio and television receivers.

SUGGESTIONS FOR FURTHER READING

Paul M. Chirlian and Armen H. Zemanian: "Electronics," McGraw-Hill Book Company, New York, 1961.

W. Ryland Hill: "Electronics in Engineering," McGraw-Hill Book Company, New York, 1961.

Jacob Millman and Samuel Seely: "Electronics," McGraw-Hill Book Company, New York, 1951.

EXERCISES

4-1 Calculate the forward resistance of a 5U4 vacuum diode at a voltage of 30 V. Assuming the reverse resistance is 10 MΩ, what is the reverse to forward resistance ratio?

Ans.: 210 Ω; 4.8×10^4

4-2 Assuming the interelectrode capacitance of a 5U4 diode is 4 pF and the reverse resistance is 10 MΩ, what is the approximate upper limit to the frequency at which the diode is useful?

Ans.: 4.0×10^3 Hz

4-3 Approximate the forward characteristic of a 5U4 diode with a straight line, using Fig. 4-4, and determine the peak rectified current in a half-wave circuit with a 5000-Ω load and an input voltage of 100 V rms. Determine the direct current by averaging the rectified half sine wave over a full cycle and compare with the peak current.

Ans.: 27.1 mA; 8.6 mA

4-4 In the voltage-doubler circuit, Fig. 4-14, the input voltage is 115 V rms at a frequency of 60 Hz. If the load resistance is 10,000 Ω, what is the minimum value of the capacitors necessary to be sure the voltage drop between charging pulses is less than 10 percent of the output voltage?

Ans.: 2.4×10^{-6} F

4-5 Calculate the output voltage and ripple factor of the circuit of Fig. 4-15 after diodes D_1 and D_4 are removed.

Ans.: 12.9 V; 0.048

4-6 Plot the dc output and the ripple voltage of the full-wave rectifier circuit given in Fig. 4-19 up to a current of 200 mA.

4-7 Determine the ripple factor of an L-section filter comprising a 10-H choke and a 8-μF capacitor used with a half-wave rectifier. Compare with a simple 8-μF capacitor filter at a load current of 50 mA and also 150 mA, assuming an output of 50 V.

Ans.: 0.042; 0.604; 1.82

4-8 In the zener diode voltage regulator of Fig. 4-22 determine the range of load resistances over which the circuit is useful if $R_s = 1500$ Ω and the supply voltage $V = 150$ V; the diode breakdown voltage is 100 V and maximum rated current is 100 mA. For a fixed $R_L = 10{,}000\ \Omega$, over what range of input voltages does the circuit regulate?

Ans.: Greater than 3 kΩ; 115 to 265 V

4-9 Using the phase-shift controlled rectifier circuit, Fig. 4-27, plot the dc load current as a function of control resistance R if the transformer secondary voltage is 100 V rms, the load resistance is 10 Ω, and $C = 0.1\ \mu$F. Assume the SCR turns on whenever the gate voltage is positive with respect to the cathode.

4-10 Analyze the peak-to-peak voltmeter circuit, Fig. 4-40, by sketching the

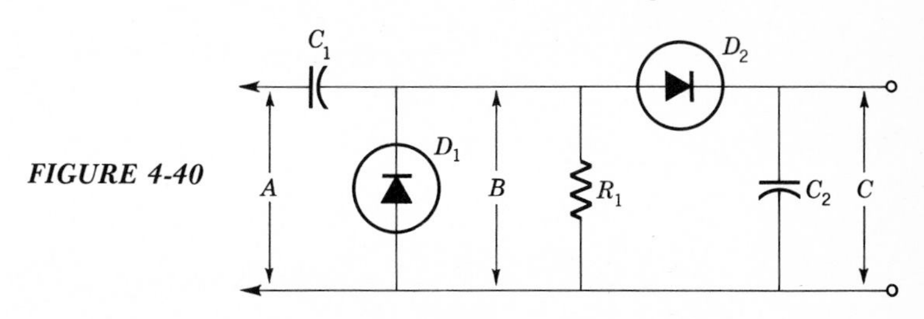

FIGURE 4-40

voltage waveforms at points *A, B,* and *C,* assuming the input voltage *A* is sinusoidal. Note that the circuit is a diode clamp followed by a peak rectifier.

LABORATORY EXERCISES

4-A The L-section Filter Understanding the operation of circuits involving nonsinusoidal signals is greatly facilitated by examination of waveforms in several parts of the circuit. Not only is a waveform picture "worth a thousand equations," but circuit analysis may proceed unfettered by mathematical complexity. In this experiment waveforms in the L-section filter are examined as a supplement to the description of circuit operation in the text.

Replace the capacitor filter in Fig. 4-15 with an L-section filter, Fig. 4-17, using $L = 0.5$ H and $C = 100\ \mu$F. In the analysis it is permissible to ignore the internal resistances of the diodes, but it may be necessary to consider the 10-Ω winding resistance of the transformer secondary in certain cases. To begin, quantitatively sketch waveforms of the transformer voltage, diode current, and capacitor voltage for the condition that the dc load current is one-half the critical current given by Eq. (4-15). A quantitative sketch is one in which the waveforms are not plotted by point-by-point calculation but are hand drawn with all pertinent features such as peak value, average value, zero points, etc., quantitatively presented. For simplicity, consider only the first three terms in the Fourier expansion of the diode current waveform. Note that your waveforms resemble those in Fig. 4-16.

A measure of your understanding of these waveforms is to sketch them for the first several cycles after the power is applied. Note that because of the impedance of the inductor, the capacitor cannot charge to its final value in one cycle. Therefore, several cycles (how many?) are needed before the waveforms reach the steady-state condition you first sketched above.

Now quantitatively sketch the waveforms for the load-current condition represented by Eq. (4-15). How well does the minimum value of load current calculated from Eq. (4-15) agree with the value you determine by waveform analysis? Which is more correct? Why?

Using quantitative waveform sketches prepare a plot of the dc output voltage, the rms value of the ac component, and the ripple factor, as a function of load current from 0 to 100 mA. Do not forget to include the effect of the winding resistances. Compare these "experimental" results with those predicted by the equations developed in the text.

4-B Zener Regulator The results of Laboratory Exercise 4-A illustrate the variations in output voltage of a practical power supply even with an L-section filter. Voltage regulator circuits are commonly used to stabilize the output voltage of power supplies against variations in load current and also against changes in the input supply voltage. The zener diode regulator is particularly appropriate and may be well analyzed by solving circuit relations graphically using load lines plotted in conjunction with the current-voltage characteristic of the diode.

The object of this experiment is to examine the regulation properties of the zener diode circuit shown in Fig. 4-22. Plot the dc output voltage of the regulator as a function of load current from 0 to 75 mA. Do this using the current-voltage characteristic given in Fig. 4-21 and by varying the load resistance to move intersection point *B*. Over what current range is regulation effective? Repeat for a 600-Ω series resistor. Is the regulation range greater or smaller? What is the major limitation to choosing a value of series resistance such that the regulation range is very large?

Now ($R_s = 300\ \Omega$ again) plot the dc output voltage and load current as a function of source voltage from 0 to 30 V for a fixed 600-Ω load. Over what input voltage range does the circuit regulate? How does a larger or smaller value of R_s influence this range?

chapter five

THE PHYSICS OF ACTIVE DEVICES

The versatility inherent in nonlinear electronic components is enhanced immeasurably by the capability to influence current in the device in accordance with signals introduced at a third control electrode. Both the vacuum triode, so named because of the third electrode, and the transistor exhibit this feature. These are considered to be active devices, rather than passive components, because the control electrode produces an active interaction with signal currents in the device.

Electrical properties of active devices are described by current-voltage characteristics as for other nonlinear components. It is often useful to develop an appreciation for the origin of these current-voltage characteristics in terms of basic physical processes. In this way it is possible to optimize performance of a nonlinear element in any application, and also to design improved components having desirable electrical properties. It turns out that the current-voltage characteristics of electronic devices depend primarily upon the motions of free electrons in them.

The electronic functions of transistors take place within solid materials called semiconductors. Accordingly, the properties of transistors and other semiconductor devices, such as the junction diode, stem directly from the behavior of electrons in semiconductor crystals. The great variety of devices that have been developed since the discovery of the transistor are possible because of the versatility of semiconductor materials.

SEMICONDUCTORS

Energy Bands The properties of any solid material, including semiconductors, depend upon the nature of the constituent atoms and upon the way in which the atoms are grouped together. That is, the properties are a function of both the atomic structure of the atoms and the crystal structure of the solid. Experiments have shown that an atom consists of a positively charged nucleus surrounded by electrons located in discrete orbits. Actually electrons can exist in stable orbits near the nucleus only for certain discrete values of energy called *energy levels* of the atom.

The allowed energies of electrons in an atom are depicted by horizontal lines on an *energy-level diagram,* Fig. 5-1. The curved lines in the diagram

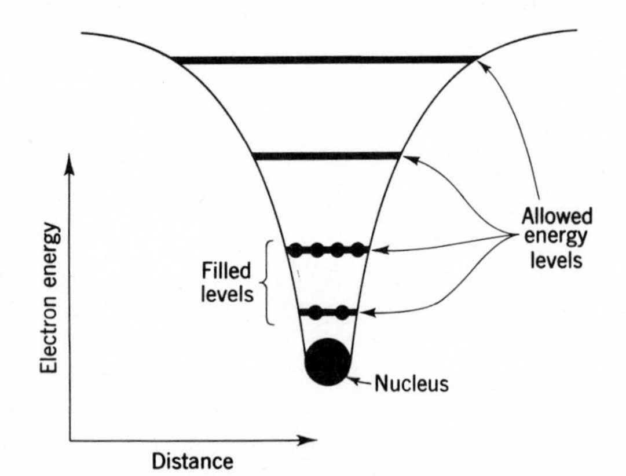

FIGURE 5-1 *Energy-level diagram for an atom.*

represent the potential energy of an electron near the nucleus, as given by Coulomb's law. As a consequence of the *Pauli exclusion principle,* only a certain maximum number of electrons can occupy a given energy level. The result is that in any atom, electrons fill up the lowest possible levels first. Electrons in occupied levels are indicated by a solid dot in the energy-level diagram.

When atoms come close together to form a solid crystal, electrons in the upper levels of adjacent atoms interact to bind the atoms together. Because of the strong interaction between these outer, or *valence,* electrons, the upper energy levels are drastically altered. This can be illustrated by an energy-level diagram for the entire crystal. Consider first two isolated atoms, each with an energy-level diagram pertaining to the outer electrons as in Fig. 5-2*a*. When these are brought close together, Fig. 5-2*b*, the valence electrons in both atoms are attracted by both nuclei. The result is that the energy required to remove an electron from one nucleus and place it on the other is reduced. This means that an outer electron is equally likely to be located near either nucleus. The appropriate energy-level diagram for the combination of two atoms has two

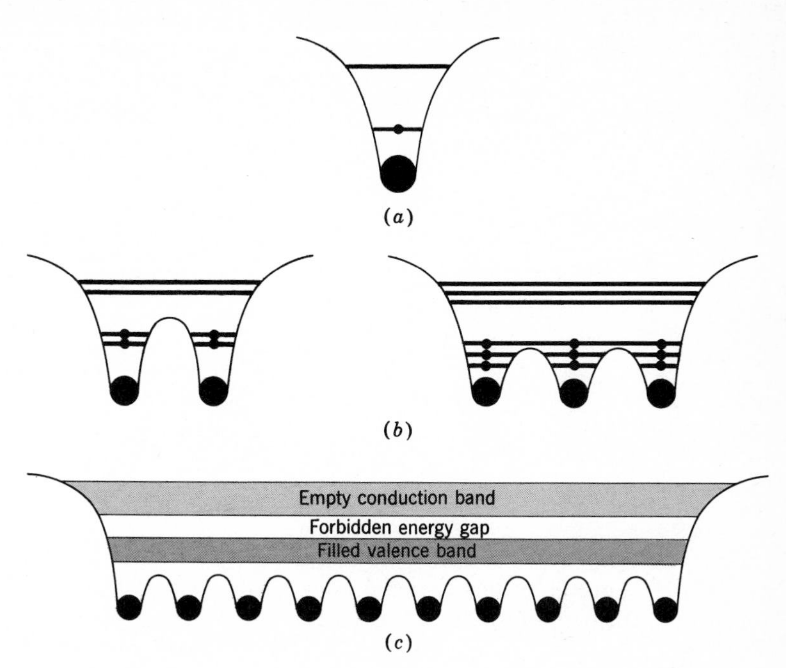

FIGURE 5-2 *Energy-level diagram for (a) isolated atom, (b) two and three atoms close together, and (c) solid crystal. In the crystal, energy levels are broadened into bands.*

energy levels near each atom core. The higher, unoccupied, levels are similarly split, indicating that these levels, too, can each contain two electrons. When three atoms are brought together, Fig. 5-2*b*, the outer electrons of all three atoms can be associated with any of the three nuclei. Consequently, three energy levels are available.

Even the tiniest crystal contains many hundreds of millions of atoms, so that very many energy levels are associated with each nucleus. Since it is impossible to show each of the many millions of levels separately, the energy-level diagram appropriate for the entire crystal has a band of levels. The lowest energy band, called the *valence band* (Fig. 5-2*c*), is completely filled with electrons for there is one electron for each of the available energy levels. Conversely, the upper energy band is empty of electrons because it corresponds to the unoccupied higher levels in the isolated atom. It is called the *conduction band,* for reasons explained below. The energy region between the valence band and the conduction band is called the *forbidden energy gap* since no electrons with such energies exist in the crystal. The forbidden energy gap corresponds to the energy region between energy levels in the isolated atom, as can be seen by comparing the energy-level diagrams in Fig. 5-2.

This picture of the electronic energy levels in a crystal is known as the *energy-band model* of a crystal. It is very useful in determining the electrical properties of any solid, since it shows how electrons can move in the crystal. While the general features of the band model for any solid are as described, many important details depend upon the specific atomic and crystal structure. In particular, the differences between metals, semiconductors, and insulators are reflected in their energy-band models.

The atomic and crystal structures of metals are such that the valence and conduction bands overlap, as indicated in the conventional energy-band model for a metal, Fig. 5-3*a*. Since there is no forbidden energy

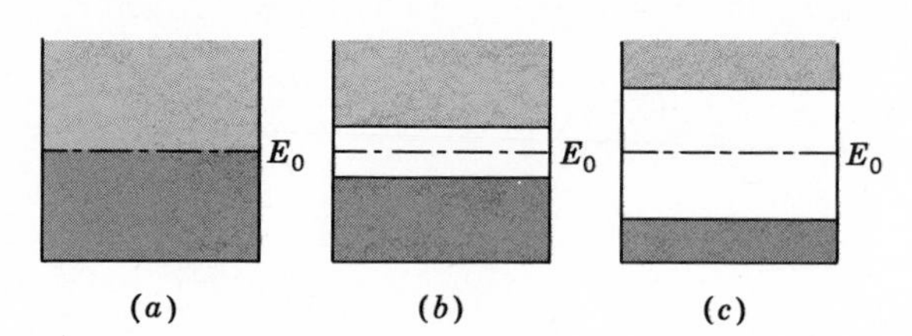

FIGURE 5-3 *Energy-band models for (a) metal, (b) semiconductor, and (c) insulator. Fermi energy level is indicated by* E_0.

gap in a metal crystal, any of the many valence electrons are free to roam throughout the solid and to move in response to an electric field. Therefore, metals are excellent electric conductors. Electrons in the bands are distributed in accordance with the Pauli exclusion principle. At absolute zero all electrons fill up the lowest levels, and the highest filled level is called the *Fermi level.* At normal temperatures some electrons at the highest energies are excited to levels slightly above the Fermi level by virtue of the heat energy in the crystal. Consequently, a few energy levels below the Fermi level are empty and a few energy levels above the Fermi level are filled. In this situation the Fermi level represents the energy at which the levels are half-filled and half-empty. The position of the Fermi level is indicated on the energy-level diagram of Fig. 5-3*a* by E_0.

An insulating crystal has a wide forbidden energy gap, Fig. 5-3*c*. The valence band is completely filled with electrons and the conduction band is completely empty. Obviously the upper band cannot contribute to electric conductivity since no electrons are present to act as carriers. It may seem paradoxical at first, but electrons in the completely filled valence band also cannot conduct electricity. When an electron moves in response to an electric field, it must move slightly faster than before. Consequently, it has greater energy and must find an empty level at a slightly higher energy. Every nearby level is filled, however, so that it is impossible for any electron in the filled valence band to be accelerated by the electric field. The crystal is therefore an insulator.

The energy-band model of a semiconductor. Fig. 5-3*b*, is similar to that of an insulator except that the forbidden energy gap is comparatively narrow. A few electrons can be promoted from the valence band to the conduction band across the forbidden energy gap by virtue of the thermal energy of the crystal at room temperature. Electrons promoted to the conduction band can conduct electricity. The corresponding electron vacancies in the valence band make it possible for electrons in this band to contribute to conductivity as well. Since the number of carriers is much fewer than in the case of a metal, semiconductors are poorer conductors than metals but better than insulators. A little thought shows that at very low temperatures a semiconductor becomes an insulator. Thermal energies at very low temperatures are insufficient to excite electrons across the forbidden energy gap. Conversely, at sufficiently high temperatures, even insulators conduct electricity because some electrons can be promoted from the valence band to the conduction band.

The width of the forbidden energy gap of semiconductors is of the order of 1 electronvolt, as shown in Table 5-1 for several typical semiconductor

TABLE 5-1 FORBIDDEN ENERGY GAPS OF TYPICAL SEMICONDUCTORS

Name	*Chemical symbol*	*Forbidden energy gap, eV*
Silicon	Si	1.1
Germanium	Ge	0.72
Gallium arsenide	GaAs	1.34
Indium antimonide	InSb	0.18
Cadmium sulfide	CdS	2.45
Zinc oxide	ZnO	3.3

crystals. The electronvolt, abbreviated *eV*, is equal to the kinetic energy gained by an electron in traversing a potential difference of 1 V. It is a convenient energy unit in semiconductor studies. In general, materials with a wide forbidden energy gap are desirable for semiconductor devices. The number of electrons promoted to the conduction band at high temperatures is small and the change in device characteristics with temperature is less severe when the forbidden energy gap is wide. For this reason, silicon crystals are more widely used than germanium crystals even though the latter are easier to prepare and less expensive.

Electrons and Holes According to the preceding discussion, the net current resulting from electrons in a filled valence band is zero. Formally, this may be written in terms of the current density using Eq. (1-13)

$$J = nev = 0 \tag{5-1}$$

where n is the density of electrons, e is the electronic charge, and v is the average velocity of the electrons in the band. Writing the average velocity explicitly as the average of the velocities of individual electrons,

$$J = ne\frac{1}{n}\sum_{i=1}^{n} v_i = e\sum_{i=1}^{n} v_i = 0 \tag{5-2}$$

Electrons in the valence band of a semiconductor at room temperature can conduct current because of the few vacant levels left behind by electrons excited to the conduction band. To show how this comes about, focus attention upon the jth electron in Eq. (5-2),

$$J = e\sum_{i=1}^{n} v_i = e\sum_{\substack{i=1\\ i\neq j}}^{n} v_i + ev_j = 0 \tag{5-3}$$

Rearranging,

$$e\sum_{\substack{i=1\\ i\neq j}}^{n} v_i = -ev_j \tag{5-4}$$

The left side of Eq. (5-4) is the current resulting from all the electrons in the valence band except for the jth one. The right side represents the current density due to one electron but of opposite electric charge. This vacant level in the valence band is called a *hole.* According to Eq. (5-4), holes in the valence band can be treated as positively charged carriers fully analogous to the negatively charged electrons in the conduction band.

The use of holes to represent the behavior of electrons in a nearly filled valence band introduces a considerable conceptual simplification. For example, the conductivity of a semiconductor crystal containing both electrons in the conduction band and holes in the valence band is, by analogy with Eq. (1-16),

$$\sigma = ne\mu_e + pe\mu_h \tag{5-5}$$

where n is the density of conduction-band electrons, p is the density of valence-band holes, and μ_e and μ_h are the respective carrier mobilities.

The energy-band model, Fig. 5-3*b*, refers to a perfect crystal structure which contains no chemical impurities and in which no atoms are displaced from their proper sites. The properties of the solid are therefore characteristic of an ideal structure, and the crystal is called an *intrinsic semiconductor.* Although it is not possible to achieve perfect structures in real crystals, this ideal may be approached and intrinsic behavior observed experimentally. The number of electrons in the conduction band of an intrinsic semiconductor is equal to the number of holes in the valence band, since both are the result of electron transitions across the forbidden energy gap. This means that the Fermi level is located in the center of the forbidden energy gap, as already indicated on the energy-band diagram. Note that the number of electrons and holes increases at high temperatures, and, according to Eq. (5-5), the conductivity also increases. This is characteristic of semiconductors.

Extrinsic Semiconductors The electrical properties of a semiconductor are drastically altered when foreign, or *impurity*, atoms are incorporated into the crystal. Since the properties now depend strongly upon the impurity content, the solid is called an *extrinsic* semiconductor. Consider, for example, a single crystal of the important semiconductor material silicon. Each silicon atom has four valence electrons which are part of the filled valence band of a silicon crystal. Suppose now a pentavalent atom, such as phosphorus, arsenic, or antimony, substitutes for a silicon atom in the crystal. Four of the impurity atom's electrons play the same role as the four valence electrons of the replaced silicon atom and become part of the valence band. The fifth valence electron is easily detached by thermal energy and moves freely in the conduction band.

Phosphorus, arsenic, or antimony impurity atoms in silicon donate electrons to the conduction band and are called *donor* impurities. The normal energy level for the extra electron is located at the impurity atom and slightly below the conduction band, Fig. 5-4. One electron is present

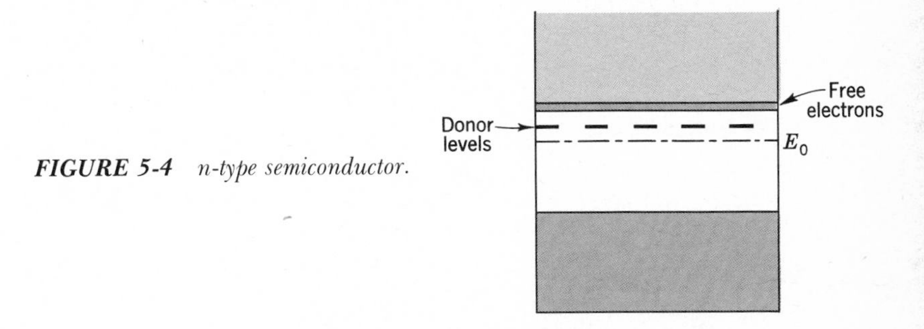

FIGURE 5-4 *n-type semiconductor.*

in the conduction band for each donor atom in the crystal. Note that there is not an equivalent number of holes in the valence band. The crystal therefore conducts electricity mainly by virtue of electrons in the conduction band. Such a crystal is called an *n-type* semiconductor because of the negative charge on the current carriers. The increased density of electrons in the conduction band means that the Fermi level is located near the donor levels, as indicated in Fig. 5-4.

By comparison, a trivalent atom like boron, aluminum, gallium, or indium substituted for a silicon atom produces a hole in the valence band. Since only three electrons are available, an electron from an adjacent silicon atom transfers to the impurity atom. The foreign atom is said to have accepted a valence electron, and such impurities are termed *acceptors*. Acceptors in a crystal create holes in the valence band and produce a *p-type* semiconductor because of the effective positive charge on each hole. Acceptor energy levels are located just above the valence band, Fig. 5-5. The Fermi level is located nearby since the number of holes in the valence band is greater than the number of electrons in the conduction band.

A crystal containing both donor and acceptor impurities is either *n*-

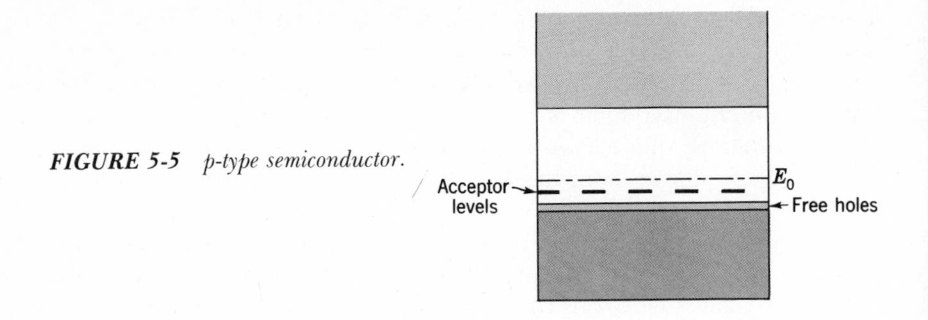

FIGURE 5-5 *p-type semiconductor.*

type or *p*-type, depending upon which impurity concentration is greater, because electrons from donor atoms fill up all available acceptor levels. Intrinsic crystals can be produced by including equal concentrations of donor and acceptor impurities, and such crystals are said to be *compensated.* Intrinsic characteristics can also be observed by heating an extrinsic crystal to a high temperature. A great many electrons are excited across the forbidden energy gap from the valence band to the conduction band so there are a great many holes and electrons. Furthermore, the numbers of holes and electrons are nearly equal and the crystal properties become essentially intrinsic.

Note that a few holes are present in the valence band of an *n*-type crystal since some electrons are excited across the forbidden energy gap. These holes are called *minority* carriers and the electrons are called *majority* carriers because of their relative concentration. Conversely, holes are majority carriers in a *p*-type semiconductor, while electrons in the conduction band are minority carriers. Intentionally introducing impurity atoms in semiconductors to obtain a desired concentration of majority carriers is called *doping.*

SEMICONDUCTOR DIODES

The pn Junction The junction between a *p*-type region and an *n*-type region in the same semiconductor single crystal, a basic structure in many devices, is called a *pn junction.* Several different techniques for producing *pn* junctions are described in a later section. All result in a transition from acceptor impurities to donor impurities at a given place in the crystal. Electrons in the *n* region tend to diffuse into the *p* region at the junction. At equilibrium, this is compensated by an equal flow of electrons in the reverse direction. The concentration of electrons is much larger in the *n* material, however, and the electron current from the *n* region would be greater except for a potential rise at the junction. This potential difference reduces the electron current from the *n* side. An identical argument applies to the hole currents diffusing across the junction. The magnitude and polarity of the internal potential rise at the junction make the two currents equal. The built-in potential rise leads to excellent rectification in the *pn* junction, as already noted in Chap. 4.

The Fermi level is continuous throughout the crystal in equilibrium, as shown in the energy-band model for a *pn* junction, Fig. 5-6. Notice that

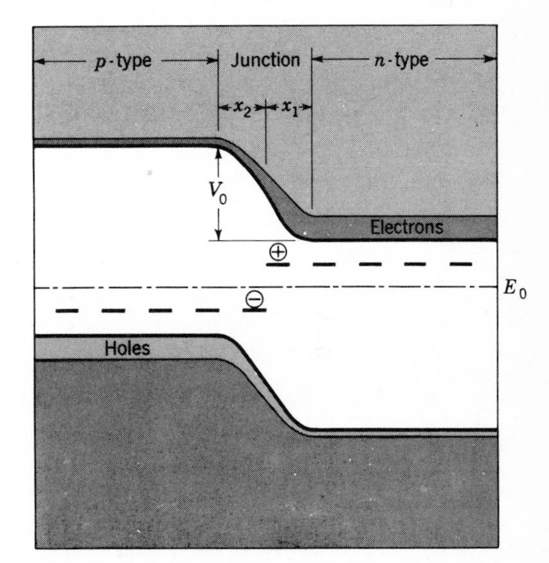

FIGURE 5-6 *Energy-band model of pn junction.*

the polarity of the potential rise tends to keep electrons in the *n* region and holes in the *p* region, as discussed above. The magnitude of the potential step is equal to the difference between the Fermi-level locations on the two sides of the junction. Therefore, the built-in potential of a *pn* junction depends upon the forbidden energy gap, the impurity concentration, and the temperature. In particular, at high temperatures, where the semiconductor becomes intrinsic and the Fermi level is at the center of the forbidden energy gap, the potential rise disappears. Proper device operation is impossible under these conditions.

The potential rise V_0 results from donor electrons at the junction that transfer into nearby acceptor levels. In the case of an *abrupt* junction, one in which the transition from *p*- to *n*-type occurs very suddenly in the crystal, the equality of charge transferred from donors to acceptors means that

$$N_d x_1 = N_a x_2 \tag{5-6}$$

where x_1 is the width of the junction in the *n* region, x_2 is the extent of the junction in the *p* region, N_d is the donor concentration, and N_a is the acceptor concentration. The magnitude of the potential rise in the *n* region

V_1 is determined by *Poisson's equation,* which is basically a reformulation of Coulomb's law relating electric field to charge distribution,

$$\frac{d^2V_1}{dx^2} = \frac{eN_d}{\epsilon_0 \kappa_e} \tag{5-7}$$

where κ_e is the dielectric constant of the semiconductor. The solution is

$$V_1 = \frac{eN_d}{2\epsilon_0 \kappa_e} x_1{}^2 \tag{5-8}$$

A similar expression

$$V_2 = \frac{eN_a}{2\epsilon_0 \kappa_e} x_2{}^2 \tag{5-9}$$

applies to the p region. The total width of the pn junction is then

$$d = x_1 + x_2 = \left(\frac{2\epsilon_0 \kappa_e}{e}\right)^{1/2} \left[\left(\frac{V_1}{N_d}\right)^{1/2} + \left(\frac{V_2}{N_a}\right)^{1/2}\right] \tag{5-10}$$

Using Eqs. (5-6) to (5-8),

$$\frac{V_1}{V_2} = \left(\frac{x_1}{x_2}\right)^2 \frac{N_d}{N_a} = \frac{N_a}{N_d} \tag{5-11}$$

Introducing $V_0 = V_1 + V_2$ and Eq. (5-11) into Eq. (5-10),

$$d = \left[\frac{2\epsilon_0 \kappa_e V_0}{e(N_a + N_d)}\right]^{1/2} \left[\left(\frac{N_a}{N_d}\right)^{1/2} + \left(\frac{N_d}{N_a}\right)^{1/2}\right] \tag{5-12}$$

If the donor concentration on the n side is much greater than the acceptor concentration on the p side, $N_d \gg N_a$, Eq. (5-12) reduces to

$$d = \left(\frac{2\epsilon_0 \kappa_e V_0}{eN_a}\right)^{1/2} \tag{5-13}$$

Furthermore, the potential rise is almost entirely confined to the p region, as can be seen from Eq. (5-11).

According to Eq. (5-13), the junction is narrow for high impurity concentrations. This principle is used to set the breakdown voltage of zener diodes at the desired potential. The avalanche breakdown process discussed in Chap. 4 depends upon the electric field in the junction. Therefore, the voltage at which breakdown occurs is small in narrow junctions and large in wide junctions.

When the pn junction is in equilibrium the current I_1 resulting from electrons diffusing from the n side is equal to the current I_2 which arises from electrons leaving the p side, Fig. 5-7*a*. Suppose now an external potential is applied to the junction so as to increase the internal barrier, as in Fig. 5-7*b*. The number of electrons diffusing across the junction from the n region is much reduced since very few electrons have sufficient energy

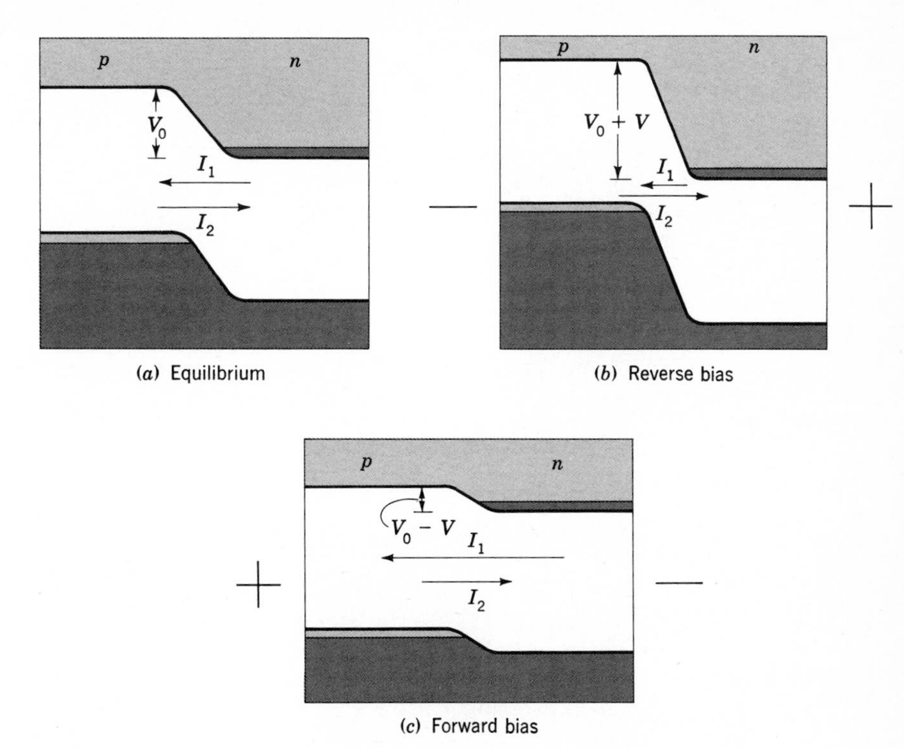

FIGURE 5-7 *Current across (a) pn junction depends upon height of potential rise at junction, which is (b) increased under reverse bias and (c) decreased by forward bias.*

to surmount the larger potential barrier. On the other hand, the number moving from the p to the n side is not affected because these electrons encounter no barrier. Thus a net current exists, but it is limited by the small number of electrons in the p region. If the polarity of the external potential is reversed, Fig. 5-7c, the internal barrier is reduced and I_1 is large because the number of electrons in the n region is so great. Again, the electron current I_2 from p-type to n-type remains unaffected. The net current in this case is large and corresponds to the forward direction. The reverse polarity increases the potential barrier and results in only a small current.

In determining the current-voltage characteristic of the junction diode, the same considerations apply to the current carried by the positive carriers as to that carried by electrons, and the total current is the sum of the two. Focusing attention on the electrons first, the current from the p region to the n region is proportional to the electron concentration in the p region n_p, so that

$$I_2 = -C_1 n_p \tag{5-14}$$

where C_1 is a constant involving the junction area and properties of the semiconductor crystal which are not of direct interest here. The minus sign in Eq. (5-14) accounts for the negative charge on the electron. According to the preceding discussion, I_2 is independent of the applied potential V. The current I_1, from the n side to the p side, is proportional to the number of electrons in the n region with sufficient energy to surmount the barrier. This number may be determined from the Boltzmann distribution, which relates the concentrations in two regions having different potential energies. That is,

$$I_1 = -C_1 n_0 e^{-e(V_0 - V)/kT} \tag{5-15}$$

where n_0 is the concentration of electrons in the n region, and the exponential represents the Boltzmann relation. When the applied potential is zero, $I_1 = I_2$. From Eqs. (5-14) and (5-15)

$$n_p = n_0 e^{-eV_0/kT} \tag{5-16}$$

The net electron current is therefore

$$\begin{aligned} I_n &= I_2 - I_1 \\ &= C_1 n_0 (e^{eV/kT} - 1) e^{-eV_0/kT} \\ &= C_1 n_p (e^{eV/kT} - 1) \end{aligned} \tag{5-17}$$

An identical expression can be derived for the positive carrier current. The result is

$$I_p = C_2 p_n (e^{eV/kT} - 1) \tag{5-18}$$

where p_n is the concentration of positive carriers in the n region and C_2 is a constant analogous to C_1. The total current is the sum of Eqs. (5-17) and (5-18),

$$\begin{aligned} I &= I_n + I_p \\ &= (C_1 n_p + C_2 p_n)(e^{eV/kT} - 1) \end{aligned} \tag{5-19}$$

$$I = I_0 (e^{eV/kT} - 1) \tag{5-20}$$

where I_0 is called the *saturation current*. Equation (5-20) is known as the *rectifier equation*.

The polarity of the applied potential is such that the p region is positive for forward bias. According to Eq. (5-20) the current increases exponentially in the forward direction. In contrast, the reverse current is essentially equal to I_0, independent of reverse potentials greater than a few volts. A plot of the rectifier equation for small values of applied voltage is shown in Fig. 5-8. It turns out that experimental current-voltage characteristics of practical junction diodes are in good agreement with the rectifier equation. The similarity between the current-voltage characteristic of a junction diode, Fig. 5-8, and that of an ideal diode, Fig. 4-1, is quite apparent.

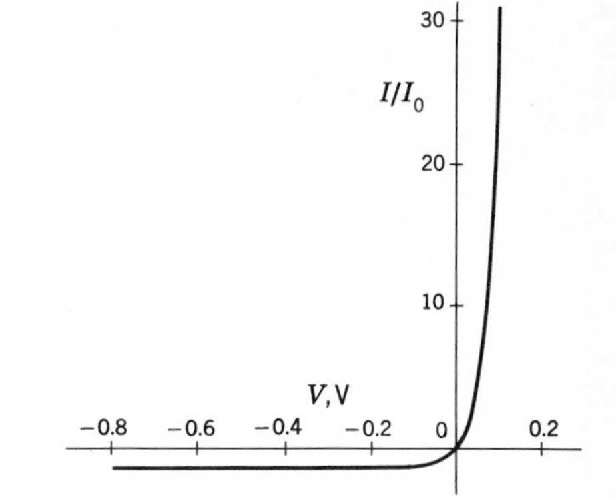

FIGURE 5-8 *Current-voltage characteristic of pn junction according to rectifier equation.*

Varicap Diodes A *pn* junction is a double layer of opposite charges separated by a small distance and so has the properties of a capacitance. Furthermore, the capacitance can be varied by applying a reverse-bias voltage. This is so because reverse bias increases the magnitude of the potential rise at the junction, and the width of the junction increases, according to Eq. (5-13). The junction capacitance is calculated from the expression for a parallel-plate capacitor, Eq. (2-21),

$$C = \frac{\epsilon_0 \kappa_e A}{d} \tag{5-21}$$

Substituting for d from Eq. (5-13),

$$C = A\left(\frac{\epsilon_0 \kappa_e e N_a/2}{V_0 + V}\right)^{1/2} \tag{5-22}$$

where the reverse-bias potential V is explicitly indicated. According to Eq. (5-22) the junction capacitance decreases as the reverse-bias potential is increased.

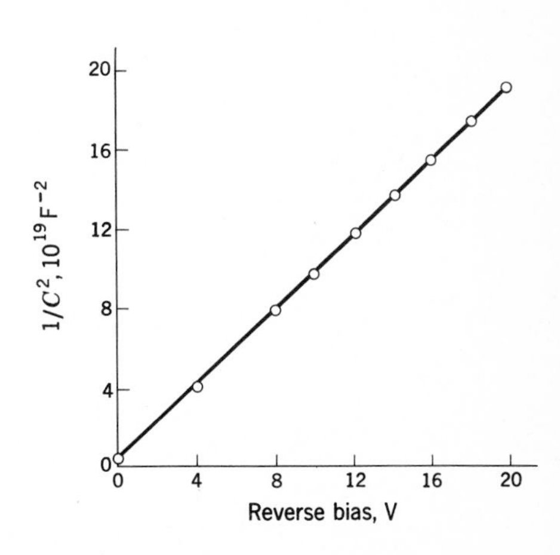

FIGURE 5-9 *Variation of capacitance of varicap diode with reverse-bias potential. Linear trend is in agreement with Eq. (5-22).*

The variation of capacitance with reverse bias for a typical *varicap diode,* Fig. 5-9, is in good agreement with Eq. (5-22). The name of this useful semiconductor device describes the ability of the *pn* junction to adjust one of its electrical parameters, capacitance, in response to an applied voltage. Parametric diodes are used as electrically controlled tuning capacitors in radio receivers to replace conventional variable capacitors, particularly in automatically or remotely tuned receivers, and also in so-called parametric amplifiers, which are very sensitive microwave amplifiers.

Tunnel Diodes A very useful effect occurs in *pn* junctions in which the impurity concentrations on both the *n* and *p* sides are very great. The carrier concentrations are large and the Fermi level lies in the valence and conduction bands on the two sides of the junction. The energy-band model of the junction is shown in Fig. 5-10*a* for the case of zero applied voltage.

FIGURE 5-10 *Energy-band model of tunnel diode (a) in equilibrium, (b) at small forward bias, and (c) at increased forward bias. Note current is greater at (b) than at (c) because of electron tunneling.*

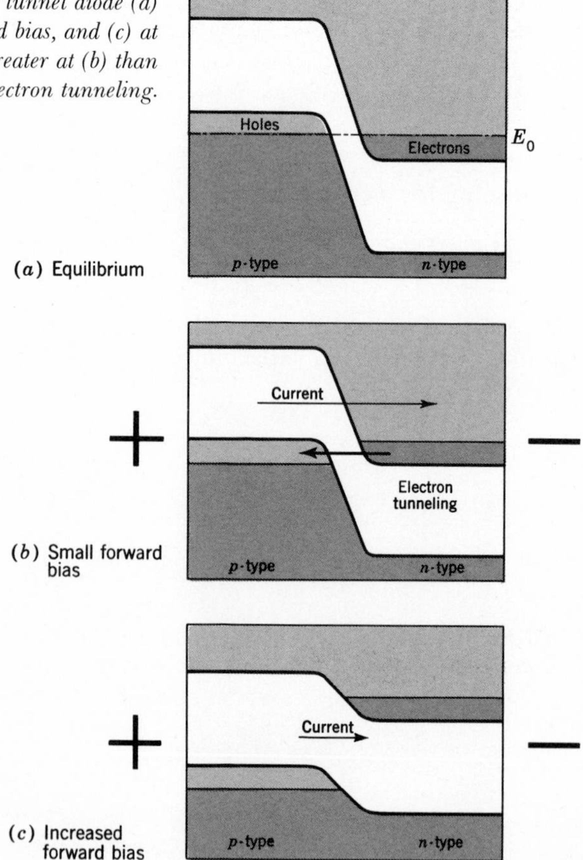

Very large impurity concentrations also mean that the junction is very narrow, according to Eq. (5-12). Widths of the order of 10 to 100 atomic diameters are easily achieved in germanium diodes and similar devices. In this situation it is possible for an electron in the conduction band on the n side to jump to the valence band on the p side by a process called electron tunneling. The tunneling transition takes place with no change in the energy of the electron. The ability of an electron to tunnel through the potential barrier of the junction is a result of the wavelike nature of the electron. The probability that an electron will penetrate a potential barrier can be calculated from the principles of quantum mechanics. The tunneling current depends upon the junction width, upon the number of electrons capable of tunneling, and upon the number of empty energy levels into which they can transfer. In equilibrium, Fig. 5-10*a*, the tunneling currents across the junction in the two directions are equal and the net current is zero.

At a small forward-bias potential, electrons tunnel from the conduction band on the n side to empty levels in the valence band on the p side and a forward current results. The current increases with voltage until the electrons on the n side are in line with the holes on the p side, Fig. 5-10*b*. As the bias is increased beyond this point, electrons in the conduction band are raised above the valence-band states and the tunnel current is reduced. In this range of forward bias each increment in voltage causes a decrease in current. Finally, Fig. 5-10*c*, the electrons and holes are completely out of line and the junction current corresponds to the normal forward *pn*-junction current.

The current-voltage characteristic of a *tunnel diode* (also referred to as an *Esaki diode* after its discoverer), Fig. 5-11, shows how the current rises to a maximum, corresponding to the condition of Fig. 5-10*b*, and then decreases as the forward bias is increased. The interval between the peak and valley of the characteristic curve exhibits a negative resistance effect since each voltage increase reduces the current. The usefulness of this effect

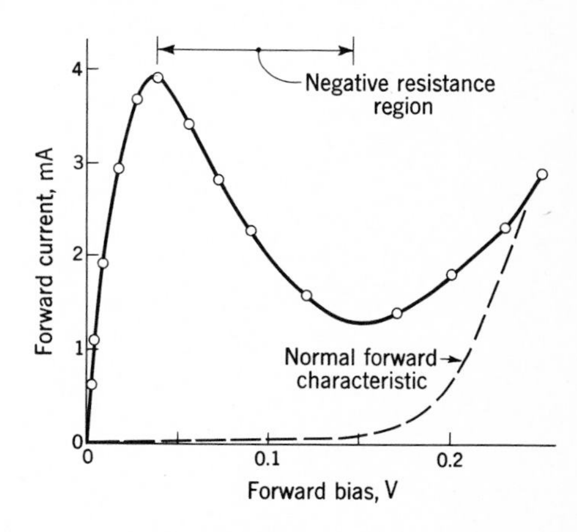

FIGURE 5-11 *Experimental current-voltage characteristic of tunnel diode. Negative resistance region exists between peak and valley currents.*

stems from the fact that all conventional electronic circuit components have positive resistance and therefore dissipate power. If a tunnel diode is placed in resonant circuit so that the net resistance vanishes, no power loss results. The circuit therefore oscillates at its resonant frequency, as analyzed in Chap. 9.

The tunnel diode makes possible very simple ac generator circuits which are particularly useful at extremely high frequencies. Tunneling takes place essentially instantaneously so that excellent high-frequency performance is achieved. Circuits that oscillate at frequencies as high as 10^{11} Hz have been designed. Furthermore, the input power requirements are extremely small, as can be seen from Fig. 5-11.

THE VACUUM DIODE

Thermionic Emission The basis of the vacuum diode considered in the previous chapter is the emission of electrons from a hot cathode. Such *thermionic emission* may be understood in the following way. As indicated schematically in Fig. 5-2, a potential-energy barrier that prevents electrons from escaping is present at the surface of any material. The origin of the barrier is the electrostatic attraction of the positively charged atomic nuclei for negatively charged electrons. The potential-energy difference between the Fermi level and an electron's energy outside of the solid, that is, the height of the potential-energy barrier at the surface, is called the *work function*, V_w of the material. This is the minimum amount of energy necessary to cause an electron to escape from the solid when the material is at a temperature of absolute zero. At higher temperatures some electrons have energies above the Fermi level, and at sufficiently high temperatures some electrons may acquire energies greater than the work function and so escape.

The relation between the concentration of electrons immediately outside of the solid and the density of the most energetic electrons inside, n_0, is given by the Boltzmann distribution,

$$n = n_0 e^{-eV_w/kT} \tag{5-23}$$

In the case of metals, n_0 can be calculated from first principles. From this calculation, and accounting for the fact only those electrons in the solid with surface-directed velocities can escape, the thermionic-emission current density is calculated to be

$$J = A_0 T^2 e^{-eV_w/kT} \tag{5-24}$$

where the quantity A_0 is a universal constant for metals with the value

$$A_0 = \frac{4\pi mek^2}{h^3} = 1.2 \times 10^5 \text{ A/(m}^2\text{)(K}^2\text{)} \tag{5-25}$$

Equation (5-24) is the *Richardson-Dushman equation* for thermionic emission. The emission current depends exponentially upon the work function and the absolute temperature, which means that the variation of emission with these quantities is exceedingly rapid. Experimental measurements of thermionic emission agree quantitatively with these results.

Typical values of work function of the practical cathode materials used in vacuum diodes are listed in Table 5-2. These values represent about

TABLE 5-2 TYPICAL WORK FUNCTIONS

Cathode material	*Work function, eV*
Tungsten	4.5
Thoriated tungsten	2.6
Oxide cathode	1.0

the extremes found for most solids. In particular the very low work function in the case of the oxide cathode makes it an efficient emitter of electrons at relatively modest temperatures.

Child's Law The Richardson-Dushman equation gives the maximum electron current emitted from a cathode at a given temperature. In the case of the vacuum diode the emitted electrons are attracted to the positive anode and constitute the anode current. Most often, the number of electrons emitted from the cathode is so great that the electric fields due to the electron charges drastically alter the electric field between the cathode and anode produced by the anode potential and a much smaller anode current results. The actual current under these conditions is determined in the following manner.

For simplicity, consider a plane cathode and a plane parallel anode separated by a distance d, as in Fig. 5-12. Assume that the potential dif-

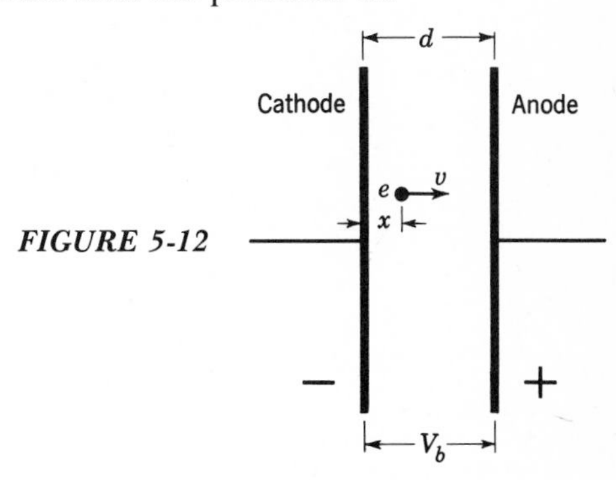

FIGURE 5-12

ference between anode and cathode is V_b and that electrons are emitted from the cathode with zero velocity, which is sufficiently true for present purposes. The effect of all the electrons is determined by applying Poisson's equation. The solution of Poisson's equation gives the potential V at any point in a region containing a volume density of charge ne. Here, n is the number of electrons per unit volume, and e is the charge on each electron. In one dimension, suitable for Fig. 5-12, Poisson's equation is written

$$\frac{d^2V}{dx^2} = -\frac{ne}{\epsilon_0} \tag{5-26}$$

The current density between cathode and anode is, from Eq. (1-13),

$$J = -nev \tag{5-27}$$

where v is the velocity of the electrons and the minus sign accounts for the negative charge of the electrons. According to the definition of potential (work per unit charge), the kinetic energy of an electron at any point is related to the potential at that point by

$$\tfrac{1}{2}mv^2 = eV \tag{5-28}$$

where m is the electronic mass. Introducing the current density from Eq. (5-27) and v from Eq. (5-28), Poisson's equation becomes

$$\frac{d^2V}{dx^2} = \frac{J}{\epsilon_0 v} = \frac{J}{\epsilon_0}\left(\frac{m}{2eV}\right)^{1/2} \tag{5-29}$$

The solution to this differential equation for V is

$$V^{3/4} = \frac{3}{2}\left(\frac{J}{\epsilon_0}\sqrt{\frac{m}{2e}}\right)^{1/2} x \tag{5-30}$$

which can be verified by direct substitution. In arriving at Eq. (5-30) it has been assumed that the cathode is at zero potential and that the electric field at the cathode is zero. These approximations are quite accurate in practical diodes.

Since we are interested in the current between cathode and anode, put $x = d$ and $V = V_b$ in Eq. (5-30) and solve for the current density. The result is

$$J = \frac{4\epsilon_0}{9d^2}\left(\frac{2e}{m}\right)^{1/2} V_b{}^{3/2} \tag{5-31}$$

This equation, known as *Child's law*, shows that the current varies as the three-halves power of the voltage. The current-voltage characteristic of a vacuum diode is therefore a horizontal line in the reverse direction and Eq. (5-31) in the forward direction, as shown in Fig. 5-13.

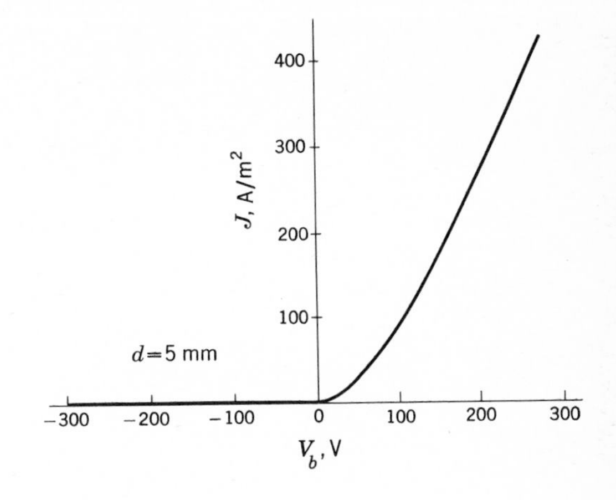

FIGURE 5-13 Current-voltage characteristic of vacuum diode according to Child's law. Anode-cathode distance is chosen to be 5 mm.

Actually the forward characteristics of practical vacuum diodes depart somewhat from Eq. (5-31) because the electrodes are normally cylindrical rather than planar and because of the simplifications introduced in deriving Child's law. The important features of the current-voltage characteristic of vacuum diodes are adequately described by Child's law, however, as is illustrated by comparing Fig. 5-13 with the characteristics of practical diodes given in Fig. 4-4.

VACUUM TUBES

The current in a vacuum diode is controlled by introducing a third electrode, the grid, between the anode and the cathode. Changes in grid voltage alter the anode current independently of the anode potential. This action makes the triode vacuum tube a useful amplifier.

The Grid The grid in a vacuum triode usually consists of a wire helix surrounding the cathode, as sketched in Fig. 5-14*a*. If the grid potential is always negative, electrons are repelled and the grid current is negligible. This means that the power expended in the grid circuit to control the anode current is very small. It is desirable to minimize the area of the grid, so the number of electrons intercepted on their way to the anode is negligible. On the other hand, if the wires are spaced too widely, the grid's ability to control the anode current is reduced. Practical triodes are designed to strike a useful compromise between these conflicting requirements. The anode in vacuum tubes is commonly termed the *plate*, because of its shape in early tube designs. The conventional circuit symbol for a triode is shown in Fig. 5-14*b*.

The grid potential alters the electric-field configuration in the space between the cathode and plate from that corresponding to the vacuum

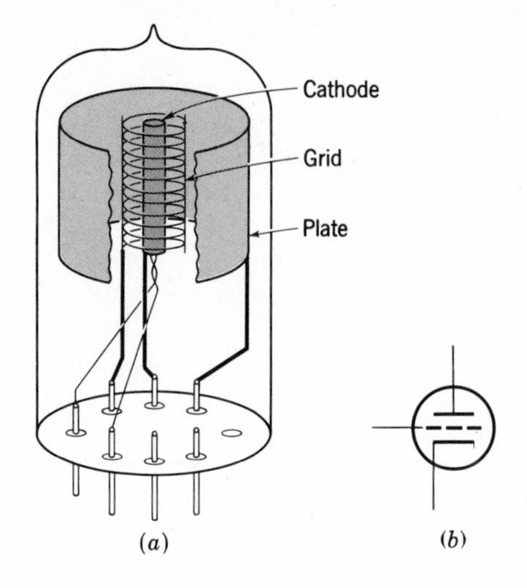

FIGURE 5-14 *(a) Vacuum triode and (b) circuit symbol.*

diode. The manner in which this happens can be described in the following way. According to the derivation of Child's law, the plate voltage sets the current so that the electric field near the cathode is very small. The grid potential has a similar effect. Under the combined influence of the grid potential and the plate potential, the tube current is such that the electric field at the cathode remains small. Since the grid is much closer to the cathode, its potential is relatively more effective in controlling the current than is the plate voltage. Therefore, using Child's law, the current in the tube may be written[1]

$$I_b = A(\mu V_c + V_b)^{3/2} \tag{5-32}$$

where A is a constant involving the tube geometry, V_c is the grid voltage, V_b is the plate voltage, and μ is a constant called the *amplification factor*. The amplification factor accounts for the greater effect of the grid voltage compared with the plate voltage. This equation agrees reasonably well with experimental current-voltage characteristics of practical triodes. It is difficult to evaluate A from first principles, however, and the exponent is often not exactly three-halves. Therefore, it is common practice to display the characteristics of practical tubes graphically, rather than attempt an accurate mathematical representation.

The action of V_c in Eq. (5-32) opposes that of the plate voltage V_b since the grid potential is negative. An expression for the amplification factor is obtained by noting that these opposing actions cancel out if the electric charge induced on the cathode by the grid potential is equal and opposite to the charge produced by the plate voltage, or

[1]It has become commonly accepted to associate the subscripts p and b with the plate circuit and g and c with the grid circuit of a vacuum tube. Similarly, the subscript k refers to the cathode.

$$-C_{gk}V_c + C_{pk}V_b = 0 \tag{5-33}$$

where C_{gk} is the capacitance between the grid and cathode. Solving for the amplification factor

$$\mu = \frac{V_b}{V_c} = \frac{C_{gk}}{C_{pk}} \tag{5-34}$$

According to Eq. (5-34) the amplification factor is increased if the grid is close to the cathode since the grid-cathode capacitance is increased. The plate is further away, and is shielded from the cathode by the grid, so that μ is always greater than unity. Actually, triodes with amplification factors ranging from 10 to 100 are commercially available.

Of the several graphical ways to represent the current-voltage characteristics of a triode, the most useful is a plot of the plate current as a function of plate voltage for fixed values of grid voltage. The curves for several grid potentials are called the *plate characteristics* of the tube. A typical set of plate characteristics, as provided by the tube manufacturer, is illustrated in Fig. 5-15. Note that each curve is similar to the current-voltage curve of a vacuum diode. Furthermore, the current at a given plate voltage is

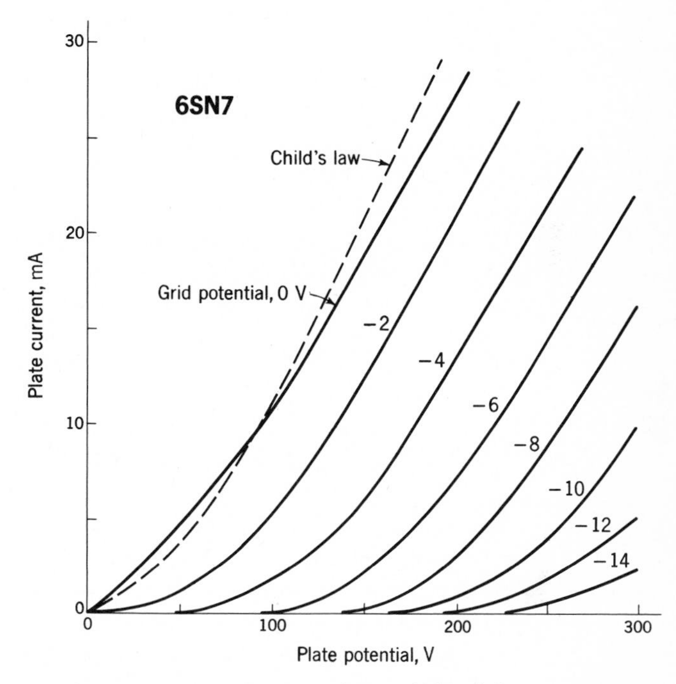

FIGURE 5-15 *Plate characteristics of type 6SN7 triode. Curve for zero grid voltage is in approximate agreement with Child's law.*

reduced as the grid is made increasingly negative. The plate current-voltage curves are displaced to the right with little change in shape for each negative increment in grid potential, in conformity with Eq. (5-32). Comparing Eq. (5-32) with the experimental curves shows why it is necessary to use graphical data: although Child's law represents the general behavior of the plate characteristics, it is not sufficiently accurate to yield satisfactory quantitative results.

According to Fig. 5-15, the plate current is essentially zero at sufficiently large negative grid potentials. The negative grid voltage necessary to put the tube in this *cutoff* condition depends upon the plate voltage. In effect, the tube is an open circuit when cut off. If the grid is at a positive potential, there is appreciable grid current. Then the grid is the anode of a diode biased in the forward direction and represents a much lower resistance than is the case for a negative grid voltage. In most circuits this effect prevents the grid from ever becoming positive. Therefore, at zero grid voltage the tube is said to be *saturated* since it is in its maximum conducting condition. The range between cutoff and saturation is the normal grid-voltage range.

Pentodes The capacitance between the grid and plate in a vacuum triode introduces serious difficulties when the tube is used as an amplifier at high frequencies. The ac plate signal introduced into the grid circuit through the grid-plate capacitance interferes with the proper operation of the circuit. Another grid is interposed in the space between the grid and plate to circumvent this difficulty. This *screen grid* is an effective electrostatic shield that reduces the grid-plate capacitance by a factor of 1000 or more. The screen grid is at a positive potential with respect to the cathode, and the current of electrons from cathode to plate is maintained. The helical grid winding is much more open than is the case for the control grid, so that the screen current is smaller than the plate current.

Two-grid four-electrode tubes called *tetrodes* are virtually obsolete except for certain special-purpose types. The reason for this is that electrons striking the plate dislodge other electrons. These may be attracted to the screen, particularly when the screen voltage is greater than the instantaneous plate potential. This introduces serious irregularities in the plate characteristics at low plate voltages. A *suppressor grid* is introduced between the screen and plate to eliminate this effect. The suppressor is held at cathode potential and effectively prevents all electrons dislodged from the plate from reaching the screen. The pitch of the suppressor grid helix is even larger than that of the screen grid. Therefore, the suppressor does not interfere with electrons passing from cathode to plate.

A tube with three grids is called a *pentode* because there are a total of five electrodes. The conventional circuit symbol of a pentode is shown in Fig. 5-16. In normal operation the screen and suppressor grids are

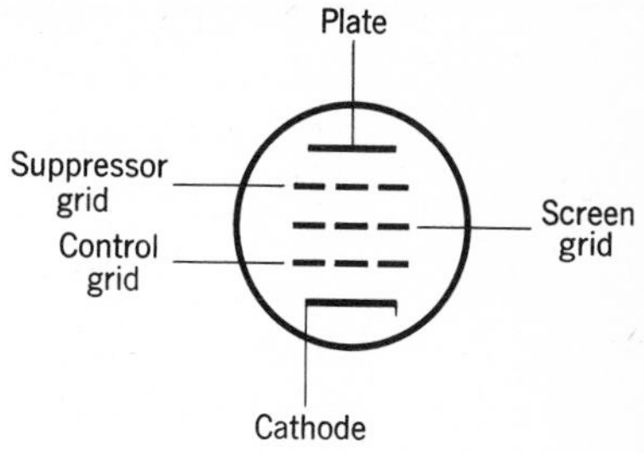

FIGURE 5-16 *Circuit symbol for pentode.*

maintained at fixed dc potentials. The influence of the plate potential on the electric field near the cathode is practically zero in pentodes, because of the shielding action of the screen and suppressor. This means that plate-voltage changes cause little or no change in the plate current. Since the plate current is almost independent of the plate voltage, the plate characteristics are nearly straight lines parallel to the voltage axis. The action of the control grid is essentially the same as in a triode, however.

Typical plate characteristics of a pentode for a given positive-screen potential and for the suppressor connected to the cathode are illustrated in Fig. 5-17. Note that the plate current is relatively independent of plate

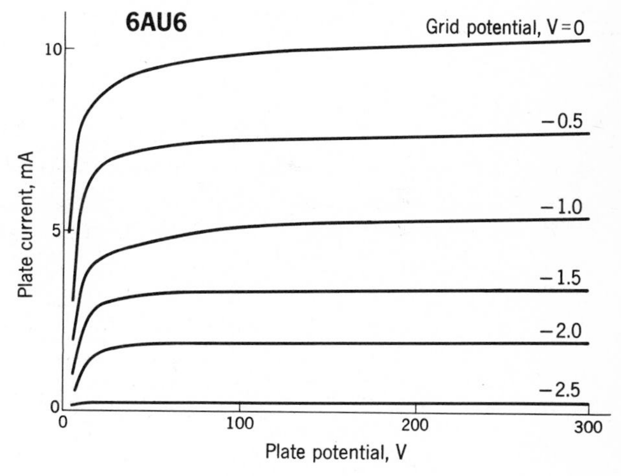

FIGURE 5-17 *Plate characteristics of type 6AU6 pentode.*

voltage, as anticipated above. Actually, however, the plate characteristics also depend upon the fixed screen potential. Higher screen voltages shift the curves in Fig. 5-17 upward to higher current values with little change in shape. It is common practice to specify pentode characteristics at two or three specific values of screen potential.

The screen voltage is maintained below the plate potential in most circuits, and it turns out that the screen current is about 0.2 to 0.4 of the plate current. As mentioned above, it is almost universal to connect the suppressor grid to the cathode; in many tubes this connection is made internally.

The pentode is extensively used in amplifier circuits because of its very high amplification factor. It surpasses the triode as a high-frequency amplifier where small grid-plate capacitance is important. Plate-voltage excursions can be nearly as large as the plate supply voltage without introducing excessive distortion, so high-power operation is possible. Finally, the pentode is a useful constant-current source, because the plate current is essentially independent of the plate voltage.

Other Multigrid Tubes Many other vacuum-tube designs suitable for special circuit applications have been developed. A pentode having a control grid wound with a helix of varying pitch has an amplification factor that depends markedly upon the grid bias. This is so because electrons are controlled best in the region of the grid where the spacing is small. At sufficiently large values of negative grid bias the electron stream is cut off at this portion of the grid. Therefore, the amplification factor of the tube corresponds to that of a vacuum tube whose grid wires are widely spaced. At less negative biases the amplification factor is that of a tube with narrow grid spacing and the amplification factor is larger. These *variable-*μ tubes are used in automatic volume-control circuits where the amplification of the tube is automatically adjusted by controlling the dc grid voltage. The voltage is adjusted to maintain the output signal constant against variations in input signal. These designs are also called *remote-cutoff tubes,* since a very large negative grid voltage is required to reduce the plate current to a small value.

More than one electrode structure may be included within a single envelope where two tube types are often used together in circuits. An example of this is the dual vacuum diode for full-wave rectifier circuits discussed in the previous chapter. Other common structures are the dual triode and the dual diode-triode. Many other combinations are possible and have been constructed. The symbol for a dual triode is indicated in Fig. 5-18*a*. This symbol also illustrates the method commonly employed to indicate the tube base pin connections to various electrodes. The base pins are numbered clockwise when viewed from the bottom. This convention facilitates identification of the various electrodes when the tube socket is examined in an actual circuit.

Vacuum tubes with more than three grids have been designed for special purposes. An example is the *pentagrid converter,* Fig. 5-18*b*. This tube has two control grids, g_1 and g_3, shielded from each other by two screen grids, g_2 and g_4. The fifth grid is the suppressor, g_5. This tube is used in frequency-converter circuits where two signals of different frequencies are

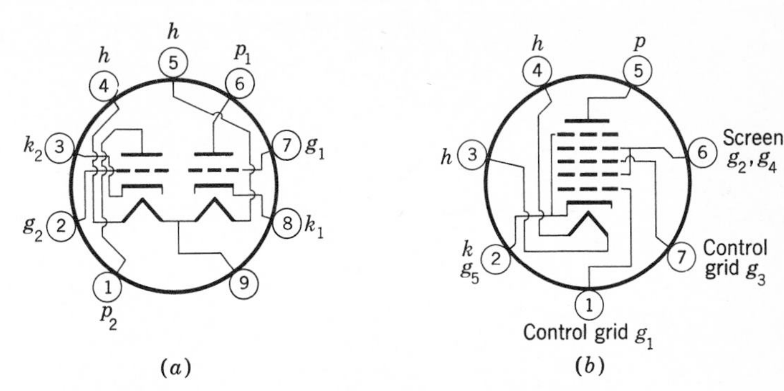

FIGURE 5-18 (a) Base diagram of 12AX7 dual triode; (b) base diagram of 6BE6 pentagrid converter.

applied to the two control grids. Nonlinearities of the tube characteristics result in sum and difference frequencies analogous to the diode first detector described in Chap. 4. Actually, the pentagrid converter can be used also to generate one of the signals within the tube itself, so only the input signal need be supplied externally.

In certain *beam power tubes,* most notably the type 6L6, the screen-grid wires are aligned with wires of the control grid. This forms the electron stream into sheets and reduces the screen current by a factor of 5 or so below that of conventional pentodes. There is no suppressor grid as such, but beam-forming plates at cathode potential further shape the electron stream. This specific electrode design causes the electron charges in the beam to produce an effective suppressor action. The result is that the plate characteristics are straight nearer to the $V_b = 0$ axis than is the case for a standard pentode. Accordingly, the beam power tube is useful as a power amplifier since the allowable plate-voltage range is nearly as large as the plate supply voltage. In addition, such tubes are often constructed with sturdy cathodes and anodes, so high-current operation is possible.

FIELD-EFFECT TRANSISTORS

Drain Characteristics Through the function of a suitable control electrode it proves as possible to influence the motion of free electrons inside semiconductors as it is in the empty space of a vacuum triode. Very effective action is obtained in the *field-effect transistor,* or *FET,* in which the flow of majority carriers is controlled by signal voltages applied to a reverse-biased *pn* junction. Consider the *n*-type semiconductor bar (most often of silicon) in Fig. 5-19 which has an ohmic contact, the *source,* at one end and a similar contact, the *drain,* at the other. Electrons moving from the source to the drain in response to the drain voltage V_d pass through a channel between two *p*-regions. This *pn* junction is called the *gate* because

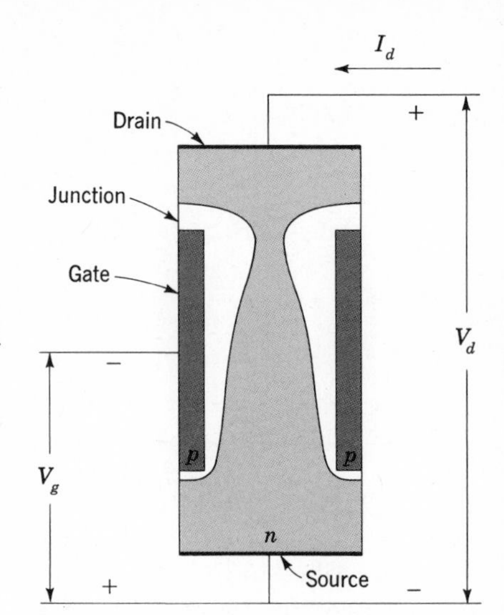

FIGURE 5-19 *n-channel junction FET.*

the width of the reverse-biased gate junction determines the width of the channel and consequently the magnitude of the current between source and drain. Signal voltages applied to the gate result in corresponding variations in drain current.

In effect the FET is a variable resistor which is controllable by the electric fields associated with the *pn* junction, whence the quite descriptive name. Very little power is expended by the applied signal because of the small reverse current in a *pn* junction. This means that the FET is an effective amplifier. Compared to the vacuum triode the FET is much smaller and more rugged because the entire structure is contained within one solid bar. Since no heater power is required to produce free electrons, the device is also much more efficient. Note that the unit described is an *n*-channel FET. Clearly a *p*-channel FET is equally possible. In this case the potentials applied to the drain and gate have the opposite polarity so that the gate junction is again under reverse bias. The circuit symbol for an *n*-channel FET is shown in Fig. 5-20. The corresponding symbol

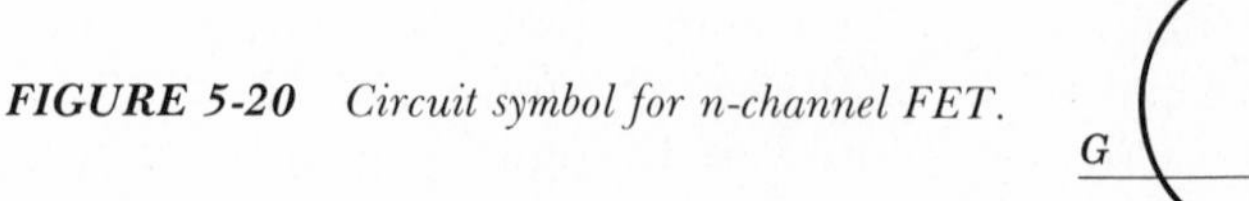

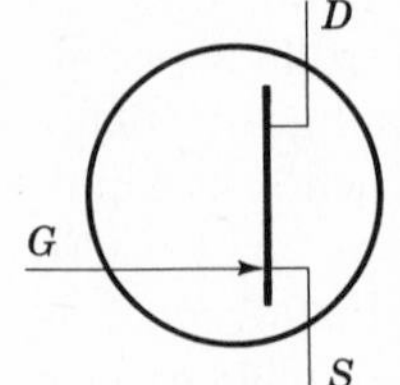

FIGURE 5-20 *Circuit symbol for n-channel FET.*

for a *p*-channel unit has the arrow representing the gate junction pointing in the other direction. Note that the direction of the arrow is consistent with the direction of conventional current flow in a junction diode.

The current-voltage characteristic of a FET is derived from an analysis of current flow in the channel. As sketched in Fig. 5-19 the width of the junction is greatest near the drain end of the gate because the reverse bias here is the sum of the gate potential plus the drain potential. The junction is narrowest at the source end because the voltage drop along the channel means that the reverse bias is essentially due to the gate potential alone. The sum of V_g and V_d may be large enough for the junction to completely pinch off the channel. It turns out, in fact, that this is the normal operating condition and the reverse bias required for pinchoff is labelled V_p.

The nonuniform voltage drop along the channel and the nonlinear variation in channel width combine to make an exact analysis complicated. It is most convenient to adopt a simplified approach and write the current in the channel in terms of the total charge of the carriers in the channel and their transit time from source to drain. Using Eq. (1-10),

$$I_d = \frac{Q}{\tau} \tag{5-35}$$

The transit time can be expressed as the length of the channel divided by the average carrier velocity using the definition of mobility from Eq. (1-14),

$$\tau = \frac{L}{\mu(V_d/L)} = \frac{L^2}{\mu V_d} \tag{5-36}$$

If the drain voltage is zero, the total charge is zero when the channel is pinched off and $V_g = V_p$. Therefore,

$$Q = C(V_g - V_p) \tag{5-37}$$

where C is the junction capacitance. A finite value of drain voltage results in additional reverse bias varying from zero at the source end to $-V_d$ at the drain end. As an approximation, the added reverse bias is taken to be the simple average value $-V_d/2$. The drain current is then, from Eqs. (5-35) and (5-37),

$$I_d = \frac{C}{\tau}\left(V_g - V_p - \frac{V_d}{2}\right)$$

Using Eq. (5-36),

$$I_d = \frac{\mu C V_d}{L^2}\left(V_g - V_p - \frac{V_d}{2}\right) \tag{5-38}$$

According to Eq. (5-38) the drain current depends upon both the gate voltage and the drain voltage, analogous to Child's law for the triode. Below pinchoff the drain current increases nonlinearly with V_d.

At pinchoff the sum of the gate voltage and the drain voltage is equal to V_p. Therefore,

$$V_d = V_g - V_p \tag{5-39}$$

Inserting Eq. (5-39) into Eq. (5-38),

$$\begin{aligned} I_d &= \frac{\mu C}{L^2}(V_g - V_p)\left(V_g - V_p - \frac{V_g - V_p}{2}\right) \\ &= \frac{\mu C}{2L^2}(V_g - V_p)^2 = \frac{\mu C V_p^2}{2L^2}\left(1 - \frac{V_g}{V_p}\right)^2 \\ &= I_{dss}\left(1 - \frac{V_g}{V_p}\right)^2 \end{aligned} \tag{5-40}$$

Evidently the drain current does not depend upon the drain voltage above pinchoff. Equations (5-38) and (5-40) are a good representation of the *drain characteristics* of practical FETs, as illustrated in Fig. 5-21.

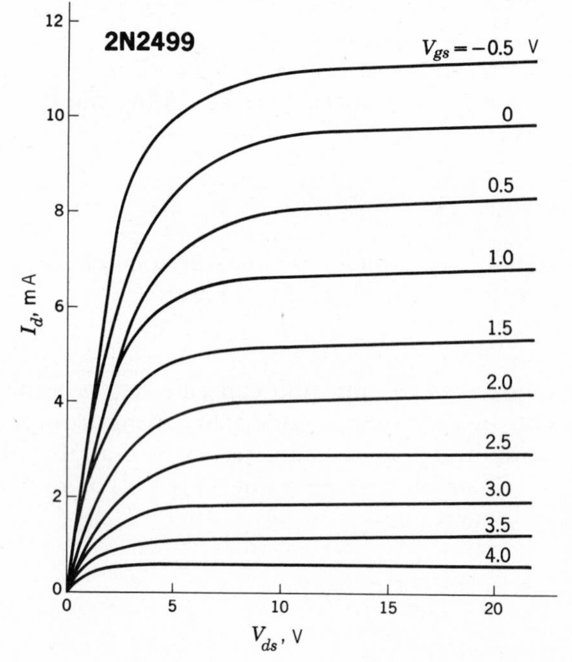

FIGURE 5-21 *Current-voltage characteristics of type 2N2499 FET.*

The drain current is quite independent of the drain voltage above pinchoff and the characteristics are reminiscent of the plate characteristics of a pentode. The maximum source-drain voltage is limited, however, by reverse breakdown at the gate junction. The quantities V_p and I_{dss} are parameters of a given geometrical design and I_{dss}, in particular, is often quoted in manufacturer's specifications.

IGFETs and MOSFETs Quite analogous control over current carriers in a semiconductor is possible without a *pn* junction by using a suitable metal gate electrode insulated from the semiconductor. Such insulated-gate field-effect transistors are known as *IGFETs*. The most successful practical units are made of silicon and employ a thin silicon dioxide layer as the insulator. The descriptive terminology for a metal-oxide-semiconductor transistor is *MOSFET*.

The MOSFET illustrated in Fig. 5-22*a* has *n*-type source and drain

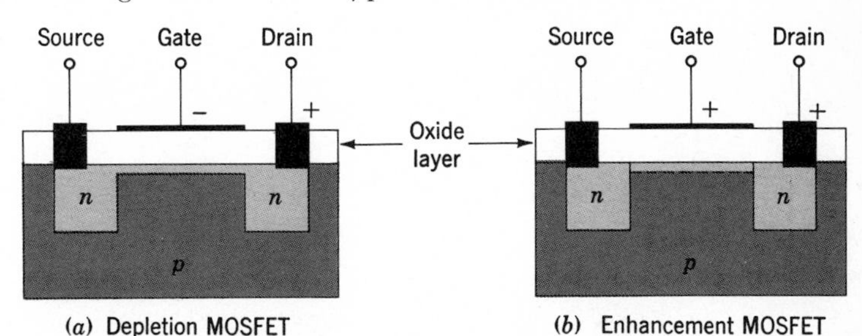

FIGURE 5-22 *n-channel MOSFETS. (a) Depletion and (b) enhancement versions.*

regions imbedded in a *p*-type layer. The metal gate electrode is deposited on top of a thin oxide layer on the surface. This structure is fabricated by methods described in a later section and very small-sized units can be manufactured. Usually the *p*-substrate is connected to the source and a negative potential is applied to the gate.

The negative gate potential repels electrons in the *n*-type channel and a *depletion* layer devoid of carriers forms. This depletion layer is quite analogous to the region between the *n*-type and *p*-type sides of a *pn* junction. Creation of the depletion layer by the gate voltage decreases the conductivity of the *n*-channel and the drain current is reduced. Pinchoff occurs when the depletion layer extends entirely across the *n*-layer and the drain characteristics of the *depletion* MOSFET are quite similar to those of the junction FET.

If the *n*-channel between source and drain is eliminated, Fig. 5-22*b*, successful operation is still obtained when the gate voltage is made positive with respect to the source and the *p*-type substrate. The drain current is zero without gate bias because of the *pn* junctions at the source and drain ends. As the gate potential is increased positively from zero, holes in the *p*-substrate immediately under the gate are repelled and the surface layer tends to become *n*-type. At the threshold gate voltage, V_T, an *n*-channel forms and current passes between source and drain. Greater gate voltage increases the drain current further. Since the gate voltage increases the *n*-channel conductivity, this design is called an *enhancement* MOSFET.

Actually, a depletion MOSFET can operate in the enhancement mode as well, since the channel can be widened as well as narrowed by the gate potential. A convenient comparison between the several possible FET types is provided by their *transfer characteristics.* These are plots of drain current as a function of gate voltage, Fig. 5-23. The junction FET employs

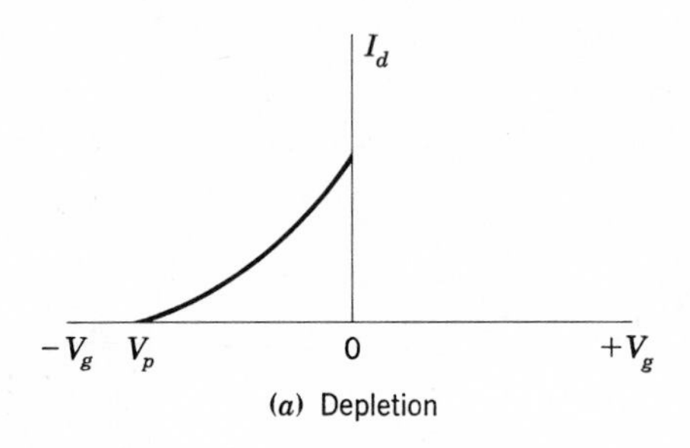

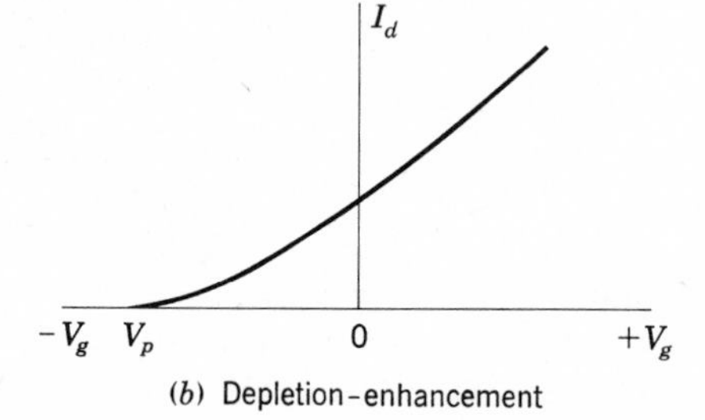

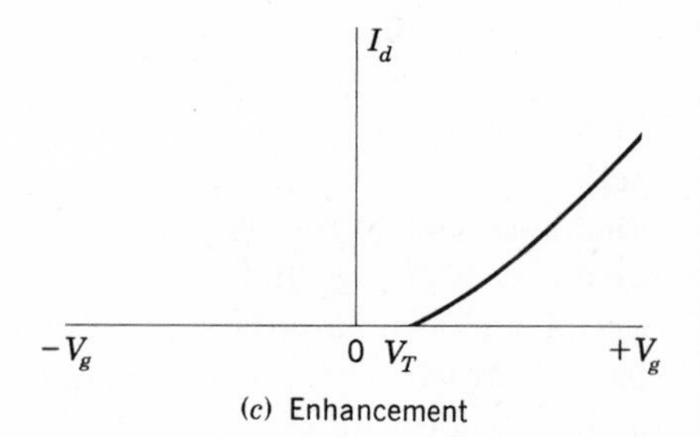

FIGURE 5-23 *Transfer characteristics of (a) depletion, (b) depletion-enhancement, and (c) enhancement FETs.*

the depletion mode, Fig. 5-23*a*, since the gate *pn* junction cannot have forward bias. The appropriate transfer characteristic is given by Eq. (5-40). The depletion MOSFET can be used with either gate polarity, Fig. 5-23*b* while the enhancement MOSFET requires positive gate bias, Fig. 5-23*c*. Note, however, that except for lateral displacement, all three transfer characteristics are quite similar.

The circuit symbol for *n*-channel MOSFETs, Fig. 5-24, suggests the insulated-gate structure. As might be anticipated, *p*-channel MOSFETs are indicated by reversing the direction of the arrowhead that suggests the direction of conventional current in the channel. Further evidence

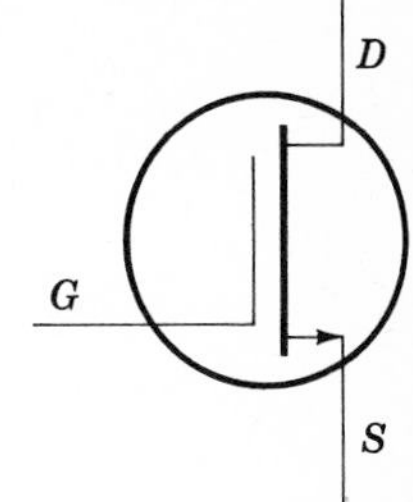

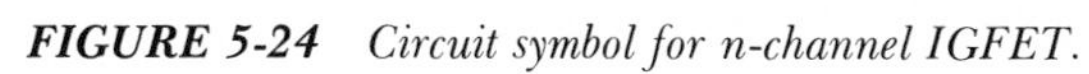
FIGURE 5-24 *Circuit symbol for n-channel IGFET.*

for the versatility of the MOSFET design is the *dual-gate* MOSFET in which two separate gate electrodes are disposed along the channel. A large n-type region similar to the source is placed between the two gates but is not externally connected. The gate closest to the drain isolates the control gate from the drain by reducing the gate-drain capacitance, performing the same function as the screen grid in a pentode. Much improved high-frequency performance is thereby obtained.

THE JUNCTION TRANSISTOR

Minority Carrier Injection The internal potential barrier in a pn junction is reduced when the junction is biased in the forward direction. The forward current results from holes diffusing across the junction from the p-type side and electrons diffusing across from the n-type side. The result is that holes are injected into the n region and electrons are injected into the p region, where in each case they are minority carriers. This *minority-carrier injection* at a pn junction is the basis for junction transistors, as described in the next section. It is usually desirable that the forward current be carried predominantly by either holes or electrons in order to enhance the injection effect. The way this is accomplished can be seen from the rectifier equation, Eq. (5-19),

$$I = (C_1 n_p + C_2 p_n)(e^{eV/kT} - 1) \tag{5-41}$$

where, it is recalled, n_p is the equilibrium concentration of electrons in the p region and p_n is the equilibrium concentration of holes in the n region. If the n region is lightly doped and the p region is heavily doped, $n_p \ll p_n$, Eq. (5-41) reduces to

$$I = C_2 p_n (e^{eV/kT} - 1) \tag{5-42}$$

This means that the forward current is carried by holes and, consequently, a large excess hole concentration is injected into the n region. If the doping ratio is interchanged, the reverse situation is true and electrons are injected into the p region.

The concentration of holes injected into the n region, $p(0)$, is simply the number in the p region with sufficient energy to surmount the reduced barrier. This is, from the Boltzmann distribution,

$$p(0) + p_n = \frac{I_2}{C_2} = p_0 e^{-e(V_0 - V)/kT} \tag{5-43}$$

where p_0 is the concentration of holes in the p region. The injected holes diffuse away from the junction because of their concentration gradient. As they diffuse, they combine with the majority carriers, electrons, so that far from the junction the hole concentration is characteristic of the n-type semiconductor. The excess hole concentration $p(x)$ decreases in moving away from the junction according to

$$p(x) = p(0)e^{-x/Lp} \tag{5-44}$$

where the constant L_p is called the *diffusion length*. The concentration of holes in the various regions is indicated schematically in Fig. 5-25.

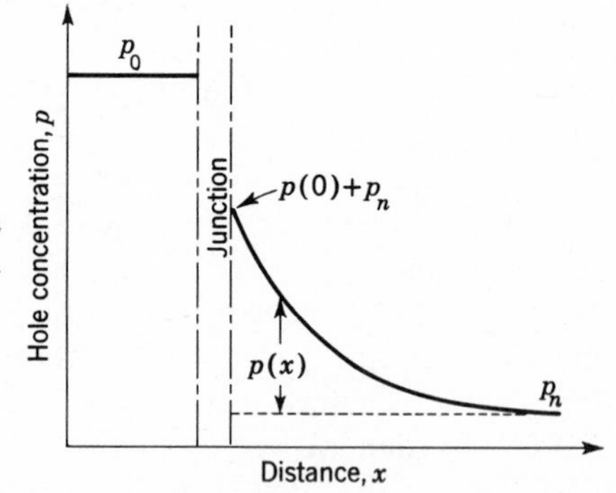

FIGURE 5-25 *Hole concentration near pn junction under forward bias.*

The current carried across the junction by the diffusing holes is simply

$$I_p = -AD_p e \left(\frac{dp}{dx}\right)_{x=0} \tag{5-45}$$

where A is the junction area and D_p is called the *diffusion constant.* The current may be calculated by inserting Eq. (5-44) into Eq. (5-45). Comparing the result with Eq. (5-42) permits the constant C_2 to be evaluated,

$$C_2 = eA\,\frac{D_p}{L_p} \tag{5-46}$$

A similar expression may be developed for the constant C_1. In this way the dependence of the reverse saturation current upon material parameters is determined.

A junction transistor consists of two parallel pn junctions juxtaposed in the same single crystal and separated by less than a minority-carrier diffusion length. Two distinct types are possible, the *pnp* transistor and the *npn* transistor, depending upon the conductivity type of the common region. The operation of the two is conceptually identical except for the interchange of minority and majority carrier types and the polarity of the bias potentials, so that it suffices to discuss the *pnp* transistor. Examples of structures having this dual junction arrangement are described in a following section.

The operation of a junction transistor can be derived from the properties of *pn* junctions. The energy-band model for a *pnp* structure in the absence of applied bias voltages, Fig. 5-26*a*, is simply that of two *pn* junctions

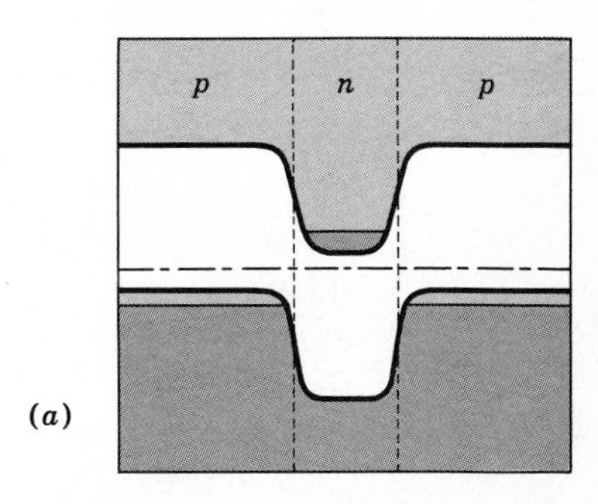

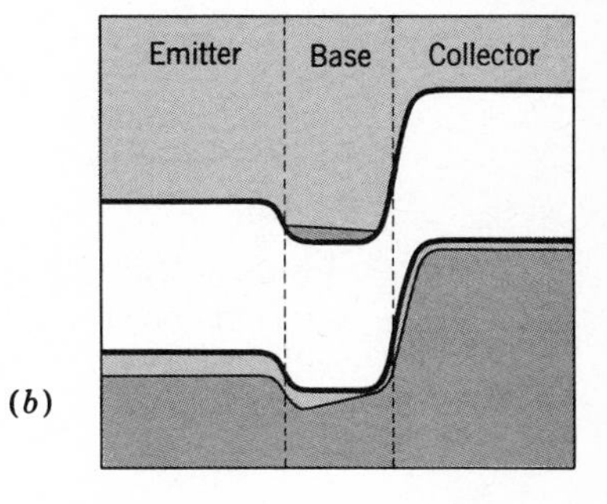

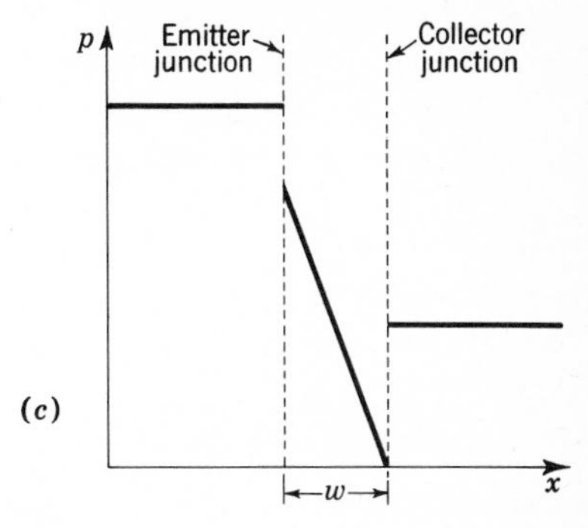

FIGURE 5-26 *(a) Energy-band diagram of pnp transistor in equilibrium, and (b) under operating bias; (c) hole concentration for case (b).*

placed back to back. In operation, one junction, called the *emitter*, is biased in the forward direction and the other, the *collector*, is biased in the reverse direction, as in Fig. 5-26*b*. Holes injected into the *n*-type *base* region at the emitter junction diffuse across to the collector junction where they are collected by the electric field at the junction. The hole concentration in the various regions is sketched in Fig. 5-26*c*.

Variations in the emitter-base bias voltage change the injected current correspondingly, and this signal is observed at the collector junction. The forward-biased emitter represents a small resistance and the reverse-biased collector a large resistance. Since nearly the same current is in both, a large power gain results. Thus the transistor is basically a power amplifier. For maximum amplification, it is desirable that the collector current be as large a fraction of the emitter current as possible. The current through the emitter junction should be carried primarily by holes, since electrons injected from the base to the emitter cannot influence the collector current. According to Eq. (5-41), this can be accomplished if the *n*-type base region is lightly doped. Secondly, the base region must be thin compared with a minority-carrier diffusion length so that few holes are lost before reaching the collector junction. It is convenient to dope the collector region lightly in order to increase the junction width. This reduces the capacitance of the collector junction and also increases the reverse breakdown voltage.

A useful figure of merit for a transistor is the *current-gain factor* α, which is the ratio of the change in collector current to the change in emitter current for constant collector voltage. Alpha is the product of two terms, the emitter efficiency γ and the base transport efficiency ϵ, so that

$$\alpha = \epsilon\gamma \tag{5-47}$$

The emitter efficiency is defined as the fraction of the emitter-junction current carried by holes (for a *pnp* transistor), while ϵ is the ratio of the collector current to the hole current injected into the base at the emitter. In terms of these definitions, the carrier currents in the various regions of a transistor are given in Fig. 5-27, using the emitter current I_e as a starting point. The current I_e in the emitter junction results in a hole current γI_e injected into the base, and a collector current $\epsilon\gamma I_e$. The difference between the emitter and collector currents $(1 - \alpha)I_e$ appears at the external connection to the base. Obviously, when the current gain is unity, all the emitter current appears in the collector circuit and the base current is zero.

The emitter efficiency in a transistor structure is larger than that for an isolated *pn* junction because the collector acts as a sink for holes in the base. The hole concentration in the base falls more rapidly with distance, Fig. 5-26*c*, compared with the exponential decrease near an isolated junction, Fig. 5-25. Therefore, the concentration gradient of holes at the base

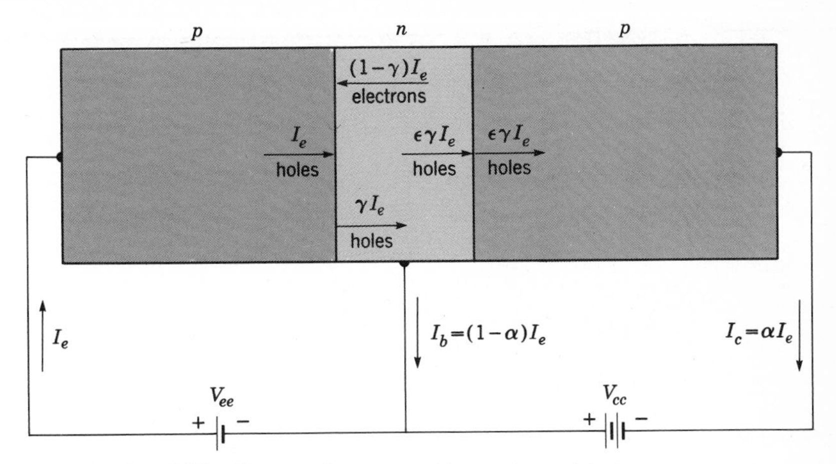

FIGURE 5-27 *Currents in pnp transistor expressed in terms of emitter current, emitter efficiency, and current gain.*

side of the emitter junction is greater because of the presence of the collector. According to Eq. (5-45), the forward hole current is increased correspondingly. Using Eqs. (5-44) and (5-45), the hole gradient for an isolated junction is $-p(0)/L_p$. Since the hole concentration in the base region of a transistor is nearly linear (for a thin-base region), the gradient in this case is $-p(0)/W$, where W is the width of the base. Thus the forward hole current in the emitter is increased by the factor L_p/W and the emitter efficiency is similarly larger. The emitter efficiency can easily be made to exceed 0.995 when the doping ratio between the emitter and base is large and when $W \ll L_p$.

With a thin-base region the base transport efficiency can be 0.999, so that the current gain α is typically of the order of $0.995 \times 0.999 = 0.994$. Actually, since α is very nearly unity for most devices the base-collector current gain β is a more sensitive measure of transistor quality. The relation between the base-collector current gain and α is, from Fig. 5-27,

$$\beta = \frac{I_c}{I_b} = \frac{\alpha}{1-\alpha} \tag{5-48}$$

The base-collector current gain is large when α approaches unity. Typical values of β range from 20 to 10^3 in practical transistors.

Minority carriers injected into the base move to the collector by diffusion, and the finite time taken by the carriers to cross the base limits the high-frequency usefulness of transistors. By ingenious fabrication techniques it is possible to achieve very narrow base widths ($\sim 5 \times 10^{-7}$ m) and obtain useful amplification at frequencies of 10^9 Hz. It is also necessary to reduce the area of the junctions in high-frequency transistors in order to minimize adverse effects of the junction capacitances.

Collector Characteristics A convenient way to represent the current-voltage characteristics of a transistor is the collector characteristics, Fig. 5-28,

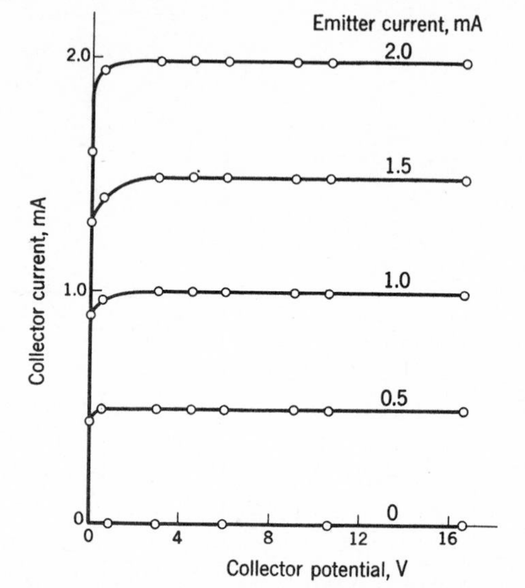

FIGURE 5-28 *Experimental grounded-base collector characteristics of pnp transistor.*

for different values of emitter current. When $I_e = 0$ the characteristic is simply the reverse saturation curve of the collector junction. Emitter current translates the curve along the current axis. These characteristics follow directly from the preceding discussion and pertain to the *grounded-base* circuit configuration already considered in Fig. 5-27. In the grounded-base connection the base terminal is common to both input and output circuits. As is considered in greater detail in the following chapter, the *grounded-emitter* configuration in which the emitter electrode is common to both input and output circuits, exhibits more favorable properties.

The quality of practical transistors is so satisfactory that the grounded-base current gain α is very nearly equal to unity and the reverse leakage current at the collector junction is negligible for all practical purposes. Consequently, the grounded-base collector characteristics are very flat, straight, and uniformly spaced. As a matter of fact, so little new information is presented in such curves that they are almost never measured. The grounded-emitter collector characteristics, Fig. 5-29, are much more illustrative. In particular, a value for the grounded-emitter current gain β may be estimated directly from the curves by noting the collector current for a given base current.

Note that either form of collector-characteristic presentation is fundamentally different from the plate characteristics of a triode, Fig. 5-15, and the drain characteristics of a FET, Fig. 5-21. The difference is that

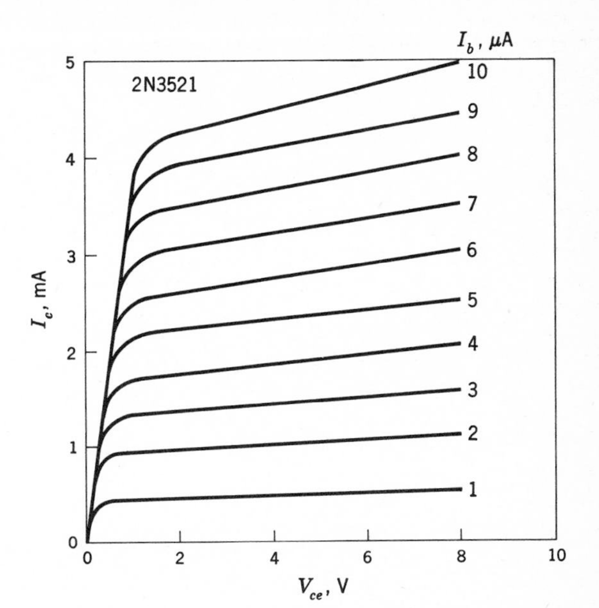

FIGURE 5-29 Grounded-emitter collector characteristics for type 2N3521 npn silicon transistor.

the control parameter is input current in the case of a transistor whereas it is input voltage in the case of a vacuum tube or FET. The consequences of this difference are explored in various ways in succeeding chapters. Actually, both types of control lead to quite effective amplifying action, but the circuit configurations and the operating properties of junction transistors are different from those of FETs. It is also worthy of note that according to Figs. 5-28 and 5-29 junction transistors perform effectively at operating potentials less than 1 V.

The operation of a junction transistor results from the properties of the emitter junction under forward bias and the collector junction under reverse bias. The properties of both the emitter junction and the collector junction depend sensitively upon temperature as may be shown in the following way. The reverse saturation current at the collector junction is given by Eq. (5-19),

$$I = C_1 n_p + C_2 p_n \tag{5-49}$$

Substituting from Eq. (5-16) and its equivalent for p_n.

$$I_0 = (C_1 n_0 + C_2 p_0) e^{-eV_0/kT} \tag{5-50}$$

According to Eq. (5-50), the reverse current at the collector junction increases exponentially with temperature. Similarly, the hole current injected at the emitter junction is, from Eq. (5-42),

$$I_e = C_2 p_0 e^{-eV_0/kT} (e^{eV/kT} - 1)$$

Since $e^{eV/kT} \gg 1$,

$$I_e = C_2 p_0 e^{e(V-V_0)/kT} \tag{5-51}$$

Under normal operating conditions, $V < V_0$, so that the current injected into the base region also increases exponentially with temperature.

The rapid increase in current with temperature indicated by these results means that transistors are quite sensitive to temperature. Comparing collector characteristics at an elevated temperature, Fig. 5-30,

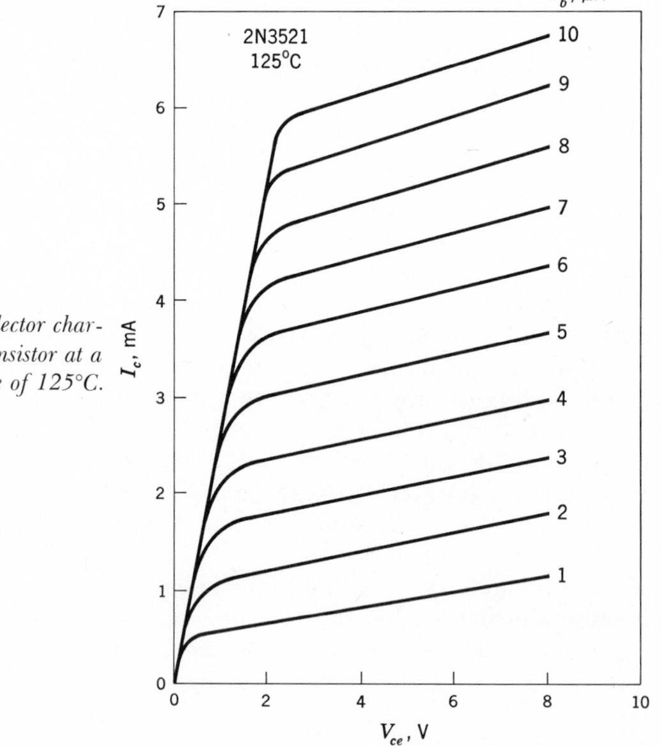

FIGURE 5-30 *Grounded-emitter collector characteristics for type 2N3521 transistor at a temperature of 125°C.*

with those for the same transistor at room temperature, Fig. 5-29, shows that currents are indeed increased considerably. The general shape of the curves is preserved, however, so that successful transistor action is still achieved.

Silicon Controlled Rectifier It is now possible to describe in greater detail the *silicon controlled rectifier* (SCR) originally discussed in Chap. 4. The useful performance of this device is obtained by adding a third *pn* junction to the junction transistor structure. The resulting *pnpn* four-layer device,

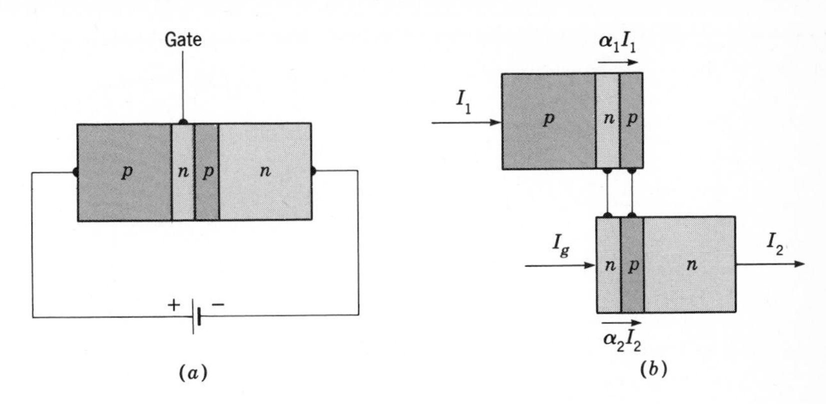

FIGURE 5-31 *(a) Sketch of SCR and (b) its interpretation in terms of pnp transistor coupled to npn transistor.*

Fig. 5-31*a*, is also called a *four-layer diode* when connections are made only to the two outer layers. A positive potential applied to the *p*-type terminal puts the center *pn* junction under reverse bias while the two outer junctions are forward-biased. The device may be looked upon as the back-to-back combination of a *pnp* transistor together with an *npn* transistor, Fig. 5-31*b*, the two transistors having a common collector junction.

Note that I_1 is the emitter current in the *pnp* transistor and that $\alpha_1 I_1$ is the collector current. Similarly, I_2 is the emitter current of the *npn* unit and $\alpha_2 I_2$ is the collector current. Using Kirchhoff's current rule at the collector junction,

$$I_2 = \alpha_1 I_1 + \alpha_2 I_2 \tag{5-52}$$

Considering the overall current input to the device

$$I_2 = I_g + I_1 \tag{5-53}$$

where I_g is the current into the gate terminal. Substituting for I_1 in Eq. (5-52) and solving the result for I_2 yields

$$I_2 = \frac{-\alpha_1}{1 - (\alpha_1 + \alpha_2)} I_g \tag{5-54}$$

According to Eq. (5-54), if the sum of the current gains $\alpha_1 + \alpha_2$ is near unity, the current I_2 can be very large even though the gate current is small. In fact, if $\alpha_1 + \alpha_2 = 1$, the current is large and limited only by the ohmic resistance of the semiconductor, even if there is no gate current.

The current-voltage characteristics of a four-layer diode, Fig. 4-24, are interpreted in terms of Eq. (5-54) as follows. For low terminal voltages the current corresponds to the minuscule reverse current of the collector junction and the current-gain factors are small. At a critical applied voltage, avalanche breakdown at the collector junction increases the current through the device and the α's increase. When $\alpha_1 + \alpha_2$ equals unity, the device switches to the high-conductance state. According to Eq. (5-54) the current can also be increased by introducing a current into the gate terminal. Thus the transition can be made to occur at a lower terminal voltage than that corresponding to current avalanche at the collector junction. Once triggered into the conducting state by gate current, the gate current may be reduced to zero since the device current itself maintains $\alpha_1 + \alpha_2$ equal to unity.

The SCR current-voltage characteristic, Fig. 4-24, may be interpreted to indicate negative resistance properties since a voltage decrease results in a current increase when a transition from the off state to the on state occurs. This negative resistance effect is inherently different from that of the tunnel diode, Fig. 5-11. The difference between the two may be identified by noting that a line through the negative-resistance region parallel to the voltage axis in Fig. 5-11 intercepts the characteristic curve at another point. By contrast, the line must be parallel to the current axis if it is to intercept the curve again in Fig. 4-24. The shape of the tunnel-diode curve is often called *N-type* negative-resistance characteristic and the SCR curve called an *S-type* negative-resistance characteristic because of the relative shapes of the two types.

An S-type negative-resistance characteristic somewhat similar to that of the SCR is produced by the *unijunction* transistor. The circuit symbol for a unijunction transistor, Fig. 5-32, shows schematically that the device

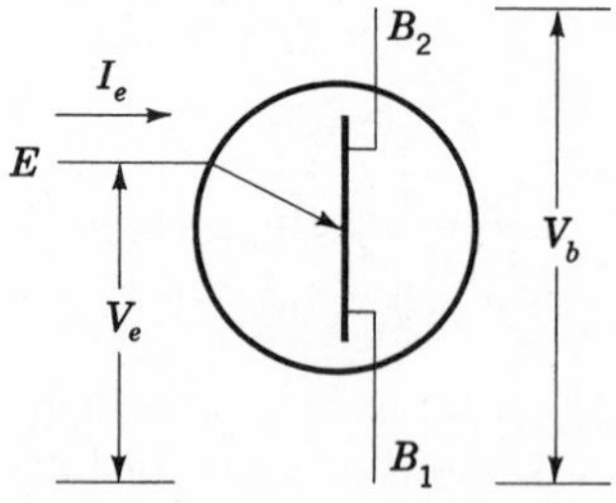

FIGURE 5-32 *Circuit symbol of unijunction transistor.*

consists of a semiconductor bar with two ohmic contacts and a single, small-area, emitter *pn* junction positioned between them. If no emitter current is present, a fraction of the voltage V_b applied between the base contacts appears at the emitter junction because of the voltage-divider action of the semiconductor bar. Only reverse current is present in the emitter circuit so long as the emitter voltage is less than this fraction and the emitter junction remains under reverse bias. If, however, the junction voltage V_e is increased such that the emitter becomes biased in the forward direction, emitter current increases and carriers are injected into the semiconductor in the region between the emitter and B_1. The resistance of this portion decreases correspondingly and the voltage divider action reduces the fraction of V_b at the junction. Thus the emitter current increases even if the emitter voltage decreases. This results in a very stable negative-resistance characteristic, as illustrated in Fig. 5-33. Circuit

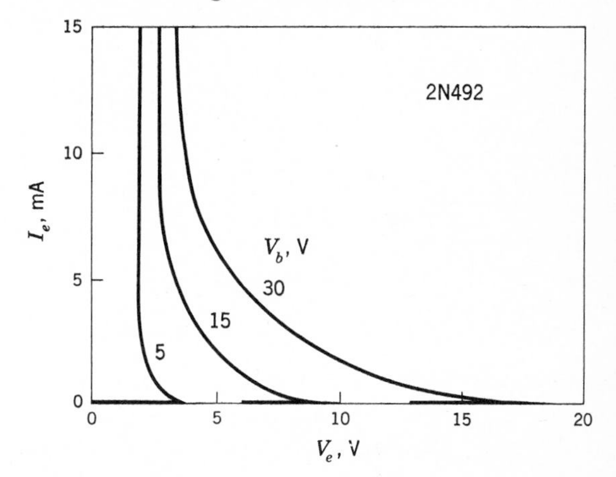

FIGURE 5-33 *Current-voltage characteristics of type 2N492 unijunction transistor.*

applications of the unijunction transistor are considered in a subsequent chapter.

Fabrication of Transistors Of the several ways to produce a semiconductor single crystal containing two *pn* junctions, four types are most popular: the grown-junction, alloy-junction, diffused-mesa, and planar transistors, Fig. 5-34. Many modifications of these four types have been devised, but they differ only in minor details from the ones shown. The grown-junction transistor was the first junction type to be fabricated and has now been supplanted by the other three. It is prepared during crystal growth from the melt by altering the impurity content of the melt as the solidified crystal is slowly withdrawn. A typical crystal may be 1 to 2 cm in diameter, while a single transistor is only about 1 mm square and many transistors can be obtained by cutting up the original crystal. Connections are soldered to the emitter, base, and collector regions. Solder containing a small amount of impurity corresponding to the conductivity type being soldered is used in order to provide good electric contacts. Thus, indium is introduced into the solder for the contacts attached to the *p* regions and antimony is added to the solder for the base. This is particularly important in making the connection to the very narrow base region because the *n*-type impurity produces a good electric connection to the *n*-type base and at the same time forms a *pn* junction with the *p* regions. This means that the soldered base contact is effectively electrically isolated from the emitter and collector regions, even though it may physically extend beyond the thin base region.

The *pnp* alloy-junction transistor is produced from a thin wafer of *n*-type

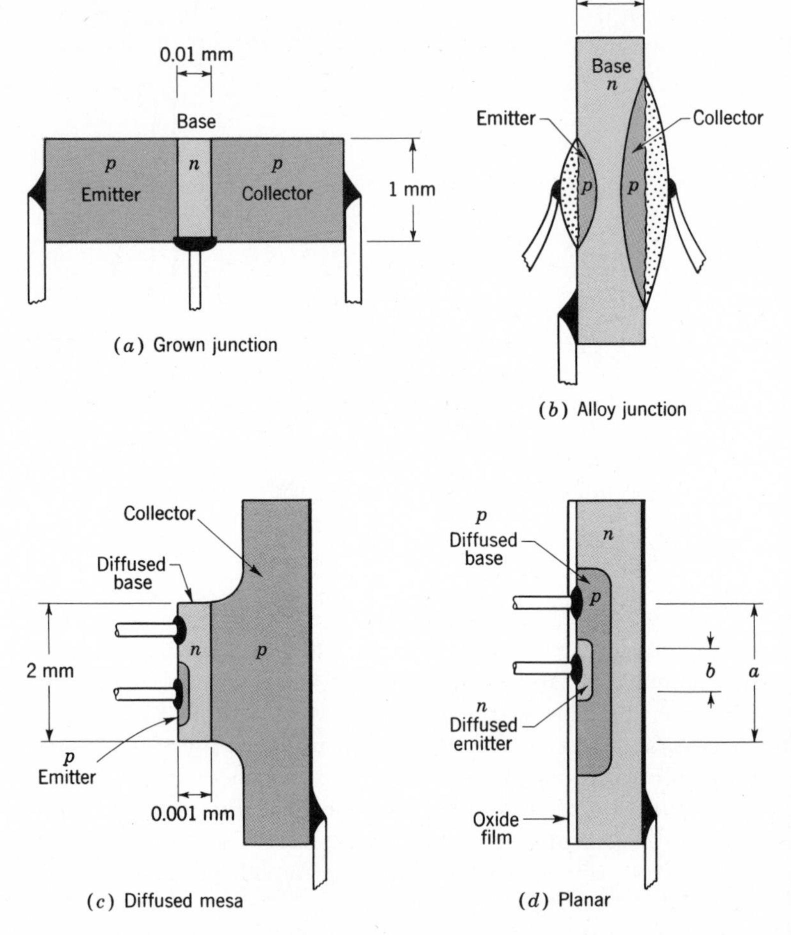

FIGURE 5-34 *Structures of junction transistors.*

single-crystal material by placing indium pellets on opposite surfaces and heating. As the indium melts it dissolves some of the germanium beneath it. During subsequent cooling the dissolved germanium recrystallizes upon the base crystal and incorporates many indium atoms into its structure. The recrystallized material is therefore *p*-type, and a transistor structure results. It is convenient to make the collector pellet larger than the emitter pellet, Fig. 5-34*b*, for in this way carriers injected at the emitter junction are collected more efficiently. Thus, the alloy junction is inherently a more satisfactory transistor geometry than the simple grown-junction shape. A micrograph of the cross section of an actual alloy-junction transistor is shown in Fig. 5-35. The emitter and collector junctions are made visible by chemical etching and the recrystallized regions and indium pellets are clearly apparent.

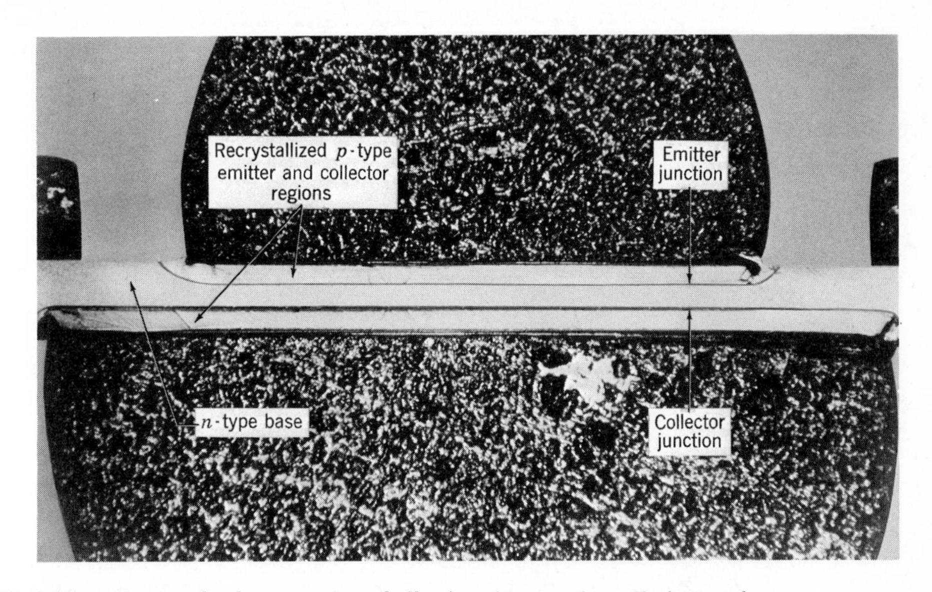

FIGURE 5-35 *Micrograph of cross section of alloy-junction transistor. Emitter and collector junctions, revealed by chemical etching, are clearly visible.* (Amperex Electronic Corporation.)

As discussed earlier, it is necessary to make the dimensions of the transistor region as small as possible in order to achieve satisfactory high-frequency performance. This is difficult in the case of the alloy-junction transistor, although many ingenious variations of the above process have been developed to accomplish this goal. The diffused-mesa transistor, Fig. 5-34c, however, is much more adaptable to high-frequency transistors because the junction fabrication process can be precisely controlled. The collector junction is formed by placing a p-type single-crystal wafer in a hot gas of, say, antimony atoms. As the wafer is heated, antimony atoms diffuse into the crystal to a depth of approximately 10^{-3} mm. The crystal is subsequently masked and chemically etched to produce a small elevated region, or mesa, about 2 mm in diameter. The mesa top is then suitably masked so that two regions 0.3 mm in diameter are exposed and appropriate metals are deposited by evaporation in high vacuum. Next, the slab is maintained at an elevated temperature while the two metal deposits diffuse into the n region. This results in the formation of a p-type emitter region by one of the metal deposits while the other one forms a contact to the base region.

The area of the collector junction is defined exactly by the size of the mesa while the emitter junction is set by the area of the vapor-deposited metal. Both processes can be precisely controlled and result in an extremely tiny volume. The width of the base region is determined by the gaseous diffusion process and subsequent diffusion of the deposited metal. Both processes can be adjusted easily by the temperature of the wafer to yield a very thin, yet accurately defined, base region. Therefore, the entire fabrication technique lends itself to precise control of transistor geometry. Very tiny active regions are produced on a wafer which is large enough to handle conveniently.

The mesa transistor, like the grown-junction and alloy-junction transistors, is so constructed that the emitter and collector junctions are exposed at the surface of the semiconductor. The electrical properties of the junction are extremely sensitive to chemical impurities from the surrounding atmosphere because of the high electric fields at the junction. Minute traces of foreign atoms on the surface in the junction region greatly degrade the performance of actual transistors compared with theoretical expectations. For this reason, transistors are hermetically sealed in carefully cleansed metal enclosures after fabrication.

A more satisfactory solution to the surface-contamination problem is to grow a thin oxide film on the surface of the semiconductor by heating in an oxidizing atmosphere. This technique, applicable primarily to silicon because of the favorable properties of silicon oxide, results in a semiconductor device completely protected from atmospheric contamination. In addition, the oxide film is a barrier to gaseous diffusion of impurities so the film itself acts as a mask during fabrication. This technique makes possible the planar transistor, Fig. 5-34*d*, so called because the entire transistor appears to be a simple plane wafer of silicon.

Construction of a planar *npn* transistor begins with a single-crystal wafer of *n*-type silicon. A thin oxide layer is grown on the top surface by heating in an oxygen atmosphere. The oxide is chemically etched away in a circular region of diameter *a*. The wafer is then exposed to a hot boron gas atmosphere. Boron atoms diffuse into the exposed silicon, forming a *p*-type base region and the collector junction. Boron atoms also diffuse laterally under the oxide film so that the base region is somewhat larger than the diameter *a* and the collector junction is protected at the surface by the oxide layer. Next, the wafer is reoxidized to produce a film covering the entire area once again. After the oxide is etched away in an area of diameter *b*, the wafer is again exposed to a hot impurity gas, this time phosphorus. These *n*-type impurity atoms diffuse into the silicon, forming the emitter region and the emitter junction. Lateral diffusion of phosphorus atoms under the oxide film again ensures that the emitter junction is protected by the oxide layer.

Finally, electric contacts are provided by vapor deposition and alloying as in the case of the mesa transistor. The result is an *npn* transistor entirely contained in one plane wafer and completely encased in an impervious oxide film. Geometric definition of the junction positions and the base width is extremely accurate because of the easily controlled gaseous diffusion doping technique. Several hundred transistors can be fabricated simultaneously from one silicon wafer a centimeter or so in diameter. These are subsequently cut into individual units by sawing up the wafer.

INTEGRATED CIRCUITS

Principles of Integrated Circuits The properties of semiconductors, most particularly silicon, make it possible to produce an entire electronic circuit within one single crystal. Such an *integrated circuit* miniaturizes electronic networks and also reduces the number of individual components in complex electronic circuits. This is so because an entire integrated circuit represents, in effect, only one component. The basis for the integrated-circuit development is that many electronic components, such as transistors, diodes, and resistors, can be made of silicon having suitably disposed *n*-type and *p*-type regions. It remains only to position the various components in a judicious geometric arrangement in order to combine them all into one single-crystal slab.

The close juxtaposition of components within a single slab leads to electrical coupling between them. It proves possible, however, to isolate components by suitable arrangement of the *p*-type and *n*-type regions. Consider the case of a resistor of *n*-type silicon in a *p*-type crystal, for example. Clearly the resistor is isolated from the crystal by the *pn* junction between resistor and the slab. This isolation is most effective only at *dc* because of *pn*-junction capacitance. The stray coupling introduced by all the junction capacitances is taken into account in the design of the circuit.

Actually, the capacitance of a *pn* junction may also serve as a capacitor in an integrated circuit. Inductors of small inductance in the form of a spiral conductor are also possible. Most often, however, the integrated circuit per se is composed of transistors, diodes, and resistors, and the necessary capacitors and inductors are connected as separate, discrete components.

Fabrication Processes Silicon integrated circuits are fabricated by processes similar to those used in the case of planar transistors. Appropriate patterns in the oxide layer prior to impurity diffusions are produced by first coating the oxide layer with a special photopolymeric material. A photopolymer becomes resistant to chemical etching when exposed to strong light. Thus, when the photopolymer is exposed to light through a suitable mask, the unexposed portions can be subsequently dissolved away. This exposes a corresponding pattern in the oxide layer which is etched away with a different chemical, in turn exposing the underlying silicon. After gaseous diffusion of *n*-type or *p*-type impurities, the wafer may be reoxidized and the process repeated to develop an additional array of doped regions.

The photographic process produces integrated circuits of extremely small size. Accurate, large-scale masks are optically demagnified to microscopic dimensions. This also means that many integrated circuits may be fabricated simultaneously from the same silicon wafer simply by constructing a mask having the desired circuit repeated many times. Subse-

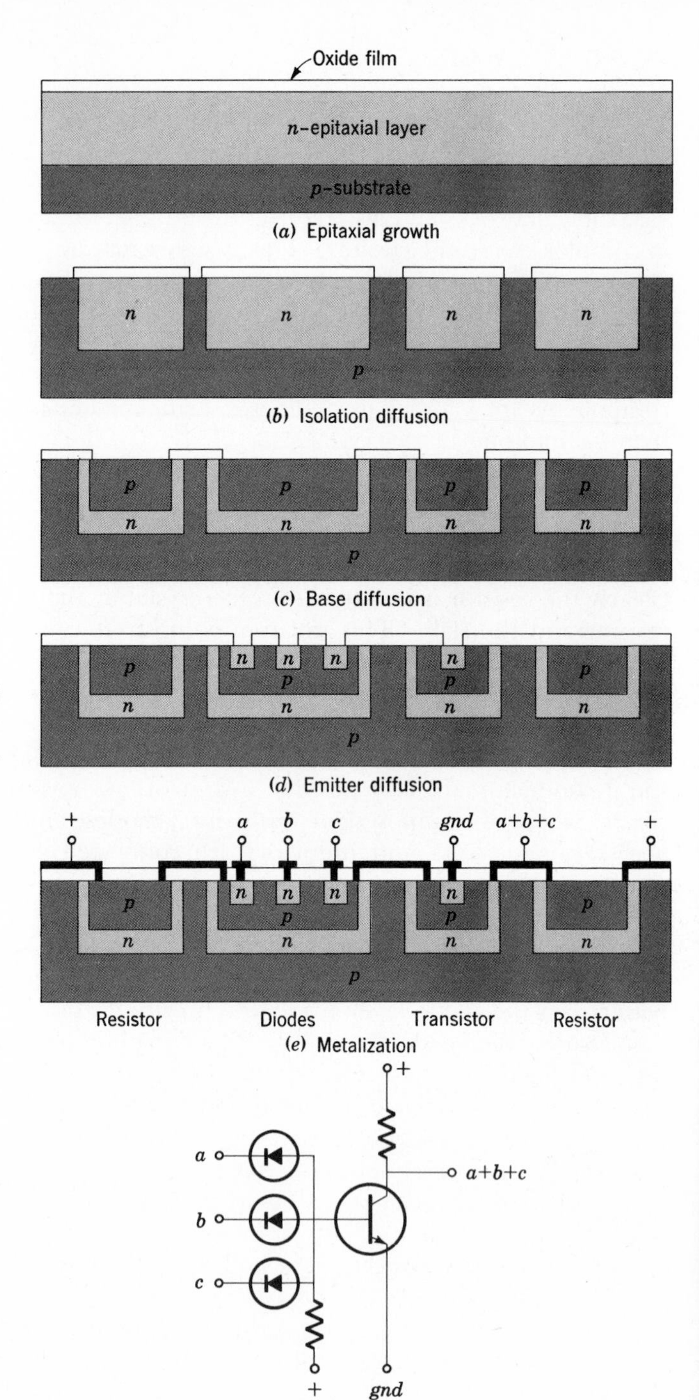

FIGURE 5-36 *Processing steps (a) through (e) result in complete integrated circuit (f).*

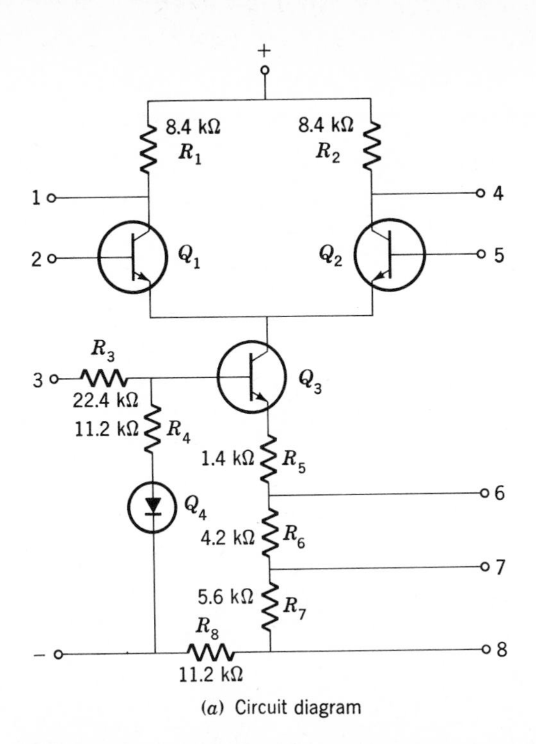

(*a*) Circuit diagram

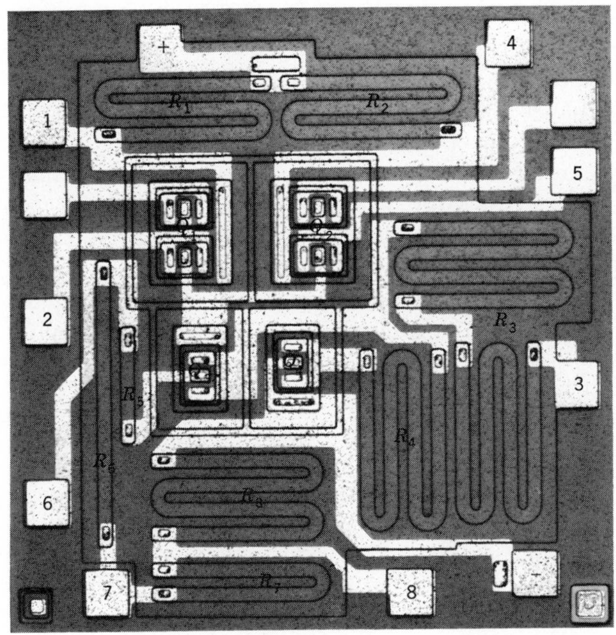

5 mm

(*b*) Integrated circuit

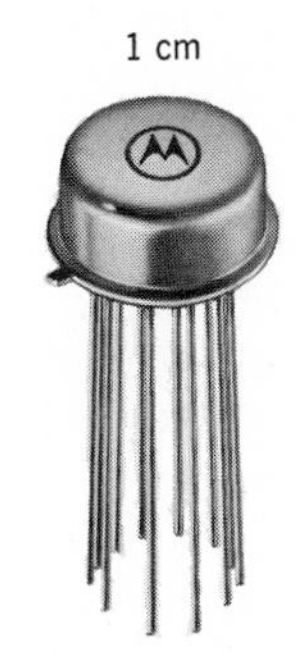

(*c*) Case

FIGURE 5-37 *(a) Entire circuit is (b) fabricated as an integrated circuit and (c) mounted in one container.* (Motorola Semiconductor Products, Inc.)

quent to the last impurity-diffusion step, a thin metal layer is deposited (also through a photopolymeric layer) to interconnect the various components and to provide terminals for external connections.

This process is best illustrated by following actual steps in the fabrication of a simple circuit, Fig. 5-36. To begin, a *p*-type single-crystal slab is exposed to silicon vapor containing *n*-type impurity atoms. The result, Fig. 5-36*a*, is an *n*-type *epitaxial* layer. Epitaxial means that the *n*-type layer has the same crystal structure as the *p*-type substrate. The surface is oxidized, masked, etched, and a *p*-type diffusion isolates several *n*-regions, Fig. 5-36*b*. These steps are repeated in Fig. 5-36*c* and Fig. 5-36*d*. Finally, Fig. 5-36*e*, the metallic interconnections are added.

This procedure results in the circuit diagramed in Fig. 5-36*f*, as can be appreciated from a detailed comparison with Fig. 5-36*e*. Note, in particular, how the resistors and diodes are isolated from the *p*-type substrate in each instance by a *pn* junction. Note also that the conventional circuit symbol for an *npn* transistor is used in the circuit diagram. The circuit, incidentally, is termed a DTL NAND gate, which is considered in greater detail in Chap. 12.

Practical Circuits It is possible to recognize the various components in the micrograph of a practical integrated circuit, Fig. 5-37*b*. This so-called *difference-amplifier* circuit is diagramed in Fig. 5-37*a* and is analyzed in the next chapter. Note that the metallic interconnections appear lighter in the micrograph and that it is often necessary to zigzag the resistors in order to achieve the desired resistance. Although this is a relatively simple circuit, it is noteworthy that all 12 components are contained in one small case about the size of a single transistor, Fig. 5-37*c*.

Much more complicated circuits are available as integrated circuits and these are discussed in subsequent sections. Each such circuit usually can be employed in several ways and is often associated with other integrated circuits to perform an entire complex task. In certain instances, most notably in digital computers, the combination of several integrated circuits can itself be made as an integrated circuit. Such *large-scale integration,* or *LSI,* further miniaturizes electronic circuits and yields the further advantage that signal propagation throughout the circuit is very rapid. LSI circuits containing 1000 or more transistors are feasible.

SUGGESTIONS FOR FURTHER READING

E. James Angelo, Jr.: "Electronics: BJTs, FETs, and Microcircuits," McGraw-Hill Book Company, New York, 1969.

Leonid V. Azároff and James J. Brophy: "Electronic Processes in Materials," McGraw-Hill Book Company, New York, 1963.

James J. Brophy: "Semiconductor Devices," McGraw-Hill Book Company, New York, 1964.
"General Electric Transistor Manual," latest edition, General Electric Company, Semiconductor Products Department, Syracuse, N.Y.

EXERCISES

5-1 Determine the impurity concentration and internal-potential rise of the parametric-diode *pn* junction corresponding to Fig. 5-9 from the slope and intercept of the line and with the aid of Eq. (5-22). The area of the junction is 4.5×10^{-7} m^2 and the dielectric constant of silicon is 11. Also calculate the width of the junction and the electric field in the junction at a reverse potential of 10 V.

Ans.: 6.7×10^{16} cm^{-3}; 0.32 V; 4.3×10^{-7} m; 2.3×10^{7} V/m

5-2 Sketch the energy-band model for an *npn* transistor in equilibrium and biased for transistor operation.

5-3 Plot Child's law, Eq. (5-32), for a type 12AU7 triode and compare with experimental plate characteristics in Appendix 1. Do this by evaluating the constant A using the experimental curve for $V_c = 0$ and evaluating μ for the $V_c = -10$-V curve.

Ans.: 1.1×10^{-5}; 17.2

5-4 By differentiating the rectifier equation for a junction diode, Eq. (5-20), determine an expression for the junction resistance $R = (dI/dV)^{-1}$. Given that $e/kT = 38$ V^{-1} at room temperature, calculate the rectification ratio at a potential of 1 V.

Ans.: 10^{33}

5-5 Calculate the ratio of the thermionic-emission current from a thoriated tungsten cathode to that of a similar tungsten cathode at a temperature of 1500 K.

Ans.: 1.8×10^{6}

5-6 Noting that Child's law only applies up to the maximum emission current given by the Richardson-Dushman equation, complete the curve of Fig. 5-13 if the oxide cathode is at a temperature of 1000 K. Repeat for a temperature of 800 K.

5-7 Calculate and plot the drain characteristics of a type 2N2499 FET using Eqs. (5-38) and (5-40). Do this by evaluating I_{dss} and V_p from the $V_g = 0$ curve in Fig. 5-21. Compare with Fig. 5-21.

Ans.: 9 mA; 6.6 V

5-8 Calculate and plot the grounded-base collector characteristics of a silicon transistor at room temperature and for collector potentials up to a maximum of 0.5 V. Assume the current gain is unity and that $I_0 = 1$ μA.

5-9 Repeat Exercise 5-8 for a temperature of 150°C. Assume that I_0 increases according to Eq. (5-50) and take $V_0 = 0.7$ V.

5-10 Plot the forward characteristics of a germanium and a silicon *pn* junction up to a current of 10 mA. Use the rectifier equation and take $I_0 = 10^{-6}$ A for germanium and $I_0 = 10^{-12}$ A for silicon. How much more forward bias is required in the case of the silicon diode to make the current 10 mA? Repeat for 1 mA.

Ans.: 0.4 V; 0.4 V

LABORATORY EXERCISES

5-A Experimental Drain Characteristics The most useful general description of a nonlinear electronic component is the relationship between current and voltage in the device. This is particularly so in the case of transistors wherein the current is controlled by the signals applied to the third electrode. The current-voltage properties of a FET are best displayed by the drain characteristics, which are curves of drain current as a function of drain voltage with the gate potential as a parameter. In this experiment the drain characteristics of a typical *n*-channel FET are determined.

Plot the drain characteristics of a FET for drain voltages from pinchoff to 20 V using Eq. (5-40) knowing that $I_{dss} = 5$ mA and $V_p = 5$ V. $\frac{1}{2}$-V increments of the gate potential are appropriate. Use Eq. (5-38) to complete the characteristics below pinchoff, which is often called the triode region. In this connection observe that the transition points between the two regions are found by inserting Eq. (5-39) into Eq. (5-38), which yields

$$I'_d = \frac{I_{dss}}{V_p^2} V_d^2 \qquad (5\text{-}55)$$

Connect the transition points on your curves with a dotted line. Is the result a parabola as predicted by Eq. (5-55)? In general, are your "experimental" drain characteristics similar to those of Fig. 5-21? Wherein do they differ?

Now prepare a transfer characteristic at a drain potential of 20 V and another one at a drain potential of 10 V. Can you tell before you determine them what differences in the two transfer characteristics to expect? What is the useful gate-voltage range in this device? What is the change in drain current per unit change in gate voltage at the middle of the range? Why is this an important parameter?

5-B Transfer Characteristics The single most important feature of active devices is the control of the output current by signals applied to the control electrode. This feature is best displayed in the transfer characteristics of the devices. Two types of transfer characteristics can be distinguished: one in which the control signal is a voltage, as in the case of the triode and FET, and the other in which the control signal is a current, as in the junction transistor. The purpose of this experiment is to examine the range and type of transfer characteristics exhibited by practical devices.

Plot the appropriate transfer characteristics of all the vacuum tubes, transistors, and FETs for which the current-voltage characteristics are given in Figs. 5-15, 5-17, 5-21, 5-28, 5-29, 5-30, 6-5, 6-24, 7-17, 7-18, and in Appendix 1. In each case, select an appropriate value of the plate, drain, or collector potential at which to plot the transfer characteristic. Note the range of currents and voltages exhibited in this data. Contrast the linearity of the transfer characteristics of the several types of devices. Comment on the effect of temperature upon transistor transfer characteristics from the data pertaining to Figs. 5-29 and 5-30.

Calculate the change in output current per change in input signal near the middle of the range over which each transfer characteristic is given. This quantity has the dimensions of conductance for voltage-controlled devices and is dimensionless in the case of current-controlled devices. Is a large value or a small value of the ratio desirable in general? Can you distinguish any trend in the magnitude of the ratio associated with the power-handling capabilities of the various units?

A direct comparison between the amplifying properties of a triode or FET with that of a transistor is not particularly appropriate since one type is voltage-controlled and the other current-controlled and each has its sphere of application. Nevertheless, some appreciation of the relative magnitudes of the two can be achieved by converting the current-gain ratio of a transistor into a ratio of the change in output current per unit change in input voltage. This is accomplished by noting that the input current signal is produced by a small voltage applied to the emitter junction. The forward resistance of the emitter junction as given by the rectifier equation, Eq. (5-20), multiplied by the change in input current is the change in input voltage which produces the observed change in output current.

Compare the ratio of the change in output current divided by the change in input voltage of the transistors with the same quantity for vacuum tubes and FETs by dividing the calculated values of current gain by the emitter junction resistance in each case. Do this using the simple expression for the junction resistance derived from the rectifier equation in the next chapter, Eq. (6-61). Comment on your comparisons.

chapter six

VACUUM TUBE AND TRANSISTOR AMPLIFIERS

The advent of electronics is reckoned from the discovery that the current in a vacuum diode can be controlled by including a third electrode, the grid, in the space between the anode and cathode. Very little power is consumed by the grid in controlling the anode current so that the triode is an exceedingly effective amplifier. The possibilities inherent in electronic circuits were greatly expanded by the invention of the transistor in 1948. Since then a wide variety of semiconductor devices have to a large extent supplanted vacuum tubes in many applications.

Both vacuum tubes and transistors are nonlinear devices and their operation in any circuit is determined by graphical analysis using the description of their electrical properties given by current-voltage characteristics. The analysis differs in detail for voltage-controlled devices such as the triode and field-effect transistor compared to current-controlled devices such as the junction transistor, but is not different in principle. Furthermore, in either case it proves possible to develop useful equivalent-circuit representations which are most satisfactory for circuit analysis.

The versatility of the triode is enhanced by the availability of multigrid tubes which outperform triodes in many applications. Correspondingly, the variety of semiconductor devices rest upon the versatility inherent in semiconductor materials. Transistors are tiny and make very efficient amplifiers since little power is required for proper operation. The major disadvantage of transistors, as with all semiconductor devices, is poor high-temperature performance, which limits the operating temperature to a few hundred degrees Celsius.

THE OPERATING POINT

Load Lines The operation of the triode as an amplifier is examined most easily with the aid of the circuit in Fig. 6-1. In this circuit the plate is

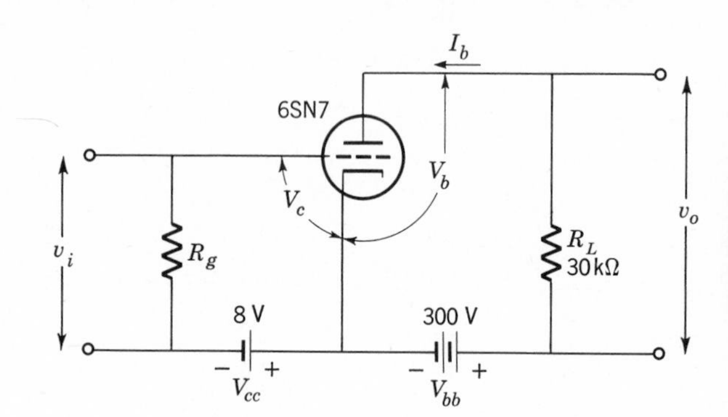

FIGURE 6-1 *Simple triode amplifier circuit.*

maintained positive with respect to the cathode by the battery V_{bb} while the grid is *biased* negatively by the grid bias battery, V_{cc}. Variations in grid voltage resulting from an input signal v_i produce changes in the plate current which are observed as a voltage signal across R_L.

The circuit is analyzed using Kirchhoff's rules in the usual way except that, since the current-voltage properties of the triode are nonlinear, the analysis must be carried out graphically. First, the current-voltage characteristic of R_L, which is called the *load line*, is plotted on the plate characteristics. This is done by writing the current as

$$I_b = \frac{V_{bb} - V_b}{R_L} = \frac{V_{bb}}{R_L} - \frac{1}{R_L} V_b \tag{6-1}$$

As noted previously, this is the equation of a straight line with slope $-1/R_L$ and with intercepts at ($V_b = 0$; $I_b = V_{bb}/R_L$) and ($V_b = V_{bb}$; $I_b = 0$), Fig. 6-2. The intersection of the load line with each curve of the plate characteristic gives the plate current for the given grid voltage.

The plate current and voltage corresponding to the dc grid bias V_{cc}, found by this procedure, is called the *operating point.* As the grid voltage varies in accordance with an applied ac input signal v_i, the plate-current excursions move back and forth along the load line, so Eq. (6-1) is satisfied at every instant. The corresponding changes in plate current give rise to an output signal across the load resistor. Suppose, for example, that the input signal is sinusoidal, as in Fig. 6-2. The output voltage is then also nearly sinusoidal but of much greater amplitude, indicating that the circuit amplifies the input signal.

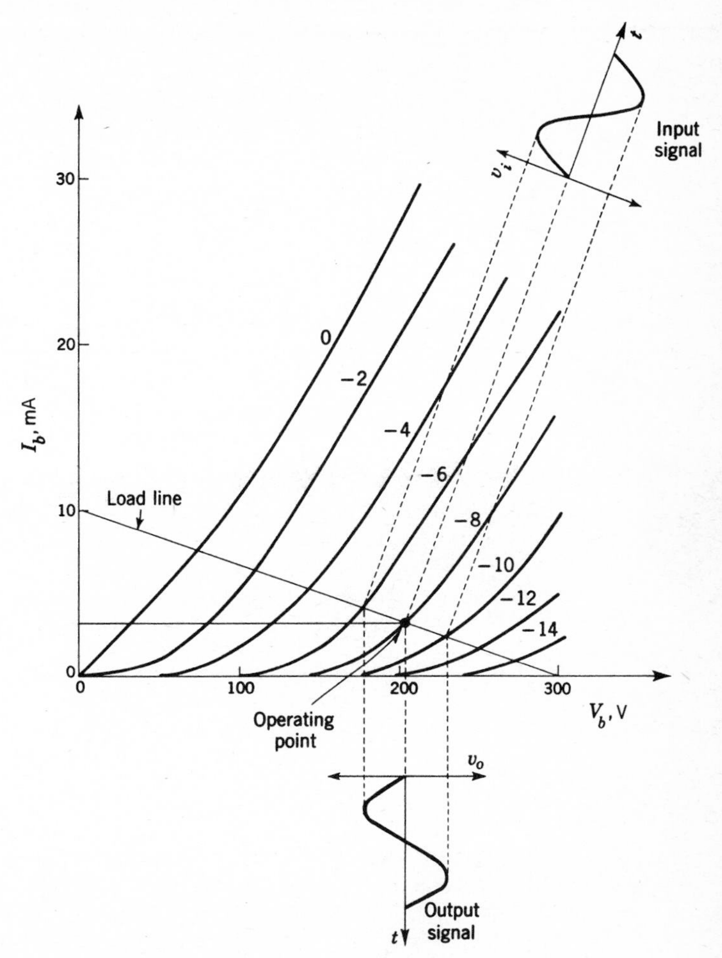

FIGURE 6-2 Analysis of triode amplifier of Fig. 6-1 using load line. Note 4-V peak-to-peak input signal results in a 40-V peak-to-peak output signal.

Note that the input power is very small, because the grid current is negligible. In contrast, the output power, which is equal to the square of the ac plate current times the load resistance, may be appreciable. This power is derived from the plate voltage supply V_{bb} and is controlled by the valvelike action of the grid. The output voltage waveform is not an exact amplified replica of the input signal, because of curvature of the plate characteristics. This *distortion* is minimized by proper circuit design and by suitable choice of the operating point. Note also in Fig. 6-2 that as the grid voltage increases, the output voltage is reduced. This means that the triode amplifier introduces a 180° phase shift between the input and output signals. It is often convenient to account for this phase shift by writing the amplification factor as $-\mu$. The minus sign signifies that the output signal lags the input by a phase difference of 180°.

It is easier to determine the output waveform corresponding to a given input signal by using the *transfer characteristics* of a triode. These are curves of plate current as a function of grid voltage for fixed values of plate voltage, Fig. 6-3. Careful comparison of Figs. 6-2 and 6-3 reveals

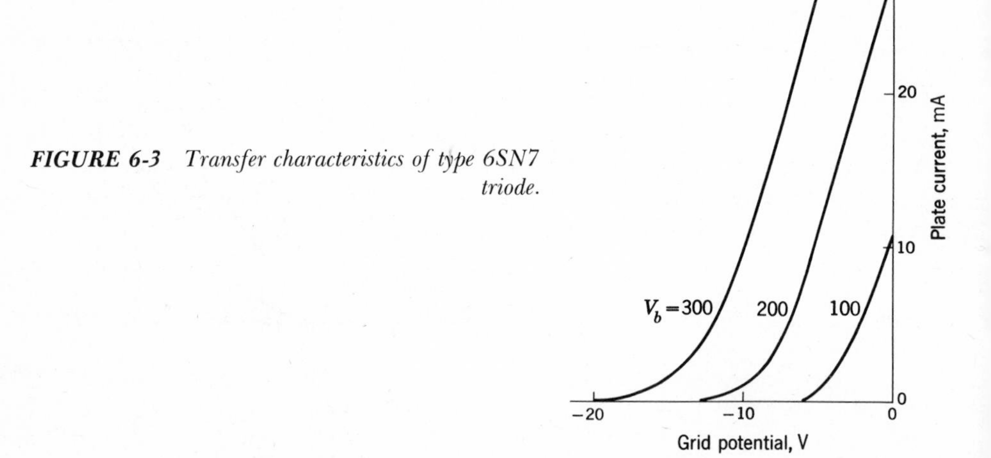

FIGURE 6-3 *Transfer characteristics of type 6SN7 triode.*

that the transfer characteristics contain the same basic information as the plate characteristics, so that one set of curves may be determined from the other. Actually, the dynamic transfer characteristic must be used to study the input-output properties because of the voltage drop across the load resistance. This is found by the same procedure developed for the vacuum diode in Chap. 4.

Cathode and Source Bias A grid-bias battery is economically impractical in most circuits. Instead, negative grid bias is obtained by inserting a resistor in series with cathode, Fig. 6-4. The *IR* drop across this resistor is $I_b \times R_k$, and the polarity makes the grid negative with respect to the cathode. The quiescent operating point may be found by first drawing the load line corresponding to

$$I_b = \frac{V_{bb} - V_b}{R_L + R_k} \tag{6-2}$$

According to Eq. (6-2) the intercepts of the load line are at V_{bb} and at

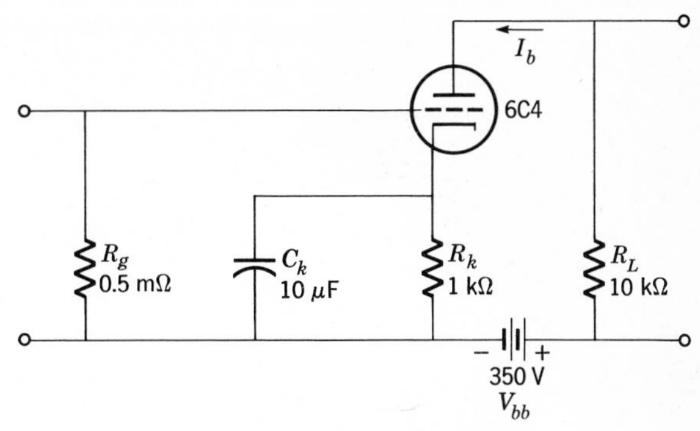

FIGURE 6-4 *Circuit diagram of a practical triode amplifier using cathode bias.*

$V_{bb}/(R_L + R_k)$, Fig. 6-5. By definition, the operating point must be located somewhere along the load line. Since the grid bias now depends upon the plate current, it is necessary to draw a second line connecting the points $I_b = -V_c/R_k$ on every plate-characteristic curve. The intersection of this curve with the load line is the operating point.

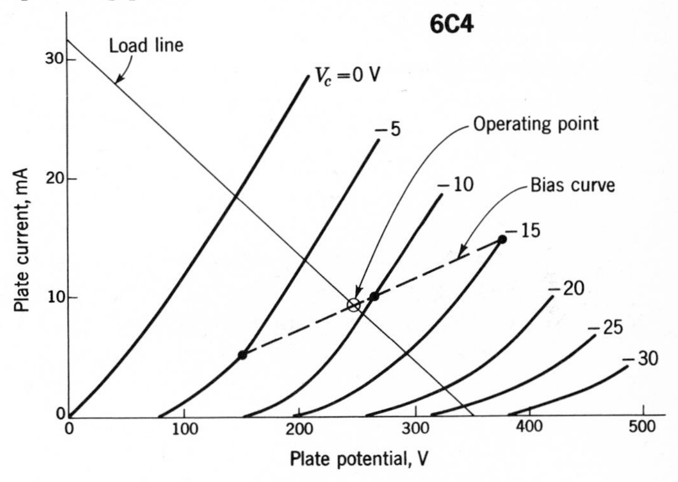

FIGURE 6-5 *Determination of operating point for 6C4 amplifier in Fig. 6-4.*

Capacitor C_k shunting the cathode-bias resistor R_k in Fig. 6-4 prevents ac signals caused by the ac plate current in the cathode resistor from appearing in the grid circuit. This is considered in greater detail in a later section.

A source-bias circuit quite analogous to Fig. 6-4 is appropriate for FET amplifiers also. To reduce changes in operating point with temperature, the gate potential is often biased positively with respect to ground, Fig.

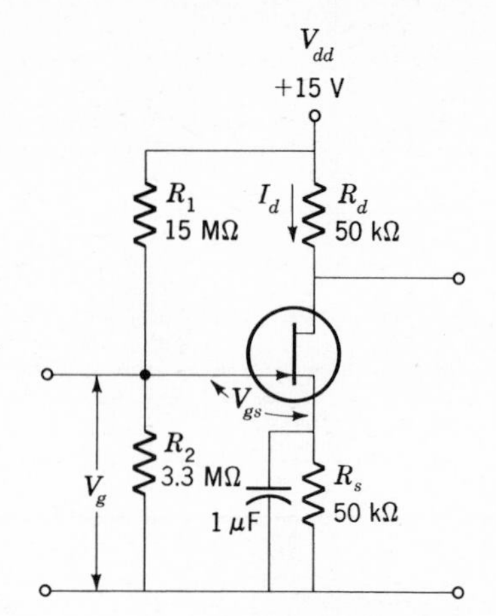

FIGURE 6-6 *Improved FET bias circuit.*

6-6, so that a larger source resistor may be used. The operating point may be found by the same technique used for triodes, although the following alternate procedure is more direct. The input circuit characteristic

$$I_d = \frac{V_g}{R_s} - \frac{1}{R_s} V_{gs} \tag{6-3}$$

is plotted on the transfer characteristic. The intersection of Eq. (6-3) with the transfer characteristic curve is the operating point. In Eq. (6-3) V_g is set by the voltage-divider action of R_1 and R_2. The net result of V_g and the drop across R_s develops the proper polarity gate-source voltage to keep the gate junction under reverse bias.

Similar considerations apply to the depletion MOSFET since the gate-bias polarity is the same as for the FET. In the enhancement MOSFET the gate-source voltage has the same polarity as the drain-source potential, according to Fig. 5-22*b*. This is achieved very simply by connecting the

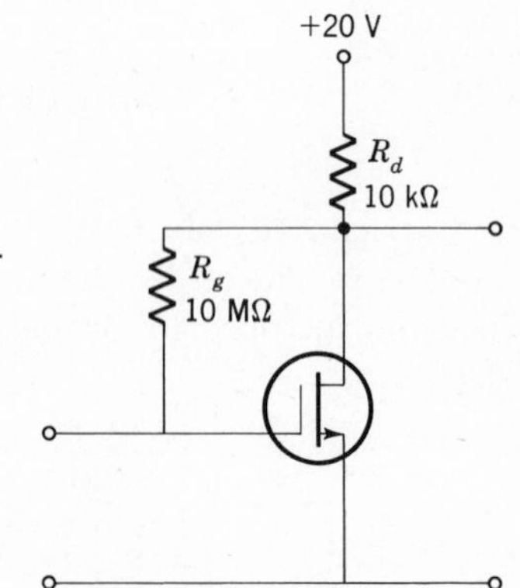

FIGURE 6-7 *Simple enhancement MOSFET amplifier.*

gate to the drain through a large resistor, Fig. 6-7, so that the gate voltage equals the drain voltage. The operating point is determined by noting the point on the load line for which the drain voltage and gate voltage are equal.

Screen Bias It is advantageous to derive the screen potential for a pentode from the plate-supply voltage. This can be accomplished by means of a resistive voltage divider across the plate-supply voltage. A simple series dropping resistor, Fig. 6-8, is even more convenient, however. The value

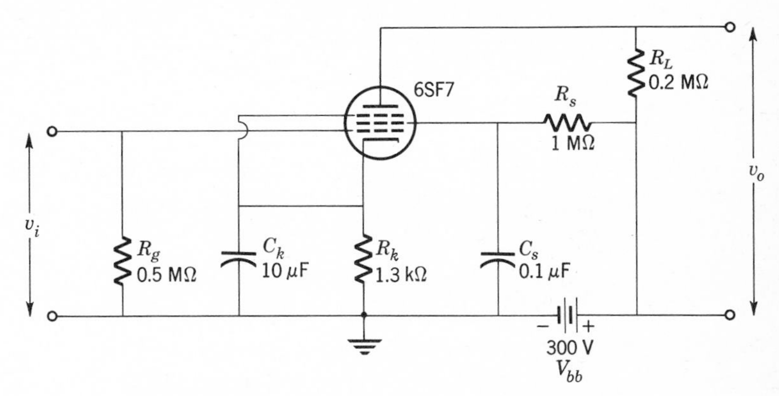

FIGURE 6-8 *Practical pentode amplifier circuit.*

of R_s is selected to yield the desired screen voltage at the given screen current and plate-supply potential. Values in the range from 0.05 to 1.0 MΩ are typical. The screen resistor is bypassed with capacitor C_s to maintain the screen voltage constant independent of variations caused by signal voltages.

Determining the operating point for cathode-biased pentodes is slightly more involved than for triodes because the bias voltage depends upon the screen current as well as the plate current. It is usually satisfactory to assume the screen current is a fixed fraction of the plate current and proceed as in the case of the triode. The correct fraction to choose depends somewhat upon the tube type in question and may be estimated from the tube's screen-current characteristics. For most pentodes the total cathode current is approximately $1.3I_b$ at the operating point. This approximation usually yields a sufficiently accurate determination of the quiescent condition.

In many cases it is simpler to resort to the following "cut-and try" procedure for determining the operating point. Choose a point on the load line corresponding to an arbitrary grid-bias voltage and plate current. The screen current is then determined from the screen characteristics for these values of V_c and V_b. The product $-(I_b + I_s)R_k$ is compared with the chosen value of V_c. If these two values are equal, the original choice is satisfactory. If they are not, the process is repeated until the desired accuracy is obtained.

Aside from the obvious economy achieved by providing grid, plate, and screen potentials from one voltage source, these bias techniques also result in more stable quiescent operation than does fixed bias. Suppose, for example, that the plate current tends to increase because of tube aging. This increases the negative grid bias and this, in turn, tends to lower the plate current. The net change in operating point is much less than is the case for fixed bias. The same situation exists with regard to the screen voltage.

It is inconvenient to measure electrode voltages with respect to the cathode in circuits employing a cathode-bias resistor. This is particularly true when more than one tube is used in a circuit. The usual practice is to refer all potentials to a common point called the *ground.* The ground point is considered to be electrically neutral so that, for example, the ground points of two different circuits may be connected with no influence upon the operation of either circuit. A typical use of the ground symbol, Fig. 6-9, is

FIGURE 6-9 *Ground symbol.*

illustrated in the pentode-amplifier diagram of Fig. 6-8. Practical electronic circuits are often constructed on a metal base, or *chassis,* which serves as the ground.

SMALL SIGNALS

Small-signal Parameters Very frequently the signal amplitudes applied to a vacuum tube are small compared with the full range of voltages covered by the plate characteristics. In this situation graphical analysis of tube performance is inaccurate because the plate characteristics are not given with sufficient precision. It is possible to represent the tube characteristics using an approximation to Child's law which is quite accurate for small signals. Once the operating point is established graphically, small departures about the operating point caused by small ac signals are treated by assuming the triode is a linear device. This approach may be illustrated in the following way. According to Child's law derived in Chap. 5, the plate current is given by

$$I_b = A(\mu V_c + V_b)^{3/2} \tag{6-4}$$

where A is a geometrical constant, μ is the *amplification factor,* V_c is the grid voltage, and V_b is the plate voltage. The change in plate current caused by a change in plate voltage is

$$\Delta I_b = \frac{3A}{2} (\mu V_c + V_b)^{1/2} \Delta V_b$$

$$= \tfrac{3}{2} A^{2/3} I_b^{1/3} \Delta V_b \tag{6-5}$$

where Eq. (6-4) has been used. The ratio $\Delta I_b / \Delta V_b$ can be identified as the reciprocal of an equivalent resistance called the *plate resistance*

$$\frac{1}{r_p} = \frac{\Delta I_b}{\Delta V_b} \tag{6-6}$$

The plate current also changes with a change in grid voltage,

$$\Delta I_b = \frac{3A}{2} (\mu V_c + V_b)^{1/2} \mu \, \Delta V_c$$

$$= \frac{3\mu}{2} A^{2/3} I_b^{1/3} \Delta V_c \tag{6-7}$$

The ratio $\Delta I_b / \Delta V_c$ is the change in current in one circuit resulting from a change in voltage in another and has the dimensions of reciprocal resistance, or conductance. Accordingly, this ratio is called the *mutual transconductance,*

$$g_m = \frac{\Delta I_b}{\Delta V_c} \tag{6-8}$$

The total change in plate current, from Eqs. (6-6) and (6-8), is written

$$\Delta I_b = \frac{1}{r_p} \Delta V_b + g_m \Delta V_c \tag{6-9}$$

A useful relation between r_p and g_m is obtained by noting that if the net change in plate current is zero, Eq. (6-9) becomes

$$0 = \frac{1}{r_p} \Delta V_b + g_m \Delta V_c$$

$$g_m r_p = -\frac{\Delta V_b}{\Delta V_c} \tag{6-10}$$

The ratio $-\Delta V_b / \Delta V_c$ is just the definition of the amplification factor. The minus sign signifies that the plate voltage decreases when the grid potential increases, as discussed previously. Therefore, Eq. (6-10) is written

$$\mu = r_p g_m \tag{6-11}$$

This relation is useful in obtaining one of the three parameters if the other two are known.

According to Eq. (6-9) small changes in plate current around the operating point caused by changes in grid voltage and plate voltage can be calculated if suitable values of the *small-signal parameters* μ, r_p, and g_m appropriate to the operating point are known. The plate resistance is just the reciprocal of the slope of the plate characteristic at the operating point and g_m is the slope of the transfer characteristic at the operating point so that quantitative values can be obtained from the current-voltage characteristics of the tube. Typical values of small-signal parameters for several triodes are given in Table 6-1.

TABLE 6-1 TRIODE SMALL-SIGNAL PARAMETERS

Type	μ	r_p, $10^3\ \Omega$	g_m, 10^{-3} mho
6C4	20	6.3	3.1
6CW4	68	5.4	12.5
6SL7	70	44	1.6
12AT7	55	5.5	10
12AU7	17	7.7	2.2
12AX7	100	62	1.6
7895	74	7.3	10.9

The magnitudes of the small-signal parameters depend upon the operating point. For this reason it is usually necessary to evaluate μ, r_p, and g_m graphically at the point determined by the dc potentials and tube characteristics. To a first approximation, the amplification factor is independent of the operating point, since it depends only upon the ratio of the interelectrode capacitances, Eq. (5-34). The plate resistance depends upon the plate current, according to (6-6) and (6-5),

$$r_p = \tfrac{2}{3}A^{-2/3}I_b^{\ -1/3} \tag{6-12}$$

Similarly for the mutual transconductance, using (6-7) and (6-8),

$$g_m = \tfrac{3}{2}A^{2/3}\mu I_b^{\ 1/3} \tag{6-13}$$

According to (6-12) and (6-13), the plate resistance decreases slowly as the plate current increases while the transconductance increases.

Since Child's law is not useful as a quantitative description of triode behavior, it is not expected that Eqs. (6-12) and (6-13) exactly represent the variation of the small-signal parameters with tube current. Nevertheless, experimental data for a practical triode, Fig. 6-10, are in surprisingly good agreement with the analytical results. Note that the amplification factor is sensibly constant, except at the smallest plate currents, while g_m and r_p respectively increase and decrease with increasing plate current. According to Fig. 6-10, it is possible to obtain considerable variation in

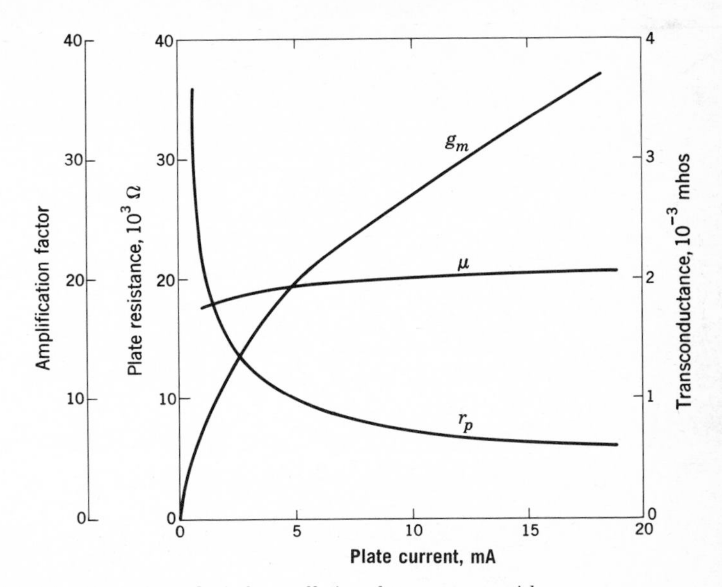

FIGURE 6-10 *Variation of triode small-signal parameters with plate current.*

the small-signal parameters by selecting different operating points. This is important in the design of vacuum-tube circuits for specific applications.

Triode Equivalent Circuit The small-signal parameters of a triode are used to calculate the performance of the tube in any circuit, so long as the signal voltages and currents are small compared with the quiescent dc values. According to Eq. (6-9), the ac component of the plate current is a function of the ac grid potential and the ac plate voltage so that

$$i_p = g_m v_g + \frac{1}{r_p} v_p \tag{6-14}$$

Henceforth the subscripts g and p signify the ac components of the grid and plate voltages (and currents), respectively. It must be kept in mind that the total electrode current or voltage has, in addition, a dc component corresponding to the operating point. After multiplying Eq. (6-14) by r_p and using Eq. (6-11), the ac plate voltage is given by

$$v_p = -\mu v_g + i_p r_p \tag{6-15}$$

Equation (6-15) may be interpreted as the series combination of a generator $-\mu v_g$ in series with a resistor r_p, as shown in Fig. 6-11. This configuration is reminiscent of the Thévenin equivalent circuit discussed in Chap. 1. Its validity in representing the performance of a triode rests on the concept of small-signal parameters for a vacuum tube. Since Eq. (6-14) refers only to the ac components of the tube currents and voltages, Fig. 6-11 is called the *triode ac equivalent circuit.*

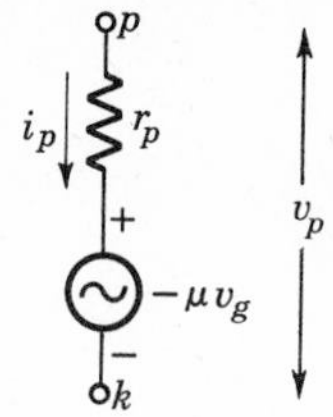

FIGURE 6-11 *Triode ac equivalent circuit.*

Consider the simple triode amplifier illustrated in Fig. 6-12*a*. According to Fig. 6-11, the equivalent circuit of the amplifier is obtained by replacing the triode with a generator and a resistor, as in Fig. 6-12. Note that the dc battery supplies are omitted since the equivalent circuit refers only to ac quantities and the ac impedance of a battery is negligible. All other circuit components are included, however. According to Fig. 6-12*b*, there is no direct electrical connection between the input circuit and the output circuit. This illustrates the valve action of the grid in controlling the plate current with essentially zero power expenditure. Note that the minus sign associated with the generator accounts for the 180° phase difference between the grid voltage and the plate voltage.

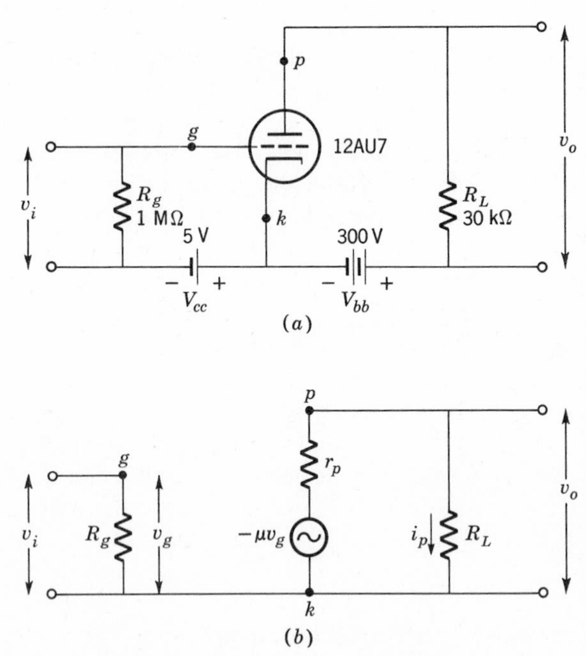

FIGURE 6-12 *(a) Triode amplifier and (b) its ac equivalent circuit.*

The output voltage of the amplifier circuit is calculated immediately from the equivalent circuit, Fig. 6-12*b*. The result is

$$v_o = i_p R_L = \frac{-\mu v_g}{r_p + R_L} R_L = \frac{-\mu v_g}{1 + r_p/R_L} \tag{6-16}$$

The ratio of the output signal to the input signal is called the *gain* of the amplifier. Suppressing the minus sign, since it represents only a 180° phase shift, and noticing that $v_g = v_i$, the gain of the amplifier is

$$a = \frac{v_o}{v_i} = \frac{\mu}{1 + r_p/R_L} \tag{6-17}$$

According to Eq. (6-17), the maximum gain of the circuit is equal to the amplification factor of the triode. This value is achieved when the load resistance is much greater than the plate resistance.

It is not always desirable to make the load resistance much larger than the triode plate resistance in a practical amplifier. A large load resistance introduces considerable dc power loss and requires a large plate supply voltage V_{bb} to put the tube at its optimum operating point. The amplification factor of most tubes is sufficiently large to yield appreciable gains with lower values of load resistance. It is also possible to use a transformer in the plate circuit. The impedance-matching properties of a transformer reflect a large ac impedance from the secondary, as explained in Chap. 3. In this case the dc resistance corresponds to the primary-winding resistance, which may be quite small. The expense and limited frequency range of transformers restrict their use to rather special applications.

The magnitude of the grid resistor R_g in this circuit is limited only by the voltage drop caused by residual grid current. The corresponding *IR* drop across R_g can change the grid bias from the value set by the grid bias battery V_{cc}. Since grid current is normally an uncontrolled quantity and varies considerably even between tubes of the same type, this is an undesirable condition. Grid resistances in the 0.5- to 10-MΩ range are satisfactory for most triodes. These values are large enough that the loading effect of the grid resistor on the input voltage source may be safely ignored in most applications.

The cathode bias triode amplifier is analyzed in identical fashion. The ac equivalent circuit corresponding to Fig. 6-4 is shown in Fig. 6-13. The impedance of R_k and C_k in parallel is

$$\mathbf{Z}_k = \frac{R_k}{1 + jR_k \omega C_k} \tag{6-18}$$

The ac signal current is given by

$$i_p = \frac{-\mu v_g}{r_p + R_L + \mathbf{Z}_k} \tag{6-19}$$

According to Fig. 6-4, the grid-cathode voltage is equal to the input voltage plus the ac voltage across $\mathbf{Z}_k$, or

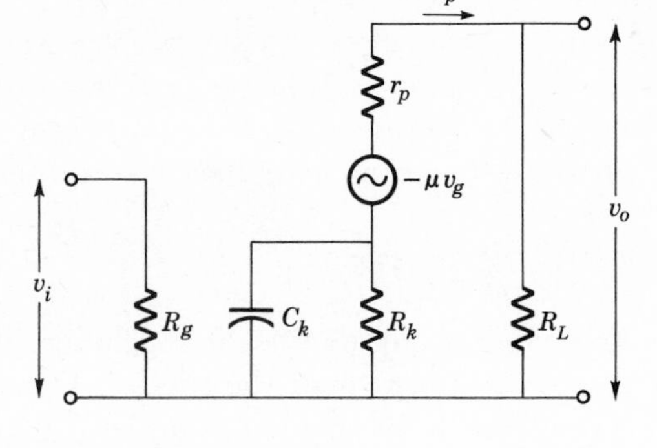

FIGURE 6-13 *The ac equivalent circuit of cathode-bias amplifier.*

$$v_g = v_i + i_p\mathbf{Z}_k \tag{6-20}$$

Inserting Eq. (6-20) into Eq. (6-19) and calculating the gain of the amplifier,

$$a = \mu \frac{R_L}{r_p + R_L + (1 + \mu)\mathbf{Z}_k} \tag{6-21}$$

Suppose now the capacitor is omitted, so $\mathbf{Z}_k = R_k$. Comparing Eq. (6-21) with Eq. (6-17) shows that the gain is reduced, because of the factor $(1 + \mu)R_k$ in the denominator. If, on the other hand, the capacitor is large enough that $(1 + \mu)\mathbf{Z}_k$ is small compared with the plate resistance plus load resistance, the gain is restored to its original value. In effect the capacitor provides an ac path around R_k and, accordingly, it is called a *cathode bypass* capacitor.

Norton Equivalent Circuit Pentodes may be represented by the same equivalent circuit as triodes so long as appropriate values of the small-signal parameters are used. These parameters depend upon the operating point and therefore vary with the screen and suppressor potentials, in addition to the grid bias and plate voltage. Because of the constant-current properties of pentodes, it is generally most useful to use the Norton representation in preference to the Thévenin circuit. According to Eq. (1-85), the equivalent current generator is given by the ratio of the Thévenin equivalent voltage generator divided by the equivalent internal resistance.

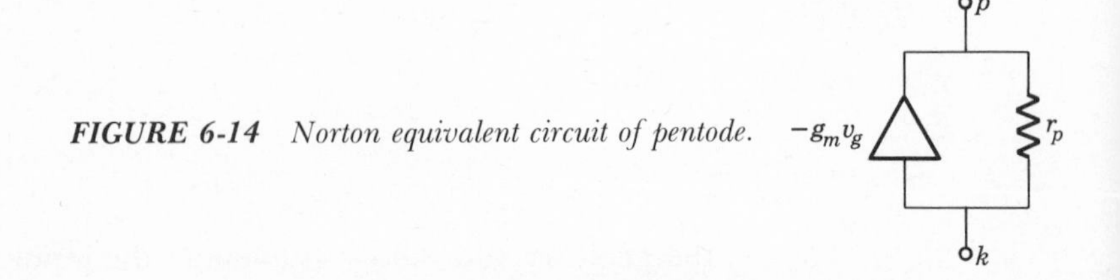

FIGURE 6-14 *Norton equivalent circuit of pentode.*

Therefore, using Fig. 6-11,

$$I_{eq} = \frac{-\mu v_g}{r_p} = -g_m v_g \tag{6-22}$$

The Norton equivalent circuit for a pentode is shown in Fig. 6-14.

A pentode amplifier stage, Fig. 6-8, is represented by the equivalent circuit shown in Fig. 6-15. The output voltage is simply the constant-

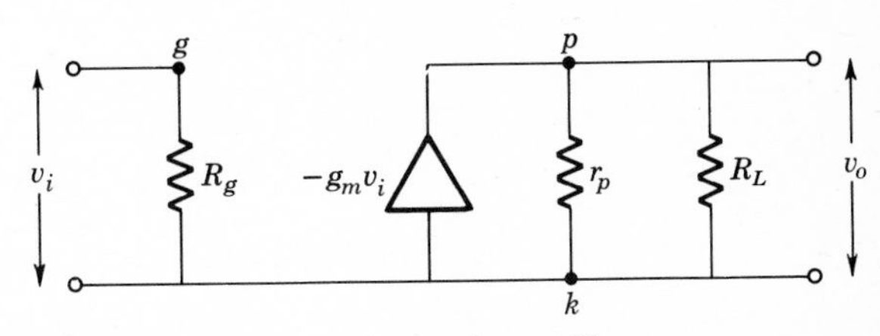

FIGURE 6-15 *Equivalent circuit of pentode amplifier.*

current source times the parallel combination of the plate resistance and load resistance,

$$v_o = -g_m v_i \frac{r_p R_L}{r_p + R_L} \tag{6-23}$$

Neglecting the minus sign, the gain is

$$a = \frac{v_o}{v_i} = g_m \frac{R_L}{1 + R_L/r_p} \tag{6-24}$$

In most cases, the plate resistance is much greater than the load resistance so that a satisfactory approximation to Eq. (6-24) is just

$$a = g_m R_L \tag{6-25}$$

Plate load resistors much in excess of 0.1 MΩ or so are inconvenient, because the excessive dc voltage drop caused by the plate current requires a very large plate-voltage source. Nevertheless, amplifications as great as 1000 are obtainable from a single pentode amplifier stage.

The similarity between the current-voltage characteristics of pentodes and FETs indicates that the Norton equivalent circuit is equally appropriate in the latter case as well. The variation of the mutual transconductance of an FET with drain current is determined with the aid of Eq. (5-40),

$$\Delta I_d = \frac{2I_{dss}}{V_p}\left(\frac{V_g}{V_p} - 1\right)\Delta V_g \tag{6-26}$$

In terms of I_d, using Eq. (5-40),

$$g_m = \frac{2I_{dss}^{1/2}}{V_p} I_d^{1/2} \tag{6-27}$$

Apparently, the transconductance of a FET increases somewhat more rapidly with current than is the case for a triode. Typical values range from 5 to 10 × 10^{-3} mho.

SINGLE-STAGE AMPLIFIERS

Many different amplifier circuits are considered in subsequent chapters, but it is useful to analyze a few special examples at this point. These circuits have important applications and the analysis illustrates the equivalent-circuit approach.

Cathode Follower The cathode is common to the input and output terminals in the simple amplifier circuit, insofar as ac signals are concerned. This configuration is used most often and is sometimes referred to as the *grounded-cathode connection.* Another configuration is the *grounded-plate* amplifier, Fig. 6-16. The plate resistor is omitted, and the output

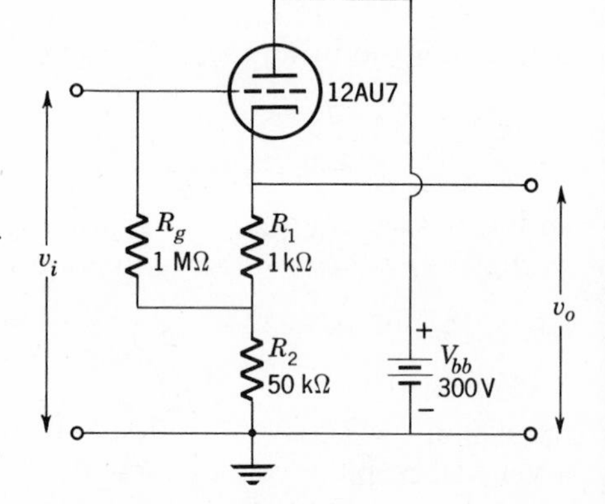

FIGURE 6-16 *Practical cathode-follower circuit.*

signal is developed across the cathode resistor. Note that in this circuit the plate is common to both input and output terminals and is at ground potential so far as ac signals are concerned. It is maintained at a high dc potential, however. Grid bias is developed across R_1 and the operating point is determined in the same fashion as for a grounded cathode amplifier. This circuit is commonly called a *cathode follower* because the cathode potential follows that of the grid. The equivalent circuit of the cathode follower is given in Fig. 6-17 assuming $R_k = R_2 \gg R_1$. We proceed to find the output signal by first calculating the current

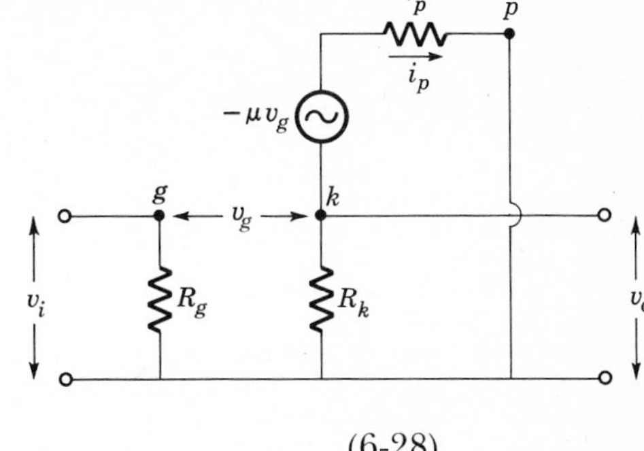

FIGURE 6-17 Equivalent circuit of a cathode follower.

$$i_p = \frac{-\mu v_g}{r_p + R_k} \tag{6-28}$$

The grid-cathode voltage is equal to the input voltage plus the voltage across the cathode resistor, or

$$v_g = v_i + i_p R_k \tag{6-29}$$

Inserting Eq. (6-29) into Eq. (6-28), the current in the circuit is

$$i_p = \frac{-\mu v_i}{(1 + \mu)R_k + r_p} \tag{6-30}$$

The drop across R_k caused by this current represents the output voltage. Note, however, that the output voltage is the negative of $i_p R_k$, because current always enters the positive terminal of a load. Therefore

$$v_o = \frac{\mu v_i R_k}{(1 + \mu)R_k + r_p} \tag{6-31}$$

The gain of the amplifier,

$$a = \frac{v_o}{v_i} = \frac{\mu}{(1 + \mu) + r_p/R_k} \tag{6-32}$$

is always less than unity. In fact, if $(1 + \mu)R_k \gg r_p$, which is usually the case,

$$a = \frac{\mu}{\mu + 1} \approx 1 \tag{6-33}$$

The voltage gain of the cathode-follower circuit is essentially equal to unity, and, according to Eq. (6-31), the output voltage is in phase with the input signal. Therefore, the cathode voltage follows the input voltage very closely.

The virtue of the cathode follower is that the cathode voltage replicates the input signal in a much lower impedance circuit. Therefore, the internal impedance of the circuit as a power source is very low. This means the tube can deliver appreciable power to low-impedance loads.

The equivalent internal impedance is determined using Thévenin's theorem. The theorem cannot be applied directly to Fig. 6-17, because the generator voltage $-\mu v_g$ depends upon the load current through R_k, according to Eq. (6-29). To find a more suitable circuit, Eq. (6-30) is put in the form

$$i_p = \frac{-v_i\mu/(1+\mu)}{R_k + r_p/(1+\mu)} \tag{6-34}$$

This expression is the ratio of an emf, $-v_i\mu/(1+\mu)$, divided by a resistance, $R_k + r_p/(1+\mu)$, which can be represented by the diagram of Fig. 6-18. Applying Thévenin's theorem to this circuit, the equivalent internal impedance is the parallel combination of R_k and $r_p/(1+\mu)$,

$$R_o = \frac{R_k r_p/(1+\mu)}{R_k + r_p/(1+\mu)} = \frac{R_k r_p}{(1+\mu)R_k + r_p} \tag{6-35}$$

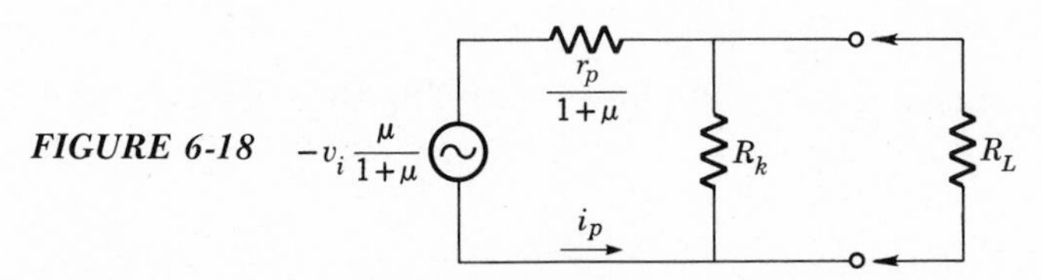

FIGURE 6-18

The *output impedance* of any circuit is the ratio of the output voltage divided by the output current when the input signal is zero. This is simply the internal impedance in the Thévenin equivalent representation. According to Eq. (6-35), the output impedance of the cathode follower is

$$R_o = \frac{r_p}{(1+\mu) + r_p/R_k} \approx \frac{r_p}{1+\mu} \approx \frac{r_p}{\mu} \tag{6-36}$$

This means that the output impedance is small. Note that the circuit results in an output impedance much smaller than R_k, since $R_k \gg r_p/\mu$.

The input impedance is the ratio of the input voltage to the current in the input circuit

$$R_i = \frac{v_i}{i_i} \tag{6-37}$$

Applying Kirchhoff's voltage rule to the input circuit, the input current is

$$i_i = \frac{v_i + i_p R_2}{R_g + R_2} \tag{6-38}$$

The current i_p is given by Eq. (6-30). Substituting into Eq. (6-38),

$$i_i = \frac{v_i}{R_g + R_2}\left[1 - \frac{\mu R_2}{(1+\mu)\,R_k + r_p}\right]$$

$$= \frac{v_i}{R_g + R_2}\left[1 - \frac{\mu}{(1+\mu)\,R_k/R_2 + r_p/R_2}\right] \tag{6-39}$$

Most often, $R_2 \sim R_k$ and $(1+\mu)R_k \gg r_p$, so that

$$i_i = \frac{v_i}{R_g + R_2}\left(1 - \frac{\mu}{1+\mu}\right) = \frac{v_i}{R_g + R_2}\,\frac{1}{1+\mu} \tag{6-40}$$

Comparing Eq. (6-40) with Eq. (6-37), the input impedance is

$$R_i = (1+\mu)\,(R_g + R_2) \tag{6-41}$$

which is quite large. This is an advantage since the cathode follower presents a very light load to preceding circuits. The reason for the high input impedance is, of course, the opposing voltage introduced into the grid circuit from the cathode resistor.

The cathode follower is an impedance-matching device with a very high input impedance and a very low output impedance. Although the voltage gain is less than unity, the power gain may be appreciable. This is so because the signal voltage is the same in both the high-impedance input circuit and the low-impedance output circuit. Since power is proportional to V^2/R, the power amplification is large. In addition, the circuit performance is stable and relatively independent of changes in component values. According to Eq. (6-33), the gain depends only on the amplification factor, which is the most stable of the small-signal parameters.

Difference Amplifier The circuit diagramed in Fig. 6-19 is often used as the input stage of oscilloscope amplifiers and other laboratory instruments, because it yields a signal proportional to the difference between two input voltages. This useful property may be illustrated by analyzing the

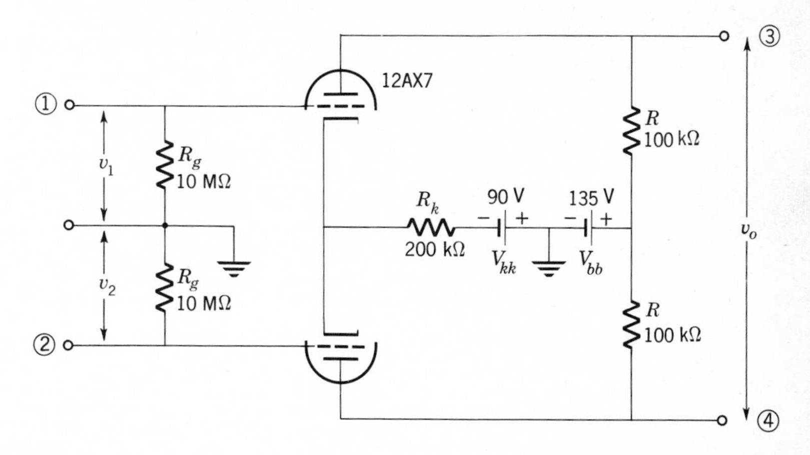

FIGURE 6-19 *Difference amplifier. Half-open tube symbol signifies a dual triode.*

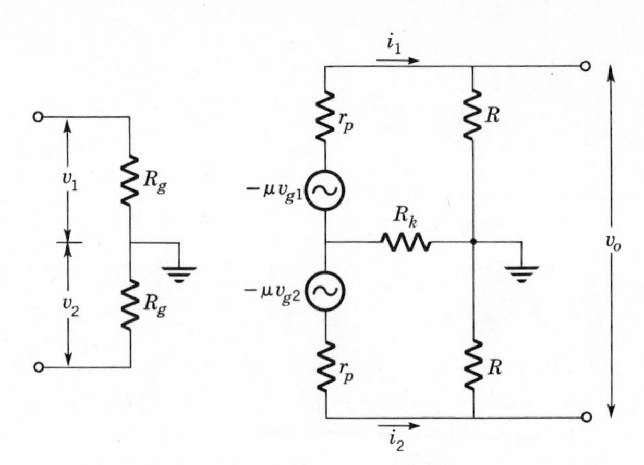

FIGURE 6-20 *Equivalent circuit of difference amplifier.*

equivalent circuit, Fig. 6-20. Kirchhoff's voltage rule applied to both loops results in

$$-\mu v_{g_1} = r_p i_1 + R i_1 + R_k(i_1 + i_2)$$
$$-\mu v_{g_2} = r_p i_2 + R i_2 + R_k(i_1 + i_2) \tag{6-42}$$

Similarly, the grid-cathode voltages are

$$v_{g_1} = v_1 + R_k(i_1 + i_2)$$
$$v_{g_2} = v_2 + R_k(i_1 + i_2) \tag{6-43}$$

Equations (6-43) are substituted into (6-42) and the result solved for the two currents. The expression for the current in the upper loop is

$$i_1 = \mu \frac{v_2 - [1 + (r_p + R)/(1 + \mu)R_k]v_1}{\{[1 + (r_p + R)/(1 + \mu)R_k]^2 - 1\}(1 + \mu)R_k} \tag{6-44}$$

Because of the circuit symmetry, the result for i_2 has the same form with subscripts 1 and 2 interchanged. The plate-to-plate output voltage is

$$v_o = Ri_1 - Ri_2 = R(i_1 - i_2) \tag{6-45}$$

Substituting for i_1 and i_2 and simplifying,

$$v_o = \mu(v_2 - v_1) \frac{2 + (r_p + R)/(1 + \mu)R_k}{(1 + r_p/R)[2 + (r_p + R)/(1 + \mu)R_k]} \tag{6-46}$$

$$v_o = \frac{\mu}{1 + r_p/R}(v_2 - v_1) \tag{6-47}$$

which is similar to Eq. (6-17) for the amplification of a single triode. According to Eq. (6-47), the output voltage is a constant times the difference between the input signals.

The *difference amplifier* rejects any voltage signals that are common to both terminals, such that $v_1 = v_2$. At the same time, signals applied from terminal 1 to terminal 2 are amplified normally, since in this case

$$v_1 = \frac{v_i}{2} \qquad \text{and} \qquad v_2 = -\frac{v_i}{2} \tag{6-48}$$

The output voltage is

$$v_0 = \frac{-\mu}{1 + r_p/R}\left(\frac{v_i}{2} + \frac{v_i}{2}\right) = \frac{-\mu}{1 + r_p/R}\, v_i \tag{6-49}$$

which means the circuit simply amplifies the input signal. The same result holds if the input signal is connected between ground and either grid, so that, for example, $v_1 = v_i$ and $v_2 = 0$. Since the circuit rejects *common-mode* signals, resulting from, say, stray electrical fields caused by the 60-Hz power mains, it is useful as the input stage of a sensitive amplifier. In addition, the input may be connected to voltage sources that have neither terminal grounded, which is often very convenient.

The ability of the difference amplifier to reject common-mode signals rests on exact symmetry in the two halves of the circuit. Since this is rarely achieved in practice because of manufacturing tolerances, it is useful to examine how slight asymmetries influence performance. Suppose, for example, the triodes are slightly different so that $r_{p1} \neq r_{p2}$. The current in the upper loop is, from Eq. (6-44),

$$i_1 = \frac{\mu}{D_1}\left\{v_2 - \left[1 + \frac{r_{p1} + R}{(1 + \mu)R_k}\right] v_1\right\} \tag{6-50}$$

where D_1 represents the denominator of Eq. (6-44). A similar expression with interchanged subscripts may be written for i_2. Substituting into Eq. (6-45), the output signal is

$$v_o = \frac{2\mu R}{D_1}\left[(v_2 - v_1) + v_2\,\frac{r_{p2} + R}{2(1 + \mu)R_k} - v_1\,\frac{r_{p1} + R}{2(1 + \mu)R_k}\right] \tag{6-51}$$

where it is assumed that $D_1 \approx D_2$. When $2(1 + \mu)\,R_k \gg r_p + R$ the denominator becomes $2(r_p + R)$ and Eq. (6-51) reduces to Eq. (6-47). That is, a large value of cathode resistor reduces the effect of circuit asymmetry.

In particular, if $v_1 = v_2 = v_i$, Eq. (6-51) yields

$$v_o' = \frac{\mu v_i}{1 + r_p/R}\,\frac{r_{p2} - r_{p1}}{2(1 + \mu)R_k} \tag{6-52}$$

which shows that an output signal results from circuit imbalance. The ratio of the difference-signal output, Eq. (6-49), to the common-mode signal, Eq. (6-52),

$$\frac{v_o}{v'_o} = \frac{2(1+\mu)R_k}{r_{p2} - r_{p1}} \tag{6-53}$$

is called the *common-mode rejection ratio.* The value of this ratio, and hence the ability of the circuit to reject common-mode signals, is enhanced by large values of R_k.

FET voltmeter The amplification and large input impedance of an FET results in a sensitive electronic voltmeter when used in conjunction with a standard d'Arsonval milliammeter. A common circuit, Fig. 6-21, is a

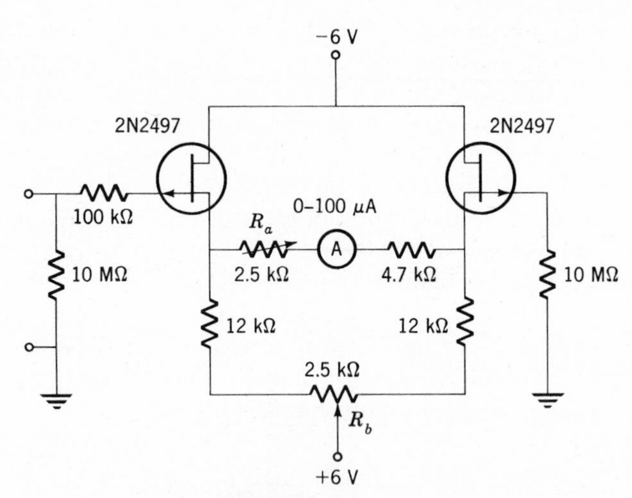

FIGURE 6-21 *FET electronic voltmeter.*

source-follower difference amplifier with the meter connected between the source terminals. The unknown voltage is applied to one of the gates while the other gate is unused and returned to ground. Note that *p*-channel FETs are used in Fig. 6-21; a voltmeter employing *n*-channel units is equally possible. Since a source follower presents a high input impedance and a low output impedance, this circuit is particularly useful in the voltmeter application.

The balanced circuit provides a convenient connection for the d'Arsonval meter which results in zero deflection with zero input signal, even though the quiescent current in each FET is nonzero. Furthermore, the symmetry of the circuit maintains performance stability against component changes caused by aging. Actually the circuit may be looked upon as a bridge comprising a FET in each of two arms together with the source resistors in the other two arms. A dc voltage applied to the gate changes the FET resistance and unbalances the bridge. The circuit may be analyzed using equivalent circuits in the usual fashion.

The adjustable resistor R_a in series with the meter is a calibration resistor in the diagram of Fig. 6-21. The potentiometer R_b is a balance adjustment which corrects for minor asymmetries in the FETs or other components. The circuit is balanced for zero current in the meter by adjusting R_b with zero voltage applied to the gate.

The input gate is connected to multiplier resistors in commercial instruments to achieve multiple-range performance, much as in the case of the VOM circuit. The resistor values are greater in the electronic instrument, however, because of the much greater sensitivity. An additional feature of interest is that the maximum meter current, which occurs when one of the FETs is completely cut off, does not damage the meter. Therefore, the meter movement is unharmed, even if the input is inadvertently connected to a voltage source much larger than the full-scale value.

Standard instruments provide for ac measurements by including a diode rectifier circuit, as discussed in the previous chapter. The high input impedance of the source-follower difference amplifier is ideally suited to measure the output of the rectifier circuit. In addition to ac and dc voltages, most instruments also measure resistance with an ohmmeter circuit similar to those discussed in Chap. 1. Again, the difference amplifier is used as the indicator. This permits a very wide range of operation. For example, quite inexpensive instruments are capable of measuring unknown resistances from 1 Ω to 100 MΩ. An earlier version of this circuit employing a dual triode is called a *vacuum-tube voltmeter,* or *VTVM.*

JUNCTION-TRANSISTOR AMPLIFIER

Many considerations of previous sections apply equally to the junction-transistor amplifier. Since, however, the transistor is a current-controlled device, there are certain notable differences as well. FET and vacuum-tube circuits have their transistor counterparts, although circuit operation is modified by the properties of transistors.

Biasing The emitter junction of a transistor requires forward bias and the collector junction reverse bias. Consider, for example, the simple grounded-emitter amplifier, Fig. 6-22, in which base bias current is supplied by the resistor R_B. Because the forward resistance of the emitter junction is very small, the base current is given by

$$I_b = \frac{V_{cc}}{R_B} \tag{6-54}$$

An approximate value of collector current is then βI_b, and the operating point is completely determined. Since the current gain depends somewhat upon the operating point, it is more accurate to plot the load line corresponding to R_L on the appropriate collector characteristics, such as

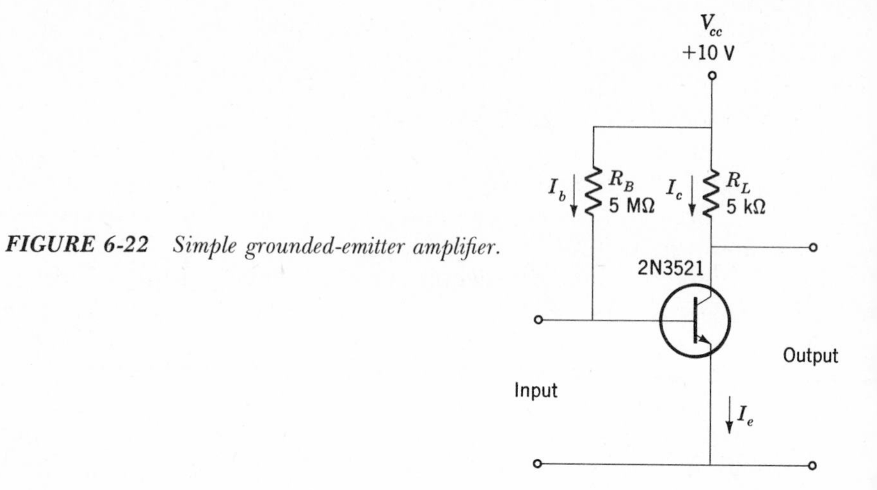

FIGURE 6-22 *Simple grounded-emitter amplifier.*

Fig. 5-29. The intersection of the load line with the base current curve given by Eq. (6-54) is the operating point.

This simple bias circuit is not generally satisfactory because the operating point shifts drastically with temperature. Comparison of collector characteristics at an elevated temperature, Fig. 5-30, with those for room temperature, Fig. 5-29, reveals that a much greater collector current exists at the higher temperature. Since the base bias current is fixed by the circuit, it is possible for the operating point to move into an unusable region of the transistor characteristics.

The most satisfactory transistor bias circuit is obtained by including a resistor in the emitter circuit, Fig. 6-23*a*. The voltage drop across R_E tends to bias the emitter junction in the reverse direction and the voltage

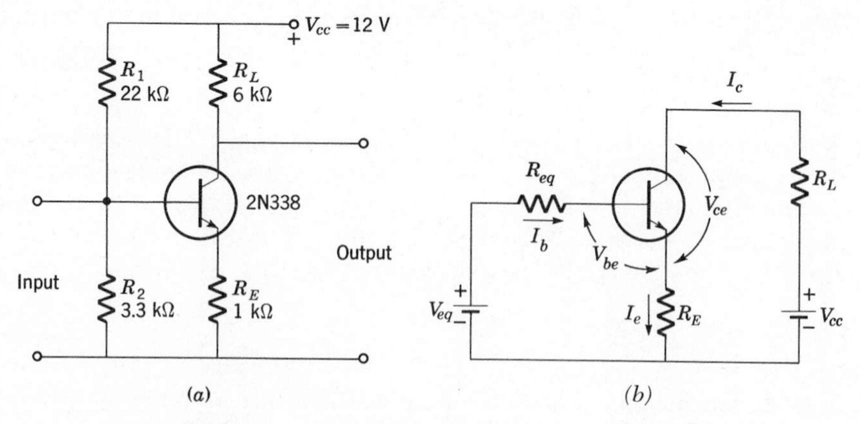

FIGURE 6-23 *(a) Practical transistor bias circuit. Note type 2N338 npn transistor is used. Circuit applies equally to pnp type if polarity of V_{cc} is reversed. (b) Dc equivalent circuit used to determine operating point.*

divider comprising R_1 and R_2 sets the base voltage so that the base-emitter potential is in the forward direction. This circuit has the advantage that an increase in transistor current changes the voltage drop across R_E such that the base-bias current is reduced, much as in the case of cathode bias of a vacuum tube.

The bias circuit of Fig. 6-23*a* is best analyzed by converting the R_1 and R_2 voltage divider to its Thévenin equivalent circuit. Using Eqs. (1-76) and (1-77), the equivalent battery and series resistance are

$$V_{eq} = \frac{R_2}{R_1 + R_2} V_{cc}$$

and

$$R_{eq} = \frac{R_1 R_2}{R_1 + R_2} \tag{6-55}$$

as shown in Fig. 6-23*b*. With this dc equivalent circuit the quiescent operating point is found as follows. First, the appropriate load line, which includes both R_L and R_E, is plotted on the transistor characteristics, Fig. 6-24. Next, writing Kirchhoff's voltage equation around the base circuit,

$$V_{eq} = I_c R_E + I_b R_{eq} + V_{be} \tag{6-56}$$

and around the collector loop,

$$V_{cc} = I_c(R_E + R_L) + V_{ce} \tag{6-57}$$

In Eqs. (6-56) and (6-57) the approximation $I_e = I_c$ has been made. These equations are solved for V_{ce} in terms of I_b. The result is

$$V_{ce} = V_{cc} - (V_{eq} - V_{be}) \left(1 + \frac{R_L}{R_E}\right) + R_{eq} \left(1 + \frac{R_L}{R_E}\right) I_b \tag{6-58}$$

In most circuits V_{be} may be considered to be a constant equal to 0.2 V for germanium transistors and 0.6 V for silicon units. The reason for this difference is that a larger forward bias is necessary when the forbidden energy gap is larger, in conformity with Eq. (5-41). The fact that V_{be} is essentially a constant is a result of the steepness of the forward characteristic of a *pn* junction (refer to Fig. 4-5).

Choosing values of I_b, the collector current is calculated from Eq. (6-58) and corresponding points plotted on the transistor characteristics. The intersection of this bias curve with the load line gives the operating point, as illustrated in Fig. 6-24.

Actually, transistor collector characteristics are so linear and evenly spaced (compare Fig. 6-24) that it is usually satisfactory to determine the operating point analytically, rather than graphically. The manufacturer usually specifies a dc current gain h_{FE}, which is the ratio of the dc collector

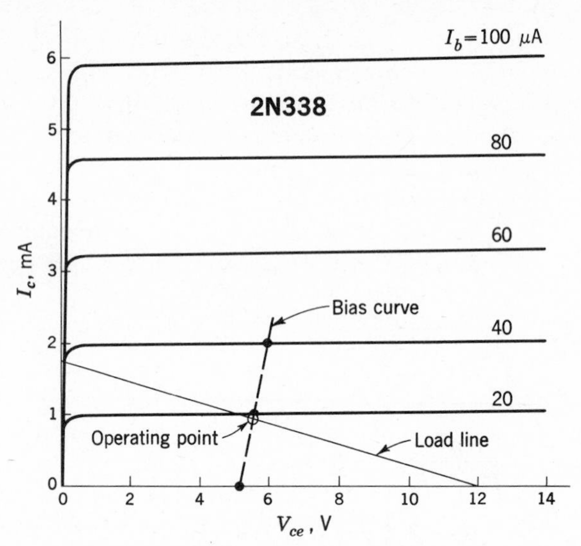

FIGURE 6-24 Determining operating point for circuit of Fig. 6-23 using common-emitter collector characteristics.

current to the dc base current. Thus, $h_{FE}I_b$ is substituted for I_c in Eq. (6-56) and the equation immediately solved for I_b. This determines the operating point since then I_c is also known.

To design a transistor amplifier circuit such as in Fig. 6-23, it is first necessary to choose the desired operating point. This is best done after consulting the transistor manufacturer's data sheets for a given transistor type. Next, a value for R_L is selected to suit gain requirements (see discussion in the next section) and output impedance considerations. In order that the potential divider made up of R_1 and R_2 may not seriously impair the amplification, R_{eq} should be larger than the transistor input impedance by a factor of 5 or 10. The emitter resistor should be chosen so that $R_E \approx R_{eq}/5$ to ensure satisfactory stability of the operating point. Then, the required dc supply voltage V_{cc} is $I_c(R_L + R_E) + V_{ce}$. The value of R_{eq} selected above, together with Eq. (6-58), yields simultaneous equations for R_1 and R_2, which completes the design. It is apparent from this brief description that the design of a transistor circuit is not completely determined by the usual requirements specified in advance. For this reason many different circuits have been developed. Each of these optimizes some particular feature of interest such as gain, minimum current requirement, or stability of operating point.

T-equivalent Circuit The electrical characteristics of a junction transistor can be represented by a *T-equivalent circuit,* Fig. 6-25, which corresponds to the grounded-base configuration already illustrated in Fig. 5-27. In the

FIGURE 6-25 *T-equivalent circuit of transistor.*

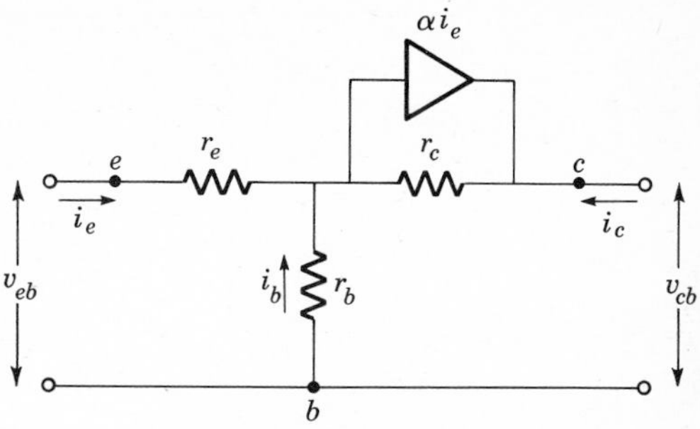

T-equivalent circuit the constant-current generator αi_e is in parallel with the collector-junction resistance r_c. The collector resistance is the reciprocal of the slope of the collector characteristic curves. An analytical expression for the collector-junction resistance can be developed by differentiating the rectifier equation, Eq. (5-20), with respect to current,

$$\frac{dI}{dI} = 1 = \frac{I_0 e}{kT} e^{eV/kT} \frac{dV}{dI} \tag{6-59}$$

$$r = \frac{kT}{eI_0} e^{-eV/kT} \tag{6-60}$$

According to Eq. (6-60) the collector reverse resistance is very large for negative collector potentials. Secondary effects such as leakage currents across the junction at the transistor surface reduce the resistance below that corresponding to an ideal *pn* junction. Therefore values of r_c must be determined experimentally and in practice are of the order of 1 to 10 MΩ.

The emitter resistance r_e is the forward resistance of the emitter junction. This is given by Eq. (6-60) with $I_0 e^{eV/kT}$ replaced by the forward emitter current I_e. This substitution is possible because commonly used values of positive forward-bias potentials make the exponential term much larger than unity in the rectifier equation. Therefore the emitter resistance in ohms is

$$r_e = \frac{kT/e}{I_e} = \frac{26}{I_e} \tag{6-61}$$

The numerical value in Eq. (6-61) is appropriate for room temperature and for the emitter current expressed in milliamperes. Accordingly, the emitter resistance of a transistor is 26 Ω at a quiescent bias current of 1 mA.

The base resistance in the T-equivalent circuit arises from two sources; the ohmic resistance of the base region (which may be appreciable since the base is so thin) and a feedback effect between the collector and the base. The origin of this effect is the decrease in base width as the collector junction widens with collector voltage. The narrower base width increases the base transport efficiency and also tends to increase the emitter efficiency. Therefore a smaller emitter-base voltage is required to maintain a constant collector current. This is exactly the effect a series resistance r_b has in the T-equivalent circuit since it reduces the influence of the input voltage. A typical value for the base resistance arising from these combined effects is 500 Ω.

The T-equivalent circuit for transistors is used in circuit analysis analogously to the vacuum-tube equivalent circuit. The T equivalent is particularly appropriate because its parameters are directly related to the basic physical structure of the transistor. Note that the T-equivalent circuit involves a direct connection between the input and output terminals, which differs from the vacuum-tube case. This may be illustrated explicitly by solving the T-equivalent circuit, Fig. 6-25, for the various currents and voltages. Using Kirchhoff's rules,

$$i_e + i_b + i_c = 0$$

$$v_{eb} = i_e r_e - i_b r_b \tag{6-62}$$

$$v_{cb} = (\alpha i_e + i_c) r_c - i_b r_b$$

Solving for i_c and v_{eb},

$$i_c = -\alpha i_e + \frac{1}{r_c} v_{cb} \tag{6-63}$$

$$v_{eb} = [r_e + (1 - \alpha) r_b] i_e + \frac{r_b}{r_c} v_{cb} \tag{6-64}$$

where the approximation $r_c \gg r_b$ has been introduced. Thus, two equations are necessary to describe the operation of a transistor, whereas a single expression is sufficient in the case of a triode or FET. Actually, Eq. (6-63) is quite analogous to the corresponding expression for a triode and should be compared with Eq. (6-15). The other relation, Eq. (6-64), indicates that the input voltage depends upon the output voltage, v_{cb}, as well as upon the input current i_e. This is a direct result of the connection between input and output in a transistor.

Hybrid Parameters Although the T-equivalent circuit is a satisfactory representation of transistor operation, it is fairly difficult to determine the various resistance parameters of the circuit by direct measurements on actual transistors. For this reason an equivalent circuit which involves so-called *hybrid parameters* is more commonly employed in circuit analysis. This representation may be illustrated by writing Eqs. (6-63) and (6-64) as

$$i_c = h_{fb} i_e + h_{ob} v_{cb} \tag{6-65}$$

$$v_{eb} = h_{ib} i_e + h_{rb} v_{cb} \tag{6-66}$$

where the h's are the hybrid parameters. The significance of the h-parameter subscripts is as follows: the subscripts i, r, f, and o refer to input, reverse, forward, and output, respectively, which may be understood by examining the meaning of each coefficient in Eqs. (6-65) and (6-66). The subscript b signifies the common-base configuration for which these equations have been developed. This designation distinguishes these parameters from those appropriate for the common-emitter and common-collector connections discussed below. A summary of the subscript notations is presented in Table 6-2.

TABLE 6-2 SUBSCRIPT NOTATION FOR *h* PARAMETERS

Subscript	*Meaning*
i	Input parameter
r	Reverse parameter
f	Forward parameter
o	Output parameter
e	Common emitter
b	Common base
c	Common collector

According to Eq. (6-66) it is possible to determine h_{ib} from the ratio of the emitter-base voltage to the emitter current v_{eb}/i_e with the collector shorted to ground so that $v_{cb} = 0$. Similarly, the ratio of the collector current to emitter current i_c/i_e with the collector shorted yields h_{fb}, using Eq. (6-65). When the emitter is open-circuited, $i_e = 0$, so that $h_{rb} = v_{eb}/v_{cb}$ and $h_{ob} = i_c/v_{cb}$. Note that the conditions $v_{cb} = 0$ and $i_e = 0$ refer to ac signals; the dc potentials are maintained at the proper operating point for the transistor. Because of the small emitter resistance and large collector resistance of a transistor it is particularly easy to achieve an open-circuited emitter or a short-circuited collector for ac signals while maintaining the desired dc potentials. In this way the small-signal h parameters can be measured using ac bridge techniques. The great virtue of the hybrid parameters is the fact that they can be measured directly with relative ease.

The *hybrid equivalent circuit* representing the grounded-base configuration is given in Fig. 6-26, as can be determined by inspecting Eqs. (6-65) and (6-66). Kirchhoff's rule applied to the input loop yields Eq. (6-66) directly. The voltage equation around the output loop gives

$$v_{cb} = \frac{1}{h_{ob}}(i_c - h_{fb}i_e) \tag{6-67}$$

Solving for i_c yields Eq. (6-65). Therefore this hybrid equivalent circuit

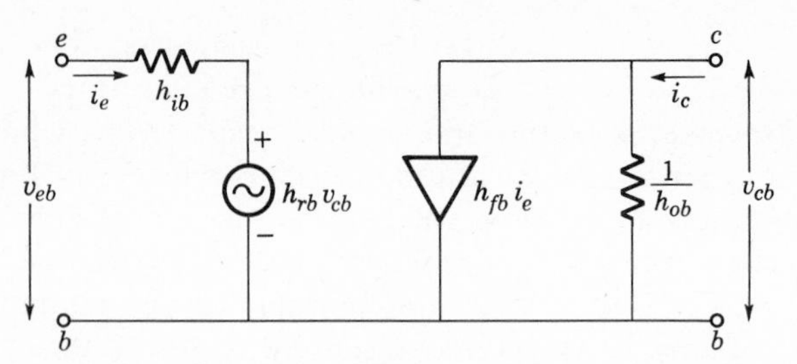

FIGURE 6-26 *Grounded-base hybrid equivalent circuit.*

represents the operation of the transistor as given by Eqs. (6-65) and (6-66). These equations, in turn, are based on the fundamental processes associated with *pn* junctions in the transistor structure.

The direct connection between input and output is not so immediately apparent in the hybrid equivalent circuit as compared with the T-equivalent. Note, however, that the voltage generator in the input circuit $h_{rb}v_{cb}$ involves the collector-base voltage and that the current generator in the output circuit $h_{fb}i_e$ includes the emitter current. Thus, the input and output circuits are indeed coupled. According to Fig. 6-26, h_{ib} represents a resistance, h_{ob} is a conductance, while h_{rb} and h_{fb} are simple numerics. These quantities represent different physical entities and are therefore termed hybrid parameters. The relation between hybrid parameters and T-equivalent parameters may be determined directly by comparing Eq. (6-65) with (6-63) and Eq. (6-66) with (6-64). These relations are summarized in Table 6-3.

The common-emitter configuration is more widely used than is the common-base circuit. While it is possible to rearrange the equivalent circuit appropriate for the common-base configuration to apply to the common-emitter case, it is much more convenient to employ the same form of equivalent circuit and adjust parameters of the circuit for the common-emitter configuration. The relations between grounded emitter h parameters and those corresponding to the grounded-base circuit are developed in the next section.

The h parameters depend upon the dc operating point of the transistor as well as upon the particular transistor type. It is common practice for transistor data sheets to specify the h parameters for the grounded-emitter or grounded-base configurations. The relations in Table 6-3 can be used to determine other parameters as needed. In addition, the variation of the parameters with operating point is usually specified so that appropriate corrections can be applied to the given values.

Illustrative values of the common-emitter h parameters and their variation with emitter current are given in Fig. 6-27. Note that the forward current gain h_{fe} and the reverse voltage amplification factor h_{re} are sensi-

TABLE 6-3 TRANSISTOR PARAMETER RELATIONS

Common-base parameters	*Common emitter*	*Common collector*	*T-equivalent*
$h_{ib} =$	$h_{ie}/(1 + h_{fe})$	$-h_{ic}/h_{fc}$	$r_e + (1 - \alpha)r_b$
$h_{rb} =$	$h_{ie}h_{oe}/(1 + h_{fe}) - h_{re}$	$h_{rc} - 1 - h_{ic}h_{oc}/h_{fc}$	r_b/r_c
$h_{fb} =$	$-h_{fe}/(1 + h_{fe})$	$-(1 + h_{fc})/h_{fc}$	$-\alpha$
$h_{ob} =$	$h_{oe}/(1 + h_{fe})$	$-h_{oc}/h_{fc}$	$1/r_c$
Common-collector parameters	*Common emitter*	*Common base*	*T-equivalent*
$h_{ic} =$	h_{ie}	$h_{ib}/(1 + h_{fb})$	$r_b + r_e/(1 - \alpha)$
$h_{rc} =$	$1 - h_{re}$	1	$1 - r_e/(1 - \alpha)r_c$
$h_{fc} =$	$-(1 + h_{fe})$	$-1/(1 + h_{fb})$	$-1/(1 - \alpha)$
$h_{oc} =$	h_{oe}	$h_{ob}/(1 + h_{fb})$	$1/(1 - \alpha)r_c$
Common-emitter parameters	*Common base*	*Common collector*	*T-equivalent*
$h_{ie} =$	$h_{ib}/(1 + h_{fb})$	h_{ic}	$r_b + r_e/(1 - \alpha)$
$h_{re} =$	$h_{ib}h_{ob}/(1 + h_{fb}) - h_{rb}$	$1 - h_{rc}$	$r_e/(1 - \alpha)r_c$
$h_{fe} =$	$-h_{fb}/(1 + h_{fb})$	$-(1 + h_{fc})$	$\alpha/(1 - \alpha)$
$h_{oe} =$	$h_{ob}/(1 + h_{fb})$	h_{oc}	$1/(1 - \alpha)r_c$
T-equivalent parameters	*Common emitter*	*Common base*	*Common collector*
α	$h_{fe}/(1 + h_{fe})$	$-h_{fb}$	$(1 + h_{fc})/h_{fc}$
r_c	$(h_{fe} + 1)/h_{oe}$	$(1 - h_{rb})/h_{ob}$	$-h_{fc}/h_{oc}$
r_e	h_{re}/h_{oe}	$h_{ib} - (1 + h_{fb})h_{rb}/h_{ob}$	$(1 - h_{rc})/h_{oc}$
r_b	$h_{ie} - h_{re}(1 + h_{fe})/h_{oe}$	h_{rb}/h_{ob}	$h_{ic} + h_{fc}(1 - h_{rc})/h_{oc}$

bly constant, while both the input resistance h_{ie} and output conductance h_{oe} vary considerably. The change in h_{ie} results mainly from the decrease of emitter-junction resistance with current. This may be noted from the relationship in Table 6-3 and Eq. (6-61).

The two most sensitive measures of transistor quality are the emitter-collector current gain h_{fe} and the grounded-base output conductance h_{ob}. Large values of current gain are desirable to achieve maximum amplification. Note that h_{fe} and β as previously defined are two commonly used symbols for the same quantity. Small values of h_{ob} are desirable since then the output resistance is large and the feedback effect small. A small

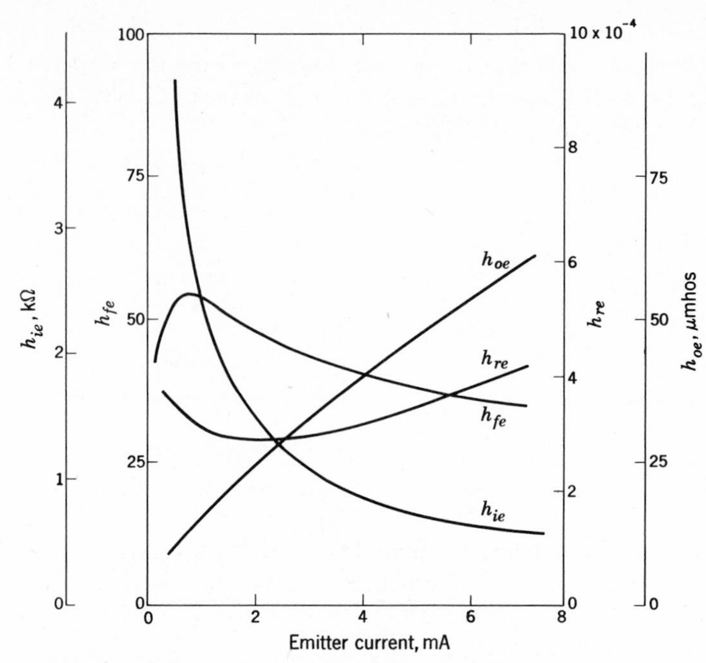

FIGURE 6-27 *Variation of hybrid parameters with emitter current.*

h_{ob} implies a large value of collector resistance (compare Table 6-3). Thus the two major parameters h_{fe} and h_{ob} reveal the quality of the emitter junction, base region, and collector junction.

TRANSISTOR CIRCUITS

Common Emitter The complete circuit of a practical common-emitter amplifier using an *npn* transistor is shown in Fig. 6-28. The input and

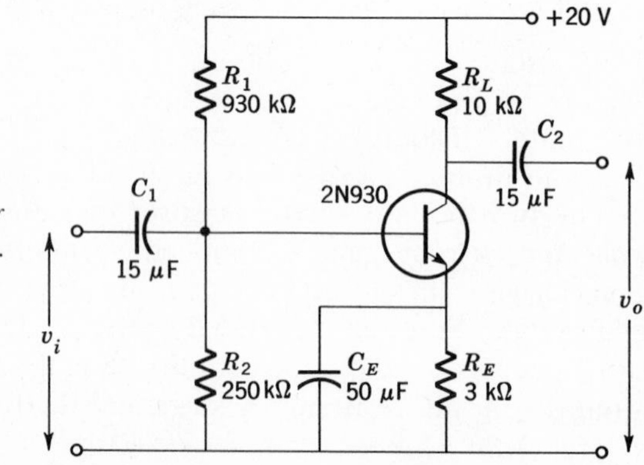

FIGURE 6-28 *Practical common-emitter amplifier using npn type 2N930 transistor.*

output coupling capacitors C_1 and C_2 pass ac signal voltages and assure that the dc operating point of the transistor is independent of the source and load conditions. The emitter bypass capacitor C_E shorts out the emitter bias resistor R_E for ac signals.

Assuming that the reactance of all three capacitors is negligible, the hybrid equivalent circuit appropriate for Fig. 6-28 is shown in Fig. 6-29.

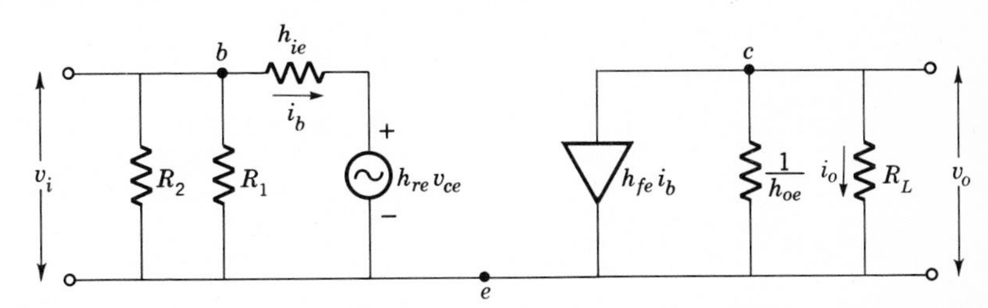

FIGURE 6-29 *Hybrid equivalent circuit of amplifier in Fig. 6-28.*

Note that it has the same form as for the grounded-base configuration, Fig. 6-26. The relation between grounded-emitter and grounded-base h parameters is developed by first writing the circuit equations pertaining to Fig. 6-29,

$$i_c = h_{fe}i_b + h_{oe}v_{ce} \tag{6-68}$$

$$v_{be} = h_{ie}i_b + h_{re}v_{ce} \tag{6-69}$$

where the h parameter subscripts are taken from Table 6-2. The corresponding relations for the grounded-base case, Eq. (6-65) and Eq. (6-66), can be put into this form with the aid of the following general relations:

$$v_{be} + v_{ec} + v_{cb} = 0 \tag{6-70}$$

$$i_b + i_e + i_c = 0 \tag{6-71}$$

which apply to all configurations. Using (6-70) and (6-71) to eliminate i_e and v_{cb}, (6-65) and (6-66) may be put in the form

$$i_c = \frac{-h_{fb}}{1 + h_{fb}}\, i_b + \frac{h_{ob}}{1 + h_{fb}}\, v_{ce} \tag{6-72}$$

$$v_{be} = \frac{h_{ib}}{1 + h_{fb}}\, i_b + \left(\frac{h_{ib}h_{ob}}{1 + h_{fb}} - h_{rb}\right) v_{ce} \tag{6-73}$$

The approximation that h_{rb} and h_{ob} are small has been introduced in arriving at (6-72) and (6-73). Comparing these equations with (6-68) and (6-69) establishes the validity of the equivalent circuit in Fig. 6-29 and also gives the relations between the common-emitter h parameters and the common-base h parameters. These relationships are summarized in Table 6-3.

The performance of the grounded-emitter amplifier is examined by analyzing the equivalent circuit as follows. The output voltage can be written directly as the current through the parallel combination of $1/h_{oe}$ and R_L

$$v_o = -h_{fe} i_b \frac{R_L}{1 + h_{oe} R_L} \tag{6-74}$$

Kirchhoff's rule applied to the input circuit yields

$$v_i = h_{ie} i_b + h_{re} v_o \tag{6-75}$$

Equation (6-75) is solved for i_b and this is substituted into Eq. (6-74). The result is arranged to give the voltage gain

$$a = \frac{v_o}{v_i} = -\frac{1}{h_{ie}(1 + h_{oe} R_L)/R_L h_{fe} - h_{re}} \tag{6-76}$$

By referring to Table 6-3 to compare the h parameters with the T-equivalent parameters, it may be seen that h_{re} is small and that $h_{oe} R_L$ may be neglected with respect to unity. Therefore Eq. (6-76) has the approximate form

$$a = -h_{fe} \frac{R_L}{h_{ie}} \tag{6-77}$$

According to Eq. (6-77), the voltage gain is approximately equal to the forward current gain of the transistor times the ratio of the load resistance to the input resistance. The voltage gain of the common-emitter amplifier is appreciable since both factors are large. The minus sign signifies that the input and output signals are 180° out of phase.

The transistor is a current-controlled device, and the current gain, which is the ratio of the output current to the input current, is also important. The output current is determined from Eq. (6-74)

$$i_o = \frac{v_o}{R_L} = -\frac{h_{fe} i_b}{1 + h_{oe} R_L} \tag{6-78}$$

For simplicity, the effect of R_{eq} in the bias network is assumed negligible, so the current gain is

$$g = \frac{i_o}{i_b} = -\frac{h_{fe}}{1 + h_{oe} R_L} \tag{6-79}$$

Here again, $h_{oe} R_L \ll 1$, so the approximate current gain is just h_{fe}.

The input resistance is the ratio of the input voltage to the input current, or

$$R_i = \frac{v_i}{i_b} = \frac{h_{ie} i_b + h_{re} v_o}{i_b}$$

Using Eq. (6-74),

$$R_i = h_{ie} - h_{re} \frac{h_{fe} R_L}{1 + h_{oe} R_L} \tag{6-80}$$

In arriving at Eq. (6-80), Eq. (6-75) has been used and the effect of R_{eq} has again been neglected. If necessary, R_{eq} may be included by calculating the parallel combination of R_{eq} and R_i. In Eq. (6-80), the second term is often negligible, so the input resistance is approximately h_{ie}. Note, however, that the exact input resistance depends upon the value of the load resistance R_L. This illustrates the coupling between input and output terminals inherent in transistors.

The output resistance of the amplifier includes the internal resistance of the input-signal source because of the coupling between input and output terminals. The effective internal resistance of the amplifier as viewed from the output terminals is determined using the Thévenin equivalent circuit. The equivalent internal resistance in the Thévenin circuit is the ratio of the open-circuit ($R_L = \infty$) voltage to the short-circuit ($R_L = 0$) current. Thus

$$R_o = \frac{(v_o)_{oc}}{(i_o)_{sc}} = -\frac{(h_{fe}/h_{oe})\,(i_b)_{oc}}{-h_{fe}(i_b)_{sc}} \tag{6-81}$$

The numerator uses Eq. (6-74) with $R_L = \infty$, and the denominator comes from the output loop of Fig. 6-29. Kirchhoff's rule applied to the input loop yields

$$v_g + i_b(R_g + h_{ie}) + h_{re} v_o = 0 \tag{6-82}$$

where v_g and R_g are the voltage and internal resistance of the signal source, respectively. Equation (6-82) may be solved for i_b and used to evaluate $(i_b)_{oc}$ by introducing v_o from Eq. (6-74). Similarly, $(i_b)_{sc}$ results from setting $v_o = 0$ in Eq. (6-82). Substituting these values into Eq. (6-81) gives the following expression for the output resistance of the amplifier,

$$R_o = \frac{1}{h_{oe} - h_{re} h_{fe}/(h_{ie} + R_g)} \tag{6-83}$$

To the same degree of approximation introduced previously, the output resistance is just $1/h_{oe}$, which is a fairly high value.

In summary, the common-emitter transistor amplifier yields both voltage and current gain. It has a modestly high input resistance and large output resistance. Because of the favorable values of these four quantities, it is the most commonly used transistor amplifier circuit. Table 6-4 illustrates the magnitude of these important parameters, calculated from the h parameters of Fig. 6-27 at an emitter current of 2 mA. The various approximations introduced above have been used in preparing this table.

TABLE 6-4 APPROXIMATE PARAMETERS OF TRANSISTOR AMPLIFIERS*

Circuit configuration	*Voltage gain*	*Current gain*	*Input resistance, Ω*	*Output resistance, Ω*
Common emitter	280	47	1700	4×10^4
Common base	280	0.98	35	1.9×10^6
Common collector	1	48	5×10^5	22

*Using Fig. 6-27 for $I_E = 2$ mA; $R_L = 10^4\ \Omega$.

Grounded Base A typical common-base amplifier circuit using an *npn* transistor is shown in Fig. 6-30. Careful comparison of this circuit with the

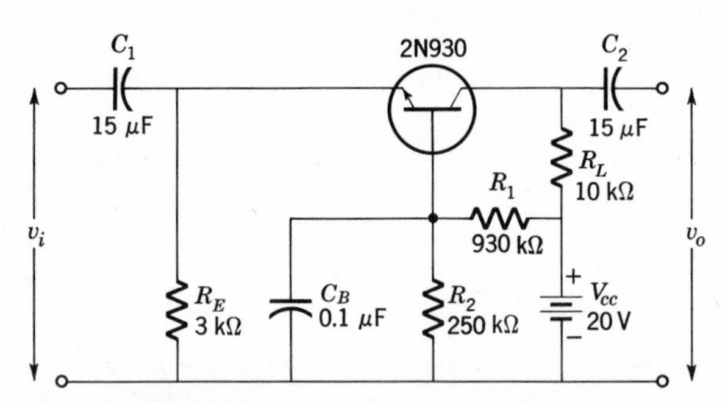

FIGURE 6-30 *Common-base transistor amplifier. Note bias circuit is identical to common-emitter case.*

common-emitter configuration, Fig. 6-28, reveals that the bias arrangements are identical. Therefore, bias considerations and the techniques for determining the operating point previously described for the common-emitter amplifier apply to the grounded-base circuit as well.

Capacitor C_B bypasses the base resistor R_2, and R_1 is shorted out for ac signals by the low impedance of the battery. Consequently, in the appropriate hybrid equivalent circuit, Fig. 6-26, these resistors are absent. The properties of this circuit can be found directly by comparing Fig. 6-26 with the hybrid equivalent of the common-emitter case, Fig. 6-29. The two are identical in form so that the previous results for the voltage and current gain, Eqs. (6-76) and (6-79), and for the input and output resistance, Eqs. (6-80) and (6-83), apply directly upon substituting the applicable h parameters. Note also that R_E replaces R_{eq} and i_e replaces i_b in converting the equations to the common-base configuration.

The approximate voltage gain in this case is $-h_{fb}R_L/h_{ib}$, which can be made large, but only if R_L is very great. This is so because the current gain is only $-h_{fb}$, which is approximately equal to unity. Note that h_{fb} is inherently a negative quantity (Table 6-3), so that the input and output voltage signals are in phase. The approximate input resistance h_{ib} is very low because it is essentially the resistance of the forward-biased emitter junction. Conversely, the output resistance $1/h_{ob}$ is the resistance of the reverse-biased collector junction and is therefore very large. This wide disparity between input and output resistance makes the grounded-base circuit less popular than the common-emitter amplifier, except for special applications. A summary of typical common-base circuit properties is presented in Table 6-4.

Emitter Follower The common-collector amplifier, Fig. 6-31, is more often termed the emitter follower because of its similarity to the cathode-

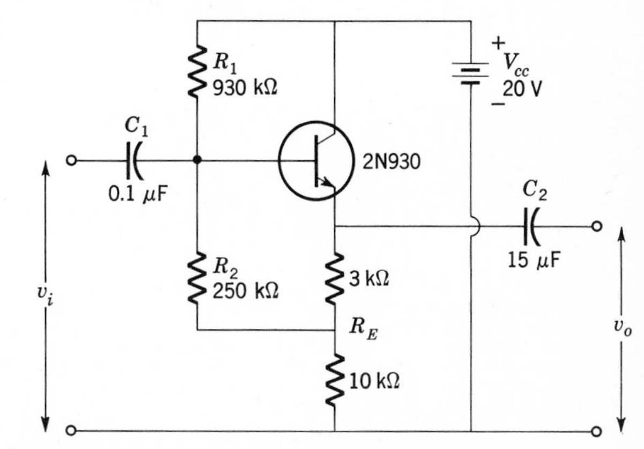

FIGURE 6-31 *Common-collector transistor amplifier. This circuit is also called emitter follower.*

follower circuit. Here again, bias considerations are identical with those previously discussed. The hybrid equivalent circuit takes the same form as the one for the common-emitter and the grounded-base circuits, Fig. 6-26 and Fig. 6-29. Relations between the common-collector h parameters and those of the other configurations are found by network analysis identical to that used above. The relationships are summarized in Table 6-3.

Circuit properties are obtained directly from the previous relations upon making the appropriate substitutions. Values of the common-collector h parameters are such that the voltage gain is unity. Furthermore, the output signal is in phase with the input. The approximate current gain $-h_{fc}$ is large. If the effect of R_{eq} can be neglected, the input resistance is simply $-h_{fc}R_L$. Thus, the input impedance is also large. The approximate output impedance reduces to $-(r_b + R_g)/h_{fc}$. Therefore, the output impedance of the amplifier itself is very small since the internal resistance R_g is associated with the signal source.

All these properties, summarized in Table 6-4 for the common-collector transistor amplifier, are similar to the cathode-follower vacuum-triode circuit. Therefore it is not surprising that the two circuits find similar applications. In particular, transistor differential amplifiers similar to the triode circuit discussed earlier are often used.

Complementary Symmetry One of the more intriguing circuit applications of transistors is based on the combination of *npn* and *pnp* transistors, with their symmetrically inverted bias and signal-voltage polarities. Consider, for example, the *complementary-symmetry* circuit, Fig. 6-32, employing

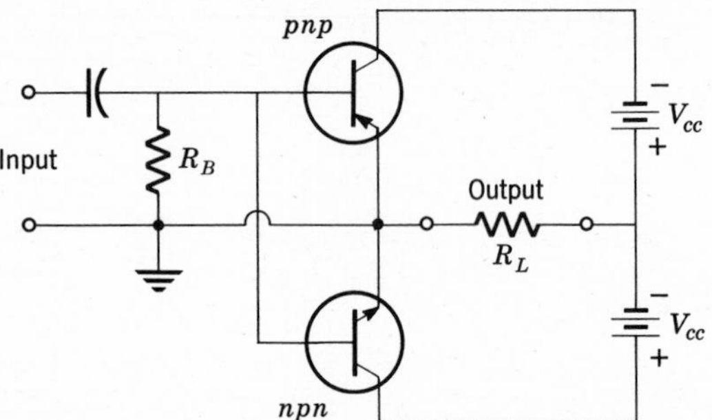

FIGURE 6-32 *Complementary-symmetry amplifier employing pnp and npn transistors.*

npn and *pnp* grounded-emitter amplifiers having common input and output connections. The base bias on both transistors is zero so in the absence of signal the transistors are cut off. Therefore, current is present in each transistor only when the input signal voltage biases its emitter junction in the forward direction. This happens on alternate half-cycles of the input voltage waveform because of the opposite polarities of the two transistors. Thus, the *npn* transistor delivers current to the load resistor when the *pnp* unit is cut off, and vice versa. The output signal is a replica of the input waveform, even though each transistor operates only half the time.

This particularly simple circuit is an efficient power amplifier since the quiescent current is zero and each transistor operates over the entire range of its characteristics. Furthermore, I^2R losses are small because the dc current in the load resistor is zero at all times. Unfortunately, it is not easy to fabricate *npn* and *pnp* transistors with identical characteristics. Also, the center-tapped collector supply voltage having neither terminal at ground is an awkward complication. For these reasons, the advantages of complementary symmetry are not widely used, although Fig. 6-32 does demonstrate the potential benefits that can be achieved in suitable circuits.

Darlington Connection The combination of two similar transistors in the so-called *Darlington connection,* Fig. 6-33, has many favorable circuit properties. Note that Q_2 is directly connected to Q_1 and the base-to-collector

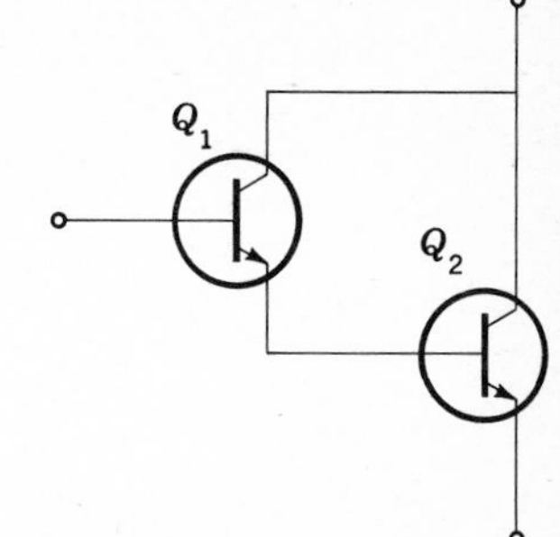

FIGURE 6-33 *Darlington-connected amplifier.*

potential of Q_2 provides the emitter-collector voltage for Q_1. Furthermore, the output emitter current of Q_1 is the base input current of Q_2.

The circuit may be viewed as an emitter follower, Q_1, followed by a grounded-emitter amplifier, Q_2. The combination produces a very large current gain, $-h_{fc} \times h_{fe} = \beta^2$. The circuit also has the voltage gain of the common emitter with the large input impedance of the emitter-follower amplifier, both of which are desirable features.

For most purposes the combination may be considered to be a single device. Indeed, units having the two transistors in one case are commercially available. Proper bias potentials are applied by a conventional transistor bias circuit such as Fig. 6-22 in which the combination of Q_1 and Q_2 replaces the single grounded-emitter transistor.

SUGGESTIONS FOR FURTHER READING

E. J. Angelo, Jr.: "Electronic Circuits," McGraw-Hill Book Company, New York, 1958.

"RCA Receiving Tube Manual," latest edition, Radio Corporation of America, Harrison, N.J.

"General Electric Transistor Manual," latest edition, General Electric Company, Semiconductor Products Department, Syracuse, N.Y.

Joseph A. Walston and John R. Miller (eds.): "Transistor Circuit Design," McGraw-Hill Book Company, New York, 1963.

EXERCISES

6-1 Calculate the small-signal parameters corresponding to Fig. 6-1 and determine the gain of the amplifier.

Ans.: 13.8

6-2 Plot the static transfer characteristics for a type 6AU6 pentode at plate voltages of 100, 200, and 300 V. Do this with the aid of the plate characteristics given in Fig. 5-17.

6-3 Find the operating point of a type 12AX7 triode amplifier in the circuit of Fig. 6-4 if $R_L = 12{,}000\ \Omega$ and $R_k = 2000\ \Omega$. Use plate characteristics given in Appendix 1.

Ans.: −2.7 V, 1.4 mA

6-4 Determine the operating point of a type 6SF7 pentode amplifier in the circuit of Fig. 6-8. Use the plate characteristics given in Appendix 1.

Ans.: −2 V, 1.5 mA

6-5 Determine the operating point of a type 12AU7 triode cathode-follower amplifier, Fig. 6-16. Calculate the input impedance, output impedance, and power gain using values of the small-signal parameters determined from the plate characteristics.

Ans.: −3 V, 3 mA; $2.3 \times 10^7\ \Omega$, 480 Ω, 4.8×10^4

6-6 Calculate an approximate value of h_{fe} for the type 2N175 transistor from the characteristic curves in Appendix 1. Repeat for h_{oe}.

Ans.: 83, 5×10^{-5} mho

6-7 Determine the operating point of a type 2N175 transistor in the grounded-emitter circuit, Fig. 6-23*a*. Draw the ac equivalent circuit of the amplifier and calculate the gain assuming $h_{ie} = 1000\ \Omega$, $h_{re} = 3 \times 10^{-4}$, and the other parameters are as in Exercise 6-6. Repeat for the circuit of Fig. 6-22 and compare the gain of the two amplifiers.

Ans.: 2.7 V, 1.4 mA, 430; 11 V, 0.24 mA, 108.

6-8 Determine the operating point of the *npn* grounded-emitter amplifier, Fig. 6-23, using a 2N338 transistor, if $R_E = 470\ \Omega$, $R_1 = 16{,}000\ \Omega$, $R_2 = 6200\ \Omega$, $R_L = 700\ \Omega$, and $V_{cc} = 12$ V. Use the characteristic curves in Appendix 1.

Ans.: 75 μA, 4 mA

6-9 Design a common-emitter amplifier corresponding to Fig. 6-23 for the type 2N35 transistor. The operating point is $V_{ce} = 6$ V, $I_c = 1$ mA, and $I_b = 20 \times 10^{-6}$ A. Assume $R_L = 10{,}000\ \Omega$ and $h_{ie} = 2000\ \Omega$. Characteristic curves of the 2N35 are in Appendix 1.

6-10 Determine the operating point of the 2N930 common-emitter amplifier, Fig. 6-28. Characteristic curves of the 2N930 are in Appendix 1.

Ans.: 4 μA, 4 mA

6-11 Calculate the voltage and current gain and the input and output impedance of the amplifier of Exercise 6-10. The *h* parameters are $h_{ie} = 3600\ \Omega$, $h_{re} = 3 \times 10^{-3}$, $h_{fe} = 150$, $h_{oe} = 1.4 \times 10^{-4}$ mho. Assume the reactances of all capacitors in the circuit are negligible and that the source resistance is 1000 Ω. Compare the calculated values with approximations introduced in the text.

Ans.: 362, 416; 62.4, 150; 1700 Ω, 3600 Ω; $2.4 \times 10^4\ \Omega$, $7.1 \times 10^3\ \Omega$

6-12 Determine the operating point of the 2N930 grounded-base amplifier, Fig. 6-30. Assuming the hybrid parameters given in Exercise 6-11, calculate the voltage and current gain and the input impedance for this circuit. Compare these with the approximations discussed in the text. Take $R_g = 100\ \Omega$.

Ans.: 200, 413; 0.99, 0.994; 27.5 Ω, 24 Ω

6-13 Determine the operating point of the type 2N930 emitter-follower amplifier, Fig. 6-31. Given the h parameters of Exercise 6-11, calculate the voltage and current gain and the input and output impedance for this circuit. Compare these results with the approximations introduced in the text. The source resistance is 10,000 Ω.

Ans.: 1, 1; 63, 151; $6.3 \times 10^5\ \Omega$, $1.5 \times 10^6\ \Omega$; 90 Ω

6-14 Draw the hybrid equivalent circuit of the transistor difference amplifier in Fig. 6-34. Plot the output signal as a function of the input signal. The appropriate h parameters are $h_{ie} = 1260\ \Omega$, $h_{re} = 1.5 \times 10^{-3}$, $h_{fe} = 44$, and $h_{oe} = 2.7 \times 10^{-5}$ mho.

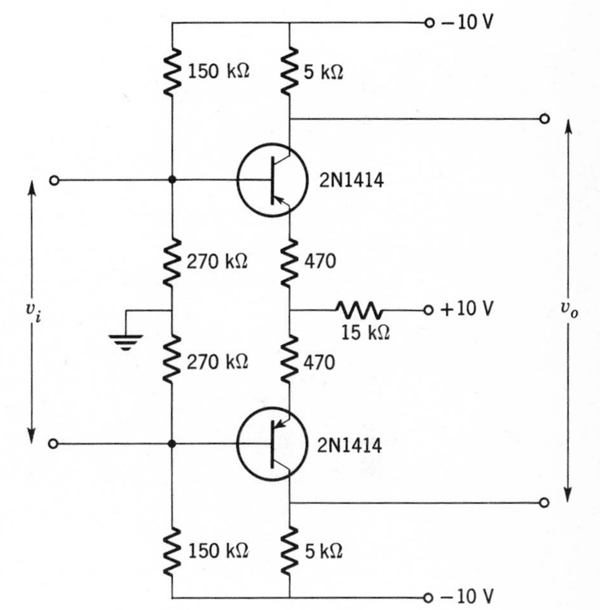

FIGURE 6-34 *Transistor difference amplifier analyzed in Exercise 6-14.*

6-15 Determine the operating point of an 2N2499 FET amplifier, Fig. 6-6, using the characteristic curves given in Appendix 1 and with $R_d = R_k = 2.5\ \text{k}\Omega$. Calculate the approximate voltage gain of the amplifier.

Ans.: −2 V, 4 mA.

LABORATORY EXERCISES

6-A The Grounded-emitter Amplifier The performance of transistor amplifiers is specified by the magnitude of the small-signal parameters of the transistor. Appropriate values of hybrid parameters can be determined by direct measurement or, alternatively, from the transistor current-voltage characteristics. To some extent amplifier performance can be adjusted by selection of the operating point, since values of the hybrid parameters depend upon the collector current. This experiment studies the properties of a grounded-emitter amplifier in terms of the small-signal parameters of the transistor.

Begin by determining values of the small-signal parameters of a 2N3521 *npn* silicon transistor at a collector potential of 5 V using the collector characteristics given in Fig. 5-29. Note that h_{oe} is calculated from the slope of the collector characteristics and h_{fe} is the slope of the transfer characteristic. The quantities h_{re} and h_{ie} may then be calculated using the forward resistance of the emitter junction, according to Table 6-3. For present purposes it is satisfactory to ignore r_b in calculating h_{ie}. Determine all four parameters for small, intermediate, and large collector currents, say, 0.5, 2, and 4 mA.

Design a suitable bias circuit similar to Fig. 6-28 so that the operating point is near a collector current of 0.5 mA and a collector potential of 5 V and such that the voltage gain is in the range 500 to 1000. Confirm the operating point by plotting the load line on the collector characteristics. Using values of the small-signal parameters previously determined calculate the voltage and current gain and the input resistance of the amplifier. By inspection, estimate the peak-to-peak magnitude of output voltage the amplifier can sustain without serious waveform distortion.

Repeat the above, this time setting the operating point near 2 mA and 5 V. Comment on the changes in amplifier performance. Prepare a dynamic transfer characteristic (output voltage as a function of input current) over the entire permissible range of input signal. Do this using the load line corresponding to R_L plotted on the collector characteristics. Is the transfer characteristic linear? Compare the permissible signal swing in this case to that possible at the low-current operating point. Now shift the operating point to 4 mA and 5 V. Calculate the gains and input and output resistances. Comment on the trends apparent in these three cases.

Using the circuit components selected to place the operating point at 2 mA and 5 V, calculate the operating point at a temperature of 125°C using the collector characteristics in Fig. 5-30. Determine values of the hybrid parameters at this operating point and calculate the amplifier-performance figures. Comment on the effect of temperature upon amplifier operation.

6-B The Emitter Follower The emitter-follower amplifier exhibits high-input resistance and low-output resistance, which makes it a useful circuit even though the voltage gain is only unity. This experiment examines the important properties of a grounded-collector amplifier using both small-signal parameters and also graphical analysis for large-signal applications.

Begin by rearranging the circuit studied in Laboratory Exercise 6-A at the 2-mA 5-V operating point into the emitter-follower configuration, Fig. 6-31. Determine suitable values of the hybrid parameters for the grounded emitter amplifier using the relations in Table 6-3 and calculate the amplifier current gain, voltage gain, and input and output resistance. Also calculate the power gain. Contrast these properties with those of the grounded-emitter amplifier.

Prepare a dynamic transfer characteristic and compare the linear range with that corresponding to the grounded-emitter case. Now convert this data into a voltage transfer characteristic (output voltage versus input voltage) and note how broad a linear portion is available.

The properties of an emitter follower are sensibly independent of transistor parameters. To illustrate, suppose a 2N930 transistor is inadvertently put in the circuit designed for the 2N3521. Determine the operating point using the collector characteristics given in Appendix 1, and estimate the small-signal parameters. Calculate the gains and resistances and compare with the 2N3521 amplifier. Are the amplifier parameters similar?

chapter seven

AMPLIFIER CIRCUITS

The principal applications of transistors and vacuum tubes are based on their ability to amplify electric signals. Some circuits amplify minute voltage signals by factors of many million, while others increase the electric power of a signal in order to operate a mechanical device such as an electric motor. Still other circuits amplify currents. In each of these applications the frequency range of the input signal is important. Different circuits have been developed for dc amplification and for use at high radio frequencies.

Most often, the signal level is increased in several successive amplifier stages to attain the desired output-signal magnitude. In this case, the interaction between amplifier stages must be considered and fairly complicated networks are involved. Fortunately, the techniques of circuit analysis developed in previous chapters, in particular the ac equivalent circuits for tubes and transistors, are sufficient for a satisfactory understanding of complete amplifier circuits.

VOLTAGE AMPLIFIERS

Cascading The transistor and vacuum-tube circuits discussed in previous chapters are ideally suited to amplify voltage signals with minimum waveform distortion. Gain factors greater than those possible with a single-stage amplifier are obtained by *cascading* several amplifier stages. The output of one amplifier stage is amplified by another stage or stages until the desired signal voltage level is achieved.

Consider, for example, the two-stage cascaded triode amplifier, Fig. 7-1.

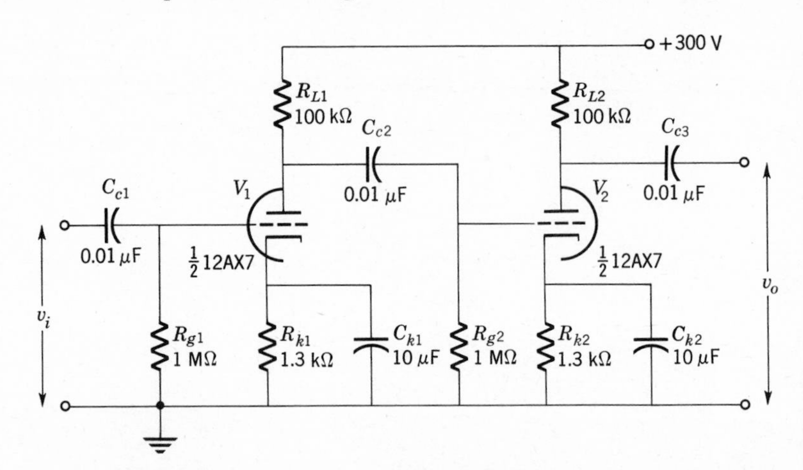

FIGURE 7-1 *Two-stage amplifier using triodes connected in cascade.*

Two individual circuits similar to those discussed in Chap. 6 are connected with the coupling capacitor C_{c2}. This capacitor passes the amplified ac signal from V_1 to the grid of V_2. At the same time it blocks the positive plate voltage of V_1 from the grid of the second triode. Similarly, capacitors C_{c1} and C_{c3} isolate the input and output circuits insofar as dc potentials are concerned.

The entire ac equivalent circuit of this amplifier may be drawn using the principles discussed in Chap. 6. The performance of the system is determined from a complete ac-circuit analysis. Actually, this procedure is unwieldy because of the number of loops in the network and is rarely attempted. Rather, the circuit is analyzed in several separate steps, each of which has a minimum of mathematical complexity. This has the further advantage that the important effects can be isolated and more clearly examined.

For example, the reactances of both cathode bypass capacitors are assumed small enough to be negligible. Accordingly, these components are absent in the ac equivalent circuit of the amplifier, Fig. 7-2. The reactances of the coupling capacitors are also ignored, even though they are included in the equivalent circuit for clarity. With these simplifications, the output voltage may be immediately written as

$$v_o = v_{g2}\frac{-\mu_2}{1 + r_{p2}/R_{L2}} = v_{g1}\frac{-\mu_1}{1 + r_{p1}/R'_{L1}}\frac{-\mu_2}{1 + r_{p2}/R_{L2}} \tag{7-1}$$

where the load resistance of the first stage R'_{L1} is the parallel combination of the plate resistor R_{L1} and the second-stage grid resistor R_{g2}.

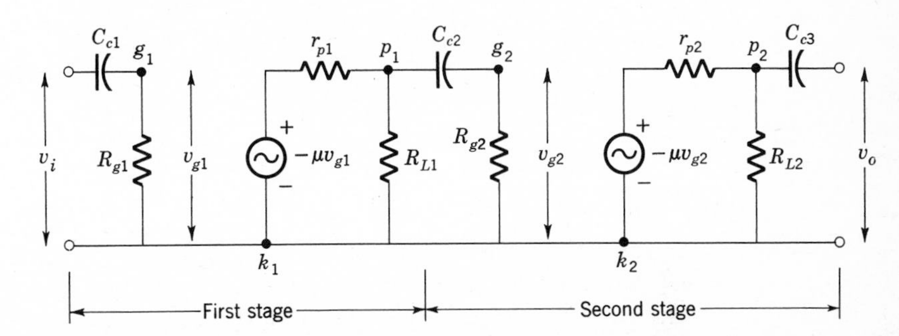

FIGURE 7-2 *Equivalent circuit of the two-stage cascaded amplifier of Fig. 7-1.*

According to (7-1), the overall gain of this two-stage amplifier is simply

$$a = a_1 a_2 \tag{7-2}$$

where a_1 and a_2 are the gains of each stage. The expression for the gain of the first stage a_1 includes the input impedance of V_2 as part of the load resistance. Usually, $R_{g2} \gg R_{L1}$, however, so a_1 is essentially the gain of the isolated V_1 stage. Equation (7-1) and (7-2) give the *midband gain* of the amplifier since the reactances of the coupling capacitors and the cathode bypass capacitors are assumed negligible. This approximation applies to signal frequencies which are neither so low that the reactances cannot be ignored nor so high that other effects reduce the gain. These other frequency regions are discussed in the following section.

Cascaded transistor voltage amplifiers most often employ the grounded-emitter configuration because of the combined voltage and current gain of this circuit. Neither the common-base nor the emitter-follower configuration achieves as great overall voltage amplification when cascaded. This is a result of the great impedance mismatch between the output impedance of one stage and the input impedance of the succeeding stage. A typical transistor voltage amplifier, Fig. 7-3, uses interstage coupling capacitors as in the vacuum-tube case to isolate the dc bias voltages of the two stages. Note that the bias resistors of the second stage are different from those of the first stage, even though both transistors are identical. The operating points are set at different places in order to obtain most favorable values of the h parameters in each stage.

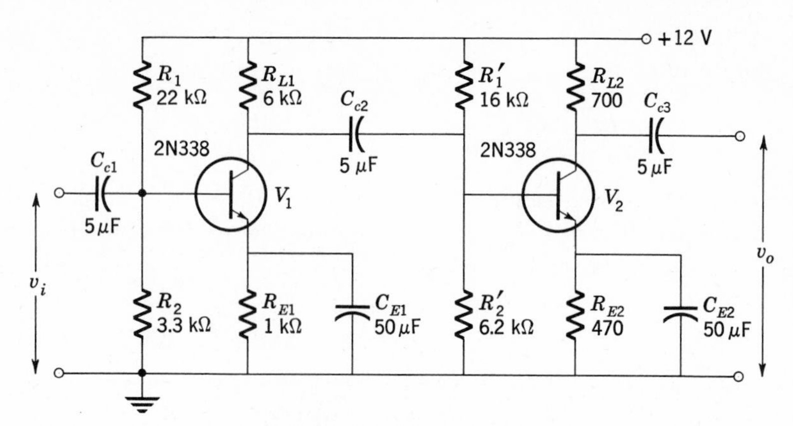

FIGURE 7-3 *Two-stage cascaded transistor amplifier.*

In the appropriate equivalent circuit of the amplifier, Fig. 7-4, the base bias resistors are replaced by their parallel combination, as explained in the previous chapter. Here again, the overall gain is the product of the

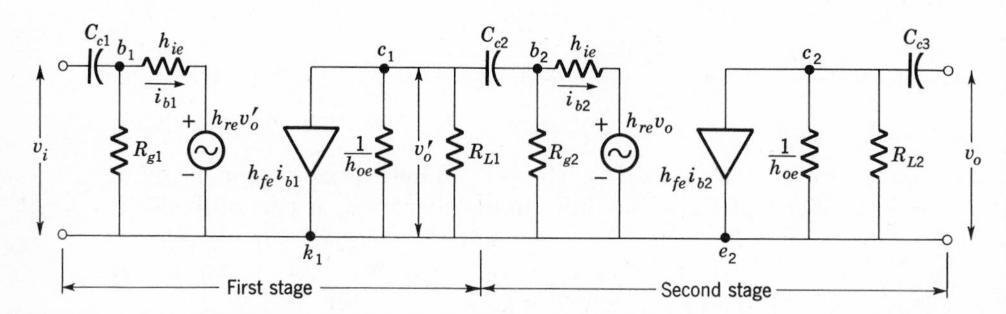

FIGURE 7-4 *Equivalent circuit of the two-stage transistor amplifier of Fig. 7-3.*

individual gain of each stage, Eq. (7-2). The loading effect of the second stage upon the output of the first stage cannot be ignored in calculating the gain because of the inherently low input impedance of transistors. In fact, it is necessary to work backward through the circuit starting at the output terminals because the input impedance of a transistor amplifier depends upon the output-load impedance. With the output load specified, the gain and input impedance of the second stage are calculated using the results developed in the previous chapter. This input impedance is part of the load for the preceding stage and its gain and input impedance are calculated accordingly. Thus, detailed analysis of transistor circuits is somewhat more complicated than is the case for vacuum-tube amplifiers, basically because of the input-output coupling in a transistor. Nevertheless, transistor circuits are treated quite satisfactorily by straightforward ac-circuit analysis of the equivalent circuit.

Low-frequency Gain At sufficiently low frequencies the capacitive reactances may no longer be neglected. The effect of the coupling capacitors is usually of greater significance than that of the cathode bypass capacitors, although both tend to reduce the gain at low frequencies. It is not practical to make the coupling capacitors large. Large values of capacitance imply increased leakage current, which upsets grid bias of vacuum tubes. This situation is aggravated by the fact that the coupling capacitor is connected between the large positive-plate potential and the low grid voltage and by the large value of grid resistance. Consequently, practical coupling capacitors are limited to values below about 0.5 μF.

No such restrictions are placed on cathode bypass capacitors since they are connected in low-impedance low-voltage circuits where leakage currents are insignificant. Electrolytic capacitors are common in this position and values ranging up to 100 μF are used. Special low-voltage electrolytics are used in transistor amplifiers as both coupling capacitors and bypass capacitors since the impedance levels are low in both places. Nevertheless, leakage currents must be minimized into the base terminal so a limit to the capacitance exists in this case as well. The low-frequency gain of cascaded transistor amplifiers is also determined primarily by the reactance of the interstage coupling capacitors.

Because the overall gain of cascaded stages is the product of individual stage gains, it is only necessary to examine the effect of the coupling capacitor reactance for an isolated amplifier stage. According to the ac equivalent circuits of both vacuum-tube and transistor amplifiers, Fig. 7-2 and 7-4, this effect can be treated by considering the simple *RC* circuit comprising the coupling capacitor and the input impedance of the amplifier.

This part of both equivalent circuits is shown separately in Fig. 7-5 for

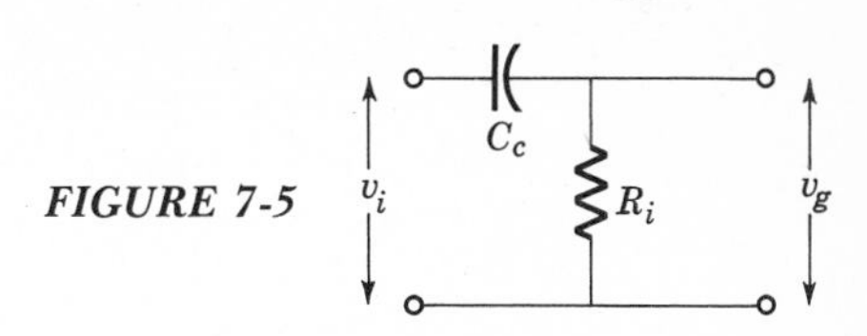

FIGURE 7-5

clarity. The input impedance R_i is simply the grid resistor in the case of the tube amplifier, but in the transistor amplifier it includes the input impedance of the transistor itself. In either case, the output voltage of the stage is

$$v_o = av_g = a\frac{v_i}{R_i + 1/j\omega C_c}R_i$$

$$= \frac{av_i}{1 - j/\omega R_i C_c} \qquad (7\text{-}3)$$

As discussed in Chap. 2, it is appropriate to define the characteristic frequency

$$2\pi f_0 = \omega_0 = \frac{1}{R_i C_c} \tag{7-4}$$

Substituting Eq. (7-4) into Eq. (7-3), the gain v_o/v_i is

$$\mathbf{a}(f) = \frac{a}{1 - jf_0/f} \tag{7-5}$$

where a is the midband gain. Note that the gain is reduced when the signal frequency is smaller than the characteristic frequency. At the same time a phase shift is introduced between the input and the output signals. Both effects are important in determining the waveform distortion of the amplifier. Recall that in the Fourier analysis of a complex signal waveform, both the amplitudes and relative phases of all frequency components must be preserved if the output wave is to be an amplified replica of the input signal.

It is convenient to rationalize Eq. (7-5),

$$a(f) = \frac{a}{\sqrt{1 + (f_0/f)^2}} \tag{7-6}$$

so that the gain at any frequency can be immediately calculated. Note that Eq. (7-6) shows that the gain is $a/\sqrt{2}$, or about 70 percent of the midband gain, when $f = f_0$. It is important to recognize that Eq. (7-6) applies to an individual stage. The low-frequency response of the entire amplifier is always poorer than that of any individual stage because the gain of cascaded stages is the product of individual stage gains.

High-frequency Gain The high-frequency gain of any amplifier is reduced by stray capacitive effects that are not purposely made part of the circuit. Referring to a simple triode amplifier, Fig. 7-6, these are the grid-cathode

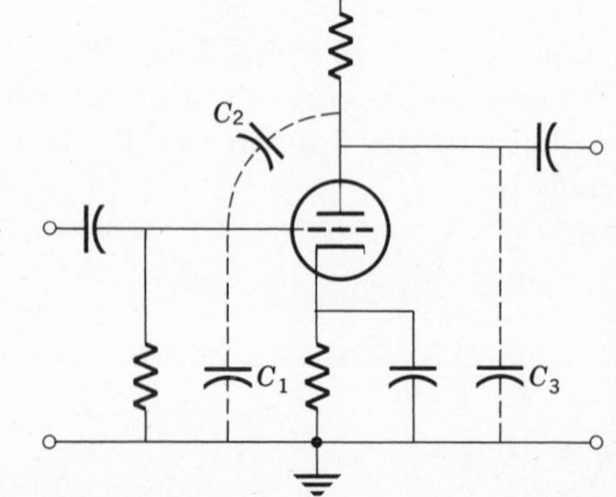

FIGURE 7-6 *Stray capacitances in a triode amplifier.*

capacitance C_1, the grid-plate capacitance C_2, and the plate-cathode capacitance C_3 of the tube itself. Also included in C_1 and C_3 are stray capacitances between the wires and components attached to the grid and plate terminals. All three of these capacitors shunt the signal at frequencies high enough that the capacitive reactances are significant.

The effect of the grid-plate capacitance is particularly important. Consider the pertinent equivalent circuit, Fig. 7-7, in which C_2 is connected

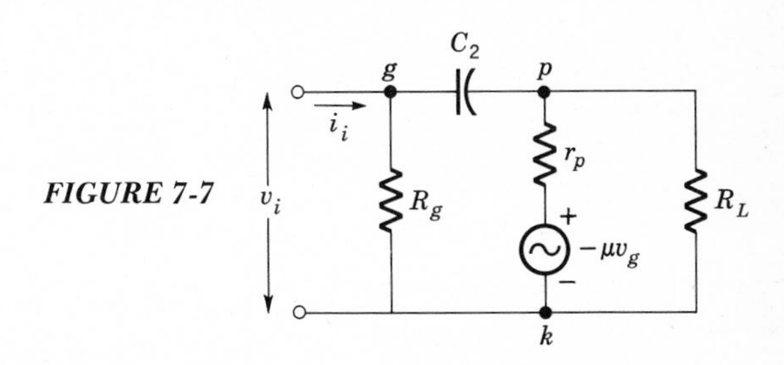

FIGURE 7-7

between the grid and plate terminals. For the moment the effect of the other capacitors is ignored. The input impedance of the amplifier is calculated by assuming that the reactance of C_2 is the controlling factor,

$$\mathbf{Z}_i = \frac{v_i}{i_i} \cong \frac{v_i}{(v_i + \mu v_i)/(1/j\omega C_2)} \tag{7-7}$$

$$\mathbf{Z}_i = \frac{1}{j\omega(1 + \mu)C_2} \tag{7-8}$$

This result indicates that the input impedance may be considered to be a capacitor $(1 + \mu)C_2$ connected from grid to ground. The increase in effective shunt capacitance caused by the amplification factor of the tube is called the *Miller effect* and is the dominating effect in determining the high-frequency response. The appropriate high-frequency equivalent circuit for the triode amplifier of Fig. 7-6 includes a shunt capacitance, as shown in Fig. 7-8. The magnitude of this capacitance,

$$C_s = C_1 + C_3 + (1 + \mu)C_2 \tag{7-9}$$

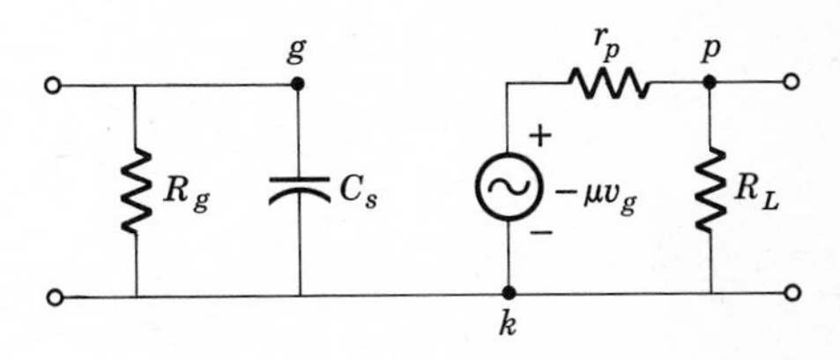

FIGURE 7-8 *High-frequency equivalent circuit of triode amplifier includes shunt capacitance C_s.*

includes the plate-cathode capacitance of the previous stage C_3 in the total shunt capacitance, as indicated by Eq. (7-9).

Shunt capacitance is less important in a transistor amplifier, Fig. 7-9, because of the small input impedance of the transistor compared with

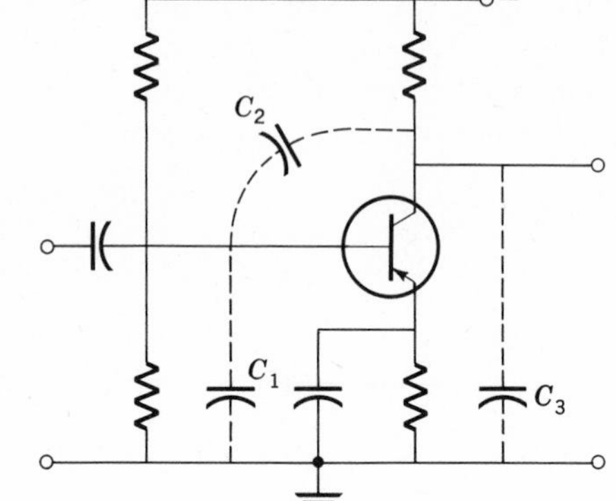

FIGURE 7-9 Stray capacitances in transistor amplifier.

the vacuum tube. Nevertheless, the collector-junction capacitance C_2 and the emitter-junction capacitance C_1 must be accounted for in assessing the high-frequency response. As in the case of the triode amplifier, capacitances C_1 and C_3 also include the effect of stray wiring capacitances.

The effect of the collector-junction capacitance is enhanced by the Miller effect of the transistor. The magnitude is found by a procedure identical to that used for the triode amplifier and results in a total shunt capacitance given by

$$C_s = C_1 + C_3 + (1 + h_{fe})C_2 \tag{7-10}$$

The third term in Eq. (7-10) is most important. The corresponding high-frequency equivalent circuit of the transistor amplifier is illustrated in Fig. 7-10.

According to Figs. 7-8 and 7-10, the high-frequency gain of both triode

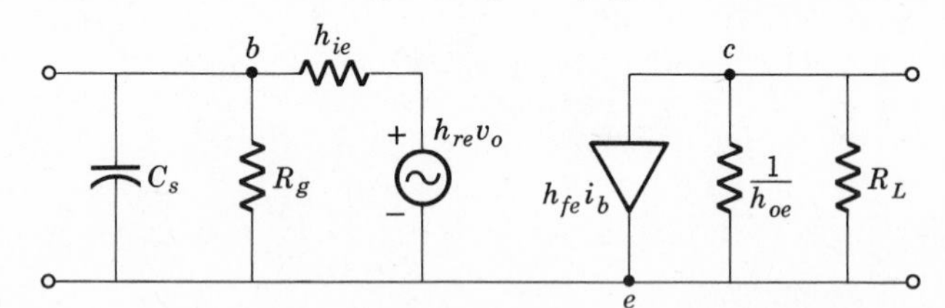

FIGURE 7-10 High-frequency equivalent circuit of a transistor amplifier includes shunt capacitance C_s.

and transistor amplifiers is accounted for by the input shunt capacitor C_s. This is evaluated using the simple circuit, Fig. 7-11, which also includes the effect of the source resistance R_s. The input resistance R_i is essentially the grid resistance R_g in the case of the vacuum tube but includes the total

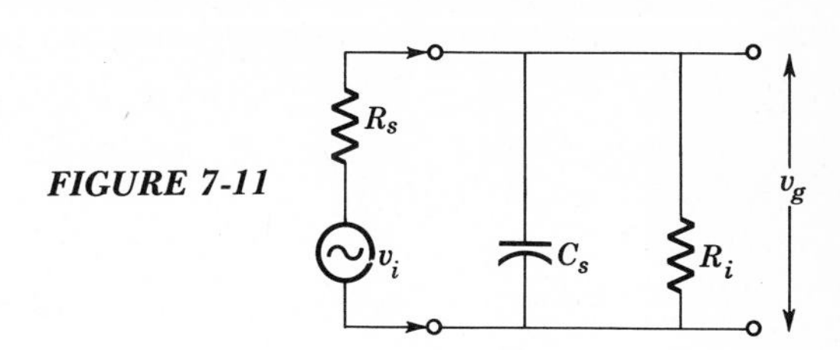

FIGURE 7-11

input impedance in the case of the transistor amplifier. The output voltage of either emplifier is found by analyzing the circuit of Fig. 7-11,

$$v_o = a'v_g = a'\frac{v_i}{R_s + \mathbf{Z}}\mathbf{Z}$$

where

$$\frac{1}{\mathbf{Z}} = \frac{1}{R_i} + j\omega C_s$$

and a' is the midband gain of the stage with no load. Substituting for Z and simplifying,

$$v_o = \frac{a'v_i}{(1 + R_s/R_i) + j\omega R_s C_s} \tag{7-11}$$

The characteristic frequency of this circuit is defined as

$$2\pi f_0 = \omega_0 = \frac{1}{C_s}\left(\frac{1}{R_s} + \frac{1}{R_i}\right) \tag{7-12}$$

Introducing Eq. (7-12) into Eq. (7-11) and solving for the gain v_o/v_i gives

$$\mathbf{a}(f) = \frac{a'}{1 + R_s/R_i}\frac{1}{1 + jf/f_0} \tag{7-13}$$

The denominator of the first term in Eq. (7-13) accounts for the loading of the amplifier input upon the previous stage. As previously discussed, this effect is usually included in the determination of the true midband gain of the entire amplifier. Accordingly, the variation of the gain at high frequencies is conveniently written, after rationalization, as

$$a(f) = \frac{a}{\sqrt{1 + (f/f_0)^2}} \tag{7-14}$$

where a is the true midband gain.

This result shows that the gain is reduced at high frequencies. Note also, Eq. (7-13), that phase shift is introduced between input and output signals, and this is equally significant in preserving the signal waveform. In the case of the vacuum tube amplifier, the input resistance is essentially equal to the grid resistor R_g. Since $R_g \gg R_s$, the high-frequency performance is controlled by the output impedance of the preceding stage, according to Eq. (7-12). Conversely, $R_i < R_s$ in the case of the transistor amplifier, so the transistor input impedance is the dominating factor. As in the low-frequency case, the overall high-frequency response of the complete amplifier is poorer than that of any individual stage.

Actually, the high-frequency amplification of many transistors is limited by the transit time of carriers diffusing across the base region. This effect results in a high-frequency gain given by an expression identical to Eq. (7-14) except that f_0 is determined by physical constants of the transistor, such as the width of the base. This characteristic frequency is usually specified by the transistor manufacturer. In specially designed high-frequency transistors the *alpha falloff frequency* is high enough that the gain is limited by the circuit parameters, as discussed above.

Using Eqs. (7-6) and (7-14), the *frequency response* of any voltage amplifier is similar to that illustrated in Fig. 7-12*a*. The *low-frequency cutoff* for each stage is determined from Eq. (7-4) while the *high-frequency cutoff* for each stage is found using Eq. (7-12). The *bandpass* of the complete amplifier is the frequency interval between the high- and low-frequency points where the gain falls to $1/\sqrt{2}$ of the midband gain. Since the power output is reduced to one-half of the midband value of these frequencies, they are referred to as the half-power points (see Chap. 2). It is conventional to employ logarithmic scales on both axes of bandpass characteristics such as Fig. 7-12*a* because the range of gains and frequencies is so great. The vertical scale is often put in terms of a unit called the *bel,* named after Alexander Graham Bell, the inventor of the telephone. Actually, a unit one-tenth as large, the *decibel,* abbreviated dB, proves more convenient in practice. The decibel is defined as

$$\text{dB} = 20 \log \frac{a(f)}{a} \tag{7-15}$$

Correspondingly, the midband gain is often quoted in terms of decibels using the definition

$$\text{dB} = 20 \log \frac{v_o}{v_i} \tag{7-16}$$

The advantage of this unit is that the total gain in dB of several amplifier stages is simply the sum of the individual gains in terms of decibels. Note that, according to Eq. (7-15), the amplifier gain is down 3 dB at the upper and lower half-power points.

The phase-shift characteristics of a single-stage amplifier are illustrated in Fig. 7-12*b*. The output signal leads the input at frequencies below the low-frequency cutoff and lags at frequencies above the high-frequency

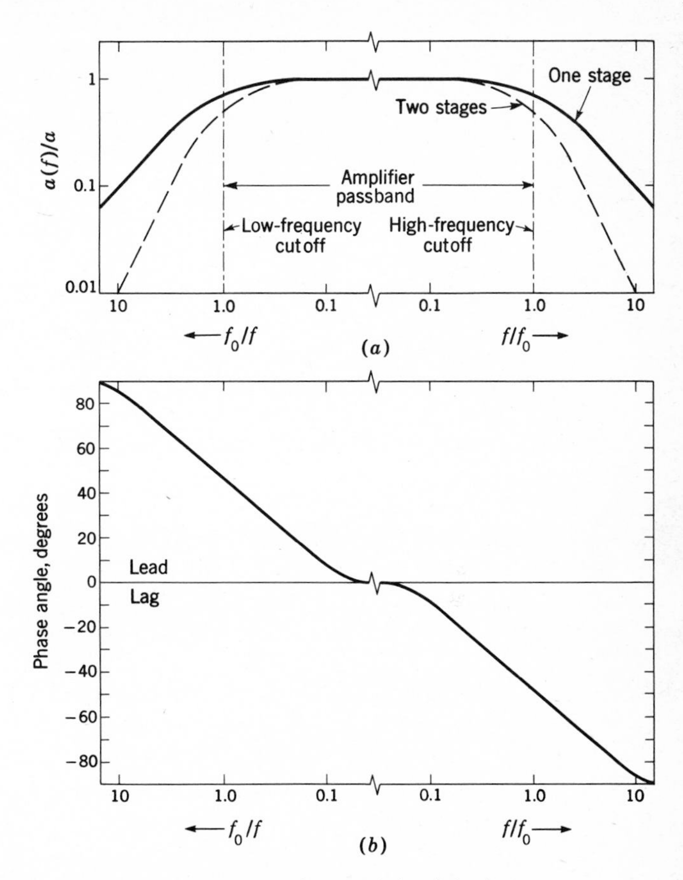

FIGURE 7-12 *(a) Frequency-response characteristic of single amplifier stage. Characteristic of two-stage amplifier is shown dashed. (b) Phase-shift characteristic of single amplifier stage.*

cutoff. The phase-shift characteristics of an entire amplifier are determined by adding the contributions from each stage.

Often an amplifier must have a wide bandpass in order to minimize waveform distortion. A number of minor circuit alterations have been developed to accomplish this end. For example, source, cathode, or emitter bypass capacitors may purposely be made small so that the capacitive reactance is appreciable except at frequencies near the high-frequency cutoff. This reduces the midband gain, according to Eq. (6-21), but increases the gain at high frequencies where the capacitive reactance becomes small. The net result is an extended high-frequency response, although at the expense of smaller overall amplification. If necessary, the loss in gain can be made up by adding another stage.

A second useful way of extending the high-frequency response is to include a small inductance as part of the load, Fig. 7-13. The load imped-

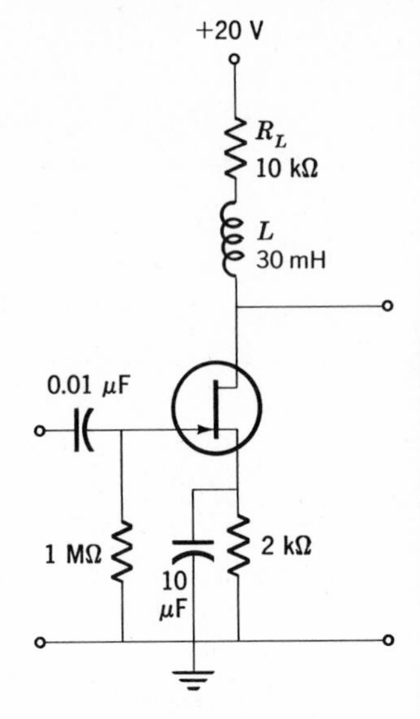

FIGURE 7-13 *Small inductance in load impedance improves high-frequency response because gain is increased by higher load impedance.*

ance increases at high frequencies and the gain of the amplifier is larger, according to Eq. (6-25). This technique is called *peaking* since the resulting frequency-response characteristic tends to be peaked at the high-frequency end.

Improved low-frequency response can be achieved by adding a series resistor-capacitor combination R_3 and C_3 to the output circuit, Fig. 7-14.

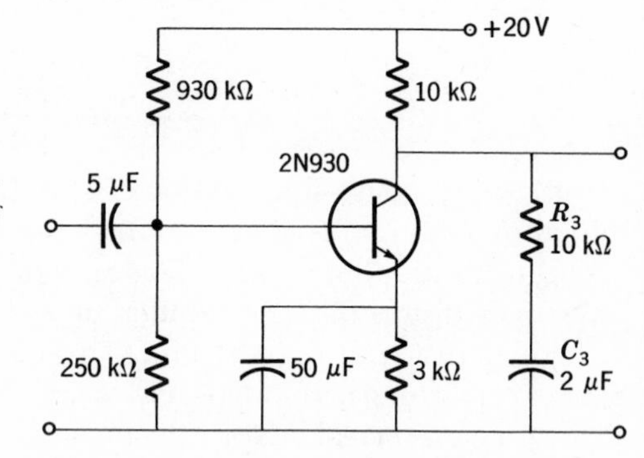

FIGURE 7-14 *Combination R_3C_3 improves low-frequency response of amplifier because reactance of C_3 becomes large.*

The gain at frequencies where the reactance of C_3 is small includes the effect of R_3 in the ac load resistance. At low frequencies the gain increases as the reactance of the capacitor increases and removes R_3 as part of the output load. The result is an extended low-frequency response at the expense of midband gain.

Decoupling When three or more stages of amplification are cascaded it is usually necessary to *decouple* the power supply of the input stage from the remainder of the amplifier. The reason for this is that the supply voltage changes with current because of the effective internal impedance of the power supply. Any small change in the power-supply voltage alters the bias on the first stage, and this change is amplified in the same fashion as an input signal. If the change in bias increases the current in the first stage, current in the second stage is reduced because of the 180° phase shift in the input amplifier. The current in the third stage is increased, however, because of the second 180° phase shift in the second stage. The change in the third stage is much larger than the original disturbance because of the gain of the amplifier. The additional load causes a decrease in the power-supply voltage. This, in turn, further alters the bias on the first stage and the process is cumulative. The changes continue until one tube or transistor is driven to cutoff or into saturation, which reduces the overall gain to zero. The power-supply voltage then returns to normal and the process repeats itself. The result of this *feedback* from output to input is that the amplifier rapidly oscillates from cutoff to saturation at a rate which is a function of the circuit components.

A low-pass *RC* filter inserted in the power-supply lead to the first stage, Fig. 7-15, circumvents this difficulty. The time constant of this decoupling

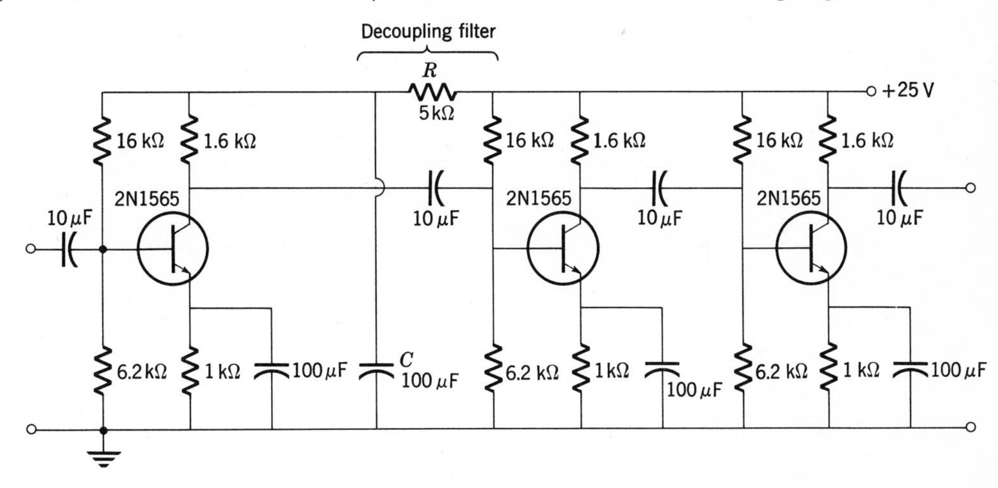

FIGURE 7-15 *Simple RC decoupling filter eliminates instability in multistage amplifier by reducing feedback effects resulting from common power supply.*

filter is selected so that power-supply variations are sufficiently attenuated and feedback is eliminated. Actually, the characteristic filter frequency is put below the low-frequency cutoff of the amplifier where the gain is insufficient to support feedback oscillations.

POWER AMPLIFIERS

Transformer Coupling When transistor or vacuum-tube amplifiers deliver appreciable amounts of power, it is no longer feasible to use resistors in the collector or plate circuit. The I^2R losses become significant at the high currents associated with large powers. Instead, a transformer couples the circuit to the load, Fig. 7-16. The dc collector current in the winding

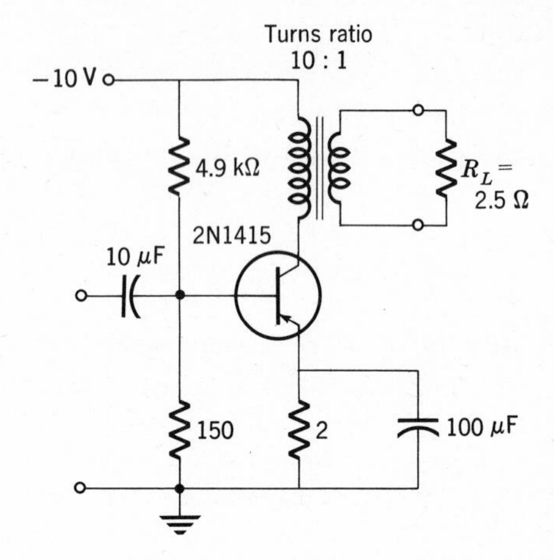

FIGURE 7-16 *Power amplifiers use transformer to couple transistor to load to reduce dc power lost in load resistance.*

resistance introduces only a small power loss, yet the reflected resistance of R_L into the primary circuit provides the proper ac load impedance for the amplifier. Furthermore, the output impedance of the amplifier is matched to the load by the transformer and the actual load resistance can be any convenient value.

The dc load line is essentially vertical on the collector characteristic curves, Fig. 7-17, because of the small winding resistance of the transformer primary. The quiescent operating point is determined exactly as outlined in Chap. 6. The ac load line corresponding to the reflected load resistance R'_L as seen from the primary side of the output transformer passes through the operating point. As usual, the slope of the ac load line is $-1/R'_L$, as shown in Fig. 7-17.

The operating point of a power amplifier is chosen to maximize the efficiency of the amplifier and to minimize the possibility of thermal runaway. Power dissipation in the transistor due to collector current is limited by the allowable temperature rise of the collector junction. If the maximum permissible temperature of the collector junction is T_M, the power dissipation must not exceed

$$P_M = KT_M \tag{7-17}$$

where K is a constant involving the thermal conductance and other geo-

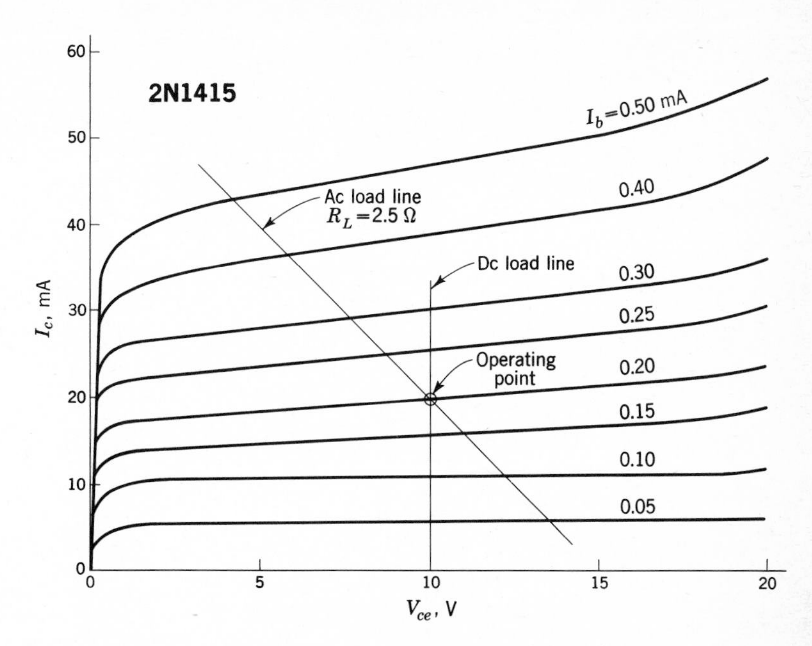

FIGURE 7-17 *Location of operating point and dc and ac load lines for power amplifier in Fig. 7-16.*

metrical factors. Writing the power dissipation in the transistor as I_cV_c, the product must never exceed KT_M. The relation

$$I_cV_c = P_M \tag{7-18}$$

is a hyperbola on the collector characteristics, as indicated by the dashed line in Fig. 7-18. The permissible operating range of collector current and voltage is to the left of this *maximum-power hyperbola.*

Power transistors firmly mounted on a good heat conductor make K in Eq. (7-17) larger. This moves the maximum-power hyperbola farther away from the origin and extends the permissible operating current and voltage range. In addition, cooling fins are often provided to maximize heat conduction away from the transistor.

The operating point is located so that the largest possible ac signals can be developed in order to maximize the power output without distortion. The maximum instantaneous collector potential is limited by reverse breakdown at the collector junction. Similarly, the maximum instantaneous transistor current corresponds to collector saturation, where the collector current no longer increases with emitter-junction current. The output waveform is badly distorted if either of these limits is exceeded because the peaks of the signal wave are clipped. Therefore, the optimum position for the operating point is in the center of the rectangle bounded by collector breakdown, collector saturation, zero collector current, and

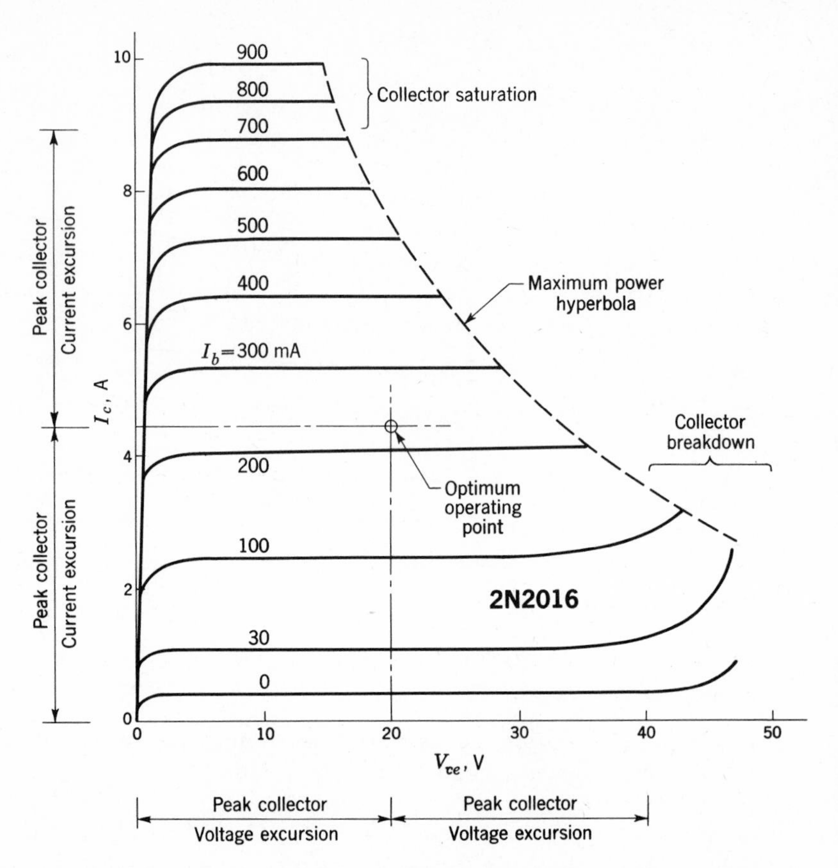

FIGURE 7-18 *Optimum location of operating point for power amplifier is determined by collector saturation and collector breakdown.*

zero collector voltage, Fig. 7-18. Here the collector current and voltage excursions on either side of the operating point are maximized without distortion.

The efficiency of a power amplifier is equal to the ratio of the ac signal power to the dc or average power from the power supply. The average power is simply the product of the quiescent current times the quiescent collector voltage I_cV_c. If the operating point is located at the optimum position, the peak output signal current is equal to I_c and the peak output signal voltage is equal to V_c. Consequently, the efficiency is

$$\eta = \frac{(I_c/\sqrt{2})\,(V_c/\sqrt{2})}{I_cV_c} = \frac{1}{2} \tag{7-19}$$

Thus, the maximum efficiency of this power amplifier is 50 percent. Practical transistor amplifiers approach the ideal quite closely, even though for minimum distortion the signal excursions must be somewhat smaller than the ideal case considered above. Efficiencies of the order of 48 percent are achieved in practice.

The plate characteristics of vacuum tubes are not nearly so ideal as are transistor collector characteristics. Curvature in the characteristics is considerably greater. Consequently, peak signal voltages and currents are smaller and the efficiency is correspondingly less. Pentodes are much more satisfactory than triodes in this respect, but the efficiency of practical circuits rarely exceeds 30 percent.

An equivalent-circuit representation of power amplifiers is not feasible because of the large signal voltage excursions. Consequently, all analyses are carried out graphically. It is most useful to determine the dynamic transfer characteristic of the amplifier, which is a plot of the collector output current as a function of base input current, Fig. 7-19. The transfer

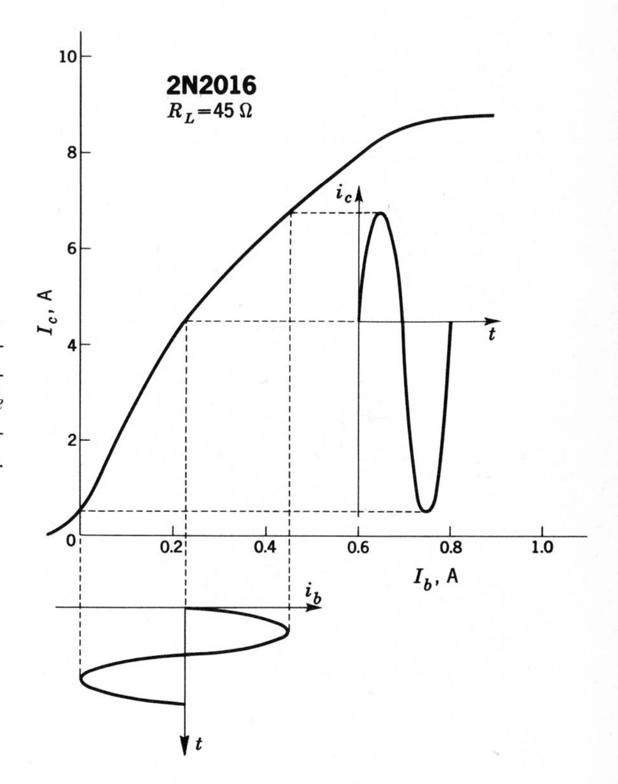

FIGURE 7-19 Dynamic transfer characteristic is used to determine amplified waveform of 2N2016 power amplifier. Note distortion in output current caused by nonlinear transfer characteristic.

characteristic is determined from intersections of the collector characteristic curves, Fig. 7-18, with the dynamic load line. For minimum distortion the transfer characteristic should be a straight line since any curvature introduces irregularities into the output waveform.

Push-Pull Amplifier The two transistor *push-pull* amplifier, Fig. 7-20, has increased power output, efficiency, and less distortion than a single-

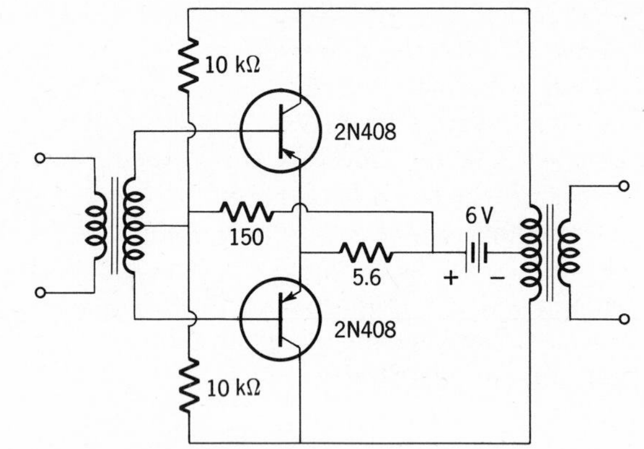

FIGURE 7-20 *Push-pull power amplifier.*

transistor circuit. The center-tapped input transformer drives each of the transistors with signals 180° out of phase, which accounts for the name of the circuit. The amplified collector currents combine in the center-tapped output transformer to produce a load current waveform that is a replica of the input signal. The input transformer also matches the driver stage to the input impedance of the amplifier.

Increased efficiency results when the push-pull amplifier is biased nearly to cutoff. This is known as *class B* operation to distinguish the circuit performance from that previously described in which current is present in the transistor during the entire input-signal cycle. Because each transistor is biased near cutoff, the quiescent current is very small and the signal-voltage and current excursions can be equal to the maximum permissible collector voltage and current, Fig. 7-21. Each tube delivers one-half of a sine-wave signal to the output transformer and the output waveform is preserved even though signal currents in each transistor represent only one-half of the input signal. This action has already been noted in the complementary-symmetry amplifier discussed in Chap. 6. In push-pull operation the peak output voltage can equal the maximum collector potential, which is the same as the dc collector supply voltage (see Fig. 7-21). Correspondingly, the peak signal current is equal to the maximum collector current. The average power of the stage is equal to the power of a half-sine wave, since only one tube conducts at a time. Therefore, the efficiency of a class B push-pull amplifier is

$$\eta = \frac{P_o}{P_{\mathrm{dc}}} = \frac{(V_c/\sqrt{2})\,(I_c\sqrt{2})}{(2/\pi)V_cI_c} = \frac{\pi}{4} \tag{7-20}$$

According to Eq. (7-20), the maximum efficiency is 78 percent, a considerable improvement over the single-transistor amplifier.

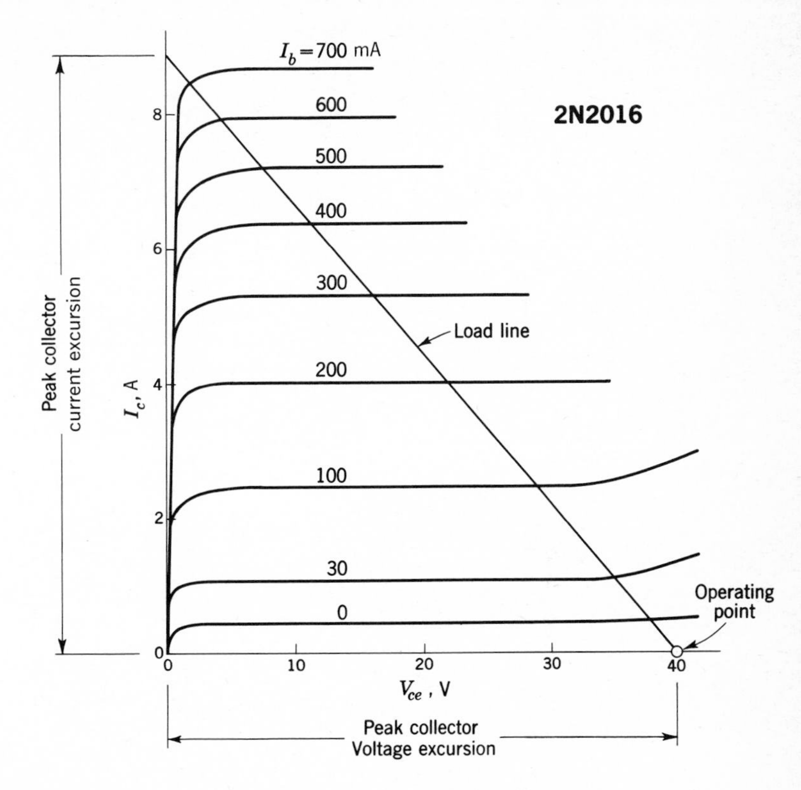

FIGURE 7-21 *Operating point for class B push-pull operation permits voltage to swing over maximum range of collector characteristics.*

The required power output P_o of any push-pull amplifier is specified by the particular application. The peak output-signal voltage is limited by the collector reverse-breakdown potential V_c, however, which means that the collector-to-collector load resistance must be

$$R_L = \frac{V_c^2}{2P_o} \tag{7-21}$$

It usually turns out that the value of R_L determined by Eq. (7-21) is smaller than the output impedance of the transistors, and maximum power-transfer conditions are not possible. Nevertheless, the turns ratio of the output transformer is selected to reflect the proper value of R_L corresponding to the actual load resistance.

A small quiescent base bias current minimizes *crossover* distortion resulting from nonlinearity in the transfer characteristic of each transistor at small currents. This is illustrated in Fig. 7-22, where the transfer characteristics of the two transistors are plotted in opposite quadrants corresponding to their reversed signal polarities. The composite transfer characteristic of the entire amplifier is the average of the individual curves

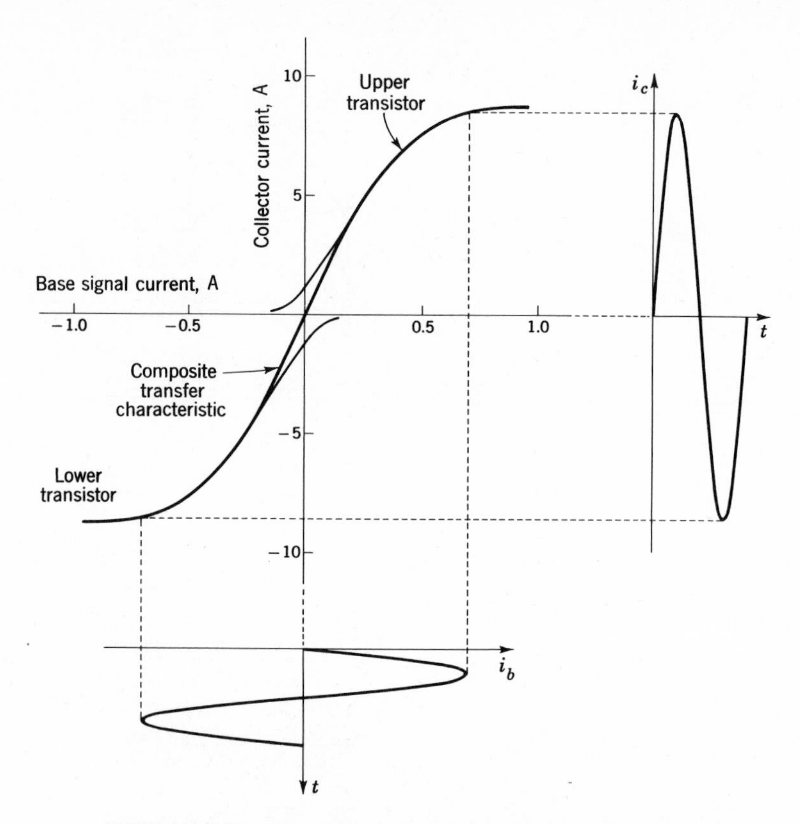

FIGURE 7-22 Composite transfer characteristic of push-pull power amplifier is more linear than that associated with each transistor. Compare output-signal current amplitude and waveform with single transistor case, Fig. 7-19.

and is much more linear than either one. In particular, the nonlinearities cancel each other near the origin where both transistors are active. The cancellation effect means that the push-pull circuit has much less distortion than a single-ended stage.

Special Circuits The frequency response of power amplifiers is seriously limited by transformer chracteristics. In particular, good low-frequency response requires large (and expensive) transformers. This implies poor high-frequency operation because of unavoidable stray capacitances. The small output impedance of the emitter-follower configuration may often be used advantageously to eliminate the output transformer and to achieve satisfactory performance over a wide frequency interval. A typical circuit, Fig. 7-23, also illustrates the flexibility inherent in transistor designs.

Transistor Q_1 is a grounded-emitter input stage which is directly coupled to a complementary-symmetry pair of transistors Q_2 and Q_3. This pair is biased for class B operation and each transistor produces an output signal on alternate halves of the input wave, as previously discussed in connection with Fig. 6-32.

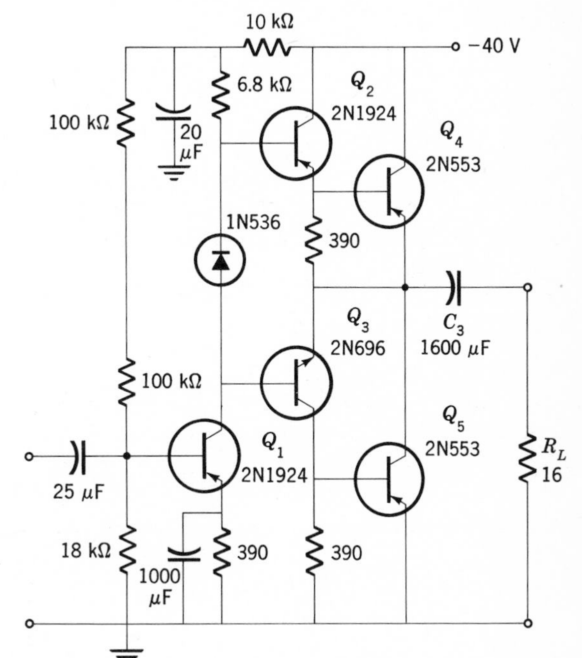

FIGURE 7-23 *Single-ended push-pull power amplifier.*

Actually, Q_2 and Q_3 have a small quiescent bias to minimize crossover distortion. This bias is set by the voltage drop across the forward-biased 1N536 silicon diode. According to Eq. (6-60) the emitter resistance of a transistor decreases significantly with increasing temperature. The forward resistance of the diode similarly decreases with temperature and reduces the base bias of Q_2 and Q_3 correspondingly. Thus the quiescent current in Q_2 and Q_3 is temperature compensated.

The complementary-symmetry pair is in the Darlington connection with respect to the output transistors Q_4 and Q_5. This produces large current gain, as described in the previous chapter. Both Q_4 and Q_5 operate class B except that a small bias provided by the voltage drop across the 390-Ω resistors minimizes crossover distortion. Note that the push-pull signals developed by Q_2 and Q_3 result in activating Q_4 and Q_5 on alternate half-cycles of the input waveform. This may be seen most clearly by assuming the input signal is a square wave. Suppose the input square wave is on the positive cycle so that Q_1 is nearly cut off. This means Q_2 and Q_4 conduct, while Q_3 and Q_5 are cut off, the output signal is negative, and C_3 is charged through R_L. On the alternate half-cycle Q_3 and Q_5 discharge C_3 through the load while Q_2 and Q_4 are cut off. Since each output transistor acts independently of the other, the circuit is, in effect, a single-ended Class B push-pull arrangement.

An integrated-circuit power amplifier is illustrated in Fig. 7-24. The cir-

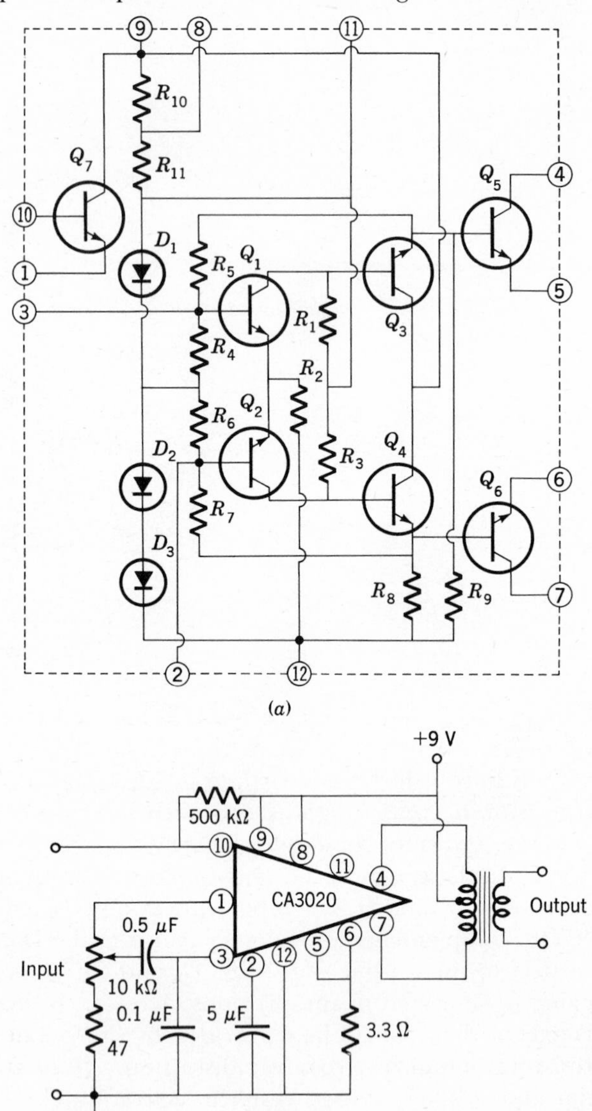

FIGURE 7-24 *(a) Wiring diagram of type CA3020 integrated amplifier circuit and (b) practical power amplifier.*

cuit produces a gain of 58 dB and a power output of $\frac{1}{2}$ W. As previously described in Chap. 5, the entire amplifier is contained within one small monolithic block of silicon.

The circuit of the amplifier, Fig. 7-24*a*, consists of an input difference amplifier, Q_1 and Q_2. The output signals from Q_1 and Q_2 are properly phased to feed Q_3 and Q_4 in push-pull. Q_3 and Q_4 are directly connected to the output transistors Q_5 and Q_6. Transistor Q_7 is a separate unit which may be connected to the remainder of the amplifier or ignored, as desired.

The combination of R_{10} and R_{11} together with the three diodes make up a temperature-compensated voltage regulator for the input stage. Since the voltage drop across a forward-biased silicon diode is approximately 0.6 V, the total base bias supply for Q_1 and Q_2 is 1.2 V, while the collector supply voltage is 1.8 V. Furthermore, the magnitudes of these voltage sources change with\ temperature to compensate for corresponding changes in transistor properties with temperature. Note that R_5 and R_7 set the bias on Q_1 and Q_2 from the base potentials of Q_5 and Q_6. Thus if the bias on the base of Q_5 increases, say, because of a temperature change, the bias on Q_1 also increases. This produces a decreased voltage on the base of Q_3 and, in turn, a decreased bias on Q_5. In effect, the original change has been partially compensated. Such *feedback* helps stabilize the operating points of the transistors against drifts caused by temperature changes and slow changes in component characteristics.

A practical power amplifier using this integrated circuit is shown in Fig. 7-24*b*. In this connection Q_7 is arranged as an emitter follower for a high impedance input and the 10-kΩ potentiometer acts as a variable gain control. The base of Q_2 is grounded for ac signals through the 5-μF capacitor, and a standard push-pull output transformer couples Q_5 and Q_6 to the load. Note how many fewer components are required in using the integrated-circuit amplifier compared to the number necessary if the entire circuit were assembled from separate components.

TUNED AMPLIFIERS

Tuned Coupling Resonant circuits couple the output of one stage to the input of the next when it is only necessary to amplify signals of a single frequency or of a narrow band of frequencies. The impedance of parallel resonant circuits is very great at resonance, as discussed in Chap. 3. Therefore, appreciable gain is achieved at the resonant frequency when a *tuned circuit* is the load impedance of a vacuum-tube or transistor amplifier. A tuned amplifier also rejects signals far from the resonant frequency, which is often a considerable advantage. In addition, stray circuit capacitances are incorporated into the resonant circuit and do not shunt the signal at high frequencies.

The elementary two-stage transistor tuned amplifier, Fig. 7-25, uses parallel resonant circuits for the input circuit and output load of each

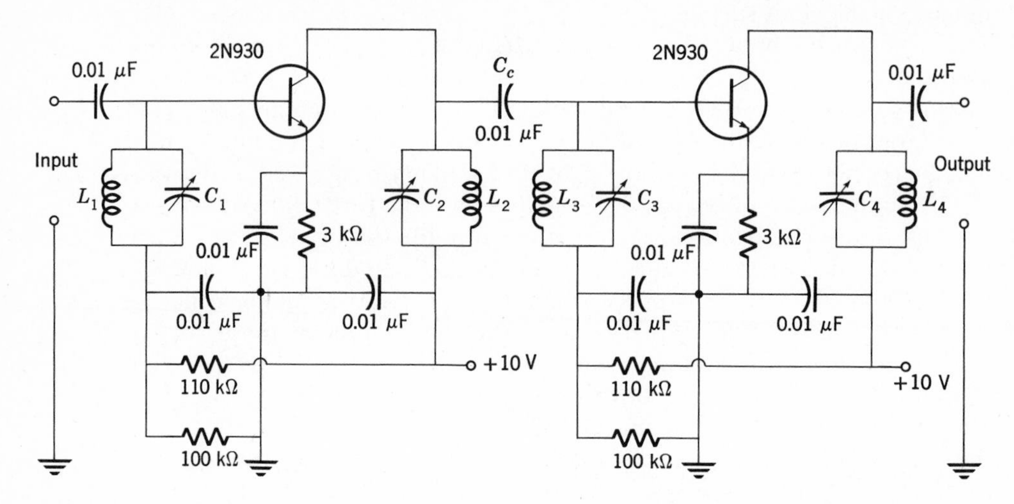

FIGURE 7-25 *Two-stage tuned amplifier.*

transistor. The coupling capacitor C_c carries the signal from one stage to the next. The operating point for each transistor is determined in the standard fashion. It is common practice to make the tuning capacitors C_1, C_2, C_3, and C_4 adjustable so that each circuit can be brought to the same resonant frequency including the effect of all stray capacitances in each stage. Tuned circuits are resistive at resonance, which means that the amplifier can be analyzed by the methods previously developed. Stray capacitances can be neglected, however, and values of the h parameters appropriate at the frequency of interest must be used. Circuit analysis at other than the resonant frequency is rather complicated because of reactance effects, but can be treated straightforwardly using the ac equivalent circuit.

If all four resonant circuits are tuned to the same frequency, the response characteristic is sharply peaked at the resonant frequency, Fig. 7-26.

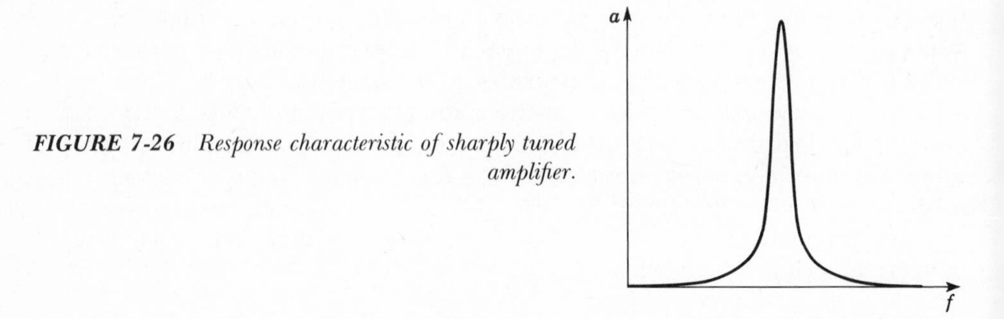

FIGURE 7-26 *Response characteristic of sharply tuned amplifier.*

Such a characteristic is useful when signals having one specific frequency are amplified: Alternatively, each circuit can be tuned to a slightly different frequency, Fig. 7-27, in which case the response characteristic

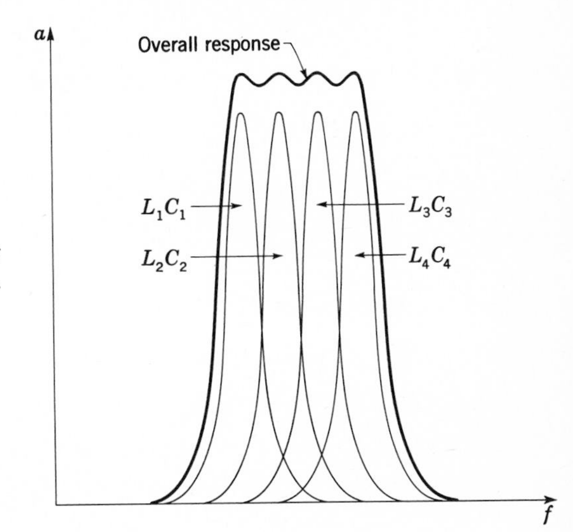

FIGURE 7-27 *Stagger-tuned amplifier has response characteristic with relatively flat top.*

becomes flat-topped. This permits amplification over a band of frequencies such as for the modulated sine wave discussed in Chap. 4. The midband gain of such a *stagger-tuned* amplifier is less than that of the single-frequency circuit since the maximum amplification of each stage occurs at different frequencies.

If L_2 and L_3 are wound on the same core the response characteristic may be double-peaked, Fig. 7-28, even though both primary and secondary

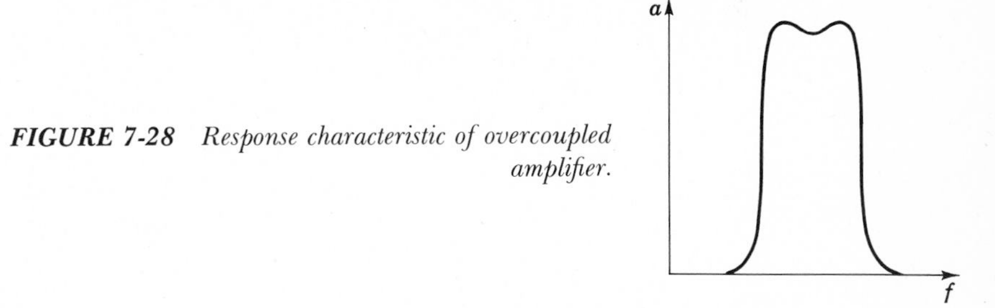

FIGURE 7-28 *Response characteristic of overcoupled amplifier.*

windings are tuned to the center frequency. This is caused by mutual inductance between the windings, which are said to be *overcoupled.* The mutual inductance is altered by changing the distance between the two coils. In this way the frequency-response curve is either sharply peaked, as in Fig. 7-26, or relatively flat-topped, as in Fig. 7-28. The overcoupled case is particularly useful because the midband gain is greater than in the stagger-tuned amplifier and the sides of the response curve rise much more steeply. The circuit thus rejects signals at frequencies immediately outside of the passband. Note that the coupling capacitor is no longer necessary since the signal is coupled from one stage to the next by the mutual inductance. In fact, the combination L_2C_2 and L_3C_3 is looked upon as a tuned transformer.

Neutralization The collector-junction capacitance in many transistors is large enough to cause an undesirable feedback effect between the collector and base. At high frequencies the capacitive reactance becomes small and the amplifier oscillates because the amplified collector signal is returned to the input where it is reamplified, etc. The circuit is useless as an amplifier when this occurs. The effect of the collector capacitance can be *neutralized* by feeding the base a signal of the same amplitude as that produced by the feedback capacitance but 180° out of phase so that the two feedback signals cancel each other. One technique for accomplishing this is illustrated in Fig. 7-29. Here, a portion of the output signal is fed back to the

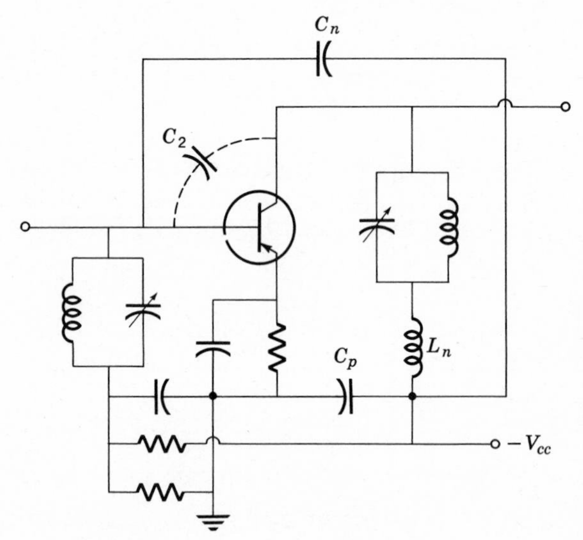

FIGURE 7-29 *Effect of collector-base capacitance is neutralized by signal returned from output circuit through C_n.*

input by the neutralizing capacitor C_n. The magnitude of the return signal is determined by the relative values of the small series inductance L_n and the capacitor C_p, as well as the value of C_n. The return signal is 180° out of phase with that due to the collector capacitance because of the inductive phase shift introduced by the series inductance. Other circuit configurations are also used. In transistor design, every effort is made to minimize collector capacitance so that neutralization is unnecessary.

A similar effect occurs in field-effect transistors because of the drain-gate capacitance. The magnitude of the drain-gate capacitance of MOSFETs is markedly reduced by interposing a second gate between the control gate and the drain, Fig. 7-30, so that neutralization becomes unnecessary.

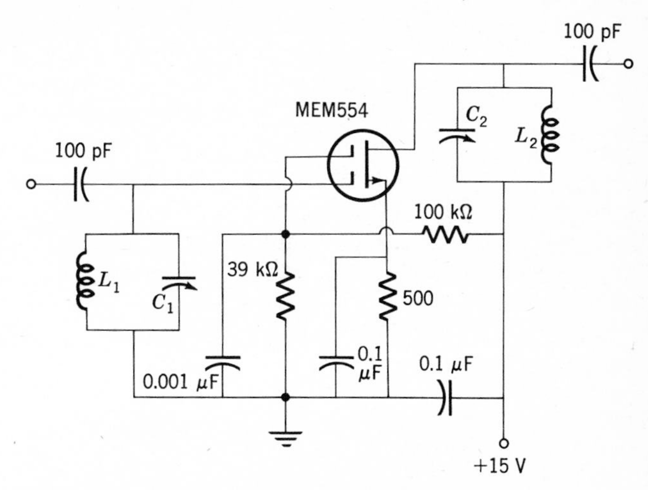

FIGURE 7-30 *Dual gate MOSFET amplifier.*

The second gate is grounded for ac signals to provide the shielding effect. The screen grid in a pentode provides a similar shielding action and for this reason pentodes are almost universally used in preference to triodes as high-frequency amplifiers.

The feedback effect in transistor amplifiers can also be circumvented by employing the grounded-base configuration, Fig. 7-31. In this circuit the

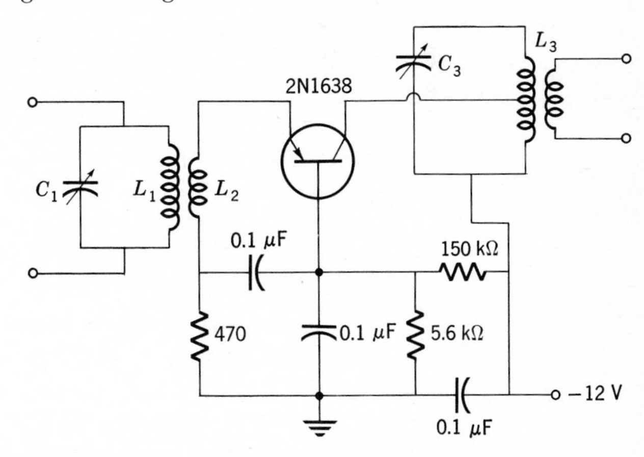

FIGURE 7-31 *Grounded-base tuned amplifier.*

collector-emitter capacitance is so small that neutralization is unnecessary. Furthermore, the grounded-base h parameters are relatively independent of frequency so that the transistor is a useful amplifier at frequencies very close to the α-cutoff frequency. The low input impedance of the grounded-base stage can be matched using a tuned transformer. The winding L_2 has only a few turns, and it is usually not advantageous to resonate the secondary of the transformer.

In the circuit of Fig. 7-31, the collector is tapped down on the inductance of the resonant circuit L_3C_3. This reduces the loading of the transistor output impedance upon the resonant circuit, thereby increasing the Q and making the resonance curve sharper. Furthermore, the collector capacitance is much less significant in determining the resonant frequency. This is important because of the change in collector capacitance with temperature. If the collector is connected across the entire resonant circuit, the collector capacitance is effectively in parallel with the tuning capacitor C_3, and the resonance frequency varies with temperature.

NOISE

Any spurious currents or voltages extraneous to the signal of interest are termed *noise* since they interfere with the signal. Noise voltages arise in the basic operation of electronic devices or are the result of improper circuit design and use. It is important to minimize noise effects in order to characterize the signals with the greatest possible precision and to permit the weakest signals to be amplified. A convenient measure of the influence of noise on any signal is the *signal-to-noise* ratio, the ratio of the signal power to the noise power at any point in a circuit.

Nyquist Noise When a number of amplifier stages are cascaded a random noise voltage appears at the output terminals, even in the absence of an input signal. This output voltage is caused by a random voltage generated in the input resistor. The noise voltages that appear across the terminals of any resistor are attributed to the random motion of the free electrons in the material of the resistance. Electrons in a conductor are free to roam about by virtue of their thermal energy and at any given instant more electrons may be directed toward one terminal of the resistor than toward the other. The result is a small potential difference between the terminals. The magnitude of the potential fluctuates rapidly as the number of electrons moving in a given direction changes from instant to instant.

Since the noise voltage across a resistor fluctuates randomly, it has Fourier components covering a wide range of frequencies. It is convenient, therefore, to specify the noise voltage in terms of the mean square noise voltage per unit cycle of bandwidth. For a resistor R this quantity is

$$\langle \Delta v^2 \rangle = 4kTR \tag{7-22}$$

where k is Boltzmann's constant, and T is the absolute temperature. The noise voltage given by Eq. (7-22) is variously called *Nyquist noise,* after the physicist who derived this equation, or *thermal noise,* since its origin is a result of the thermal agitation of free electrons.

The meaning of Eq. (7-22) is as follows. A noise voltage appears between the terminals of any resistance. The magnitude of the noise voltage actually measured with any instrument depends upon the frequency response of the instrument. For example, the rms noise voltage of a 1000-Ω resistor at room temperature as measured by a voltmeter with a bandwidth of 10,000 Hz is, using Eq. (7-22),

$$v = (1.65 \times 10^{-20} \times 10^3 \times 10^4)^{1/2} \cong 4.1 \times 10^{-7} = 0.41\ \mu\text{V} \tag{7-23}$$

This rather small voltage is not inconsequential. The output voltage of an amplifier with a 10-kHz bandpass and a gain of 10^6 is nearly $\frac{1}{2}$ V if the input resistor is 1000 Ω. This output voltage is present even when no input signal is applied.

The Nyquist expression for the noise voltage of resistances, Eq. (7-22), may be understood in the following way. Replace the actual resistor by an equivalent circuit, Fig. 7-32, containing a noise voltage generator in series

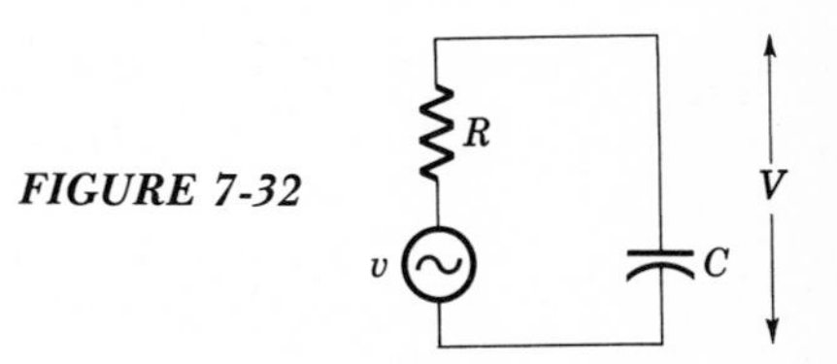

FIGURE 7-32

with a noiseless resistor and in parallel with a capacitor representing the inherent stray capacitance of the actual resistor. The square of the voltage across the capacitor is simply

$$V^2 = \frac{v^2}{1 + (\omega RC)^2} \tag{7-24}$$

The condition of the circuit is completely determined if the voltage across the capacitor is known; in thermodynamic terms the circuit is a system with 1 degree of freedom. According to the equipartition theorem in thermodynamics, the total energy of the capacitor, $\frac{1}{2}CV^2$, must equal $\frac{1}{2}kT$. Thus, using Eq. (7-24),

$$\tfrac{1}{2}kT = \tfrac{1}{2}CV^2 = \tfrac{1}{2}C \int_0^\infty \frac{v^2 df}{1 + (\omega RC)^2} \tag{7-25}$$

The integration extends over all frequencies because of the random nature of the noise voltage.

Equation (7-25) determines the magnitude of the noise voltage v^2. We proceed by assuming that the noise voltage is independent of frequency, so v^2 may be brought out from under the integral sign. Therefore, Eq. (7-25) becomes

$$kT = Cv^2 \int_0^\infty \frac{df}{1 + (\omega RC)^2} = \frac{v^2}{4R} \tag{7-26}$$

Solving Eq. (7-26) for v^2 yields the Nyquist expression.

This development indicates that the noise voltage of resistances is independent of frequency. Accordingly, Nyquist noise is called "white" noise by analogy with the uniform spectral distribution of white light energy. As indicated by Fig. 7-32, the presence of Nyquist noise in any circuit is accounted for by including a noise generator given by Eq. (7-22) in series with a noiseless resistor. In practice it is usually necessary to consider the Nyquist noise of only those resistors in the input circuit of an amplifier. The gain of the first stage makes the amplified noise of the input resistor larger than the noise of resistors in succeeding stages.

Nyquist noise is a fundamental and unavoidable property of any resistance. An amplifier should have a bandwidth only as wide as is necessary to adequately amplify all signal components in order to minimize the ever-present Nyquist noise voltages. If it is desired to amplify a single-frequency signal, for example, the frequency response of the amplifier should be sharply peaked at that frequency, as in Fig. 7-26. The total noise voltage at the output is therefore reduced since only noise components having frequencies in the amplifier passband are amplified. The signal-to-noise ratio is enhanced and weak signals can be amplified usefully.

1/f Noise Noise voltages in excess of Nyquist noise are observed experimentally in certain resistances when a direct current is present. Although the physical origins of this additional noise are not clear, many experiments have shown that the noise is largest at low frequencies and that it increases with the square of the current. An empirical expression for this effect is

$$\langle \Delta v^2 \rangle = K\frac{I^2}{f} \tag{7-27}$$

where K is an empirical constant involving the geometry of the resistor, the type of resistance material, and other factors; I is the dc current; and f is the frequency. According to Eq. (7-27) the mean square noise voltage per unit bandwidth depends inversely upon frequency and the phenomenon is therefore called *$1/f$ noise.* Since the noise also depends on I, it is sometimes referred to as *current noise.*

The magnitude of $1/f$ noise varies markedly with the material of the conductor and its physical form. It is absent entirely in bulk metals, so that only Nyquist noise is observed in wire-wound resistors. Composition resistors, on the other hand, generate a large $1/f$ noise level, Fig. 7-33, which is associated with the intergranular contacts in such resistors. Although such contacts are known to be important, $1/f$ noise is also observed in single-crystal semiconductors where contact effects are negligible.

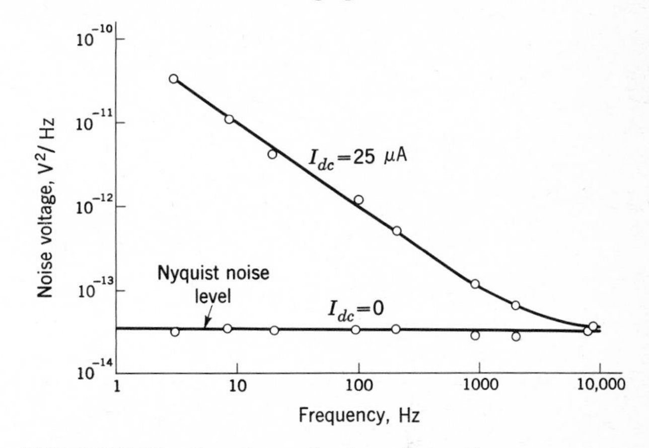

FIGURE 7-33 *Experimental noise voltage of 2.2 -MΩ composition resistor. Note spectrum is 1/f when dc current is present and white Nyquist noise in absence of current.*

To minimize the low-frequency noise level, resistor types in which the $1/f$ noise is small, such as wire-wound units, are selected. Fortunately, at high frequencies where wire-wound resistors are unsuitable because of their inductance, the $1/f$ noise of composition resistors is usually negligible compared with Nyquist noise. In other situations the direct current in noisy components is minimized to reduce the current noise generated. The total current noise voltage in any given circuit is found by integrating Eq. (7-27) over the frequency response characteristic of the amplifier.

Low-frequency $1/f$ noise is also present in vacuum tubes and transistors. In the former it is often known as *flicker noise* and originates in the semiconducting cathode material, particularly at the emitting surface. The noise in transistors results from semi-conductor properties, in which surface conditions are very important. In general, $1/f$ noise is more prevalent in germanium devices than in silicon units.

Noise in Tubes and Transistors Other noise effects are also present in tubes and transistors. *Shot noise* in vacuum tubes is a result of the random emission of electrons from the cathode. Since each electron represents an increment of current, the plate current fluctuates slightly about the dc value. This effect is analogous to the noise of raindrops on a tin roof. That is, the basic reason for shot noise is that the electron is a discrete unit of electrical charge.

An expression for the magnitude of shot noise can be developed as follows. Suppose that n is the average number of electrons emitted from the cathode in a time interval t. The direct current is then, from Eq. (1-10),

$$I = \frac{en}{t} \tag{7-28}$$

According to a general principle of statistical phenomena the variance in n is equal to its average value, so that

$$\langle \Delta n^2 \rangle = n = \frac{It}{e} \tag{7-29}$$

Therefore, the current fluctuations in the time interval t are

$$\langle \Delta I^2 \rangle = \left(\frac{e}{t}\right)^2 \langle \Delta n^2 \rangle = \frac{e}{t} I \tag{7-30}$$

Finally, it can be shown that the relation between the total fluctuations in a given time interval and the mean square fluctuation per unit bandwidth is given by

$$\langle \Delta i^2 \rangle = 2t \langle \Delta I^2 \rangle \tag{7-31}$$

Introducing Eq. (7-30), the current fluctuations are

$$\langle \Delta i^2 \rangle = 2eI \tag{7-32}$$

The quantity $\langle \Delta i^2 \rangle$ is analogous to $\langle \Delta v^2 \rangle$ in the Nyquist expression, Eq. (7-22), except that the noise is expressed here in terms of current fluctuations. The mean square noise voltage output of a tube is simply Eq. (7-32) multiplied by the square of the load resistance. Note that shot noise is a white noise since the right side of Eq. (7-32) is independent of frequency.

The basic expression for shot noise, Eq. (7-32), applies to the situation in which each electron emitted from the cathode proceeds to the anode independently of all other electrons. As discussed in Chap. 5 in connection with Child's law, this is not the case in practical vacuum tubes. Each electron is influenced by the presence of all the others. The result of this interaction is to make the electron current more uniform and thereby reduce the magnitude of shot noise. It turns out that the actual value is a function of the operating point. A full analysis of this effect is complicated because it depends upon subtle details of the emission current and electric field.

It is convenient to express the effective shot noise in terms of the Nyquist noise of an equivalent resistor in the grid circuit. The magnitude of the equivalent noise resistor in the case of a triode is given by

$$R_{neq} = \frac{2.5}{g_m} \quad \Omega \tag{7-33}$$

The meaning of R_{neq} is simply that the effective shot noise of the triode can be represented by a noise voltage generator in the grid circuit given by the Nyquist expression, Eq. (7-22), using the resistance value determined from Eq. (7-33). Therefore, the appropriate equivalent circuit, Fig. 7-34, includes two noise generators, one associated with Nyquist noise

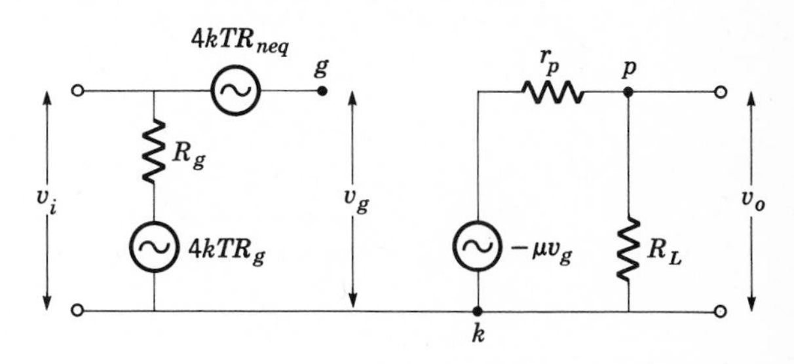

FIGURE 7-34 *Equivalent circuit of triode including noise generators.*

of the grid resistor and the other associated with the effective shot noise of the tube. According to Eq. (7-33), the effective shot noise is smallest for tubes having a large mutual transconductance and is of the order of 250 Ω for the lowest-noise triodes listed in Table 6-1. The internal noise as a function of frequency for typical triodes is illustrated in Fig. 7-35. In these

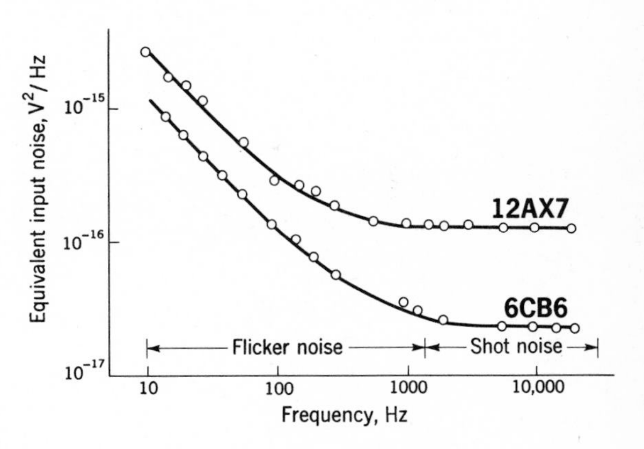

FIGURE 7-35 *Experimental internal noise levels of 12AX7 triode and triode-connected (screen tied to plate) 6CB6. Note flicker noise at low frequencies and white shot noise at high frequencies.*

curves the output noise level is referred to a noise voltage generator in the grid circuit. Any contribution due to the grid resistor has been eliminated simply by shorting the grid to ground. The flicker-noise and shot-noise regions can be easily discerned.

Noise voltages are treated by including noise voltage generators in the equivalent circuits. The effect of the amplifier passband, which may be determined by stray shunt capacitance or by tuned circuits, is included in assessing the magnitude of the noise voltages. The noise resulting from each generator may be treated separately since random noise voltages are independent. The total output noise voltage is therefore simply the sum of all noise effects. As mentioned previously, it is not necessary to consider the noise of the resistors r_p and R_L in Fig. 7-34 because the amplified noise of the input circuit is much larger than the Nyquist noise of these resistors.

An expression analogous to Eq. (7-32) applies to each electrode in pentode and other multigrid tubes where the appropriate value of current to each electrode is used. The total noise is the sum of the individual electrode noises, suitably modified by the internal conditions. The sum is known as *partition noise* since the total tube current is divided among several electrodes. Pentodes are therefore noisier than triodes and the latter are universally used when it is necessary to obtain the largest possible signal-to-noise ratio. Note that this applies principally to the input stage of an amplifier. The gain of the first stage increases the signal level sufficiently so that noise effects in succeeding stages are negligible. The signal-to-noise ratio of any circuit is determined primarily by conditions in the first amplifier stage.

The input signal to any circuit has associated with it a given signal-to-noise ratio since Nyquist noise corresponding to the source resistance is present, at least. An ideal amplifier amplifies the incoming signal and the incoming noise equally and introduces no additional noise. Therefore, the original signal-to-noise ratio is preserved at the output. Practical amplifiers are not ideal because of Nyquist noise of the input circuit and shot noise of the first stage. A useful figure of merit for any circuit is the *noise figure, NF,* which is defined as the input signal-to-noise ratio divided by the output signal-to-noise ratio. An ideal amplifier has a noise figure of unity and many practical circuits approach this value fairly closely.

The noise of transistors is a result of Nyquist noise of the semiconductor resistance, $1/f$ noise caused by current in the semiconductor crystal, and shot noise of carriers crossing the junctions. In addition, still another noise phenomenon has been observed in semiconductors. Electrons are promoted randomly from the valence band to the conduction band and also return randomly to the valence band, keeping the proper average number of carriers in each band. Random generation and recombination of carriers is caused by thermal energies and produces a conductivity fluctuation of the semiconductor. This generates a noise voltage when a direct current is present; this second type of current noise in semiconductors is termed *generation-recombination,* or *g-r noise.*

Analysis of noise phenomena in transistors is complicated by these many factors and by the inherent input-output coupling. It turns out that at intermediate frequencies the noise figure of a transistor can be expressed as

$$\text{NF} = 1 + \frac{r_b}{R_s} + \frac{r_e}{2R_s} + \frac{(R_s + r_b + r_e)^2}{2h_{fe}r_eR_s} \tag{7-34}$$

where R_s is the source resistance and the other symbols have their usual meaning. According to Eq. (7-34) the noise figure approaches unity if the emitter and base resistances are small with respect to the source resistance and if the forward current gain h_{fe} is large. Note that the internal noise of a transistor depends upon the operating point, much as is the case for a vacuum tube.

Equation (7-34) ignores $1/f$ noise, which increases the noise level at low frequencies, Fig. 7-36. Additionally, an increase in noise at frequencies

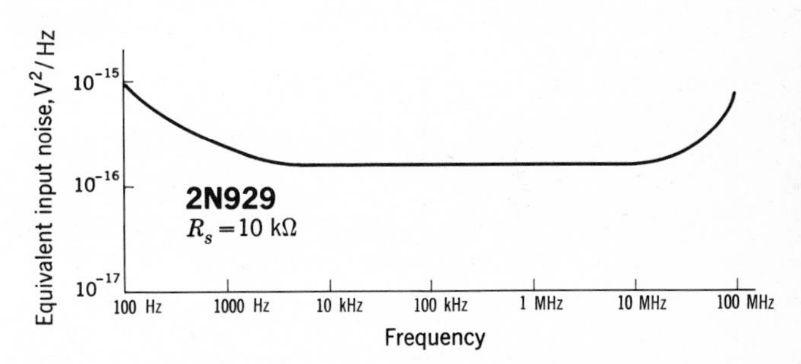

FIGURE 7-36 *Experimental internal noise level of a type 2N929 transistor with a source resistance of 10,000 Ω.*

of the order of the α-cutoff frequency is observed. The latter effect, also apparent in Fig. 7-36, can be quite well explained on the basis of the influence of current gain and the α-cutoff frequency upon *g-r* noise.

The internal noise of field-effect transistors is a result of thermal noise in the channel. Therefore, at intermediate frequencies the drain noise is represented quite well by

$$\langle \Delta i_d{}^2 \rangle = 4kTg_m \tag{7-35}$$

where g_m is the transconductance of the FET. The analogy between (7-35) and (7-22) is evident; note that the equivalent noise resistance in the drain circuit is just $1/g_m$ and, therefore, depends upon the operating point. The noise increases at low frequencies because of $1/f$ noise and also increases at high frequencies as a result of capacitive coupling between channel noise and the gate electrode. With careful design the FET may have the lowest noise figure of all active devices, particularly at high impedance levels.

Shielding and Grounding Noise is often introduced in practical circuits by extraneous voltage signals coupled into the circuit from the surroundings. The most common source of this noise comes from the 60-Hz electric and magnetic fields produced by power mains. The 60-Hz signal induced by these fields is called *hum* because it is audible as a low-frequency tone in amplifiers connected to a loudspeaker. Other *stray pickup* may result from electric fields generated by nearby electronic equipment, electric motors, lightning discharges, etc.

It is useful to *shield* those portions of a circuit where the signal level is small and, consequently, where noise voltages are most troublesome. Electric fields induce noise voltages capacitively, so it is only necessary to surround the circuit with a grounded conducting shield in order to reduce stray pickup. This is illustrated schematically in Fig. 7-37, where the

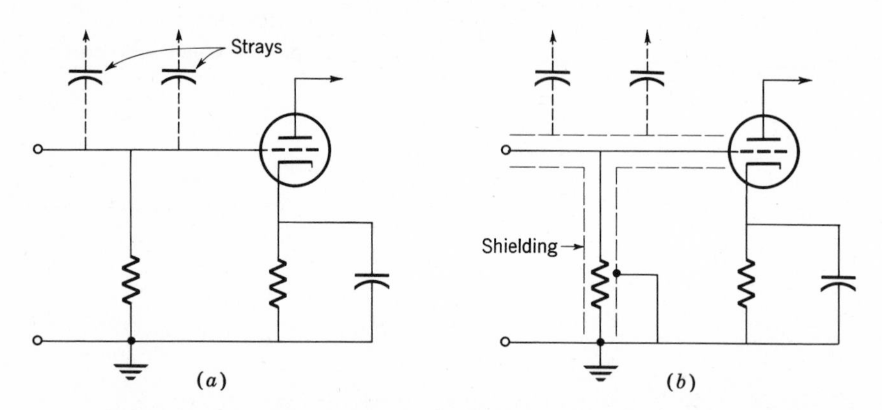

FIGURE 7-37 *(a) Stray capacitive coupling introduces noise pickup signals into sensitive circuits. (b) Grounded conductor shields the circuit from surroundings.*

capacitive coupling to external sources in Fig. 7-37*a* is interrupted by interposing a grounded conductor, Fig. 7-37*b*. Such shielding is also effective in reducing so-called *crosstalk* between different stages of the same circuit as, for example, between the input stage and the power output stage of a complete amplifier.

Additionally, it is useful to shield a circuit to minimize induced currents resulting from stray magnetic fields. This is accomplished with high-permeability ferromagnetic enclosures which reduce the intensity of the magnetic field inside. Such shielding is never complete because of the properties of ferromagnetic materials, and it is always advantageous to minimize the area of the circuit by using the shortest possible signal leads. According to Eq. (2-53) the induced voltage in any circuit resulting from changing magnetic fields decreases if the enclosed area of the circuit is reduced. Transformers are particularly troublesome with respect to inductive pickup because of their many turns of wire. They are kept well removed from all power transformers because of the strong magnetic fields generated by such units.

In general it is good practice to keep all circuits physically small in order to minimize stray pickup, crosstalk, and stray capacitance. All grounded components, such as bypass capacitors, pertaining to a given stage are returned to a single point. This reduces so-called *ground loops*, which are current paths through the metal chassis on which electronic circuits are often mounted. If all components of one stage are not grounded at the same point, the currents may cause undesirable signal coupling between stages.

Vacuum tubes are also *microphonic* in that noise voltages are generated by movement of the grid wires caused by mechanical vibration or shock. Transistors are much superior in this respect because of their simpler mechanical construction. On the other hand, the lower impedance level in transistor circuits makes them more susceptible to induced magnetic pickup because the induced currents are larger. Hum may also result from an inadequate power-supply filter or the ac heater-current wires. It is common practice to twist heater-current wires tightly together to reduce the net magnetic field from the current. In very sensitive circuits hum pickup from this source is reduced by heating the cathodes of the input stages with direct current.

When a number of individual electronic units are interconnected, shielded cable is used for all signal leads between units. The shield is used as the ground lead, as illustrated in Fig. 7-38. It is important that

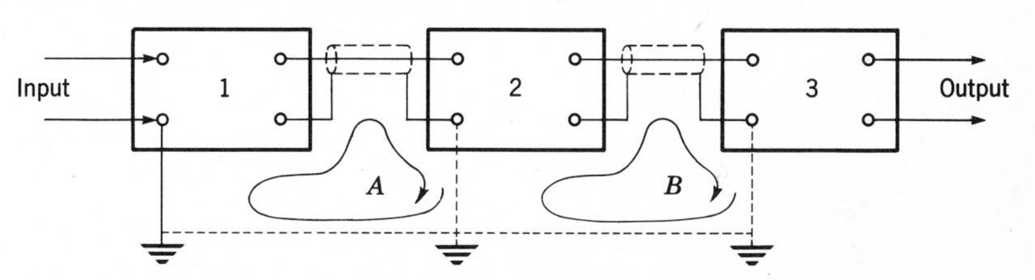

FIGURE 7-38 *When several electronic devices are connected, system must be grounded only at one point, preferably at input. Multiple grounds can lead to large ground-loop currents.*

the entire system be grounded at only one point, usually the input terminal. If each unit is grounded separately, as shown by dashed lines in Fig. 7-38, large 60-Hz currents can be induced in the circuit because of the large area encompassed. The current induced in loop *A* by stray 60-Hz magnetic fields introduces a large stray pickup signal into the amplifier.

Note that the capacitance between the central wire and its shield tends to shunt the signal at high frequencies. For this reason such cables are kept as short as possible. The circuit output impedance is also made small since this reduces the effect of shunt capacitance. Therefore, the output stage of many electronic circuits is a cathode or emitter follower.

SMALL-SIGNAL MEASUREMENTS

Direct-coupled Amplifiers All the amplifier circuits discussed to this point have zero gain for dc signals because of the infinite reactance of the coupling networks at zero frequency. Amplification of dc or very slowly varying signals is achieved by eliminating the coupling networks entirely. In addition to the dc response of such a *direct-coupled* amplifier, the high-frequency performance is also enhanced. This comes about because stray capacitances associated with the coupling networks are eliminated as well.

The Darlington connection already discussed is an elementary two-stage dc amplifier. More elaborate circuits are necessary to achieve larger values of voltage gain. Consider, for example, the three-stage direct-coupled amplifier in Fig. 7-39 which consists of two grounded-emitter transistors

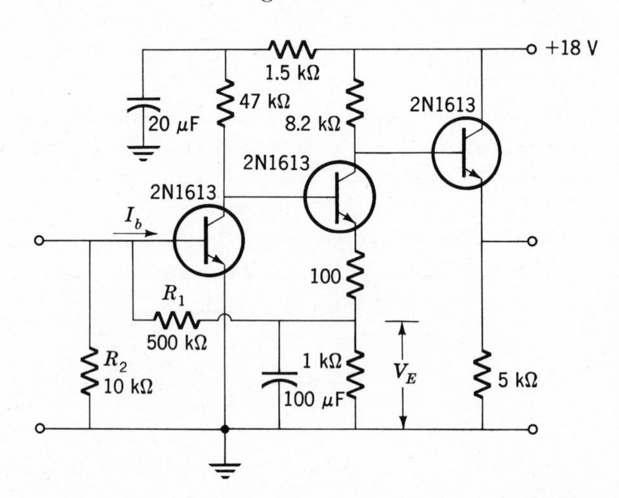

FIGURE 7-39 *Three-stage direct-coupled amplifier.*

in cascade followed by an emitter-follower output stage. The major difficulty with such cascaded dc amplifiers is *drift* caused by small variations in transistor characteristics with time or temperature and changes in the supply potentials. The input transistor is particularly susceptible because any small changes in the input are amplified by succeeding stages in the amplifier.

It is common practice to stabilize the quiescent point of direct-coupled amplifiers by returning a portion of the output voltage to the input stage in such a fashion that drifts tend to be compensated. This is accomplished in Fig. 7-39 by the resistor R_1 which sets the base bias on the input transistor. Note that if the emitter voltage at the second stage, V_E, increases for some reason the base current of the first stage increases correspondingly. This results in a decrease in base current of the second transistor which tends to return V_E to its original value. The stabilizing action of R_1 can be examined best with the equivalent circuit of the input stage base circuit, which is similar to Fig. 6-23*b*. The voltage drops around the equivalent circuit are

$$\frac{R_2}{R_1 + R_2} V_E - I_b \frac{R_1 R_2}{R_1 + R_2} - V_{be} = 0$$

where V_{be} is the emitter-base voltage. Solving for V_E

$$V_E = \left(1 + \frac{R_1}{R_2}\right) V_{be} + R_1 I_b \tag{7-36}$$

Now V_{be} is a relatively constant quantity. If the circuit is arranged so that the first term in Eq. (7-36) is much larger than the second, V_E is constant and the amplifier is stabilized. After calculating V_E from Eq. (7-36), the currents in each transistor may be determined from the transistor characteristics and the quiescent point of the amplifier is completely determined.

Note, however, that in achieving stability against drift the amplifier cannot respond to dc signals. If the second term in Eq. (7-36) is negligible, an input signal I_b does not cause any change in V_E; hence the output signal is zero. This is not so for ac signals where the 100-μF emitter bypass capacitor shorts out V_E, and no signal is returned to the input. Practical amplifiers strike a compromise between stability and dc response such that both terms in Eq. (7-36) are active.

Drifts caused by variations in device characteristics and supply voltages are minimized by using a balanced difference amplifier, Fig. 7-40. Changes in one side of the circuit tend to be compensated by similar changes on the other side. Furthermore, the output terminals are at the same potential when the input signal is zero. Note the similarity of Fig. 7-40 to Fig. 5-37*a* and Fig. 6-19.

The purpose of transistor Q_3 is to provide a constant-current source for the difference amplifier to further reduce drift tendencies. In addition, the ac collector impedance of Q_3 is very large, essentially equal to $1/h_{oe}$, which greatly reduces the effect of circuit assymetry, according to Eq. (6-51). This is more convenient than a large emitter resistor since the voltage drop across Q_3 is much smaller than that associated with a resistance equal to $1/h_{oe}$. Note that the base bias resistors for Q_1 and Q_2 are returned to the opposite collectors. This return tends to stabilize the operating point without reducing gain since the collector potentials change in opposite directions in response to an input signal.

The output signal of the difference amplifier is, using Eq. (6-47) and Eq. (6-25),

$$v_o = g_m R_L (v_2 - v_1) \tag{7-37}$$

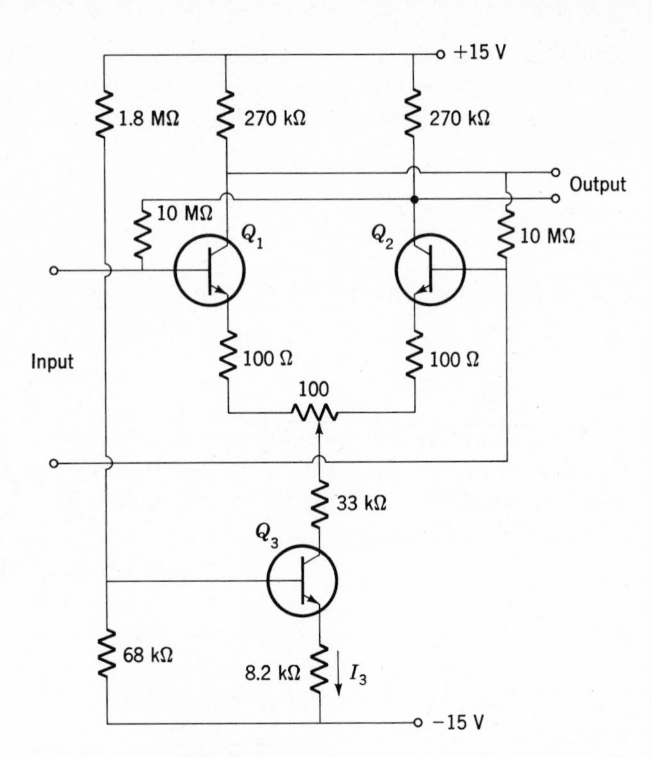

FIGURE 7-40 *Difference amplifier is balanced dc amplifier.*

It is instructive to evaluate the transconductance of the transistor using the rectifier equation, Eq. (5-20),

$$g_m = \frac{dI_c}{dV_e} = \alpha \frac{dI_e}{dV_e} = \frac{\alpha e}{kT} I_o e^{eV_e/kT}$$

$$= \frac{\alpha e}{kT} I_e \tag{7-38}$$

the circuit is balanced so that

$$I_{e1} + I_{e2} = 2I_e = I_3 \tag{7-39}$$

From Eq. (7-38) and Eq. (7-39) and Eq. (7-37), the output signal is then

$$v_o = \frac{\alpha e}{2kT} I_3 R_L (v_2 - v_1) \tag{7-40}$$

According to Eq. (7-40) the gain of the amplifier is set by the current in Q_3. This current, in turn, is determined by the base bias on Q_3 and may be adjusted to suit gain requirements.

The stability and balance of the difference amplifier is such that several stages may be cascaded to achieve large values of amplification, as considered in the following chapter. A useful two-stage amplifier which is a practical dc millivoltmeter is illustrated in Fig. 7-41. A FET stage is fol-

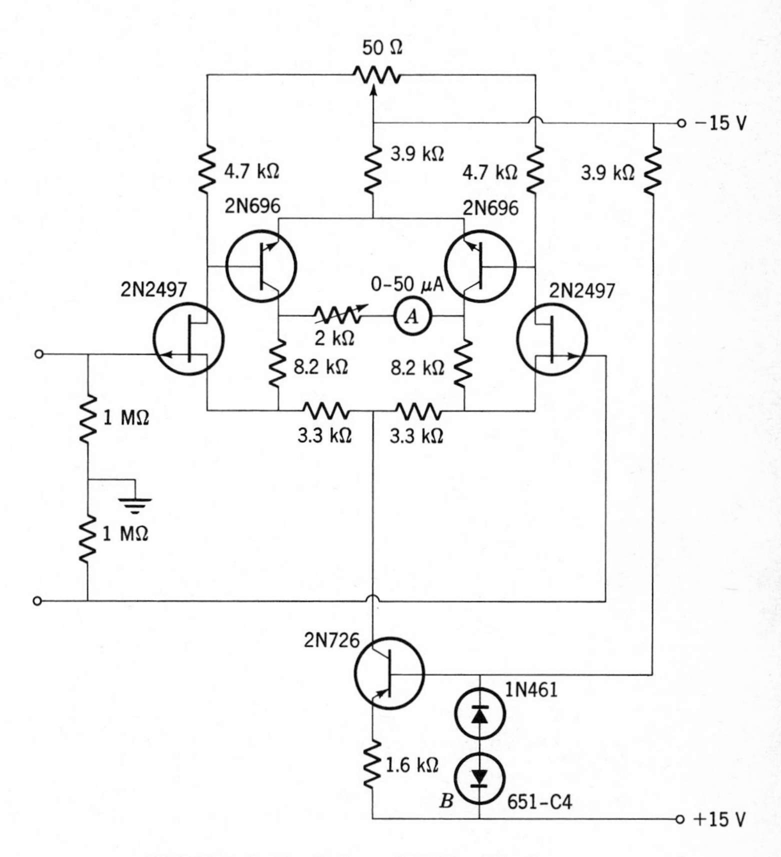

FIGURE 7-41 *Balanced FET millivoltmeter.*

lowed by a gounded-emitter transistor stage. The FET amplifier provides large input impedance and is stabilized against voltage changes through the 2N726 emitter transistor and the 651-C4 zener diode and also against temperature changes as a result of the 1N461 diode in the base bias circuit. The overall gain of the amplifier is such that full-scale deflection of the meter is achieved for an input signal of 50×10^{-3} V, so that the equivalent sensitivity is 40 MΩ/V. Alternatively, the circuit may be considered to be a microammeter having a full-scale deflection of $(50 \times 10^{-3})/(2 \times 10^{6}) =$ 25 nA.

Chopper Amplifiers To circumvent the drift and instabilities inherent in direct-coupled amplifiers, it is useful to convert the dc input signal to an ac voltage which can be amplified by a standard ac-coupled circuit. Subsequently, the amplifier output is rectified to recover the amplified dc input signal. The superior stability features of ac-coupled amplifiers provide much greater gains than are practical with dc amplifiers.

A convenient technique for converting a dc voltage to an ac signal is with a mechanical *chopper*, which is a rapidly vibrating switch driven by an electromagnet, Fig. 7-42. As the switch alternately closes contact A and B,

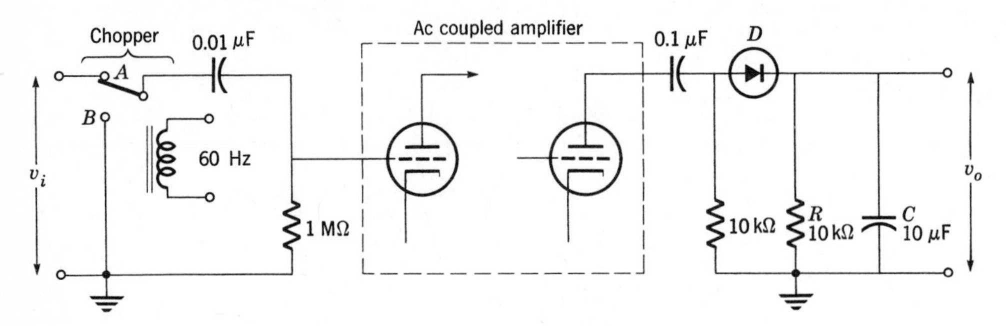

FIGURE 7-42 *Chopper amplifier using electromagnetically driven switch to convert input voltage to ac signal.*

a square wave with an amplitude equal to the dc signal is produced. The frequency of the square wave corresponds to the chopping frequency. This is most often 60 Hz, since it is convenient to drive the chopper from the power line. The amplifier is broadly tuned to the chopping frequency, which minimizes broad-band noise effects. A simple rectifier-filter combination attached to the amplified output results in a dc voltage corresponding to the amplified input signal.

Nonmechanical choppers employing diodes, transistors, vacuum tubes, SCRs, etc., are also used in various circuits. Consider, for example, the *balanced modulator* circuit, Fig. 7-43, which consists of four diodes arranged in a bridge. The chopping signal may be either a square-wave or a sine-wave signal at some convenient frequency. When the chopping voltage makes terminal A positive with respect to terminal B, diodes D_1 and D_2 are biased in the forward direction and diodes D_3 and D_4 are biased in the reverse direction. The small forward resistance of D_1 and D_2 means that the lower terminal of the input is effectively connected to the upper output terminal. On the alternate cycle of the chopping voltage, D_3 and D_4 conduct and the lower input terminal is connected to the lower output terminal. The result is a square wave with a peak-to-peak amplitude equal to twice the dc input signal.

In many applications the chopper amplifier must amplify a range of frequencies from dc to some high-frequency cutoff. The output voltage can then accurately reflect changes in the input signal. As a rule of thumb, the upper frequency limit of a chopper amplifier is about one-fourth of

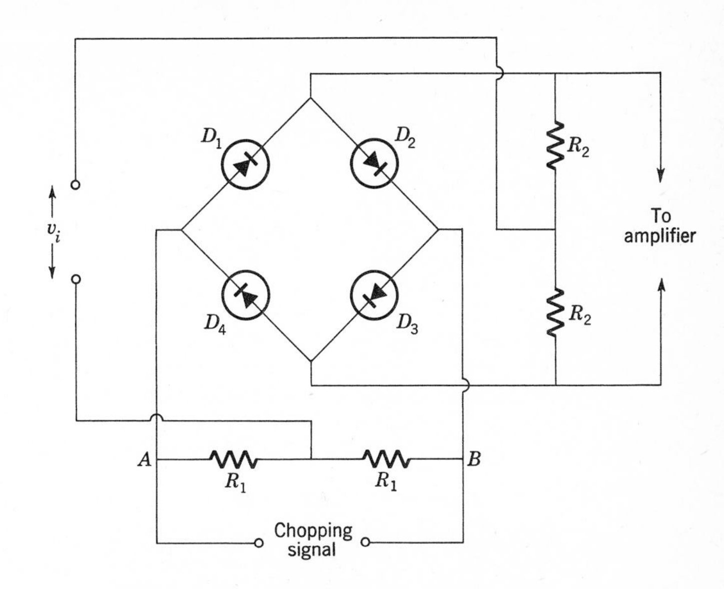

FIGURE 7-43 *Balanced modulator electronic chopper for use at high chopping frequencies.*

the chopping frequency. Mechanical choppers are limited to 60 Hz, although 400-Hz units are occasionally used. Nonmechanical choppers, such as the balanced modulator, are useful because the chopping frequency can be much higher.

Mechanical choppers are most satisfactory for very-high-gain dc amplification because they introduce minimum noise into the circuit. Suppose, for example, that one of the four diodes of the balanced modulator is slightly different from the other three. The bridge is therefore slightly unbalanced and a portion of the chopping signal appears at the amplifier terminals even when the dc input signal is zero. This signal is amplified by the circuit and appears as a noise voltage at the output. Similarly, other semiconductor and tube devices introduce switching noises to a greater or lesser extent. The mechanical chopper has very small resistance when the contacts are closed and very high resistance when the contacts are open. It is nearly ideal in this respect. Nevertheless, chopping noise limits the amplifier sensitivity at the very highest gain applications. One source of noise is stray coupling between the driving solenoid and the signal circuits.

A major advantage of chopper amplifiers is their very low effective internal random-noise level. The reason for this is the rectifier-filter combination in the output circuit. The output voltage signal may be made as noise-free as desired by increasing the time constant of the filter. In effect, the overall bandwidth of the system is equal to the frequency interval from dc (zero Hz) to $1/2\pi\tau_0$, where τ_0 is the filter time constant. If, for example, the time constant is 10 s, the effective amplifier bandwidth is

0.016 Hz, a very small value indeed. Since the total noise voltage increases with bandwidth [compare Eqs. (7-22) and (7-27)], the total noise is very low. Of course, when τ_0 is 10 s, a time interval of approximately 30 s is required for the output voltage to reach its final value. Thus the amplifier responds very slowly to changes in the amplitude of the input signal. This reciprocity between bandwidth and response time is a general property of all measuring systems.

A major difficulty with the simple chopper amplifier is that the output voltage is independent of the polarity of the input signal. That is, the ac-coupled amplifier yields an ac output signal for either polarity of input voltage. This situation is corrected in chopper amplifiers which employ a second set of contacts to rectify the amplifier output, Fig. 7-44. The

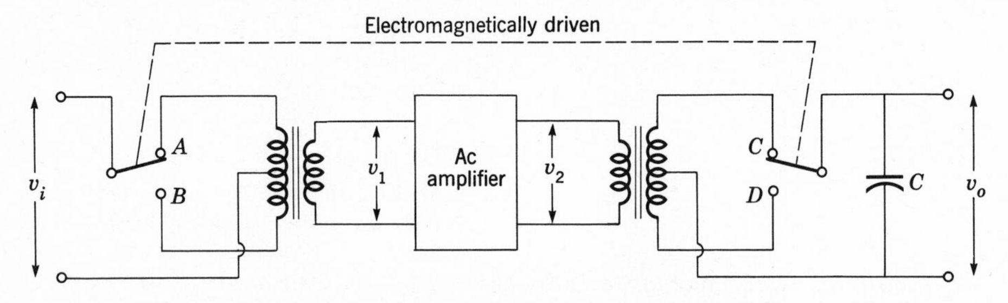

FIGURE 7-44 *Synchronous chopper amplifier preserves polarity of input voltage.*

contacts are arranged to close synchronously so that as the input chopper converts the dc signal to a double-ended square wave, Fig. 7-45*a* and *b*,

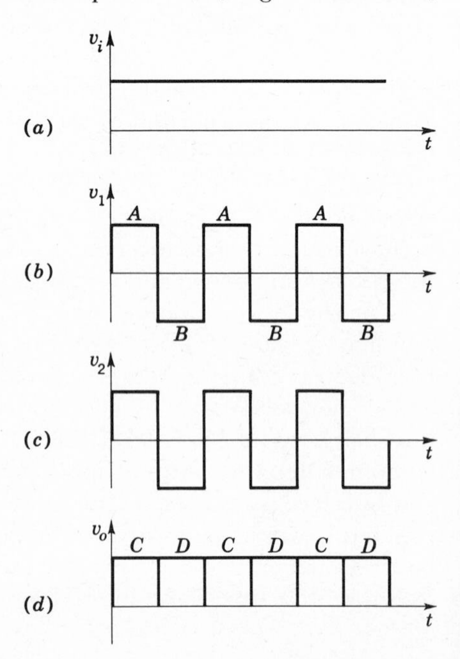

FIGURE 7-45 *Waveforms in synchronous chopper amplifier: (a) dc input signal, (b) square-wave input to amplifier, (c) amplified square-wave output, and (d) rectified square wave produced by second chopper.*

the output chopper reconverts the amplified square wave back to a dc signal, Fig. 7-45*c* and *d*. Comparing the waveforms in Fig. 7-45 shows that if v_i is positive, v_o is also positive. Similarly, if the input signal is negative, the output signal is also negative. Both sets of contacts are put on the same vibrating arm in practical synchronous choppers to assure that they open and close simultaneously. An electronic version of this *synchronous rectifier* is analyzed in the next section.

Lock-in Amplifiers The principles of the synchronous chopper amplifier are employed in an electronic circuit which has found wide use in instrumentation systems. The version illustrated in Fig. 7-46 uses a diode modu-

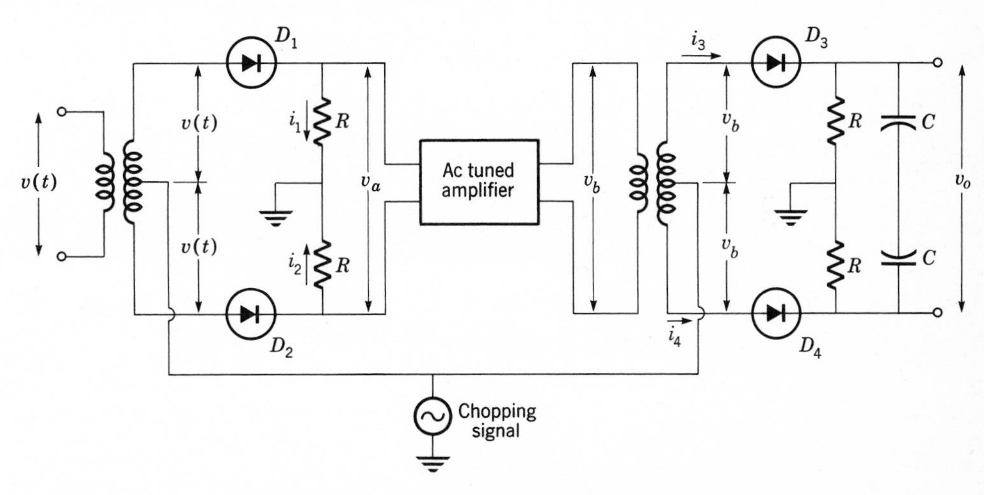

FIGURE 7-46 *Electronic synchronous chopper amplifier.*

lator to convert the input signal to a form suitable for the amplifier and a diode demodulator to recover the amplified version of the input signal. For generality, we consider the input signal to be a slowly varying signal; if dc signals are important the input transformer can be replaced by a center-tapped resistor.

The input circuit is analyzed using techniques discussed in Chap. 4. It is assumed that the diode characteristic can be represented by a quadratic expression. Then the current in the diode D_1 is, from Eq. (4-30),

$$i_1 = a_1 v(t) + a_1 V_2 \sin \omega_2 t + a_2 v^2(t) + a_2 V_2^2 \sin^2 \omega_2 t + 2a_2 v(t) V_2 \sin \omega_2 t \qquad (7\text{-}41)$$

where a_1 and a_2 are constants related to the diode characteristic, $v_2 = V_2 \sin \omega_2 t$ corresponds to the chopping signal discussed earlier, and $v(t)$ is the input signal. An identical expression applies to the current in D_2, except that the polarity of $v(t)$ is reversed with respect to v_2. The voltage signal applied to the amplifier is

$$v_a = R(i_1 - i_2) \tag{7-42}$$

Substituting for i_1 and i_2, many terms cancel because of the reversed polarity of $v(t)$ in D_2. The result is

$$v_a = 2Ra_1v(t) + 4Ra_2v(t)V_2 \sin \omega_2 t \tag{7-43}$$

The first term in Eq. (7-43) is not transmitted by the amplifier since it is assumed that ω_2 is much larger than any of the frequency components associated with $v(t)$. Note that the second term is simply a modulated sine wave of frequency ω_2. The amplitude variations correspond to the input signal.

If the gain of the amplifier is k, the signal applied to the demodulator circuit is

$$v_b = 4kRa_2v(t)V_2 \sin \omega_2 t \tag{7-44}$$

Ignoring the capacitors C for a moment, the current in D_3 is, again using Eq. (4-30),

$$i_3 = a_1v_b + a_1V_2 \sin \omega_2 t + a_2v_b^2 + a_2V_2^2 \sin^2 \omega_2 t + 2a_2v_bV_2 \sin \omega_2 t \tag{7-45}$$

The current in D_4 is similar except that the polarity of v_b with respect to v_2 is reversed. Consequently, the output voltage

$$v_o = R(i_3 - i_4) \tag{7-46}$$

reduces to

$$\begin{aligned} v_o &= 2a_1Rv_b + 4a_2Rv_bV_2 \sin \omega_2 t \\ &= 2a_1Rv_b + 16a_2^2kR^2v(t)V_2^2 \sin^2 \omega_2 t \end{aligned} \tag{7-47}$$

Inserting the standard trigonometric identity $2 \sin^2 \omega t = 1 - \cos 2\,\omega t$,

$$v_o = 2a_1Rv_b + 8a_2^2kR^2V_2^2v(t)\,(1 - \cos 2\omega_2 t) \tag{7-48}$$

The filter capacitors eliminate the high-frequency terms in ω_2 and $2\omega_2$. Therefore, the filtered output signal is simply

$$v_o = (8a_2^2kR^2V_2^2)v(t) \tag{7-49}$$

According to Eq.(7-49) the output voltage is an amplified replica of the input signal.

The amplifier passband is conveniently peaked at ω_2 to minimize noise effects. Actually, however, the output filter circuit determines the effective bandwidth in the same way as for the chopper amplifier, and very-low-noise performance is possible. The minimum usable bandwidth depends upon the frequency components present in the signal $v(t)$.

In instrumentation applications it is often possible to produce the initial modulation in some way associated with the physical quantity being measured. For example, the infrared beam of an infrared spectrometer is chopped by means of a rotating shutter before it strikes the infrared detector. The amplified signal from the detector is demodulated by a circuit similar to Fig. 7-46 using a voltage v_2 derived from the shaft of the rotating shutter. The low-noise performance of the circuit permits extremely weak infrared signals to be detected. In this form the circuit is usually called a *lock-in amplifier*, since the detector is locked in step with the input signal. The system is also called a *phase-sensitive detector* because the circuit can recognize the phase of the input signal.

A specific example is the *vibrating-reed electrometer* which uses an electromagnetically driven vibrating capacitor to measure feeble electric currents. The vibrating capacitor develops an ac signal. This signal is amplified, rectified, and fed back to the input circuit. The system is exactly analogous to the lock-in amplifier and has the stability and sensitivity advantages of this circuit. Most often, the vibrating capacitor C_e, Fig. 7-47, is a small

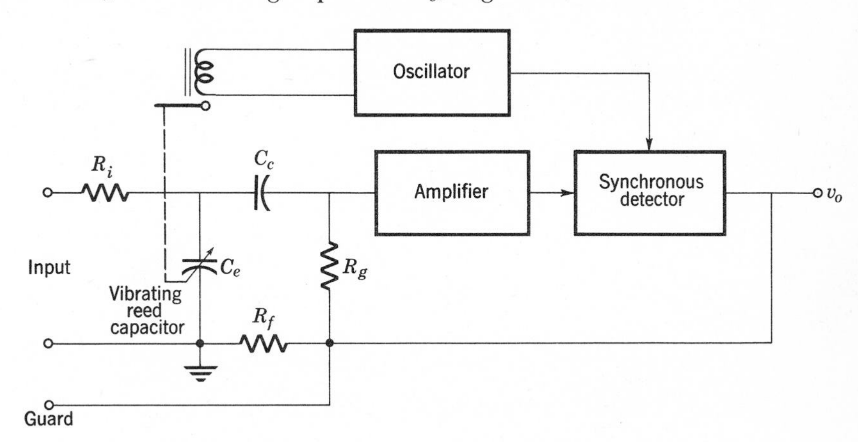

FIGURE 7-47 *Block diagram of vibrating-reed electrometer.*

metal reed electromagnetically vibrated by a sinusoidal voltage derived from an oscillator. The oscillator output also provides a signal for the synchronous detector to develop a dc output voltage in accordance with the input signal. The mechanical vibration of the reed results in a sinusoidal variation in the capacitance of the reed in combination with an adjacent stationary electrode. A charge Q_e placed on the capacitance produces an ac voltage

$$v_e = R_i Q_e \frac{dC_e}{dt} \tag{7-50}$$

which is presented to the amplifier input.

The operation of the vibrating-reed electrometer is analyzed in the following way, with reference to Fig. 7-47. A charge Q applied to the input terminals divides between the reed capacitance and the coupling capacitance C_c so that the charge on C_e is

$$Q_1 = \frac{QC_e}{C_e + C_c} \tag{7-51}$$

The ac signal from the vibrating capacitor is amplified and rectified and appears as an output voltage v_o across the feedback resistor R_f. This voltage also produces a charge on C_e,

$$Q_2 = \frac{v_o}{1/C_c + 1/C_e} = \frac{v_o C_e C_c}{C_e + C_c} \tag{7-52}$$

Therefore, the total charge on C_e is the sum of Eqs. (7-51) and (7-52), or

$$Q_e = Q_1 + Q_2 = \frac{C_e}{C_e + C_c}(Q + v_o C_c) \tag{7-53}$$

We define the overall gain of the system A as the ratio of the output voltage to the dc voltage across the vibrating capacitor, so that

$$v_o = -A\frac{Q_e}{C_e} \tag{7-54}$$

The gain involves the effective change in capacitance of the vibrating reed, Eq. (7-50), together with the amplifier gain and the rectifier efficiency. Introducing Q_e from Eq. (7-54) into Eq. (7-53) and solving for the output voltage, the result is

$$v_o = -\frac{Q}{C_c + (C_e + C_c)/A} \cong -\frac{Q}{C_c} \tag{7-55}$$

which is true when A is large.

According to Eq. (11-67), the output voltage is a direct measure of the input charge. The output voltage is measured by a conventional d'Arsonval meter. Carefully designed vibrating-reed electrometers can measure charges as small as 10^{-16} C, equivalent to a few hundred electrons. Minute currents can be measured by timing the rate at which C_c charges or by shunting the input terminals with a large known resistance. Currents of the order of 10^{-17} A, a few hundred electrons per second, can be detected.

The input resistor R_i is present to isolate the instrument from the capacitance of the source. This is necessary because any fixed capacitance in parallel with C_e effectively reduces the ac signal developed by the vibrating reed since the net change in capacitance is reduced. The time constant R_iC_e is made larger than the vibration period of the reed to accomplish this isolation. Note also that the feedback voltage is essentially equal to the input voltage. Therefore, the *guard* terminal, Fig. 7-47, can be used to reduce unwanted leakage currents between the upper input terminal and other points in the circuit. When these points are connected to the guard terminal the potential difference is negligible and stray leakage currents are eliminated.

SUGGESTIONS FOR FURTHER READING

E. J. Angelo, Jr.: "Electronic Circuits," McGraw-Hill Book Company, New York, 1964.

Jack J. Studer: "Electronic Circuits and Instrumentation Systems," John Wiley & Sons, Inc., New York, 1963.

A. van der Ziel: "Noise," Prentice-Hall, Inc., Englewood Cliffs, N.J., 1954.

EXERCISES

7-1 Plot the frequency-response characteristics of the two-stage triode amplifier in Fig. 7-1. Assume $C_1 + C_3 = 50$ pF and $C_2 = 1.7$ pF. Use the small-signal parameters listed in Table 6-1.

7-2 Plot the frequency-response characteristics of the two-stage transistor amplifier in Fig. 7-3. Assume the output load resistance is 5000 Ω. The common-base h parameters for V_1 are: $h_{ib} = 50\ \Omega$, $h_{rb} = 300 \times 10^{-6}$, $h_{fb} = -0.99$, $h_{ob} = 0.2 \times 10^{-6}$ mho; for V_2: $h_{ib} = 11\ \Omega$, $h_{rb} = 600 \times 10^{-6}$, $h_{fb} = -0.99$, $h_{ob} = 0.72 \times 10^{-6}$ mho. Assume $C_1 + C_3 = 50$ pF and $C_2 = 8$ pF.

7-3 Plot the midband and high-frequency response of the FET amplifier in Fig. 7-13. Compare with the response of the amplifier without the inductance. Assume $C_1 + C_3 = 10$ pF and $g_m = 8 \times 10^{-3}$ mho.

7-4 Plot the low-frequency response of the transistor amplifier of Fig. 7-14. Compare with the response of the amplifier with R_3 and C_3 removed. Use values of h parameters given in Exercise 6-11.

7-5 Consider the power amplifier in Fig. 7-16. Calculate the maximum power output for load resistances of 5, 10, 20, and 40 Ω, using the characteristic curves of Fig. 7-17. Calculate the power efficiency for each load.

7-6 Plot the dynamic transfer characteristic for each load resistance of the power amplifier studied in Exercise 7-5. Which load yields minimum distortion?

7-7 Determine the required reflected load impedance R_L' for a class B push-pull power amplifier using 2N1415 type transistors if the output power is 0.3 W and $V_{cc} = 6$ V.

Ans.: 60 Ω

7-8 Plot the composite transfer characteristic of the amplifier of Exercise 7-8. Repeat if each transistor is biased to a quiescent base current of 0.2 mA.

7-9 Sketch the equivalent circuit of the transistor tuned amplifier, Fig. 7-25. Calculate the peak gain assuming that the reactances of the tuned circuit are infinite at resonance. Use h parameters given in Exercise 6-11.

Ans.: 3.7×10^5

7-10 Draw the equivalent circuit of the MOSFET tuned amplifier, Fig. 7-30. Calculate the peak gain, assuming that the reactance of the tuned circuit is infinite at resonance. Take $g_m = 9500$ μmhos.

Ans.: 4.75×10^3

7-11 Suppose an amplifier with a bandpass from 10 Hz to 100 kHz and a gain of 10^5 uses a 1-MΩ input resistor. What is the rms noise output voltage?

Ans.: 4.1 V

7-12 Determine the mean square noise voltage per unit bandwidth of two 100,000-Ω resistors in parallel. Repeat for a 10,000-Ω resistor in parallel with a 100,000-Ω resistor. Can you generalize on these results?

Ans.: 8.26×10^{-16} V²/Hz; 1.5×10^{-16} V²/Hz

7-13 Repeat Exercise 7-11 if the input resistor is the carbon resistor of Fig. 7-33 and the direct current indicated is present.

Ans.: 7.34 V

7-14 With the aid of the equivalent circuit of Fig. 7-34 determine the minimum detectable input voltage (signal-to-noise ratio equal to unity) of the triode differential amplifier of Fig. 6-19. Assume the amplifier bandpass is 1 Hz to 40 kHz.

Ans.: 1.15×10^{-4} V

7-15 What is the minimum detectable input signal voltage of the chopper amplifier of Fig. 7-42, assuming chopper noise is negligible?

Ans.: 1.62×10^{-7} V

LABORATORY EXERCISES

7-A Multistage Power Amplifier Analysis of complicated electronic circuits is accomplished by isolating each subcircuit and considering its performance separately. This technique is particularly appropriate in the case of amplifier circuits since the signal is amplified sequentially in each stage. Subsequently, the performance of the entire amplifier is determined by combining the characteristics of each stage into a unified, operating system. The purpose of this experiment is to analyze the multistage power amplifier circuit in Fig. 7-23 in this fashion.

Begin by considering the input grounded-emitter stage. Determine the operating point, the input and output resistances, and the voltage gain of this stage using small-signal analysis. Take the properties of the 2N1924 to be the same as the type 2N35 in Appendix 1 and the drop across the 10,000-Ω filter resistor to be about 10 V. Because of the high input resistance of the Darlington connection of Q_2 and Q_3, it is sufficiently accurate to assume that the Q_1 load is just the 6.8 kΩ collector resistor.

To examine the temperature compensation feature involving the 1N536 diode, sketch the emitter-base portions of Q_2 and Q_3 in conjunction with the 1N536 as an isolated circuit. Note that the potential across the diode (what is the polarity and approximate magnitude?) together with the drop across the 390-Ω resistor determine the emitter-base voltage of both transistors. Set up a simple expression for the base currents and show how the change in the emitter-base resistances with temperature is compensated by the change in the voltage across the diode.

Sketch the waveform at the collector of Q_1 and show how the complementary symmetry of Q_2 and Q_3 converts this signal to a push-pull signal for Q_4 and Q_5. Isolate Q_3 and Q_5 and illustrate qualitatively how this *npn-pnp* Darlington connection is much the same as the more familiar *npn-npn* connection of Q_2 and Q_3.

Finally, consider Q_4 and Q_5 separately. Draw the combined collector characteristics (assumed to be the same as type 2N1715 in Appendix 1), and introduce the load line. It is helpful to sketch input and output waveforms to understand circuit operation. Now prepare a composite transfer characteristic. What is the maximum output voltage? Output power? What signal at the amplifier input is necessary to achieve this power output, if $\beta = 100$ for Q_2 and Q_3?

7-B Random Noise Signals Sensitive measurement of electrical signals is limited ultimately by the presence of random noise signals, since Nyquist noise, at least, is unavoidable. Many ingenious and useful techniques have been developed to detect and measure minute signals in the presence of noise. A typical example is the lock-in amplifier discussed in the text. In the final analysis, all techniques rest upon the fact that random noise signals are spread over a wide range of frequencies while desired signals are limited to a finite bandwidth. Thus, a narrow bandwidth measuring system always enhances the signal-to-noise ratio. The improvements obtainable through narrowband measurement techniques can be examined quantitatively.

In this experiment we attempt to measure a small dc voltage in the presence of random noise. For convenience the combined signal has been sampled rapidly (every 0.01 s) and regularly for a period of 1 s. The 100 consecutive voltage readings are tabulated in Table 7-1 and constitute the experimental data from which a useful measurement is derived. To obtain an appreciation of the signal represented by this data, plot the 100 points and connect each consecutive point. Can you estimate the desired dc signal in this noisy waveform?

Actually, the dc signal is just the average value of the 100 readings. Compute the average, V_{dc}. Now compute the rms value, V_{rms}. You do this, according to the definition of the rms value, by squaring each voltage reading, determining the average of these quantities (the "mean square"), and then taking the square root of the average. You may wish to prepare a simple Fortran program for these calculations, but before doing so plan the entire experiment. Later calculations can also benefit from the labor-saving digital computer. Of greater significance than the rms value is $V_{rms}^2 - V_{dc}^2$, which is called the *variance* of the noise. The variance is a measure of the departure of the individual readings from the dc value and, hence, a measure of the noise magnitude. It is often considered that a useful measurement is possible only if the ratio of the average value to the square root of the variance is equal to unity or larger. Is this criterion satisfied in the present case?

Suppose now the response of the system is such that a reading is attained only every 0.5 s, rather than every 10^{-2} s. Compute the dc (that is, the average) value of the readings during the first 0.5 s, V_{dc1} and also during the second 0.5 s, V_{dc2}. Is $V_{dc1} = V_{dc2}$? Why? Notice that the reciprocal relation between bandwidth and averaging time ($\Delta\omega = 1/\tau_0$) means that the 0.5-s readings are obtained in a system of one-fiftieth the bandwidth and hence in the presence of less noise. A quantitative measure of this effect is the variance of the two readings, which, by analogy with above, is just

$$\text{Var} = \frac{V_{dc_1}{}^2 + V_{dc_2}{}^2}{2} - V_{dc}{}^2 \tag{7-56}$$

Repeat these calculations for intervals of 0.25, 0.10, 0.05, and 0.02 s. Plot the variances of the readings as a function of $1/\tau_0$ to see if the noise tends to increase with bandwidth as expected.

It is often stated that the minimum detectable signal in the presence of noise decreases inversely with the square root of the averaging time. Is this consistent with the above results? What is the minumum dc voltage signal that can be measured in the experimental data used in this experiment?

TABLE 7-1 SUCCESSIVE VOLTAGE READINGS OF NOISY SIGNAL (sample interval: 10^{-2} s; voltage readings expressed as 10^{-8} V)

292	296	297	142	436
443	297	141	147	498
446	349	31	101	491
378	393	1	119	484
345	334	54	178	529
367	246	214	222	609
375	280	316	294	697
324	417	438	349	734
257	480	489	361	570
242	382	556	413	314
345	223	470	489	171
417	155	306	544	248
374	198	194	297	449
323	321	117	223	570
318	393	94	296	474
345	392	284	425	257
350	385	290	432	65
357	457	245	281	1
330	501	107	210	8
293	436	67	313	100

chapter eight

OPERATIONAL AMPLIFIERS

The performance of transistor and vacuum tube amplifiers is enhanced in many respects by returning a fraction of the output signal to the input terminals. This process is called feedback. The feedback signal may either augment the input signal or tend to cancel it. The latter, called negative feedback, is the primary concern of this chapter. Improved frequency-response characteristics and reduced waveform distortion are attained with negative feedback. In addition, amplifier performance is much less dependent upon changes in tube or transistor parameters caused by aging or temperature effects.

A particular form of negative feedback, known as operational feedback, is used in amplifiers which perform mathematical operations such as addition or integration on an input signal. Commercially available high-performance dc-coupled units have come to be called operational amplifiers. Operational amplifiers are widely used in measurement and control applications, as well as in electronic analog computers.

NEGATIVE FEEDBACK

Voltage Feedback The circuit alterations of a standard amplifier which return a portion of the output signal to the input may be analyzed by the techniques developed in previous chapters. It is more illustrative, however, to isolate the feedback portion of the circuit and treat it separately. Consider the feedback amplifier, Fig. 8-1, comprising a standard amplifier

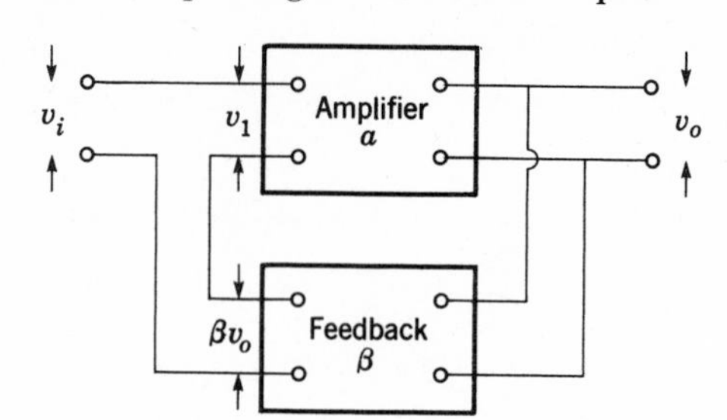

FIGURE 8-1 *Block diagram of feedback amplifier.*

with a gain a and a feedback network indicated by the box marked β. According to this circuit, a voltage βv_o is added to the input signal v_i so that the total input signal to the amplifier is

$$v_1 = v_i + \beta v_o \tag{8-1}$$

Introducing the fact that $v_o = av_1$,

$$v_o = av_1 + a\beta v_o \tag{8-2}$$

so that

$$v_o = \frac{a}{1 - \beta a} v_i \tag{8-3}$$

According to Eq. (8-3) the overall gain of the amplifier with feedback,

$$a' = \frac{a}{1 - \beta a} \tag{8-4}$$

may be greater or smalier than that of the amplifier alone, depending upon the algebraic sign of βa.

The condition of greatest interest in this chapter is *negative feedback*, when βa is a negative quantity. In this case, Eq. (8-4) shows that the overall gain is reduced because, in effect, the feedback voltage cancels a portion of the input signal. If the amplifier gain is very large, so that $-\beta a \gg 1$, the overall gain reduces to

$$a' = \frac{1}{\beta} \tag{8-5}$$

This shows that the gain depends only upon the properties of the feedback circuit. Most often, the feedback network is a simple combination of resistors and/or capacitors. Therefore, the gain is independent of variations in tube or transistor parameters in the amplifier. In addition to this desirable improvement in stability, the gain may be calculated from circuit values of the feedback network alone. Thus it is not necessary to know, for example, the h parameters of all transistors in the circuit.

Negative feedback is also effective in reducing waveform distortion in amplifiers. Waveform distortion results from a nonlinear transfer characteristic, which may be interpreted as a smaller gain where the slope of the transfer characteristic is less, and as a larger gain where the slope of the transfer characteristic is greater. According to Eq. (8-5), however, the gain of an amplifier with feedback is essentially independent of variations caused by nonlinearities in tube or transistor characteristics. Therefore, the transfer characteristic is more linear and distortion is reduced.

A quantitative measure of the reduction in distortion achieved with feedback is obtained by assuming that distortion signals can be represented by a voltage generator in the amplifier, Fig. 8-2. With this approximation

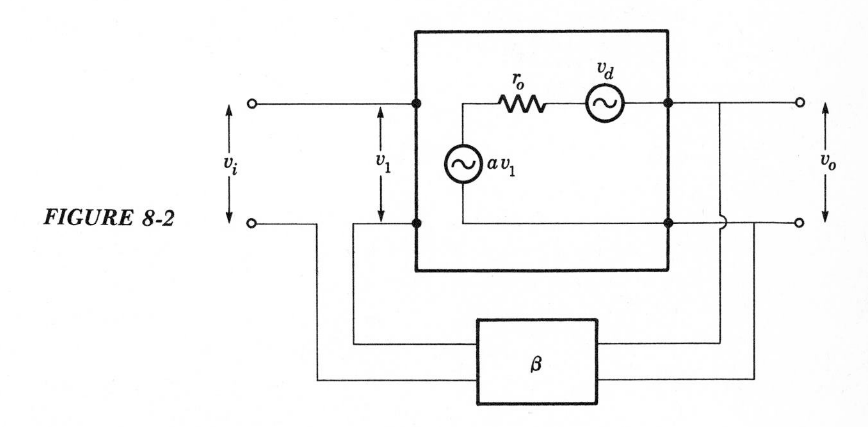

FIGURE 8-2

the amplified signal av_1 is distortionless. As usual, r_o represents the internal impedance of the circuit as viewed from the output terminals of the amplifier. Under this condition, the output voltage

$$v_o = av_1 + v_d \tag{8-6}$$

includes the distortion voltages v_d. Both the amplified signal and distortion voltages are fed back, so that the input to the amplifier is

$$v_1 = v_i + \beta v_o = v_i + \beta(av_1 + v_d) \tag{8-7}$$

Solving for v_1,

$$v_1 = \frac{v_i + \beta v_d}{1 - a\beta} \tag{8-8}$$

Equation (8-8) is inserted into Eq. (8-6),

$$v_o = \frac{a}{1 - a\beta}(v_i + \beta v_d) + v_d \tag{8-9}$$

$$v_o = \frac{av_i}{1 - a\beta} + \frac{v_d}{1 - a\beta} \tag{8-10}$$

The first term in Eq. (8-10) represents the undistorted signal output while the second term is the distortion voltage. Both decrease with the addition of feedback, but the signal output, v_s, can be kept constant by increasing v_i. Therefore, Eq. (8-10) is rewritten

$$v_o = v_s + \frac{v_d}{1 - a\beta} \tag{8-11}$$

According to Eq. (8-11), feedback reduces the relative importance of distortion signals in the output by the factor $1 - a\beta$. Since $a\beta$ is a large number, this improvement is significant. In effect, feedback results in an amplified distortion signal that cancels the original distortion voltages to a large extent. This result is particularly useful in power amplifiers where transistors or tubes are used over the full range of their characteristics.

The benefits of negative feedback are obtained at the expense of reduced gain, according to Eq. (8-4). This is not a serious loss, however, because large amplifications are easily obtained in transistor and vacuum-tube circuits. In practice, the maximum usable gain is limited by random noise effects anyway and it is not difficult to achieve the maximum amplification that can be effectively used, even with feedback included.

In the foregoing, a portion of the output voltage is returned to the input terminals, a condition referred to as *voltage feedback.* The two-stage transistor amplifier, Fig. 8-3, uses voltage feedback introduced by the resistor R_F connecting the output terminal with the input circuit of the first stage. The feedback voltage is introduced into the emitter circuit of the first stage because the output voltage of a two-stage amplifier is in phase with the input signal. The feedback factor is the result of the resistor divider made up of R_F and R_1, so that

$$\beta = \frac{R_1}{R_F + R_1} \cong \frac{R_1}{R_F} \tag{8-12}$$

Capacitor C_1 isolates the dc components of the two stages. Also, note that dc interstage coupling is used in this amplifier, which is quite independent of negative-feedback considerations. Resistor R_2 sets the bias on the first stage and also provides a dc feedback effect which helps stabilize the dc-coupled amplifier against slow drifts, as discussed in Chap. 7. It is not part of the ac feedback circuit.

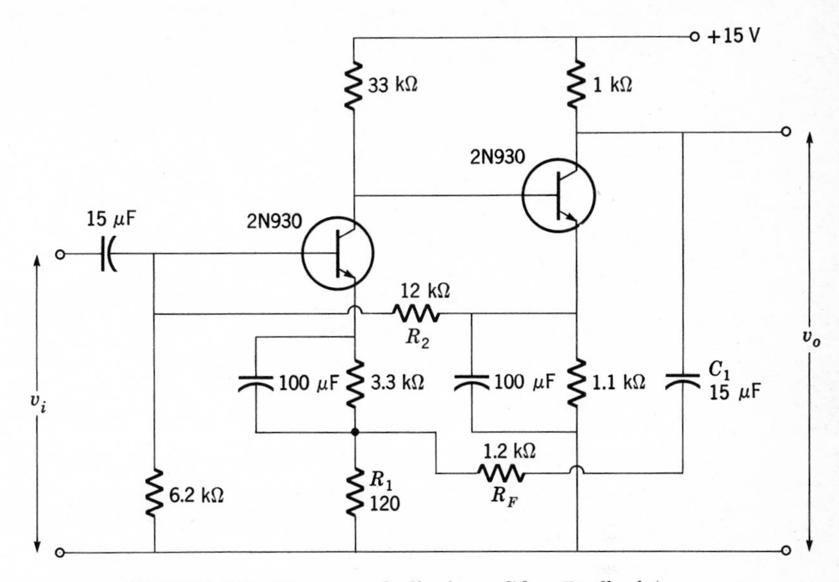

FIGURE 8-3 *Two-stage feedback amplifier. Feedback is determined by ratio R_1/R_F.*

The effect of the feedback ratio on the frequency-response characteristic is illustrated in Fig. 8-4. Without feedback (resistor R_F removed) the mid-

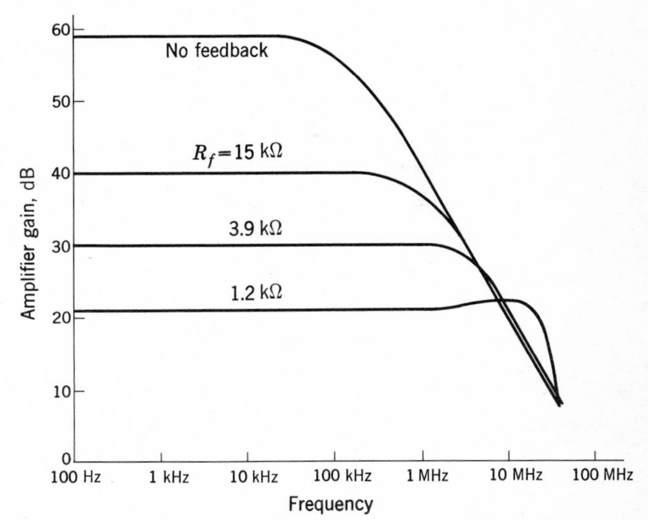

FIGURE 8-4 *Effect of feedback on response characteristic of amplifier in Fig. 8-3.*

band gain of the amplifier is 1000 and the high-frequency cutoff is 100 kHz. As feedback is increased by using smaller values of R_F the midband gain is reduced and the frequency response is extended to lower and higher frequencies. With a 1200-Ω feedback resistor the gain is 10 and the upper frequency cutoff is extended to 15 MHz. Under this condition $\beta = 120/1200 = 0.1$, so $a\beta = 100$. According to Eq. (8-5) the midband gain is $1/\beta$, in agreement with the experimental response curves, Fig. 8-4. Thus, a very stable wideband amplifier is possible through the use of feedback.

Note that in Fig. 8-4 the gain increases at the extreme of the response curve when a large feedback ratio is used. This is a result of phase shift in the amplifier at these frequencies. In effect, the negative feedback is reduced since the feedback voltage is no longer exactly 180° out of phase with the input signal. The phase characteristic of feedback amplifiers at frequencies outside of the passband is very important. Such irregularities in the response curve should be small, particularly when large feedback ratios are employed. It is possible for the phase shifts to become great enough to cause positive feedback at the frequency extremes, a condition which must be avoided if the amplifier is to remain stable. This matter is considered further in a later section.

If the feedback circuit is frequency-selective, it is possible to develop a specific frequency-response characteristic for the amplifier. Consider, for example, an amplifier having a bridged-T feedback network in one stage, Fig. 8-5. In this circuit source followers are used to provide a high

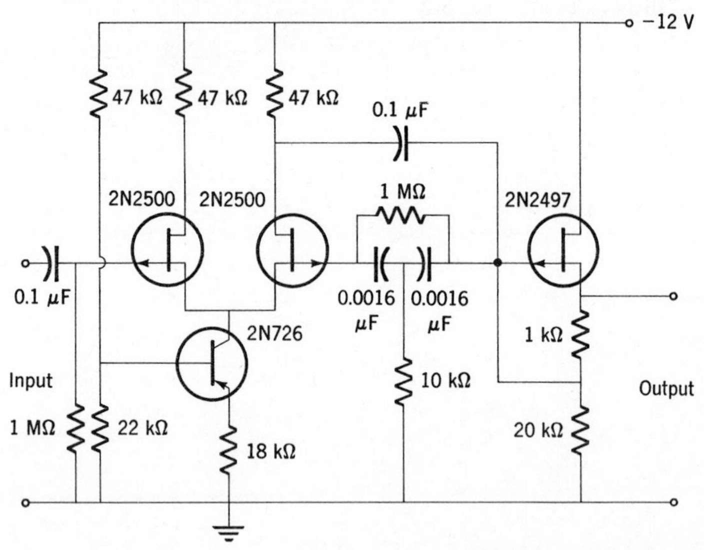

FIGURE 8-5 *Tuned amplifier using a frequency-selective feedback network consisting of bridged-T filter.*

input impedance and a low output impedance so that the frequency-selective stage is isolated from disturbing influences of input and output loads. A difference amplifier is used in the first two stages to simplify coupling between stages in the presence of the feedback network.

Since the bridged-T filter is connected from drain to gate, the feedback voltage is 180° out of phase with the input signal, as required for negative feedback. According to the response curve of the bridged-T filter, the feedback voltage is a minimum at the characteristic frequency of the filter. Accordingly, the amplifier gain is a maximum at this frequency and the result is a tuned amplifier. The feedback ratio β is a function of frequency because of the characteristics of the bridged-T filter, and the feedback effect considerably enhances the selectivity of the filter, as illustrated in Fig. 8-6.

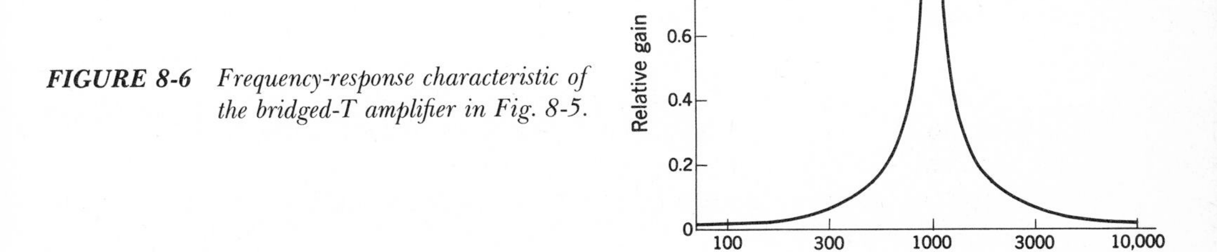

FIGURE 8-6 *Frequency-response characteristic of the bridged-T amplifier in Fig. 8-5.*

Feedback tuned amplifiers are commonly used at audio frequencies where high-Q inductances are difficult to construct because large values of inductance are required. Furthermore, it is a simple matter to tune the amplifier by making the resistors variable. This approach is also useful in integrated circuits where inductances are difficult to produce. Other frequency-selective feedback networks, such as the twin-T or Wien bridge, are also used in tuned feedback amplifiers.

Negative feedback alters the input and output impedance of an amplifier. To see how this comes about, replace the amplifier by its Thévenin equivalent, Fig. 8-7. The output impedance with feedback is determined

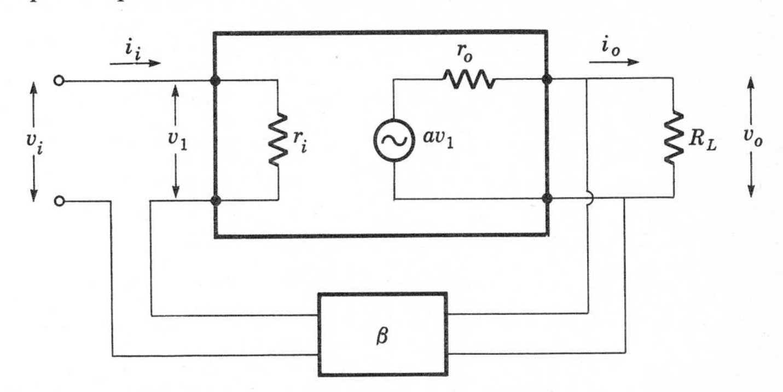

FIGURE 8-7 *Circuit to determine effective input and output impedance of amplifier with voltage feedback.*

from the ratio of the open-circuit output voltage to the short-circuit output current. The open-circuit output voltage is, from Eq. (8-3),

$$(v_o)_{oc} = \frac{av_i}{1 - a\beta} \tag{8-13}$$

The short-circuit output current is found by making $R_L = 0$ in Fig. 8-7,

$$(i_o)_{sc} = \frac{av_1}{r_o} = \frac{a(v_i + \beta v_o)}{r_o} = \frac{av_i}{r_o} \tag{8-14}$$

where the fact that $v_o = 0$ has been used. Therefore, from Eqs. (8-13) and (8-14),

$$R_o = \frac{(v_o)_{oc}}{(i_o)_{sc}} = \frac{r_o}{1 - a\beta} \tag{8-15}$$

According to this result, the effective output impedance is reduced by feedback by the same factor as the reduction in gain.

The effective input impedance is found from the ratio of the input voltage to the input current. Applying Kirchhoff's rule to the input circuit of Fig. 8-7,

$$v_i + \beta v_o - i_i r_i = 0 \tag{8-16}$$

The output voltage with no load may be written as

$$v_o = av_1 = ar_i i_i \tag{8-17}$$

Inserting Eq. (8-17) into Eq. (8-16) and solving for the ratio v_i/i_i yields

$$R_i = \frac{v_i}{i_i} = (1 - a\beta)r_i \tag{8-18}$$

which shows that the effective input impedance is increased by the same factor as the decrease in output impedance. When applied to a single stage, negative feedback is often called *bootstrapping* since the input impedance appears to be increased by the amplifier lifting itself up by its own output signal.

Both changes, increased input impedance and reduced output impedance, are desirable improvements in amplifiers, as discussed previously. Feedback is often introduced to achieve these benefits alone. Actually, the cathode, source, and emitter followers, in which the input impedance is large and the output impedance is small, may be considered to be feedback amplifiers. In this case the full output voltage is applied to the input and the feedback ratio is -1.

Current Feedback It is possible to develop a feedback signal proportional to the output current rather than to the output voltage, and this is called

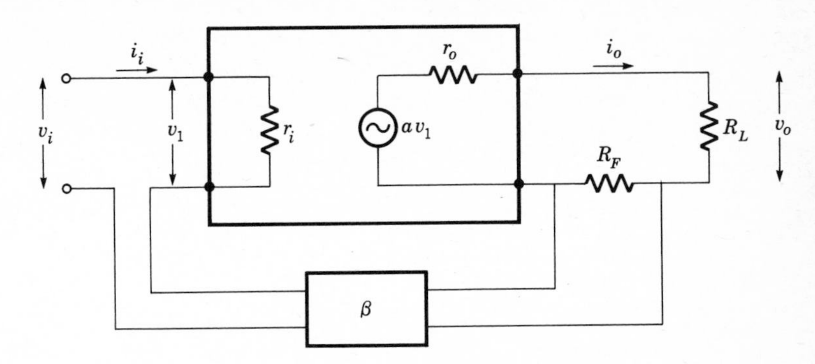

FIGURE 8-8 *Circuit to determine effective input and output impedance of amplifier with current feedback.*

current feedback. Consider the current-feedback circuit, Fig. 8-8, in which the feedback signal results from the voltage drop across a resistor R_F in series with the load. Writing Kirchhoff's voltage equation around the output circuit,

$$-av_1 + i_o(r_o + R_F) + v_o = 0 \tag{8-19}$$

and around the input circuit,

$$-v_1 + v_i + \beta i_o R_F = 0 \tag{8-20}$$

Solving Eq. (8-20) for v_1 and inserting this into Eq. (8-19) results in

$$v_o = av_i - i_o[r_o + (1 - a\beta)R_F] \tag{8-21}$$

Interpreting Eq. (8-21) in terms of a Thévenin equivalent circuit, the open-circuit voltage is av_i, and the internal impedance is the coefficient of i_o. Accordingly, the effective output impedance is increased, since

$$R_o = r_o + (1 - a\beta)R_F \tag{8-22}$$

The input impedance is also increased, as can be shown by considering the input circuit as in the case of voltage feedback.

The output current is found from Eq. (8-21) by introducing $v_o = i_o R_L$ and solving for i_o,

$$i_o = \frac{av_i}{R_L + r_o + (1 - a\beta)R_F} \cong -\frac{v_i}{\beta R_F} \tag{8-23}$$

where the approximation applies when the gain is large. Note that the output current is independent of the amplifier gain and transistor parameters. This equation is analogous to Eq. (8-5) for the case of voltage feedback. It indicates that current feedback minimizes distortion in the output current, rather than the output voltage.

Omitting the emitter or cathode bypass capacitor in power amplifiers is a commonly used form of current feedback. The load current in the emitter or cathode bias resistor introduces negative current feedback into the input circuit, with a consequent reduction in distortion. In this connection, the fact that bypass capacitors cannot be used in push-pull class B power amplifiers, as mentioned in the previous chapter, is actually advantageous, except for the concomitant loss of gain.

Both voltage and current feedback can be employed in the same amplifier, Fig. 8-9. Voltage relations in the output circuit of Fig. 8-9 are identical to Eq. (8-19), so that

$$v_o = av_1 - i_o(r_o + R_F) \tag{8-24}$$

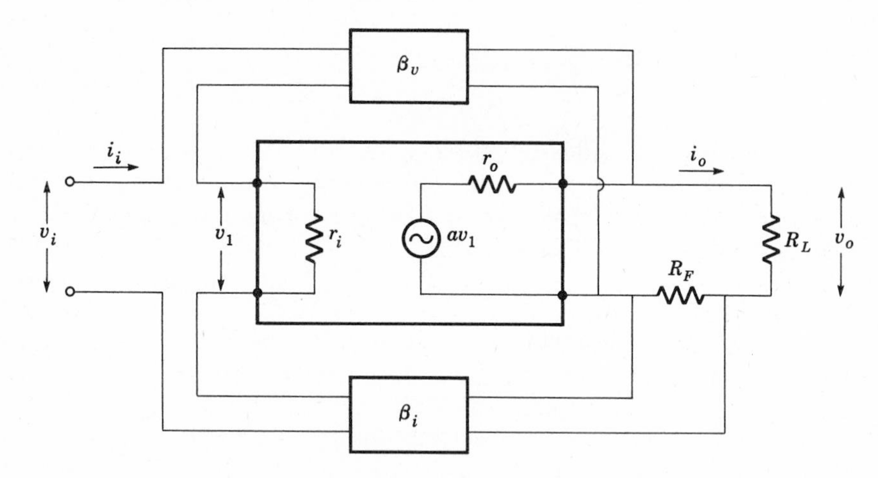

FIGURE 8-9 Combined voltage and current feedback.

Similarly, Kirchhoff's voltage equation for the input circuit yields

$$v_1 = v_i + \beta_v v_o + \beta_i i_o R_F \tag{8-25}$$

Substituting Eq. (8-25) into Eq. (8-24) and solving for the output voltage,

$$v_o = \frac{a}{1 - a\beta_v} v_i - i_o \frac{r_o + (1 - a\beta_i)R_F}{1 - a\beta_v} \tag{8-26}$$

Note that this expression incorporates both Eq. (8-3) for voltage feedback alone ($\beta_i = 0$) and also Eq. (8-21) for current feedback only ($\beta_v = 0$).

Combined current and voltage feedback permits a unique adjustment of the output impedance of the amplifier. The output impedance is given by the coefficient of i_o in Eq. (8-26) and can be made to vanish if

$$r_o + (1 - a\beta_i)R_F = 0$$

or

$$a\beta_i = 1 + \frac{r_o}{R_F} \tag{8-27}$$

If Eq. (8-27) is satisfied, the amplifier has zero internal impedance and therefore can deliver maximum power to any load impedance. Note that this requires positive current feedback, according to Eq. (8-27). Positive current feedback is permissible in this case because of the stabilizing effect of negative voltage feedback which is also present.

Stability The amplifier gain a and the feedback factor β are inherently complex numbers. That is, phase shifts associated with coupling capacitors and stray capacitance effects are present, particularly at frequencies outside of the passband of the amplifier. These phase shifts cause a departure from the 180° phase shift necessary for the feedback voltage to interfere destructively with the input signal. It can happen that the overall phase shift becomes zero (i.e., 360°) so that $a\beta$ is positive and the feedback voltage augments the input signal. This is called *positive feedback* and leads to serious instability effects in feedback amplifiers.

Note, particularly, that if $a\beta = +1$ the output voltage, Eq. (8-3), becomes very large, even in the absence of an input signal. This means that positive feedback may cause an amplifier to oscillate, as previously discussed in connection with the feedback effects of the grid-plate capacitance in triodes and the collector capacitance in transistors. Oscillation is deleterious in amplifiers since the output voltage is not a replica of the input signal. Positive feedback is, however, a useful condition in oscillator circuits, as discussed in the next chapter.

Great care is taken in the design of feedback amplifiers to make them stable in the face of phase shifts leading to positive feedback at the frequency extremes of the amplifier bandpass. Fortunately there exists a rather straightforward criterion which can be applied to establish the stability of feedback amplifiers. According to Eq. (8-4), instability occurs if $a\beta$ is positive and equal to unity. It follows that if the absolute value of $a\beta$ drops below unity before the phase shift reaches −360°, the amplifier

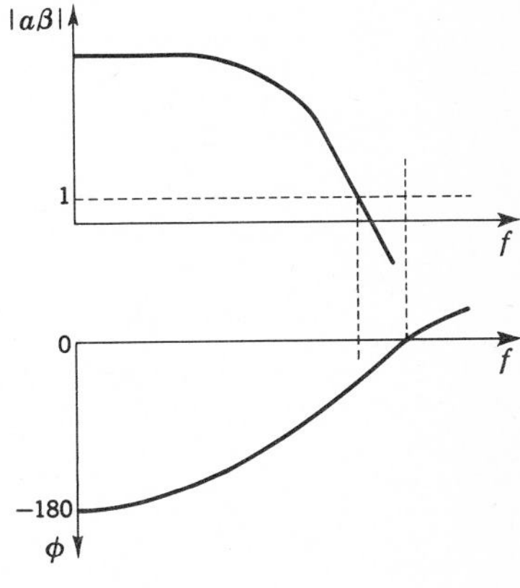

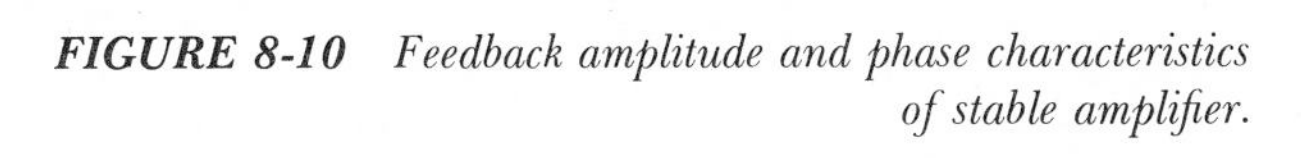
FIGURE 8-10 *Feedback amplitude and phase characteristics of stable amplifier.*

will be stable. It is convenient to plot the magnitude and phase of $a\beta$ as a function of frequency, Fig. 8-10, in order to establish the relative positions at which $a\beta = 1$ and the phase shift equals zero. By this standard, the amplifier with characteristics represented by Fig. 8-10 is stable.

In order to apply this criterion for stability, the amplifier must be stable in the absence of feedback. Also, there exist certain situations in which a feedback amplifier can be conditionally stable, that is, stable under some conditions of, say, loading, and not under others. For most practical circuit analysis, however, the stated condition is sufficient.

It is useful to consider the $a\beta$ characteristics of several amplifier types in the light of this stability requirement. Since in most cases the feedback ratio is independent of frequency, it is sufficient to examine the amplitude and phase characteristics of the amplifier gain separately. In the case of a single *RC*-coupled amplifier stage, Fig. 8-11*a* (compare Fig. 7-12), the phase shift never exceeds 90°. This means the phase of $a\beta$ is always less than $90 + 180 = 270°$. Since this is smaller than 360°, a single-stage negative feedback amplifier cannot be unstable.

Two identical *RC*-coupled stages in cascade have the ideal characteristic sketched in Fig. 8-11*b*. The phase shift reaches 180° at very low and very high frequencies, but the gain is very small at these extremes. Therefore, it is unlikely that $a\beta$ can be equal to unity, where the total phase shift is $180 + 180 = 360°$. Practical amplifiers always have stray capacitive effects that can introduce additional phase shift at high frequencies, however. It is possible that the additional phase shift caused by stray capacitances may result in unstable conditions at high frequencies if the gain is large.

The characteristics of three identical *RC*-coupled stages, Fig. 8-11*c*, are such that the extreme phase shift approaches 270°. Therefore, 180° phase shift is encountered at both low and high frequencies, and a three-stage feedback amplifier is certain to be unstable if the midband gain is large enough. It can be shown that the maximum value of $a\beta$ permitted at midband in this case is equal to 8, although even this value puts the amplifier on the verge of oscillation.

Many possibilities exist to remove this unfavorable difficulty other than minimizing the gain. For example, the bandwidth of two stages in a three-stage amplifier can be made very much greater than that of the remaining stage. This means that the phase characteristics are determined primarily by the single stage which is unconditionally stable. Alternatively, dc coupling can be used between two stages to eliminate the phase shift associated with one interstage coupling network. It is common practice to alter the feedback and gain characteristics, particularly at high frequencies, by including small capacitances to change the phase and gain characteristics in a way that improves stability. Usually such changes must be made empirically on the actual amplifier because of the unavoidable and unknown stray wiring capacitances.

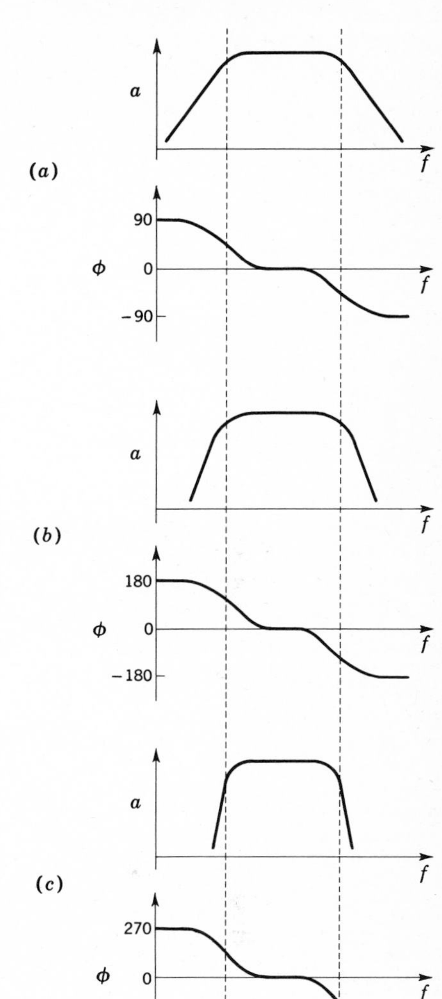

FIGURE 8-11 Gain and phase characteristics of (a) one-stage amplifier, (b) two-stage amplifier, and (c) three-stage amplifier.

OPERATIONAL FEEDBACK

The Virtual Ground A particularly versatile negative-feedback connection is called *operational feedback* because the circuit is capable of performing a number of mathematical operations on input signals. High-gain dc-coupled amplifiers are universally used in this application. Such *operational amplifiers* exhibit high input impedance, low output impedance, and conventionally introduce a 180° phase shift between the input signal and the output voltage.

To illustrate the features of an operational amplifier, consider the

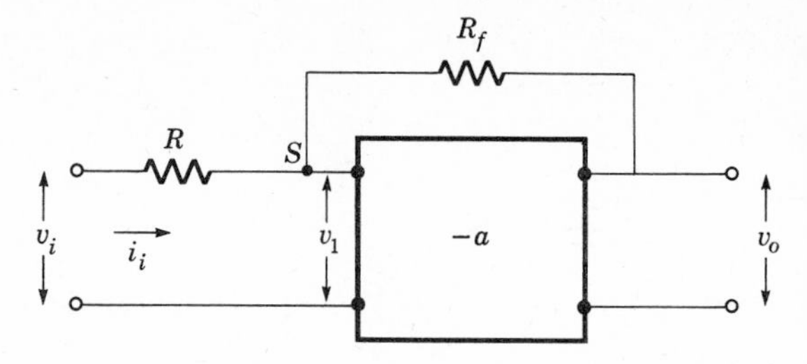

FIGURE 8-12 *Block diagram of operational amplifier.*

feedback circuit, Fig. 8-12, in which negative voltage feedback is produced by the resistor R_f connected between the output and input. Note that the feedback is negative because of the phase inversion in the amplifier. The feedback ratio in operational feedback can vary from unity for a high-impedance source to $R/(R + R_f)$ for a low-impedance source since the feedback voltage is effectively connected in parallel with the input signal source. It is convenient to analyze the operational feedback circuit by applying Kirchhoff's current rule to the branch point S. Since the amplifier input impedance is large, the current in this branch is negligible, which means the current in R equals the current in R_f, or

$$\frac{v_i - v_1}{R} = \frac{v_1 - v_o}{R_f} \tag{8-28}$$

Introducing $v_1 = -v_o/a$ and rearranging,

$$v_o\left(1 + \frac{1}{a} + \frac{R_f}{aR}\right) = -\frac{R_f}{R}\,v_i \tag{8-29}$$

Since the gain is very large

$$v_o = -\frac{R_f}{R}\,v_i \tag{8-30}$$

which means that the output voltage is just the input signal multiplied by the constant factor $-R_f/R$. If precision resistors are used for R_f and R, the accuracy of this multiplication operation is quite good.

The branch point S has a special significance in operational amplifiers. This may be illustrated by determining the effective impedance between S and ground, which is given by the ratio of v_1 to the input current,

$$Z_s = \frac{v_1}{i_i} = \frac{v_1 R_f}{v_1 - v_o} = \frac{R_f}{1 - v_o/v_1} = \frac{R_f}{1 + a} \tag{8-31}$$

where the right side of Eq. (8-28) has been inserted for the input current. According to Eq. (8-31) the impedance of S to ground is very low if the gain is large. Typical values are $R_f = 1000\ \Omega$ and $a = 10^4$, so that the impedance is 10.1 Ω. The low impedance results from the negative feedback voltage, which cancels the input signal at S and tends to keep the branch point at ground potential. For this reason the point S is called a *virtual ground.* Although S is kept at ground potential by feedback action, no current to ground exists at this point. The virtual ground at S shows immediately that the impedance viewed from the input terminals is equal to R.

Mathematical Operations The operational feedback circuit, Fig. 8-12, multiplies the input signal by the constant $-R_f/R$. It is conventional to simplify the circuit diagram, as in Fig. 8-13, by not showing the ground

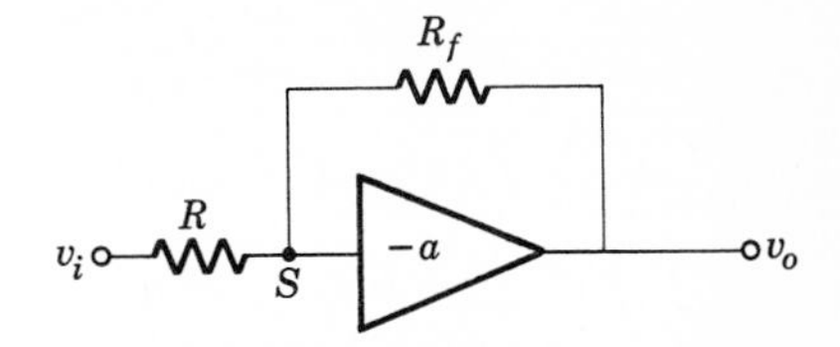

FIGURE 8-13 *Conventional circuit symbol for operational amplifier.*

terminals specifically. The operational amplifier is indicated by a triangle pointing toward the output terminal. It is understood that the simplified circuit diagram, Fig. 8-13, actually implies the corresponding complete circuit, Fig. 8-12. If $R_f = R$ in the multiplier circuit, the signal is simply multiplied by -1. This is often useful in obtaining a signal multiplied by a positive constant wherein this operational amplifier precedes one which determines the multiplier. Multiplicative factors ranging from -0.1 to -10 are possible, and the precision is determined principally by the accuracy of the two resistors.

The operational amplifier can also be used to sum several signals by connecting individual resistors to the branch point S, Fig. 8-14. In this

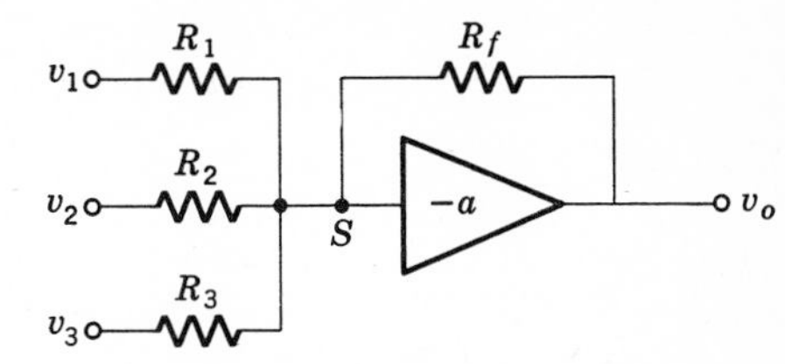

FIGURE 8-14 *Summing circuit using operational amplifier.*

circuit, the sum of the currents in R_1, R_2, and R_3 equals the current in R_f, since no current to ground exists at S. Furthermore, S is at ground potential, so that [compare Eq. (8-28)]

$$\frac{v_1}{R_1} + \frac{v_2}{R_2} + \frac{v_3}{R_3} = -\frac{v_o}{R_f} \tag{8-32}$$

$$v_o = -R_f\left(\frac{v_1}{R_1} + \frac{v_2}{R_2} + \frac{v_3}{R_3}\right) \tag{8-33}$$

Many signals can be added in this way. There is no interaction between individual signal sources since S is a virtual ground. Because the addition appears to take place at this point, S is commonly called the *summing point.* Note that each signal may be multiplied by the same factor (or -1), if $R_1 = R_2 = R_3$, etc., or that individual factors may be selected, as convenient.

If the feedback resistor is replaced by a capacitor, Fig. 8-15, the circuit

FIGURE 8-15 *Integrating circuit.*

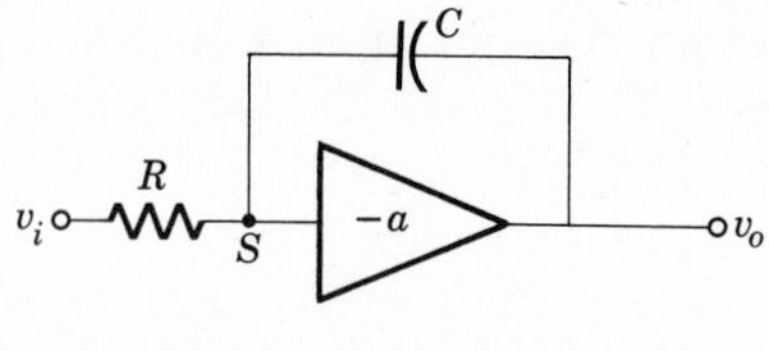

performs the operation of integration. Using the fact that S is at ground potential,

$$v_o = \frac{q}{C} = \frac{1}{C}\int_0^t i\,dt = -\frac{1}{RC}\int_0^t v_i\,dt \tag{8-34}$$

where Eq. (2-20) has been inserted for the voltage across a capacitor. According to Eq. (8-34) the output voltage is the integral of the input signal. Note that there is no restriction placed on the frequency components of the input signal, as in the case of the simple RC integrator discussed in Chap. 2. It is only necessary that the amplifier bandwidth be large enough to handle all signal frequencies. The integral of the sum of several signals is obtained by introducing several input resistors, as in Fig. 8-14.

Interchanging R and C, Fig. 8-16, results in a differentiating circuit

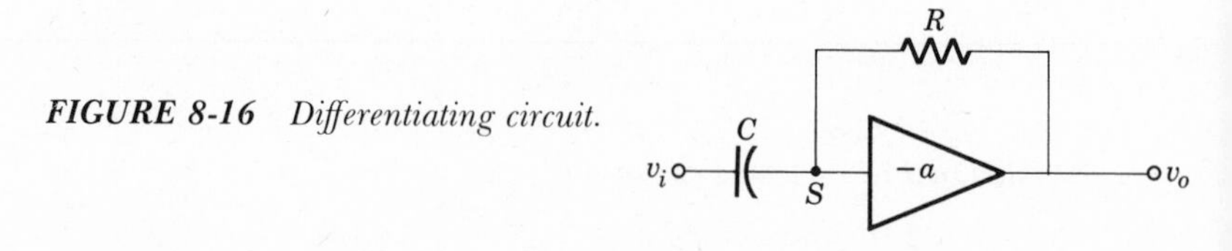

FIGURE 8-16 *Differentiating circuit.*

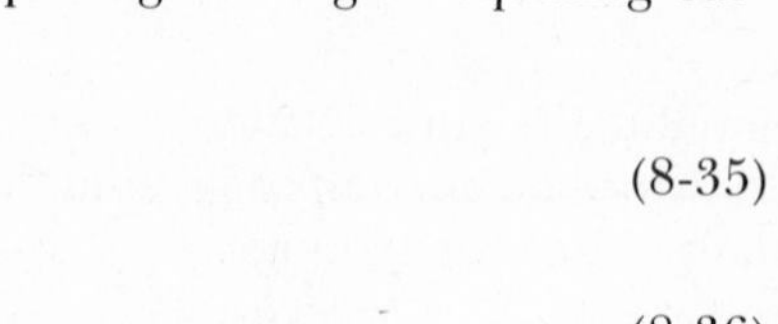

which gives the time derivative of the input signal. Again equating currents at the summing point,

$$-\frac{v_o}{R} = \frac{dq}{dt} = \frac{d}{dt}Cv_i = C\,\frac{dv_i}{dt} \tag{8-35}$$

$$v_o = -RC\,\frac{dv_i}{dt} \tag{8-36}$$

The differentiator circuit is equally independant of frequency restrictions, assuming only that the gain of the amplifier is sufficiently large at all signal frequencies.

OPERATIONAL AMPLIFIER CIRCUITS

Practical Amplifiers The circuit of a useful but relatively simple operational amplifier, Fig. 8-17, includes a FET difference-amplifier input stage

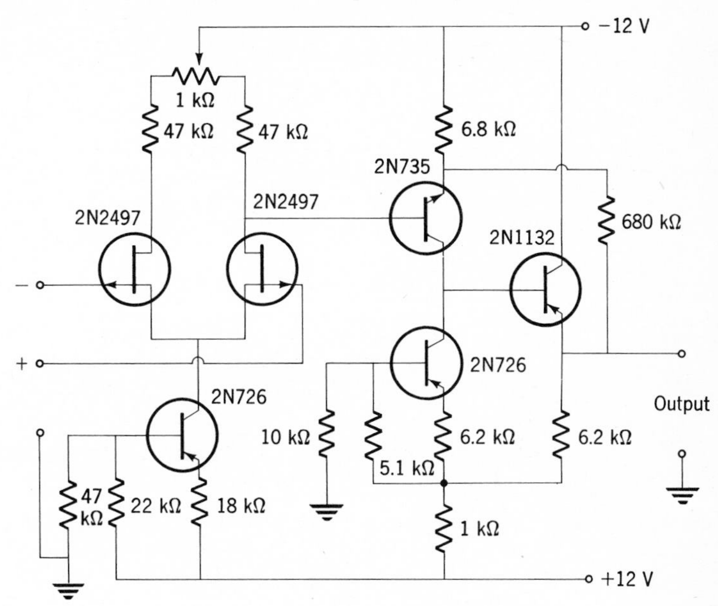

FIGURE 8-17 *Operational amplifier with FET input stage.*

employing a constant-current source transistor to improve balance, as discussed in the previous chapter. The second stage is an *npn* grounded-emitter amplifier using a *pnp* transistor as the collector load resistor. This is directly coupled to an emitter-follower output stage. Negative voltage feedback is provided by the 680-kΩ resistor connected from the output to the emitter of the second stage. In addition, positive current feedback is obtained by connecting the output emitter resistor to a portion of the emitter resistor of the collector load transistor.

This circuit has a large gain, 10^5, and a large input impedance, approximately 5 MΩ, by virtue of the FET input stage. Additionally, the output impedance is low because of the emitter-follower output and the effect of negative feedback. Positive current feedback is employed to achieve some gain in the third stage and to increase the maximum permissible output-voltage swing without distortion. Note that the differential input terminals include polarity markings. The negative terminal, usually referred to as the *inverting* input, causes an output signal of opposite polarity to the input signal. The positive terminal, called the *noninverting* input, produces an output signal in phase with the input. The inverting input is the normal input terminal since the 180° phase shift is necessary in operational feedback.

Integrated-circuit operational amplifiers are very effective because of their small size and useful properties. The circuit of a typical unit, Fig. 8-18*a*, is compared with the actual silicon wafer in Fig. 8-18*b*. The entire unit is mounted in a small package, such as Fig. 8-18*c* or Fig. 5-37*c*. The five-stage circuit consists of a Darlington connected difference amplifier with a temperature-compensated emitter transistor. This is followed by a second difference amplifier stage coupled to an emitter-follower amplifier which uses another transistor for the emitter resistor. A grounded-emitter amplifier feeds the output emitter follower. The emitter follower also employs a transistor for an emitter resistor. Note that temperature

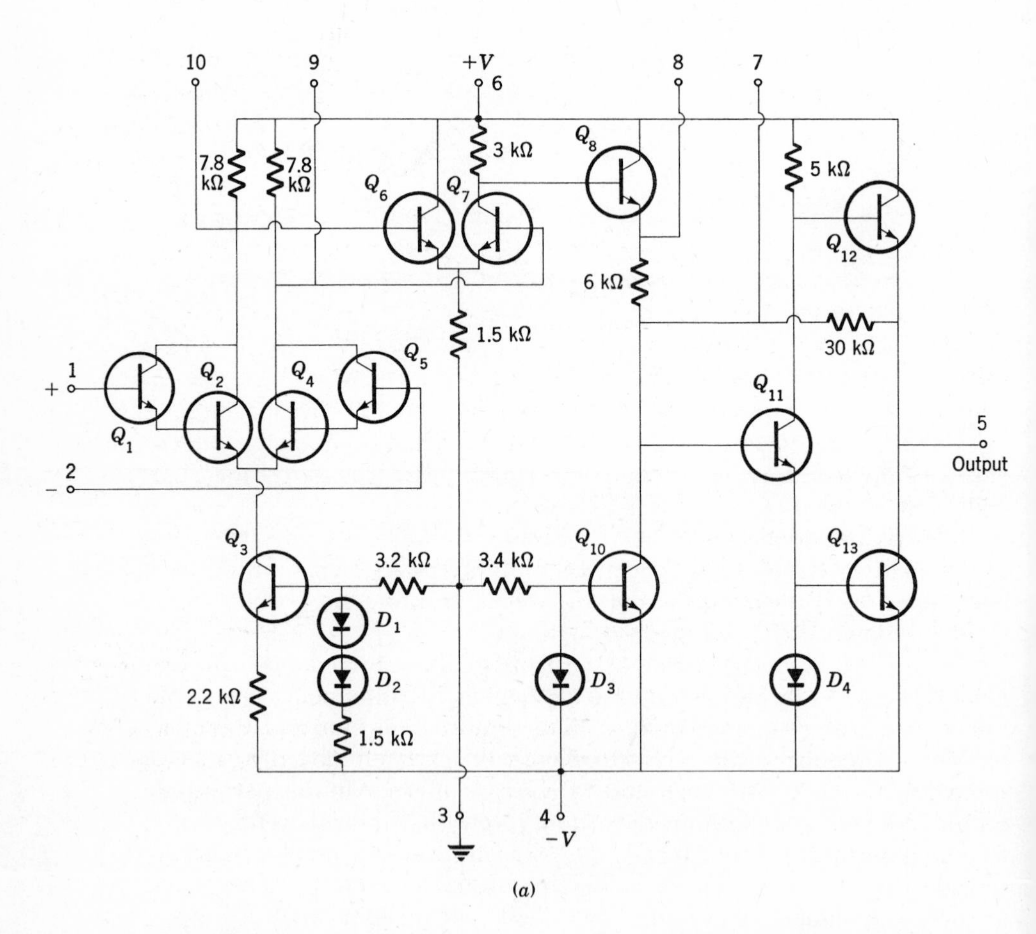

FIGURE 8-18 *(a) Integrated-circuit operational amplifier, (b) actual silicon wafer, and*

compensation is provided in the input, middle, and output stages by diodes in the base bias circuits of the transistors acting as emitter resistors.

As indicated in Fig. 8-18, access to the collectors of the input-stage transistors and to the internal negative feedback network at the third stage is provided in addition to input, output and power-supply terminals. Terminal pairs 7, 8 and 9, 10 permit the addition of external phase-compensation networks to prevent positive feedback at the extremes of the amplifier passband (see pages 313-315). The compensation required depends somewhat upon the particular operational-amplifier application and, accordingly, is not fabricated as part of the integrated circuit. Examples of phase compensation networks are illustrated in a later section.

Even with very careful design and well-stabilized supply voltages, it

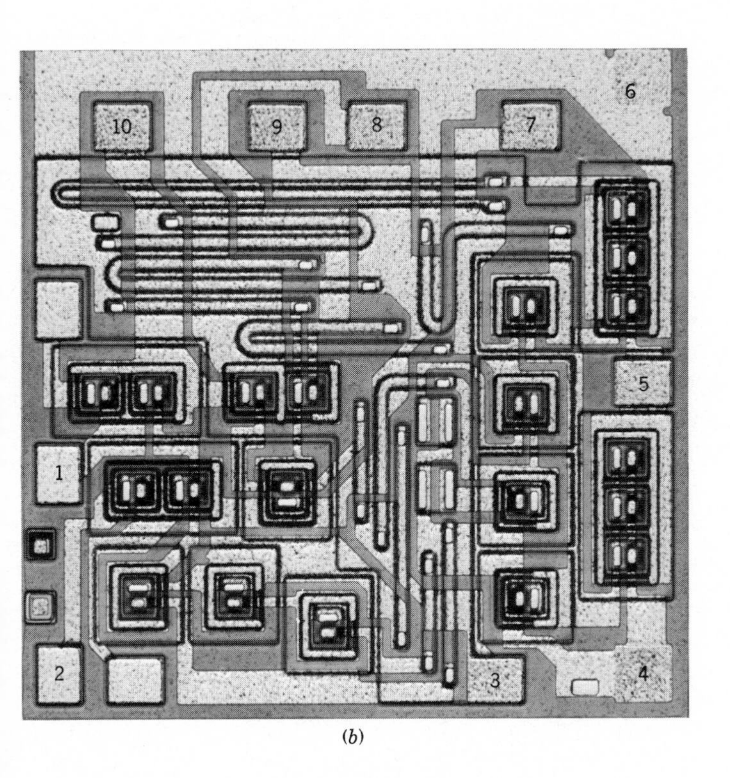

(b)

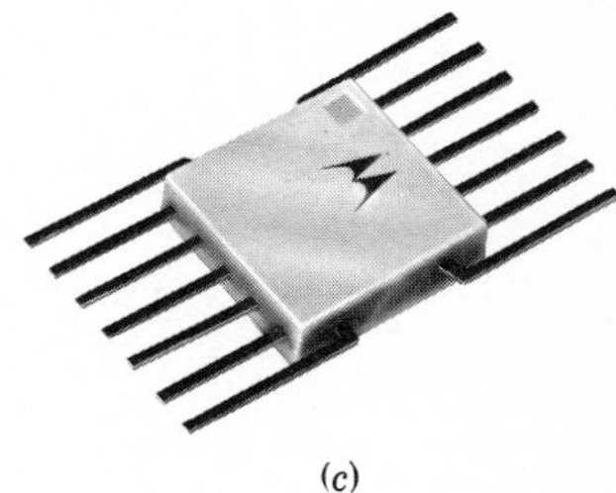
(c)

(c) flat-pack case. (Motorola Semiconductor Products, Inc.)

is difficult to prevent drifts in dc amplifiers. Drift voltages at the amplifier output mean that operational feedback no longer keeps the branch point S at virtual ground and the error in the virtual ground is referred to as *offset*. Typically, offset voltages are of the order of 1 mV, which may be serious in critical applications. Improved performance is achieved by using a chopper amplifier to stabilize the virtual ground voltage, Fig. 8-19. Here, the chopper amplifier, which is drift-free because it is ac-

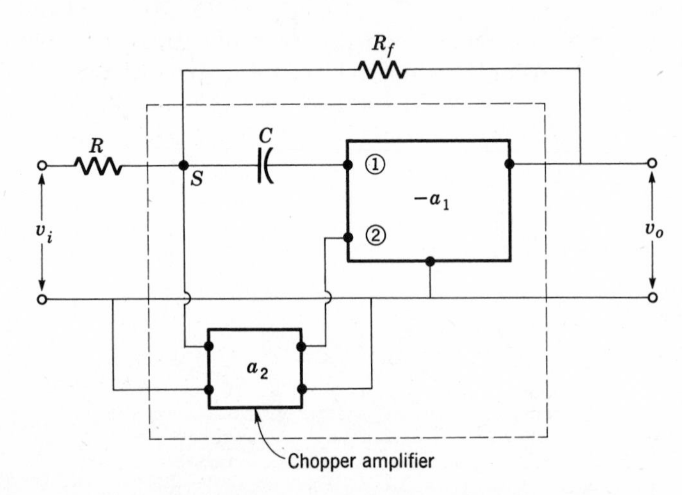

FIGURE 8-19 *Chopper-stablilized operational amplifier.*

coupled, measures the offset voltage at S and provides an amplified error signal to the positive input of the operational amplifier. This signal counteracts drifts in the operational amplifier and, in practice, offset can be reduced to about 10 μV.

Note that a capacitor is interposed between S and the operational amplifier so that offset is controlled only by the chopper amplifier. This prevents the operational amplifier itself from responding to dc signals, but the chopper amplifier does so and, in feeding its output signal to the operational amplifier, becomes part of the operational feedback circuit. In effect, the operational amplifier is not dc-coupled and handles the high-frequency signals, while the chopper amplifier handles the dc and very-low-frequency signals. The dc gain is very large since it is the product of the gains of both amplifiers. This means that the overall response curve is nonuniform in that the low-frequency gain is larger than the gain at high frequencies. According to Eq. (8-29) this is unimportant since by feedback action the gain does not appear in the expression for the output signal, so long as the gain remains large at all frequencies of interest.

The combination of an operational amplifier stabilized by a chopper amplifier is often itself referred to as an operational amplifier. Thus the dashed line in Fig. 8-19 encloses a chopper-stabilized operational amplifier. Chopper amplifiers specifically designed for stabilizing operational amplifiers are also available as separate units. A typical circuit, shown in Fig. 8-20, is a rather conventional ac-coupled amplifier and uses a synchro-

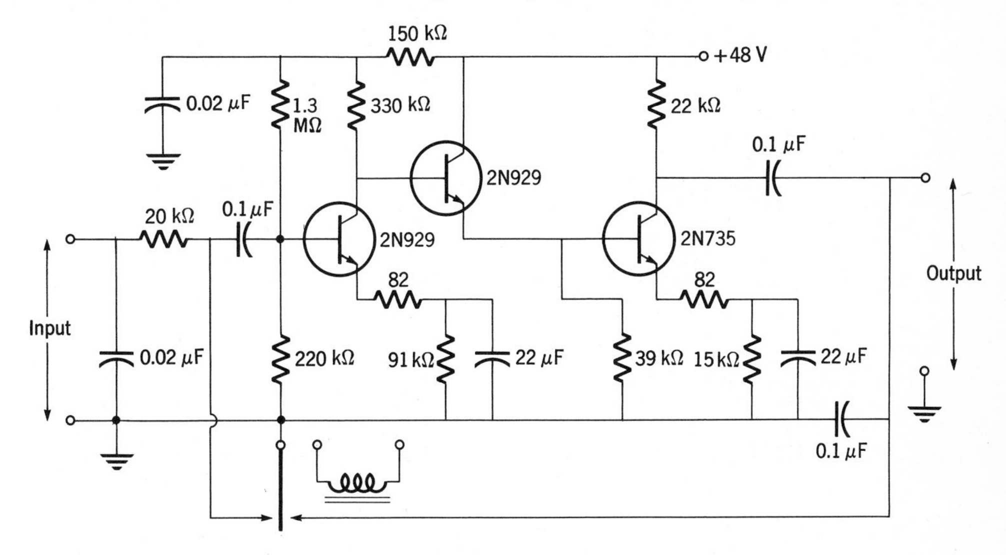

FIGURE 8-20 *Typical chopper amplifier.*

nous chopper to recover the dc signal. Note that unbypassed emitter and cathode resistors provide current feedback to improve phase characteristics so that the synchronously chopped output is in phase with the chopped input signal. In fact, dc coupling is used for convenience for the same reason. Input and output capacitors are present, so that the amplifier is actually ac-coupled.

Specifications The properties required of an operational amplifier depend upon the particular application envisioned. Certain parameters, however, are characteristic of many applications and these serve as a basis for comparison between different units. Consider the characteristics of the several types of operational amplifiers summarized in Table 8-1. This listing compares two typical transistor amplifiers assembled from discrete components (one with FET input similar to Fig. 8-17), with two integrated-circuit units similar to Fig. 8-18, and a chopper-stabilized unit. Many variations of these basic types are also available for different applications, but the range of parameters represented in the table is quite representative.

The foremost characteristic of an operational amplifier is, of course, the gain. This parameter is usually specified as the *open-loop* voltage gain, which simply means the gain of the amplifier itself without any operational

TABLE 8-1 PROPERTIES OF OPERATIONAL AMPLIFIERS

Type	*Open-loop voltage gain*	*Open-loop bandwidth, unity gain, MHz*	*Input impedance, Ω*	*Input offset voltage, mV*	*Offset drift, μV/°C*
Transistor, discrete component	3×10^4	1	2×10^5	adjust	25
FET, discrete component	10^6	10	10^{12}	adjust	2
MC1431, integrated circuit	3500	20	6×10^5	5	10
μA702A, integrated circuit	3600	30	4×10^4	0.5	2.5
Chopper-stabilized	3×10^7	15	5×10^5	0.01	0.2

feedback connection. A large value is desirable in order for Eq. (8-29) to apply accurately. According to the second column in Table 8-1, values in excess of 1000 and ranging to more than 10 million are typical.

Bandwidth is also significant since it specifies the range of signal frequencies over which appreciable gain is available. Note that a uniform frequency response is not necessary since the actual value of gain does not appear in the operational feedback expression. For this reason it is not appropriate to cite the bandwidth at the half-power frequency, as used in Chap. 7. A useful basis for comparison is the open-loop bandwidth at unity gain, which is the interval from direct current to the frequency for which the gain falls to unity. Although this parameter is useful for comparison purposes, an operational amplifier is not useful over this entire bandwidth, since the gain near the extreme frequency is not sufficient for satisfactory performance. The third column shows that very high signal frequencies can be accommodated, however.

The input impedance of an operational amplifier must be great enough for the input current to be negligible, according to the analysis leading to Eq. (8-30). This is easily achieved with an FET stage, and the Darlington connection is often employed for the same reason. Although the input impedance of a simple grounded-emitter difference amplifier is smaller than these alternatives (fifth column), it is sufficiently large for most applications.

If the input difference amplifier is not exactly balanced, an output voltage exists even in the absence of an external input signal. This spurious signal can be troublesome, as, for example, in the integrator circuit since, according to Eq. (8-34), a constant voltage results in an ever-increasing output signal which eventually drives the amplifier into saturation. Although, as previously noted, a constant-current transistor in the emitter or source circuit minimizes the effect of circuit imbalance, it usually proves necessary to provide an adjustable balance control in discrete component amplifiers (compare Fig. 8-17). The balance control may require periodic adjustments as components age. Integrated-circuit amplifiers are sufficiently well balanced so that this is unnecessary, but it is desirable to know the extent of any residual assymmetry. The voltage that must be applied to the input to obtain zero output voltage, the offset, is a convenient measure. Values of offset in the millivolt range or less are common, according to Table 8.1.

The *drift* of the input offset voltage, particularly with temperature, is significant since a slow change in offset voltage is indistinguishable from a slowly-varying signal. According to the last column in Table 8-1, temperature changes of tens of degrees are permissible before the drift signal is comparable with the offset signal.

The superior properties of the chopper-stabilized operational amplifier are clearly evident from the parameters listed in Table 8-1. Certain applications may require the very large input impedance of the FET circuit, however, while for other uses the small size and lower cost of the integrated-circuit design may be more appropriate. Thus the parameters of the several types make the amplifiers complementary rather than competitive.

In addition to those listed in Table 8.1, a number of other parameters are commonly used to specify operational amplifiers. Some of these additional properties are output impedance, input bias current and current offset, common-mode rejection ratio, and internal noise. For a specific application any one of these, as well as others, may prove more important than the common parameters indicated in the table.

Applications It is appropriate to illustrate the versatility of operational amplifiers with a few specific applications. For example, an operational amplifier makes an effective high-gain amplifier for use with small signals. If, however, the input impedance R, Fig. 8-12, is $10^5\ \Omega$ and a gain of 10^3 is desired, the feedback resistor must be $10^8\ \Omega$. Stable resistors of this magnitude are not readily available, so an alternate connection, Fig. 8-21, is used. The voltage-divider action of R_1 and R_2 returns only a fraction of the output signal to the input while maintaining the operational

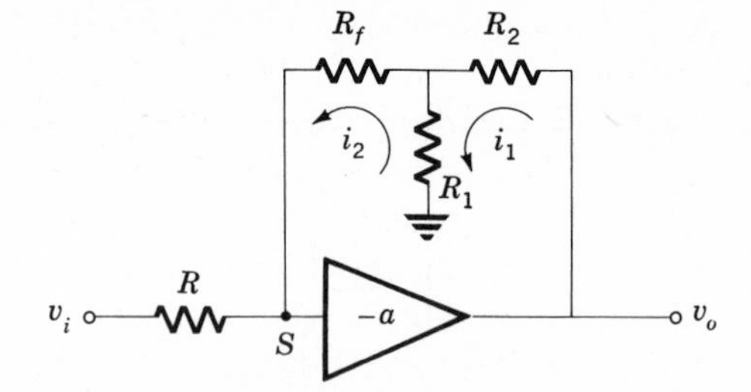

FIGURE 8-21 *High-gain operational amplifier.*

feedback connection. The circuit is analyzed by writing voltage drops around the feedback loops, remembering that S is at ground potential,

$$\begin{aligned} v_o - (R_2 + R_1)i_1 + R_1 i_2 &= 0 \\ (R_f + R_1)i_2 - R_1 i_1 &= 0 \end{aligned} \tag{8-37}$$

Since no current flows to ground at S, $v_i/R = -i_2$. This value is inserted into Eqs. (8-37) and the result solved for v_o/v_i,

$$\frac{v_o}{v_i} = -\frac{R_f}{R}\left[1 + \frac{R_2}{R_1}\left(1 + \frac{R_1}{R_f}\right)\right] \tag{8-38}$$

Most often, $R_f \gg R_1$ and $R_2 \gg R_1$ so that Eq. (8-38) reduces to the previous expression for the gain of an operational amplifier, Eq. (8-30), multiplied by the ratio R_2/R_1. A gain of 10^3 can thus be achieved at an input impedance of 10^5 Ω with a feedback resistor, $R_f = 10^6$ Ω, if, say, $R_2 = 10^4$ and $R_1 = 10^2$ Ω.

A practical integrator and differentiator circuit is illustrated in Fig. 8-22. The integrated circuit operational amplifier used here is similar

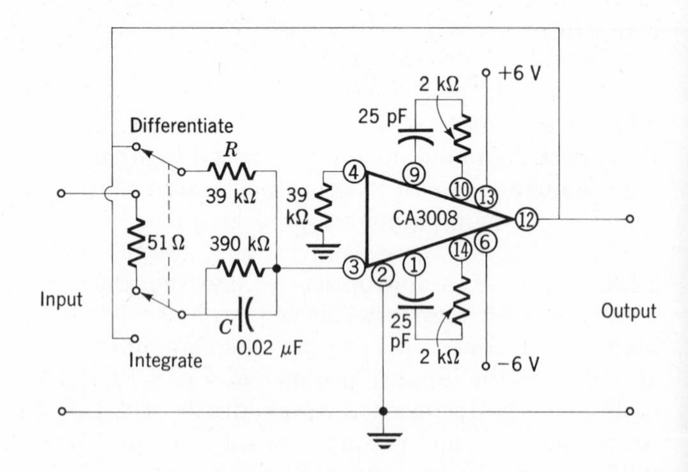

FIGURE 8-22 *Practical integrator-differentiator circuit.*

to that in Fig. 8-18. The switch in the input circuit interchanges the positions of the integrating capacitor and resistor, as may be noted by tracing through the circuit and comparing with Figs. 8-15 and 8-16. Thus the same circuit can be used either to integrate or differentiate the input signal. Typical experimental output waveforms for the case of a square-wave input signal are shown in Fig. 8-23.

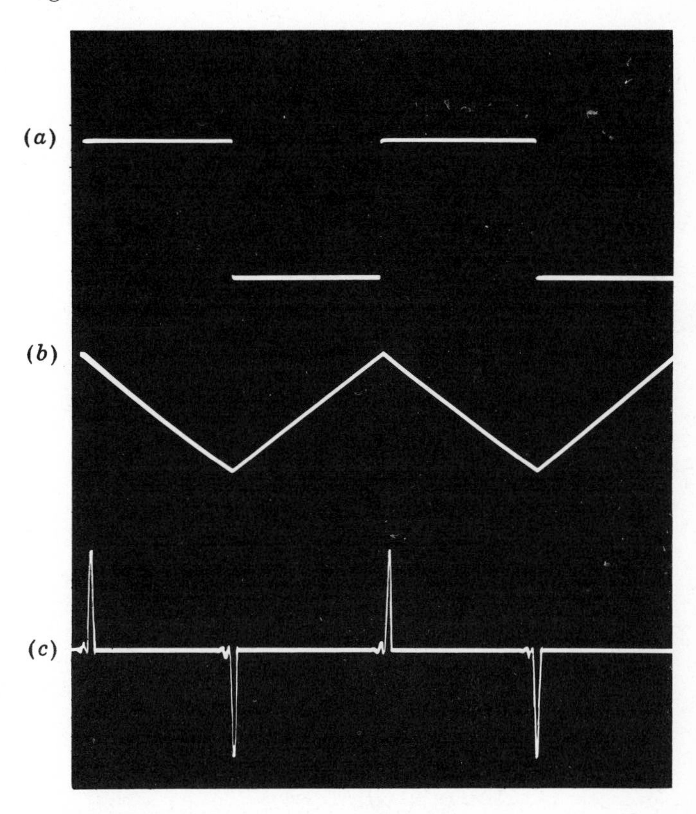

FIGURE 8-23 *(a) Square-wave input (b) integrated and (c) differentiated by operational amplifier circuit of Fig. 8-22.*

The purpose of the 390-kΩ resistor shunting the integrating capacitor in Fig. 8-22 is to eliminate the effect of long-time integration of the input offset voltage. In effect the resistor keeps the capacitor discharged for slow changes in signal and so the integrator does not integrate properly at frequencies below about $f = 1/2\pi RC = 20$ Hz. The two series combinations of a small capacitor and resistor are phase compensation networks. As previously discussed, these networks ensure that the overall phase shifts do not produce positive feedback at the high-frequency limit of the amplifier bandpass.

Nonlinear feedback components in operational-amplifier circuits are used to develop special input-output characteristics. Consider, for example, the grounded-base transistor feedback element, Fig. 8-24, that

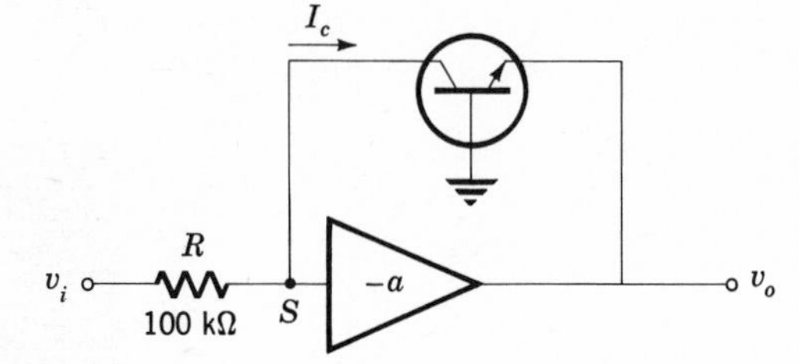

FIGURE 8-24 *Elementary logarithmic amplifier.*

results in an amplifier in which the output voltage is the logarithm of the input signal. Such a logarithmic amplifier operates over a very wide range of input-signal magnitudes without saturating.

The logarithmic input-output characteristic comes about in the following way. The collector current in a transistor is just the emitter current times the current gain, according to Fig. 5-27, and the emitter current is given by the rectifier equation. Ignoring the unity in Eq. (5-20) since the exponential term dominates, the collector current is

$$I_c = \alpha I_0 e^{eV/kT} \tag{8-39}$$

In the operational-feedback connection appropriate for Fig. 8-24, the collector current may be expressed in terms of the input signal

$$\frac{v_i}{R} = \alpha I_0 e^{ev_o/kT} \tag{8-40}$$

where the output signal v_o is the base-to-emitter voltage. Rearranging and taking the logarithm of both sides,

$$v_o = \frac{kT}{e} \ln \frac{v_i}{\alpha I_0 R} \tag{8-41}$$

Experimental input-output characteristics, Fig. 8-25, illustrate the logarithmic properties predicted by Eq. (8-41) and the extremely wide range of input signals over which the amplifier operates. In practical circuits it is usually necessary to supplement the elementary circuit in Fig. 8-24 by including a collector bias supply and to compensate for offset voltages using the noninverting input in order to obtain the logarithmic characteristic over a wide range of input voltages.

In addition to its unique signal-handling capability, the logarithmic amplifier can be used to multiply two arbitrary input signals. Consider the circuit in Fig. 8-26 in which the outputs of two logarithmic amplifiers are summed and then the output of the combination is passed through an inverse logarithmic amplifier. According to Eq. (8-41), the output of the

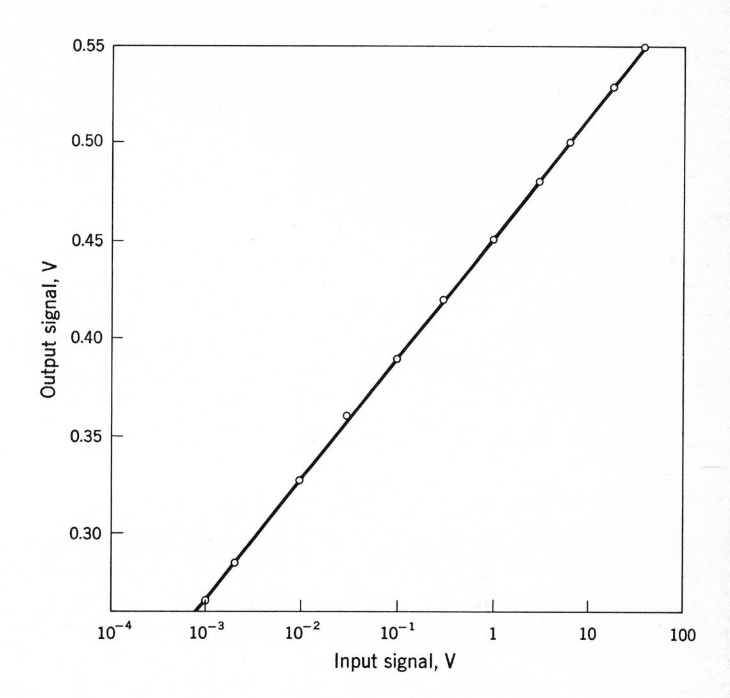

FIGURE 8-25 *Input-output characteristic of logarithmic amplifier.*

summing amplifier is

$$v_3 = (a \ln v_1 + b) + (a \ln v_2 + b) - E_c \tag{8-42}$$

where a and b are constants. The constant voltage E_c is adjusted so that

$$v_3 = a \ln v_1 v_2 \tag{8-43}$$

Therefore, the inverse logarithmic amplifier yields a signal proportional to the product $v_1 v_2$.

These examples serve to suggest the broad application range of operational amplifiers. Many additional possibilities exist. For example, eliminating the input resistor in Fig. 8-13 results in a very sensitive ammeter circuit. According to Eq. (8-30), the output voltage of an amplifier with operational feedback is

$$v_o = -R_f i_i \tag{8-44}$$

where i_i is the input current. Thus, if $R_f = 10^6\ \Omega$, the output signal is 1 V for each microampere of input current. Also, the resistance introduced into the circuit is the impedance between S and ground, which may be only a fraction of an ohm. Currents as low as $10^{-12}a$ can be measured with this circuit. Operational amplifiers are also useful in comparing the

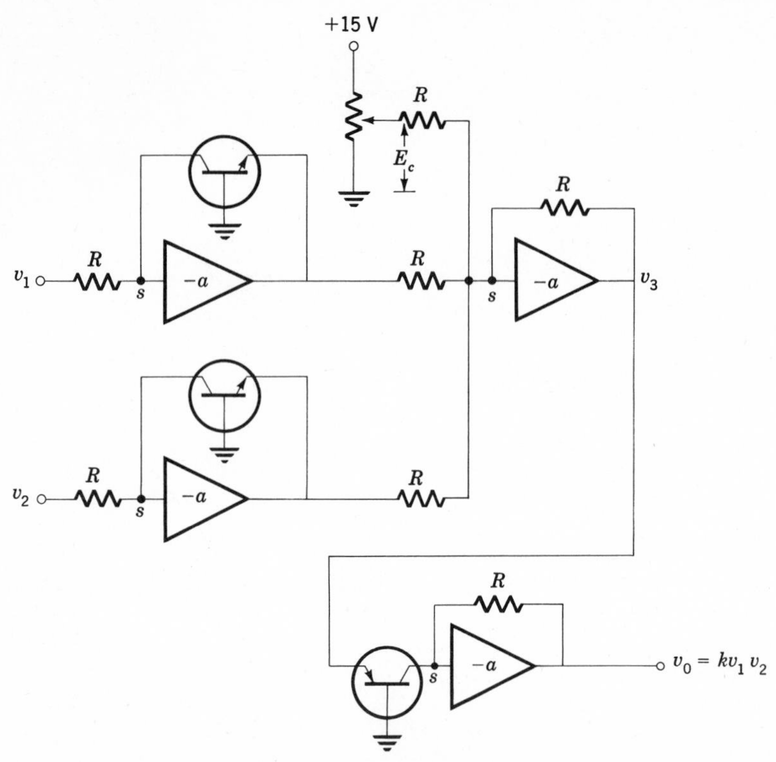

FIGURE 8-26 Product-function circuit.

magnitudes of two signals. If one signal is applied to the inverting input and the other to the noninverting input, the output of the amplifier is zero only when the two signals are equal. Such a *comparator* circuit proves to be a very accurate way of establishing the equality of two signals. The large common-mode rejection ratio and excellent input-stage balance characteristic of operational amplifiers is appropriate in this application.

ANALOG COMPUTERS

Simulation Operational feedback circuits can be assembled into *analog computers* in which voltage signals are analogous to variables in physical systems. Many physical systems can be described by mathematical equations based on the laws of nature. Analog computors are used to solve these equations and thereby to display the behavior of the system.

The voltages and other circuit parameters of an analog computer correspond to the variables and properties of the real system. The parameters of the computer are easily altered and adjusted, however, so that the behavior of the system may be examined over a wide range of conditions. The computer is, in fact, looked upon as *simulating* the physical system, and its response to external stimuli corresponds to the actual system to the extent that the mathematical description is accurate.

Damped Harmonic Oscillator A great many physical systems are described by differential equations, for example, mechanical motions in accordance with Newton's laws of motion. Consider the *damped harmonic oscillator.* This is the mechanical vibration of a body of mass m on the end of a spring with force constant k in the presence of viscous damping described by the damping constant b and driven by an arbitrary force $F(t)$. The differential equation for the position x of the body is

$$m \frac{d^2x}{dt^2} + b \frac{dx}{dt} + kx = F(t) \tag{8-45}$$

Rearranging,

$$\frac{d^2x}{dt^2} = -\frac{b}{m}\frac{dx}{dt} - \frac{k}{m}x + \frac{1}{m}F(t) \tag{8-46}$$

The design of an analog computer to solve this equation begins by assuming a voltage signal corresponding to d^2x/dt^2 is available. This is integrated to yield $-dx/dt$, where for convenience the RC time constant in Eq. (8-34)

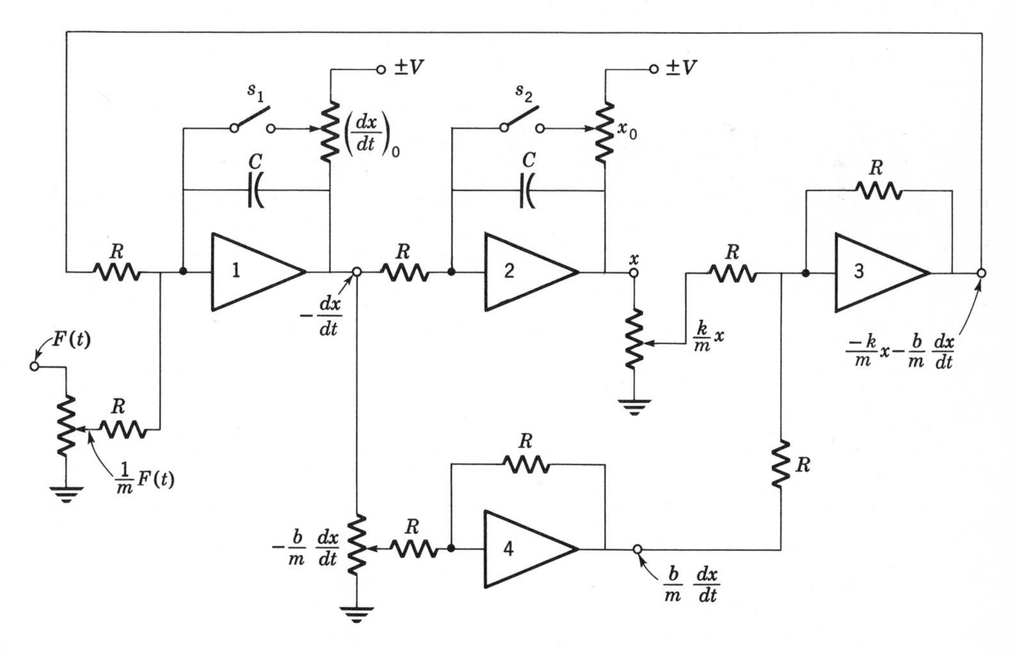

FIGURE 8-27 *Analog computer for vibrating-mass problem.*

is made equal to unity. Next, $-dx/dt$ is integrated again to obtain x. A fraction b/m of $-dx/dt$ is obtained from a voltage-divider potentiometer across the output of the first integrator; this is inverted and added to the fraction k/m of x from the output of the second integrator and to a voltage signal corresponding to $(1/m)F(t)$ (Fig. 8-27). The sum is equal to d^2x/dt^2, according to Eq. (8-46), and is returned to the input, where the second-derivative signal was assumed to be originally. This computer, therefore, continuously solves the original differential equation (8-45) and voltages corresponding to x and, if desired, dx/dt can be measured at appropriate points in the circuit.

It is necessary to set voltages corresponding to dx/dt, and x for their initial values at the time when the solution begins, as in solving any differential equation. This is most effectively accomplished by opening switches s_1 and s_2 at $t = 0$. The voltages across the integrating capacitors represent the velocity and displacement at all times, as can be seen by the fact that the summing points are at ground potential. These switches must be opened simultaneously with the beginning of $F(t)$ and in practice this is most often done with electronic switches such as diodes.

Note that Eq. (8-45) has the same form as Eq. (3-28) for the current in an *RLC* circuit. This means that the analog computer in Fig. 8-27 also can be used to investigate, say, resonance effects in this simple circuit. Although the computer is much more complicated than the circuit it simulates, study of the circuit is facilitated by the ease with which circuit parameters can be altered. Furthermore, the computer can be rewired to simulate other circuits as the need arises.

SUGGESTIONS FOR FURTHER READING

E. J. Angelo, Jr.: "Electronic Circuits," McGraw-Hill Book Company, New York, 1964.

H. V. Malmstadt, C. G. Enke, and E. C. Toren, Jr.: "Electronics for Scientists," W. A. Benjamin, Inc., New York, 1963.

R. D. Middlebrook: "Differential Amplifiers," John Wiley & Sons, Inc., New York, 1963.

EXERCISES

8-1 Draw the equivalent circuit of the bridged-T feedback amplifier, Fig. 8-5. Plot the characteristic curve of the filter alone, that is, the frequency variation of β, and determine the response curve of the feedback amplifier. Assume $g_m = 5 \times 10^{-3}$ mho for all FETs.

8-2 Consider the emitter follower, Fig. 6-31, as a feedback amplifier. Derive an expression for the gain corresponding to Eq. (6-32).

8-3 Calculate the input and output impedance of the transistor feedback amplifier in Fig. 8-3. The h parameters are $h_{ie} = 3600\ \Omega$, $h_{fe} = 150$, $h_{re} = 3 \times 10^{-3}$, $h_{oe} = 1.4 \times 10^{-4}$ mho.

Ans.: $1.61 \times 10^{6}\ \Omega$, $12\ \Omega$

8-4 Analyze the circuit of Fig. 6-28 as an amplifier with current feedback if the emitter bypass capacitor is omitted. Calculate β and the gain and compare with the corresponding expression for a vacuum-tube amplifier, Eq. (6-21).

Ans.: 1, 3.3

8-5 Derive an expression for the input impedance of a feedback amplifier with both voltage and current feedback. Use this expression to show that both current and voltage negative feedback increase the input impedance. Using Eq. (8-27) find the expression for the input impedance when the output impedance is zero.

8-6 Calculate the voltage and current feedback ratios β_v and β_i in the output stages of the operational feedback amplifier of Fig. 8-17.

Ans.: 10^{-2}, 1

8-7 Use Fig. 8-25 to develop a numerical expression corresponding to Eq. (8-41). What is the approximate value of I_0 for the transistor used in the circuit of Fig. 8-24? Take $e/kT = 38\ \text{V}^{-1}$.

Ans.: 3.8×10^{-13} A

8-8 Derive an expression analogous to Eq. (8-41) for the inverse logarithmic amplifier used in Fig. 8-26.

8-9 Using the principles employed in Fig. 8-26, sketch a circuit for which the output voltage is either the square or the cube of the input signal.

8-10 Design an analog computer to solve the falling-body problem, for which the equation is

$$\frac{d^2x}{dt^2} = g \tag{8-47}$$

where g is the acceleration due to gravity.

LABORATORY EXERCISES

8-A The Difference Amplifier Most often the input stage of a practical operational amplifier is a difference amplifier. This circuit is chosen because of its balanced configuration which minimizes drift and because of the versatility inherent in the dual inputs. This experiment examines the properties and operating characteristics of a typical integrated-circuit difference amplifier.

Consider the circuit in Fig. 5-37*a* with supply voltages + 12 V and −12 V and with terminal 3 connected to 0 V, that is, to ground. Assuming the collector characteristics of all three transistors are similar to those for type 2N338 given in Appendix 1, determine the current in each transistor and the zero signal potential of the input terminals 2 and 5. This is most easily accomplished by first calculating the current in Q_3 and noting that this value is equally divided between Q_1 and Q_2.

Draw the ac equivalent circuit of the amplifier and verify Eq. (7-37) for the output voltage. Note that Q_3 is not part of the signal circuit and may be represented by a resistor. What is the value of this resistor? Develop an expression for the common-mode rejection ratio if the transconductances of Q_1 and Q_2 differ slightly. Calculate the rejection ratio if the difference is 0.1 percent.

Quantitatively sketch the waveforms, including the dc component, at all three collectors for a sinusoidal input signal applied to terminals 2 and 5. What is a convenient magnitude for the input signal? Repeat for the same signal applied between terminal 2 and ground only. Now connect terminals 2 and 5 together, apply the signal between this point and ground and plot the waveforms.

Imagine a capacitor connected from the collector of Q_3 to ground so that the emitters of Q_1 and Q_2 are effectively at ground potential at the signal frequency. Now plot waveforms as in the previous paragraph for the three input configurations. Comment on the effect of Q_3 in the normal differential-amplifier circuit.

8-B Operational Amplifier Comparator Operational amplifiers are widely used in measurement and control applications without the operational-feedback connection. The large gain, high input impedance, and good frequency response are all attractive features. One rather unique application is the comparator circuit, Fig. 8-28, in which the inverting and noninverting inputs are used to compare the instantaneous voltages of two signals and indicate when they are equal.

Suppose that the operational amplifier in Fig. 8-28 has an open loop gain of

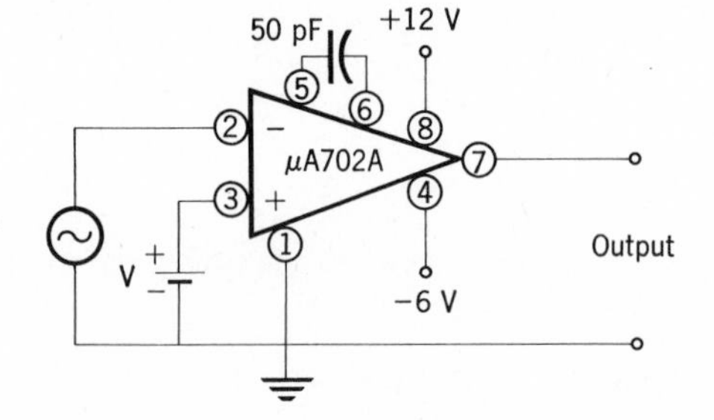

FIGURE 8-28 Operational-amplifier comparator.

4×10^3 and a maximum output signal at saturation of $\pm$ 10 V. The reference voltage V = 1.5 V. Apply a 6-V (peak-to-peak) sinusoidal signal to the input. Quantitatively sketch the output voltage waveform. Repeat for sawtooth and square-wave input signals having the same magnitude. Calculate the accuracy with which the comparator using the given operational amplifier indicates that the two voltages are equal. Your answer should be a few millivolts.

chapter nine

OSCILLATORS

Electronic circuits can generate ac signals of a variety of waveforms over a wide range of frequencies. In fact, transistor and vacuum tube oscillators are the only convenient way of generating high-frequency voltages. They are widely used in radio and TV transmitters and receivers, for dielectric and induction heating, and in electronic instruments for timing and testing purposes. An oscillator, in effect, converts power delivered by the dc supply voltages into ac power having the desired characteristics. In addition to the frequency and waveform of the oscillations, the conversion efficiency and frequency stability are important in the design of oscillator circuits.

POSITIVE FEEDBACK

Oscillation is achieved through positive feedback which produces an output signal without any input signal. According to Eq. (8-4), the gain of a feedback amplifier is given by

$$a' = \frac{a}{1 - a\beta} \tag{9-1}$$

If circuit conditions are arranged so that

$$a\beta = 1 \tag{9-2}$$

the gain becomes infinite, which means that an output signal exists even when the input signal is zero. For sinusoidal oscillations the feedback network is designed so that Eq. (9-2), called the *Barkhausen criterion,* is satisfied at only one frequency, and the circuit oscillates at that frequency. The Barkhausen criterion requires that the overall phase shift of the feedback signal is 360°, and this is the significant factor in determining the frequency of oscillation. In addition, the amplifier gain must be large enough to assure that the $a\beta$ product is equal to unity, in order for the oscillations to persist.

The amplitude of oscillation is determined indirectly by Eq. (9-2). The gain of any amplifier is reduced at large-signal amplitudes because of cutoff and saturation conditions in the transistors or vacuum tubes. Accordingly, the quiescent amplitude is such that the absolute value of gain is $1/\beta$. Since the feedback network is most often a passive circuit, the amplitude depends primarily upon amplifier characteristics.

It is not necessary to supply an input signal in order to initiate oscillations. Random noise voltages or transients accompanying application of the supply voltages are sufficient to start the feedback process. Since the amplitude of the feedback signal depends upon amplifier gain, the rapidity with which the oscillations reach the steady-state magnitude increases when the gain is large. It is usually desirable for the small-signal gain to be significantly larger than required by the Barkhausen criterion. This produces strong oscillations unaffected by minor circuit changes. On the other hand, if the gain is very great, nonsinusoidal oscillations may result from nonlinearities accompanying large-signal amplitudes.

RC OSCILLATORS

Phase-shift Oscillator A simple but useful oscillator circuit employing a conventional amplifier stage and an *RC* feedback network is the *phase-shift oscillator*, Fig. 9-1. The grounded-emitter stage has an inherent phase shift of 180°, so the three cascaded *RC* circuits shift the phase an additional

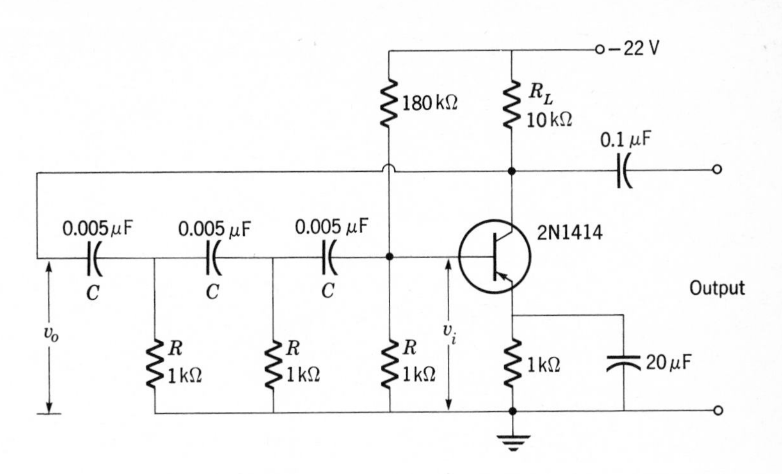

FIGURE 9-1 *Phase-shift oscillator.*

180° in order to satisfy the Barkhausen criterion. At some particular frequency the phase shift in each RC section is 60°, so the total phase shift in the feedback network is 180° and the circuit oscillates at this frequency, provided the amplification is great enough. Note that the maximum phase shift in one RC section is limited to 90°. This means that a two-section feedback network is not possible because it would require an infinite gain to overcome the attenuation in the feedback network at a total phase shift of 180°. Conversely, there is no particular advantage to having more than three RC sections in the feedback network, although it is possible to design such an oscillator.

The phase-shift oscillator is analyzed by first ignoring the loading effect of the amplifier upon the network. Considering that a voltage v_o is applied to the feedback network, the signal v_i applied to the transistor can be calculated by straightforward network analysis. The result is

$$\beta = \frac{v_i}{v_o} = \frac{1}{1 - 5/(\omega RC)^2 + j[1/(\omega RC)^3 - 6/\omega RC]} \tag{9-3}$$

In order for the phase shift of the feedback network to be 180°, the imaginary part of Eq. (9-3) must vanish, or

$$\frac{1}{(\omega_0 RC)^3} = \frac{6}{\omega_0 RC} \tag{9-4}$$

The oscillation frequency is found by solving for $f_0 = \omega_0/2\pi$,

$$f_0 = \frac{1}{2\pi\sqrt{6}}\frac{1}{RC} \tag{9-5}$$

Inserting Eq. (9-5) into Eq. (9-3), $\beta = 1/(1 - 5 \times 6) = -1/29$, which means that the gain must be

$$a = \frac{1}{\beta} = -29 \tag{9-6}$$

in order to satisfy the Barkhausen criterion. According to this result the amplification must be at least 29 or the circuit cannot oscillate.

Actually the amplifier gain must be somewhat larger than 29 in order to assure stable oscillations in the face of circuit losses and component aging effects. The amplitude of oscillations increases until limited to a value of 29 by nonlinearities in the transistor. Most often, the onset of cutoff at the peak of the wave is the limiting factor, and a peak signal amplitude nearly equal to the quiescent collector potential is expected.

In a practical circuit the loading of the amplifier upon the feedback network cannot be ignored. For example, consider the equivalent circuit of a vacuum-tube phase-shift oscillator, Fig. 9-2. Using Eq. (6-16) for the gain of a vacuum-tube amplifier,

$$v_o = \frac{-\mu v_i}{1 + r_p/\mathbf{Z}_L} = \frac{-\mu \beta v_o}{1 + r_p/\mathbf{Z}_L} \tag{9-7}$$

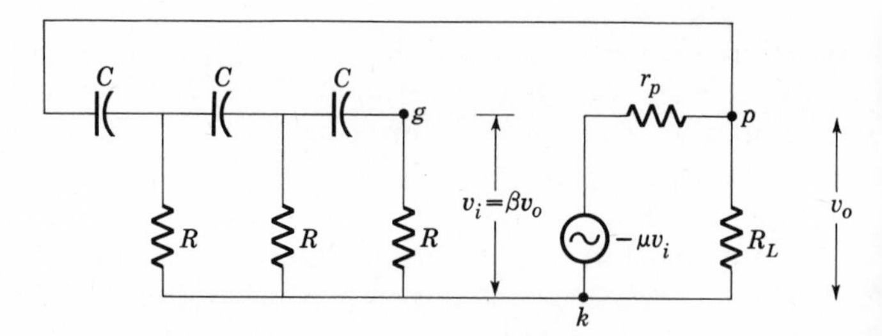

FIGURE 9-2 *Equivalent circuit of vacuum-tube phase-shift oscillator.*

where $\mathbf{Z}_L$ is the parallel combination of the plate load resistor R_L with the input impedance of the feedback network, $\mathbf{Z}_i$. Solving for β,

$$\beta = -\frac{1}{\mu}\left(1 + \frac{r_p}{\mathbf{Z}_L}\right) = -\frac{r_p}{\mu}\left(\frac{1}{r_p} + \frac{1}{\mathbf{Z}_L}\right) \tag{9-8}$$

$$\beta = -\frac{1}{g_m}\left(\frac{1}{r_p} + \frac{1}{R_L} + \frac{1}{\mathbf{Z}_i}\right) \tag{9-9}$$

This result, together with Eq. (9-6), determines the characteristics of the vacuum tube necessary for the circuit to oscillate. Note that the impedance of the feedback network is included in Eq. (9-9). A similar result applies to the transistor phase-shift oscillator except that the loading effect of the transistor input impedance must also be taken into account.

The simplicity of the phase-shift oscillator makes it attractive for noncritical applications, particularly at medium and low frequencies down to about 1 Hz. The frequency stability is not as good as can be obtained with other *RC* oscillators, however. In addition, in order to change frequency it is necessary to vary all three capacitors (or the three resistors), and this is inconvenient.

Wien-bridge Oscillator The frequency-selective properties of the Wien bridge discussed in Chap. 3, Fig. 3-14, are very appropriate for the feedback network of an oscillator. The *Wien-bridge oscillator,* Fig. 9-3, is widely

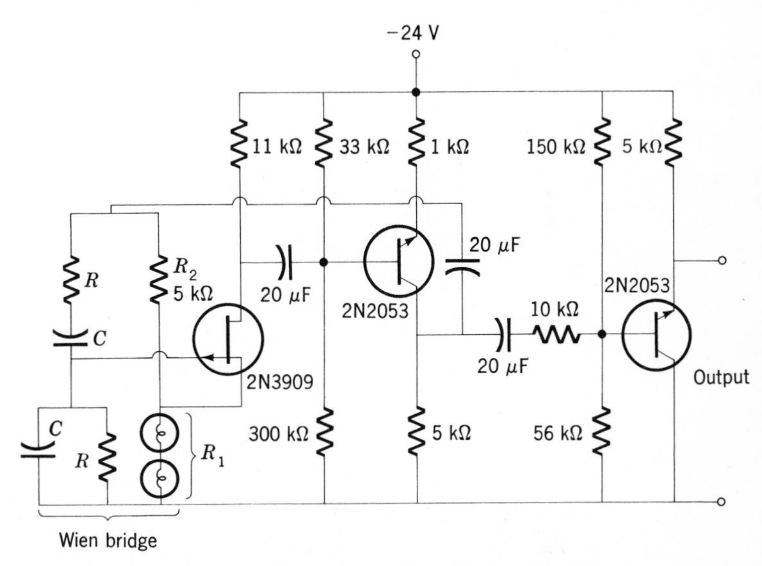

FIGURE 9-3 *Wien-bridge oscillator.*

used for variable-frequency laboratory instruments called *signal generators.* A conventional two-stage amplifier provides positive feedback at the resonant frequency of the bridge, where, according to Eq. (3-84), the phase shift in the feedback network is zero. The characteristic frequency is, from Eq. (3-80),

$$f_0 = \frac{1}{2\pi RC} \tag{9-10}$$

The output voltage of the Wien bridge is zero at exact balance when $R_2 = 2R_1$ (refer to Fig. 3-15), so it is necessary to unbalance the bridge slightly by adjusting the ratio R_2/R_1. This provides sufficient feedback voltage to maintain stable oscillations, since the feedback ratio is determined by the relative values of R_1 and R_2. It is common practice to use small tungsten-filament lamp bulbs for R_1 in order to stabilize the amplitude of the oscillations. The way this comes about is as follows. A tungsten lamp bulb filament has a positive temperature coefficient of resistance. If the output of the amplifier increases for any reason, the greater current through the lamp increases the filament temperature and therefore its resistance. Accordingly, the feedback voltage is decreased and the amplitude of oscillation returns to nearly the original value.

The excellent frequency stability of the Wien-bridge oscillator compared with the phase-shift oscillator is a result of the rapid change of phase of the feedback voltage with frequency. The feedback ratio is, from Eq. (3-84),

$$\beta = \frac{v_o}{v_i} = \frac{1}{3 + j(\omega/\omega_0 - \omega_0/\omega)} - \frac{1}{1 + R_2/R_1} \tag{9-11}$$

The phase shift is found by rationalizing Eq. (9-11) and computing the tangent of the phase angle. For frequencies near to the resonant frequency, the result is

$$\tan\theta \cong \frac{1 + R_2/R_1}{9 - 3(1 + R_2/R_1)}\left(\frac{\omega}{\omega_0} - \frac{\omega_0}{\omega}\right) \tag{9-12}$$

If, for example, $R_2 = 1.9R_1$, the coefficient of the frequency term in Eq. (9-12) is equal to 9.7. By comparison, the corresponding expression for the phase-shift oscillator is, after rationalizing Eq. (9-3),

$$\tan\theta \cong -\frac{\sqrt{6}}{5}\left(\frac{\omega}{\omega_0} - \frac{\omega_0}{\omega}\right) \tag{9-13}$$

where Eq. (9-5) has been used to introduce ω_0. The numerical factor in Eq. (9-13) is only 0.49, which is smaller by a factor of 20 than in the case of the Wien bridge.

This comparison is illustrated more clearly in Fig. 9-4 where the phase

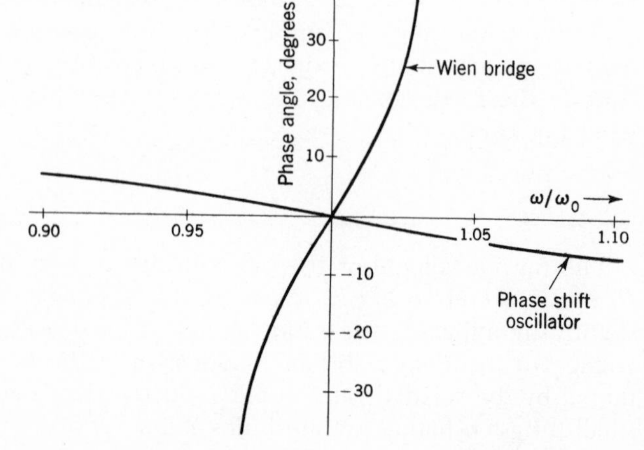

FIGURE 9-4 Phase-shift characteristics near resonant frequency for two RC oscillators.

angle of the feedback signal is plotted as a function of ω/ω_0 for both oscillators. The phase angle changes much more rapidly with frequency in the Wien-bridge feedback circuit. This means that the oscillation frequency is quite stable since only feedback signals with a near-zero phase angle are effective.

The excellent frequency stability together with the relative ease in changing frequency (only the two capacitors need be variable) makes the Wien-bridge oscillator popular. In most practical circuits the variable capacitors provide a frequency range of about 10 to 1. In addition, fixed decade values of the resistors are selected by a switch so that a wide frequency range can be covered in a single instrument. A typical commercial Wien-bridge oscillator can have a frequency range extending from 5 Hz to 1 MHz in decade steps.

RESONANT CIRCUIT OSCILLATORS

LC Oscillators Resonant *LC* circuits are often used in the feedback network of oscillators to select the frequency of oscillation. Consider the so-called *Hartley oscillator,* Fig. 9-5, in which a parallel resonant circuit is

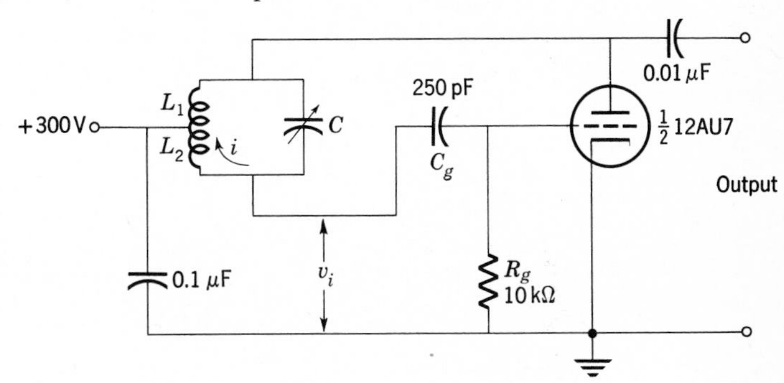

FIGURE 9-5 *Hartley oscillator.*

connected between grid and plate. The inductance is tapped so that, in effect, the portion L_1 is part of the plate load while the remainder L_2 is in the grid circuit. The resonant frequency involves the series inductance of L_1 and L_2, so that, as in Chap. 3,

$$\omega_0 = \frac{1}{\sqrt{(L_1 + L_2)C}} \tag{9-14}$$

The feedback ratio is determined by first calculating the feedback voltage across L_2,

$$v_i = ij\omega L_2 = \frac{v_o j\omega L_2}{j\omega L_2 + 1/j\omega C} \tag{9-15}$$

Introducing the resonance condition $\omega_0(L_1 + L_2) = 1/\omega_0 C$,

$$v_i = v_o \frac{\omega_0 L_2}{-\omega_0 L_1} \tag{9-16}$$

so that the feedback ratio is just

$$\beta = -\frac{L_2}{L_1} \tag{9-17}$$

Note that the ratio is negative, which means that the additional 180° phase shift of the amplifier produces positive feedback as required. Equation (9-17) and the Barkhausen criterion specify the gain of the amplifier needed to sustain oscillations.

The bias conditions in the Hartley oscillator are worthy of note. Operating bias is supplied by the R_gC_g combination in the grid circuit. When oscillations start, the grid bias is zero. As oscillations build up, the grid-cathode diode rectifies the feedback signal, thereby charging C_g to nearly the peak value of the input signal. The R_gC_g time constant is much longer than the period of oscillation so the the voltage across C_g is constant and represents the necessary dc grid bias. In effect, the grid is clamped at ground potential (compare Figs. 4-31 and 4-32). The grid bias therefore automatically adjusts itself to the amplitude of the feedback signal and this action stabilizes the amplitude of oscillation.

This form of bias means that the tube is cut off during most of the cycle, a condition labeled *class C* operation. The grid voltage produces plate current at the peak of the feedback voltage cycle. The large impedance of the resonant circuit allows only the fundamental component of the output signal to have any appreciable amplitude, however, so that the output waveform is sinusoidal. Another way of explaining the action of the resonant circuit is to say that the circuit is excited by periodic pulses from the

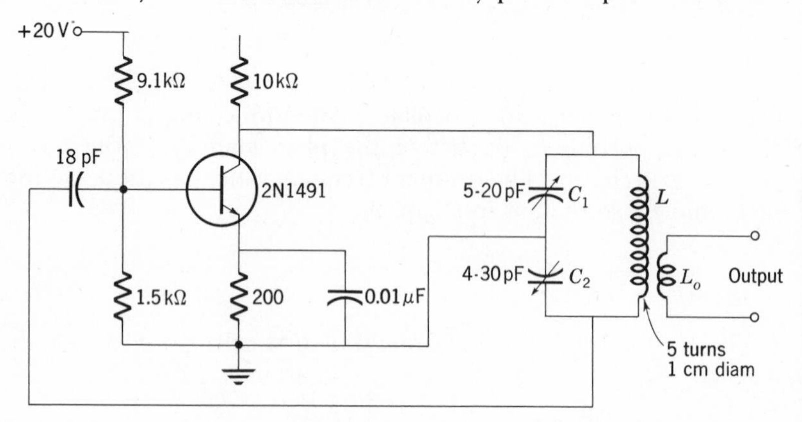

FIGURE 9-6 *70-MHz Colpitts oscillator.*

class C amplifier and that a continuous wave is produced by ringing in the resonant circuit. These two explanations are equivalent.

The *Colpitts oscillator,* Fig. 9-6, is similar to the Hartley circuit except that the feedback ratio is determined by the relative values of C_1 and C_2. Base bias in this circuit is such that the transistor operates as a class A amplifier, but class C operation equivalent to the Hartley oscillator is equally possible. The output load is coupled to the resonant circuit with a secondary winding L_o, which is particularly useful in feeding low-impedance loads.

Another way to develop feedback voltage is by including a secondary winding, or *tickler* winding, coupled to the inductance. Consider, for example, the grounded-base oscillator, Fig. 9-7*a*. Mutual inductance be-

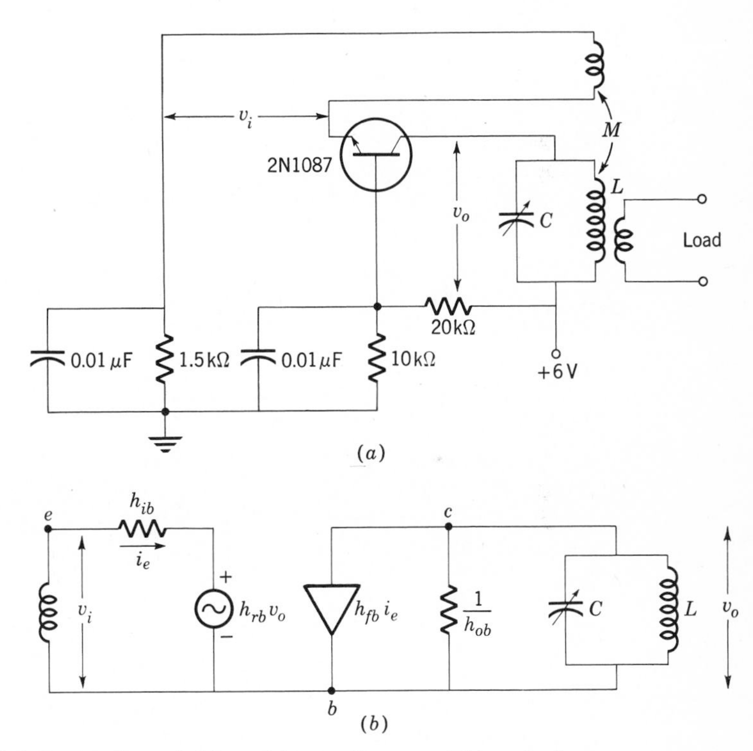

FIGURE 9-7 *(a) Grounded-base tickler oscillator and (b) equivalent circuit.*

tween L and the tickler winding induces a feedback signal of the proper amplitude and phase to sustain oscillations. From the definition of mutual inductance in Chap. 3, the feedback voltage is just the ratio of the mutual inductance M between coils to the total inductance L times the voltage across L [see Eqs. (3-91) and (2-53)]. Therefore

$$v_i = \frac{M}{L} v_o \tag{9-18}$$

or

$$\beta = \frac{v_i}{v_o} = \frac{M}{L} \tag{9-19}$$

The equivalent circuit of the oscillator, Fig. 9-7*b*, is analyzed in the following way. The impedance of the parallel resonant circuit is very large at resonance, so the gain of the grounded-base amplifier is, from Eq. (6-76),

$$a = \frac{-1}{h_{ib}h_{ob}/h_{fb} - h_{rb}} = \frac{1}{r_e/\alpha r_c + r_b/r_c}$$

$$= \frac{\alpha r_c}{r_e + \alpha r_b} \cong \frac{r_c}{r_b} = \frac{1}{h_{rb}} \tag{9-20}$$

Applying the Barkhausen criterion, the condition for oscillation is obtained by comparing Eqs. (9-20) and (9-19),

$$h_{rb} = \frac{M}{L} \tag{9-21}$$

In a good-quality transistor h_{rb} is of the order of 10^{-4}, which means that the mutual inductance may be quite small. Accordingly, a tickler winding consisting of only a few turns is sufficient. This is fortunate since the small input impedance of the amplifier is best matched with a large turns ratio between L and the feedback winding.

A major advantage of the tickler feedback circuit is that the amplitude of the feedback voltage can be easily adjusted by choosing the number of turns on the feedback winding. The proper phase is obtained by interchanging the leads of the winding, if necessary, to obtain positive feedback. Thus, although Fig. 9-7 is a grounded-base circuit, the grounded-emitter and grounded-collector configurations are equally useful. Furthermore, the resonant circuit can be put in the emitter or base circuits as well. The specific configuration illustration by Fig. 9-7 is particularly appropriate for good frequency stability, as discussed below.

Class C operation of transistor oscillators is often used in circuits designed to develop appreciable power at high frequencies. Figure 9-8 is the circuit diagram of a typical power oscillator. The overall power efficiency of an oscillator is not as good as that of a class C amplifier. Input circuit losses must be supplied by the oscillator itself, but even so, efficiencies of the order of 70 percent are possible. In this circuit, base bias is developed by the resistor-capacitor combination in the base-emitter circuit analogous to the action in the Hartley vacuum-tube oscillator.

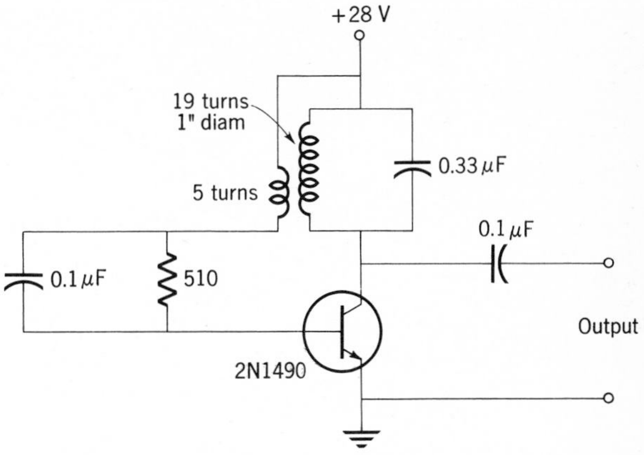

FIGURE 9-8 *100-kHz 10-W power oscillator.*

Crystal Oscillators The frequency stability of *LC* oscillators is determined primarily by the *Q* factor of the resonant circuit. The resonance curve is sharply peaked and the rate of change of phase with frequency is rapid when the *Q* is large, and both factors contribute to frequency stability of the oscillator. In this connection, any equivalent resistance connected in parallel with the resonant circuit lowers the effective *Q*. Therefore, the loading effect on the resonant circuit should be minimized to improve frequency stability. According to the discussion in Chap. 6, the output impedance of the grounded-base amplifier is larger than for any other configuration. This is why the grounded-base oscillator with the resonant combination in the collector circuit is the most satisfactory transistor oscillator configuration. Correspondingly, vacuum-tube oscillators most often have the resonant circuit in the grid circuit. In either case, practical values of *Q* from 100 to 500 can be obtained, and quite stable oscillations result.

Many applications require a higher order of frequency stability than can be obtained with *LC* resonant circuits, and *crystal oscillators* are widely used to fill this need. Certain crystalline materials, most notably quartz, exhibit piezoelectric properties, that is, they deform mechanically when subjected to an electric field. Piezoelectricity also implies that the inverse is also true: when the crystal is forcibly deformed, an electric potential is developed between opposing faces of the crystal. As a result of this piezoelectric property, a thin plate of quartz provided with conducting electrodes vibrates mechanically when the electrodes are connected to an alternating voltage source. The vibrations, in turn, produce electrical signals which interact with the voltage source. The vibrations and electrical signals are a maximum at the natural mechanical resonant frequency of the crystal.

The equation for the motion of a vibrating body has already been written in Eq. (8-45),

$$m\frac{dx^2}{dt^2} + b\frac{dx}{dt} + kx = F(t) \tag{9-22}$$

where in the present case m is the mass of the crystal, b is the internal mechanical loss coefficient, and k is the elastic constant of the crystal. As has already been noted, this expression is identical in form with that for the current in a series resonant circuit, Eq. (3-31),

$$L\frac{d^2i}{dt^2} + R\frac{di}{dt} + \frac{1}{C}i = F(t) \tag{9-23}$$

Comparing Eqs. (9-22) and (9-23) it can be seen that the vibrating mass is analogous to inductance, mechanical losses are equivalent to resistance, and the elasticity corresponds to the reciprical of capacitance. Because of the identical form of the two equations, mechanical resonance is expected, and it is useful to define a mechanical Q factor by analogy with Eq. (3-57),

$$Q = \frac{\omega m}{b} \tag{9-24}$$

It turns out that the internal losses in quartz crystals are very small and that Q values reaching as high as 100,000 can be achieved. Furthermore, the elastic constants are such that resonant frequencies ranging from 10 kHz to several tens of megahertz are possible, depending upon the mechanical size and shape of the crystal.

The piezoelectric properties of quartz result in electrode potentials corresponding to the mechanical vibrations. This suggests that the electrical characteristics can be represented by an equivalent circuit. Comparing Eq. (9-22) and (9-23), the appropriate circuit is a series combination of a resistance, inductance, and capacitance. To this must be added the electric capacitance resulting from the parallel-plate capacitance of the electrodes with the crystal as a dielectric. Therefore, the complete equiva-

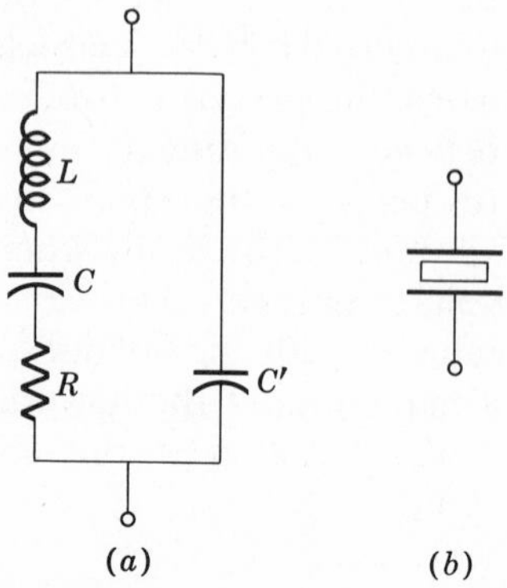

FIGURE 9-9 *(a) Equivalent circuit of a quartz crystal and (b) its circuit symbol.*

L C R C′ (a) (b)

lent circuit of a quartz crystal is the series-parallel combination shown in Fig. 9-9*a*. In this equivalent circuit L, C, and R are related to the properties of the quartz crystal and C' is the electrostatic capacitance of the electrodes. Appropriate values for a 90-kHz crystal are $L = 137$ H, $C = 0.0235$ pF, $R = 15{,}000\ \Omega$, and $C' = 3.5$ pF. The conventional circuit symbol for a crystal is a parallel-plate capacitor with the crystal between the plates, Fig. 9-9*b*.

The series-parallel equivalent circuit of a quartz crystal shows that there will be a series resonant frequency (zero impedance) and a parallel resonant frequency (infinite impedance). The frequency of series resonance is $\omega_S = 1/\sqrt{LC}$. The parallel resonance occurs when the reactance of C' equals the net inductive reactance of the combination of L and C, $\omega_P = \sqrt{1/L(1/C + 1/C')}$. Accordingly, the parallel resonant frequency is always greater than the series resonant frequency, although, since $C' \gg C$, the two are very close. The reactance is capacitive both below and above the resonant frequencies, Fig. 9-10.

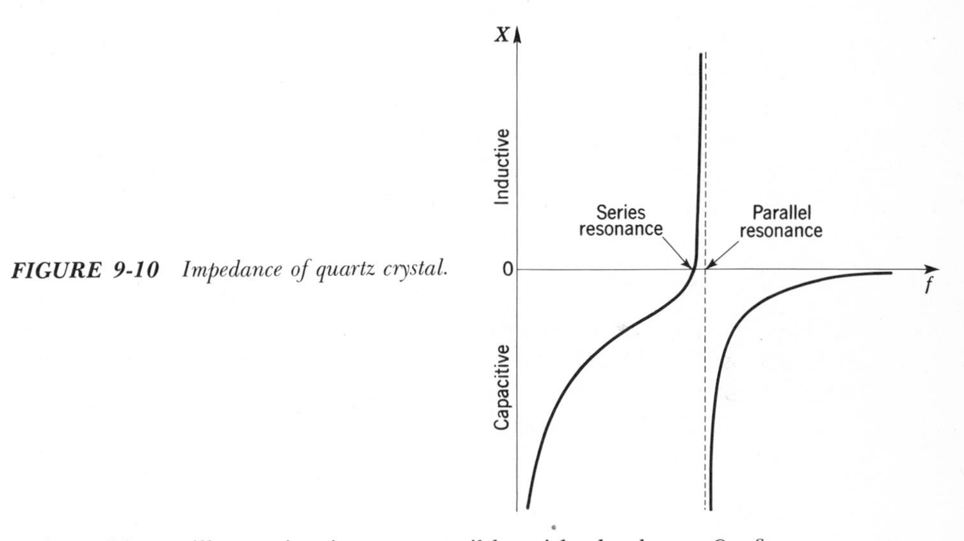

FIGURE 9-10 *Impedance of quartz crystal.*

Extremely stable oscillator circuits are possible with the large Q of a quartz resonator, and a variety of circuits have been designed. Either the series or parallel resonance frequency may be used, although parallel resonance is more common. Consider, for example, the *Pierce oscillator*, Fig. 9-11, in which the crystal is connected between base and collector. This circuit is identical to the Colpitts oscillator, Fig. 9-6, with the crystal replacing the resonant circuit. The feedback ratio is determined by the relative values of C_1 and C_2. The inductance *RFC* (for *radio-frequency choke*) in the collector lead is a useful way of applying collector potential without shorting the collector to ground at the signal frequency. It can be replaced by a 10,000-Ω resistor with some loss in circuit performance.

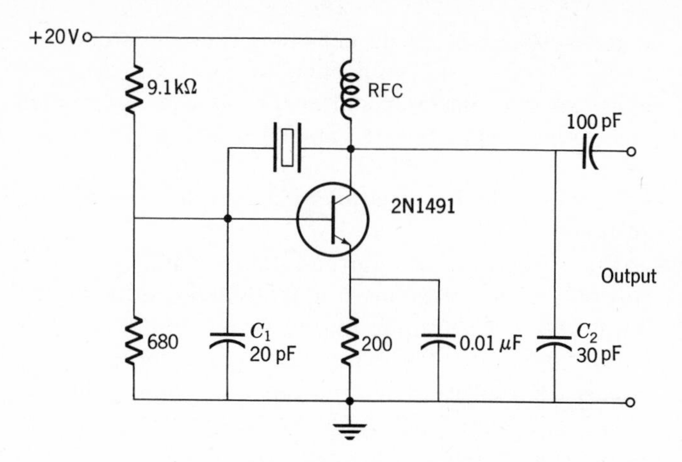

FIGURE 9-11 *Quartz-crystal oscillator.*

A parallel resonant *LC* combination can also be used with attendant gain in circuit performance. In the latter case the resonant circuit is merely a convenient collector load impedance and does not determine the oscillation frequency.

Quartz is almost universally used in crystal oscillators because it is hard, reasonably strong, and has a small temperature coefficient of expansion. Suitable orientation of the plane faces with respect to the crystalline structure makes the resonant frequency independent of temperature over a reasonable range. As a result, frequency stability of the order of 100 ppm can be achieved. Even greater accuracy is obtained by placing the crystal in a temperature-controlled oven and evacuating the crystal holder to reduce air damping forces on the vibrating crystal. It is also common practice to stabilize the temperature of the remainder of the circuit and to employ a regulated power supply for the oscillator. Amplifier stages are used to isolate the oscillator from variations in load. Such carefully designed crystal oscillators provide an extremely precise standard of time which can be accurate to 1 part in 100 million.

NEGATIVE RESISTANCE OSCILLATORS

Stability Analysis The current-voltage characteristics of several devices, most notably the tunnel diode and the unijunction transistor, exhibit *negative-resistance* properties. That is, over a portion of the characteristic the current decreases as the applied voltage increases. The physical mechanisms leading to negative resistance properties which are discussed in Chap. 5 prove to be very useful in a variety of oscillator circuits.

Analysis of negative resistance oscillation is best carried out by examining the complex impedance of the ac small-signal linear equivalent circuit of the oscillator. Consider, for example, the series resonant circuit in Fig. 9-12 where r represents the negative resistance of a tunnel diode

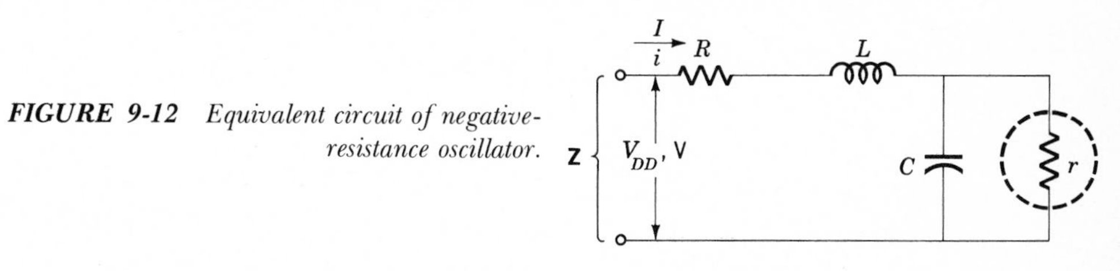

FIGURE 9-12 Equivalent circuit of negative-resistance oscillator.

at the operating point. The impedance of the circuit is

$$\mathbf{Z} = R + j\omega L + \frac{1}{1/r + j\omega C}$$

$$= R + j\omega L + \frac{r}{1 + j\omega rC}$$

$$= R + \frac{r}{1 + (\omega rC)^2} + j\omega\left[L - \frac{r^2 C}{1 + (\omega rC)^2}\right] \tag{9-25}$$

As usual, the circuit current is given by Ohm's law

$$i = \frac{v}{\mathbf{Z}} \tag{9-26}$$

Equation (9-26) predicts an infinite current under conditions for which the impedance vanishes. This situation is analogous to (9-2) and implies that the circuit oscillates. Analysis of circuit oscillation is effected, therefore, simply by searching for zeros of the impedance function.

The impedance vanishes when both the real and imaginary parts of Eq. (9-25) are equal to zero,

$$R + \frac{r}{1 + (\omega rC)^2} = 0 \tag{9-27}$$

and

$$L - \frac{r^2 C}{1 + (\omega rC)^2} = 0 \tag{9-28}$$

These relations are solved for the angular frequency

$$\omega^2 = -\left(\frac{1}{rC}\right)^2 \left(1 + \frac{r}{R}\right) \tag{9-29}$$

and

$$\omega^2 = \frac{1}{LC} - \left(\frac{1}{rC}\right)^2 = \omega_0{}^2 - \left(\frac{1}{rC}\right)^2 \tag{9-30}$$

where ω_o is the resonant frequency. Since both expressions must yield the same value for the oscillation frequency, equating (9-29) and (9-30),

$$-\left(\frac{1}{rC}\right)^2\left(1 + \frac{r}{R}\right) = \frac{1}{LC} - \left(\frac{1}{rC}\right)^2 \tag{9-31}$$

Solving for the value of negative resistance necessary to sustain oscillations,

$$r = -\frac{L}{RC} \tag{9-32}$$

Inserting this value into Eq. (9-30), the frequency of oscillation is

$$\omega^2 = \omega_0{}^2 - \left(\frac{R}{L}\right)^2 \tag{9-33}$$

This expression is very similar to the natural frequency of a series resonant circuit. In effect, the diode negative resistance cancels out the positive resistances in the circuit and the circuit rings continuously at its natural frequency.

According to Eq. (9-32) the diode resistance is negative, as expected. This means that the absolute value of r must be greater than the circuit resistance R in order for the frequency calculated from (9-29) to have a real value. Useful circuit operation is also achieved when $|r| < R$, however. The fact that (9-29) indicates that the oscillation frequency is imaginary under this condition suggests that the small-signal linear equivalent circuit analysis does not apply.

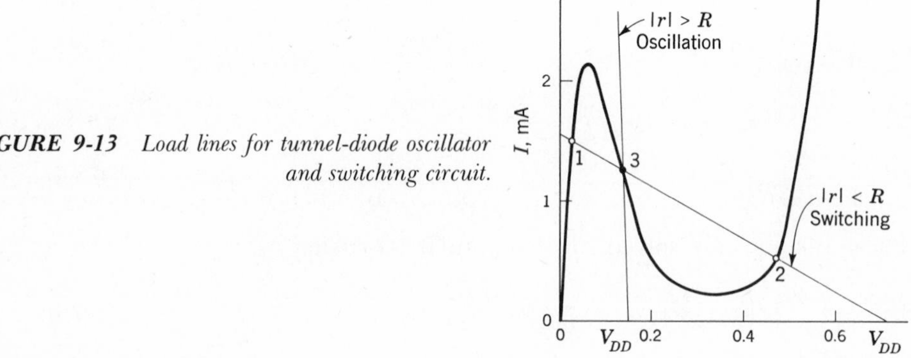

FIGURE 9-13 *Load lines for tunnel-diode oscillator and switching circuit.*

Circuit performance in the two cases is best compared with the aid of load lines plotted on the tunnel-diode current-voltage characteristic, Fig. 9-13. When $|r| > R$ the load line intersects the curve only at the operating point on the negative resistance portion and the circuit oscillates at its resonant frequency as analyzed above. By comparison, when $|r| < R$ three intersections occur. In this case both point 1 and point 2 are stable (nonoscillatory) circuit conditions. Point 3 is a condition of unstable equilibrium such that if the circuit is brought to this point the current immediately begins to change and finally settles at either point 1 or point 2. In effect the circuit has two stable states. It may be switched from one state to the other by temporary application of voltages. Suppose, for example, that the circuit is at point 1, and that a temporary increase in applied voltage increases the current to the peak. When the temporarily greater voltage is removed, the circuit traverses the *IV* characteristic and ends up at point 2. Similarly, the circuit may be switched from point 2 to point 1 by temporarily removing the applied voltage. When the voltage is reapplied, the current increases from the origin and stops at point 1. Such *switching* properties prove to be very useful in digital computors, for example.

The variety of operating conditions is not exhausted by the two cases specified by the dc load line. The oscillatory condition, $|r| > R$, actually has several different modes of operation which depend upon the ac properties of the circuit. Suppose, for example

$$r^2 < \frac{L}{C} \tag{9-34}$$

Then the frequency given by Eq. (9-30) is imaginary. This condition leads to nonsinusoidal, *relaxation* oscillations which are treated in a later section. Note that, since $|r| > R$, Eq. (9-34) leads to

$$|r|\,|r| < \frac{L}{C}$$

$$|r|\,R < \frac{L}{C} \tag{9-35}$$

$$|r| < \frac{L}{RC}$$

This condition for relaxation oscillations should be compared with the sinusoidal oscillating condition, Eq. (9-32).

Suppose, however, that while $|r| > R$ the circuit properties are also such that

$$|r| > \frac{L}{RC} \tag{9-36}$$

Then,

$$|r|\,R > \frac{L}{C}$$

but since $|r| > R$,

$$R^2 > \frac{L}{C}$$

or

$$\left(\frac{R}{L}\right)^2 > \frac{1}{LC} \tag{9-37}$$

This is the condition for complete damping, as may be noted by comparison with Eq. (2-98).

The relation between circuit conditions for sinusoidal oscillation, relaxation oscillation, and damped behavior may be appreciated from the current-voltage characteristic, Fig. 9-14, and the ac load lines associated

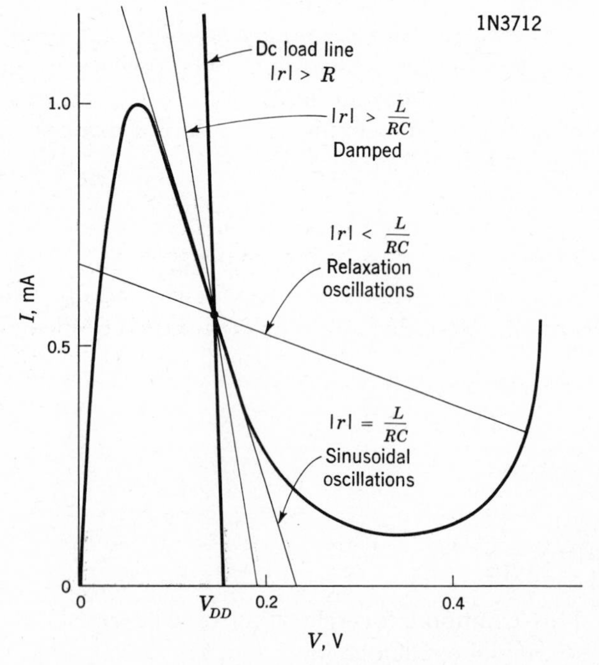

FIGURE 9-14 *Combination of dc and ac load lines determine oscillating mode of tunnel-diode oscillator.*

with (9-32), (9-35), and (9-36). According to this analysis sinusoidal oscillations at the natural frequency of the circuit are obtained when the magnitude of the negative resistance at the operating point is properly matched to circuit parameters, Eq. (9-32). A large value of inductance and small capacitance lead to relaxation oscillations, Eq. (9-35). On the other hand, large C and small L produce complete damping, Eq. (9-37). This latter condition is useful in experimentally tracing out the entire current-voltage characteristic which, clearly, cannot be accomplished with the circuit oscillating.

Tunnel-diode Oscillator The circuit of a practical tunnel-diode oscillator is shown in Fig. 9-15. In this configuration the diode is placed in series

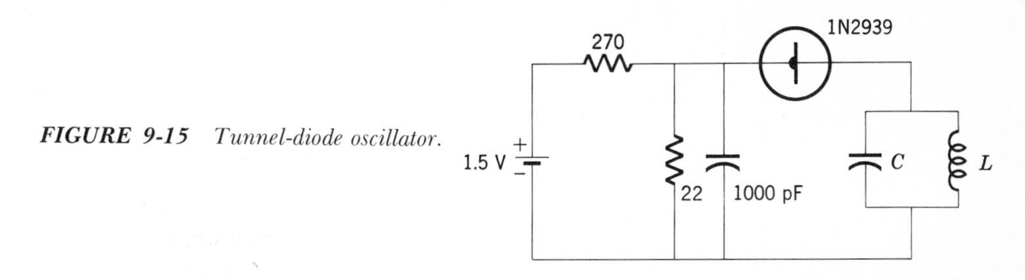

FIGURE 9-15 *Tunnel-diode oscillator.*

with the parallel resonant circuit but the principle of operation is identical to the analysis of the preceding section. The resistive voltage divider biases the diode to an operating point on the negative resistance portion of the current-voltage characteristic.

It may appear that the condition for oscillation, Eq. (9-32), is so restrictive that sinusoidal oscillations are difficult to attain. This is not the case since r is the slope of the IV characteristic at the operating point and therefore varies considerably with operating points ranging from the peak to the valley. Furthermore, the circuit tends to adjust itself to the proper value since the amplitude of oscillation changes to yield an average value of r over a complete cycle which is in agreement with Eq. (9-32).

Tunnel diodes are excellent high-frequency devices and oscillation frequencies as large as 10^{11} Hz have been achieved. Actually this virtue often proves troublesome in experimental circumstances since even tiny stray capacitances and inductances are sufficient to cause oscillations at very high frequencies.

According to the typical current voltage characteristics in Fig. 9-13 and Fig. 9-14, tunnel diodes can operate at very small power levels. Correspondingly, however, the maximum signal amplitude and ac power output of tunnel-diode oscillators are limited.

RELAXATION OSCILLATORS

Oscillator circuits considered to this point can be analyzed in terms of linear elements. Circuits which employ highly nonlinear active elements are termed *relaxation oscillators* for reasons which become clear in the following discussion. Very often relaxation oscillators are based upon negative resistance properties of the active element. Although, as discussed in the previous section, it is possible to generate sinusoidal waveforms by means of negative resistance characteristics, relaxation oscillators characteristically produce nonsinusoidal signals.

Sawtooth Generators Consider the unijunction relaxation-oscillator circuit in Fig. 9-16. The operation of this circuit can be understood after

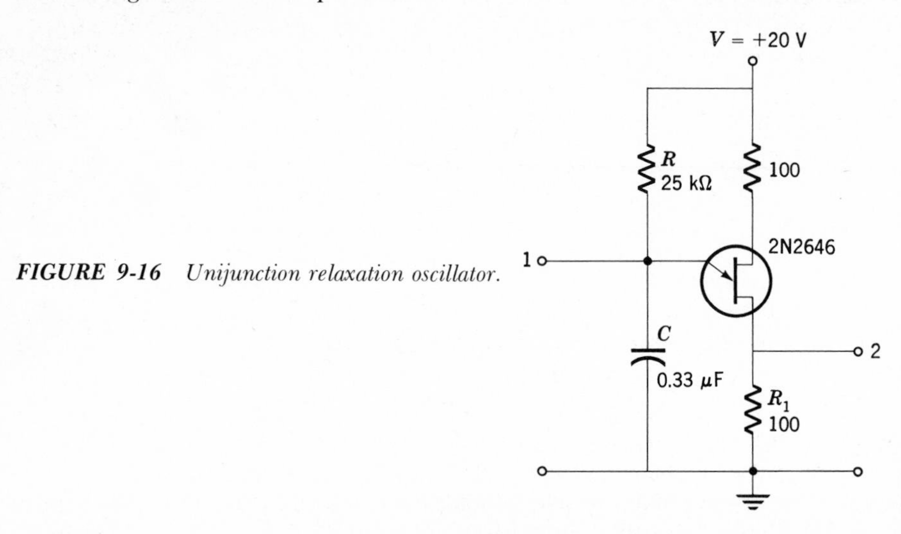

FIGURE 9-16 *Unijunction relaxation oscillator.*

examining the current-voltage characteristic of a unijunction transistor, Fig. 9-17. This characteristic exhibits a negative resistance region between the peak voltage V_p and the valley point V_v, as discussed in Chap. 5.

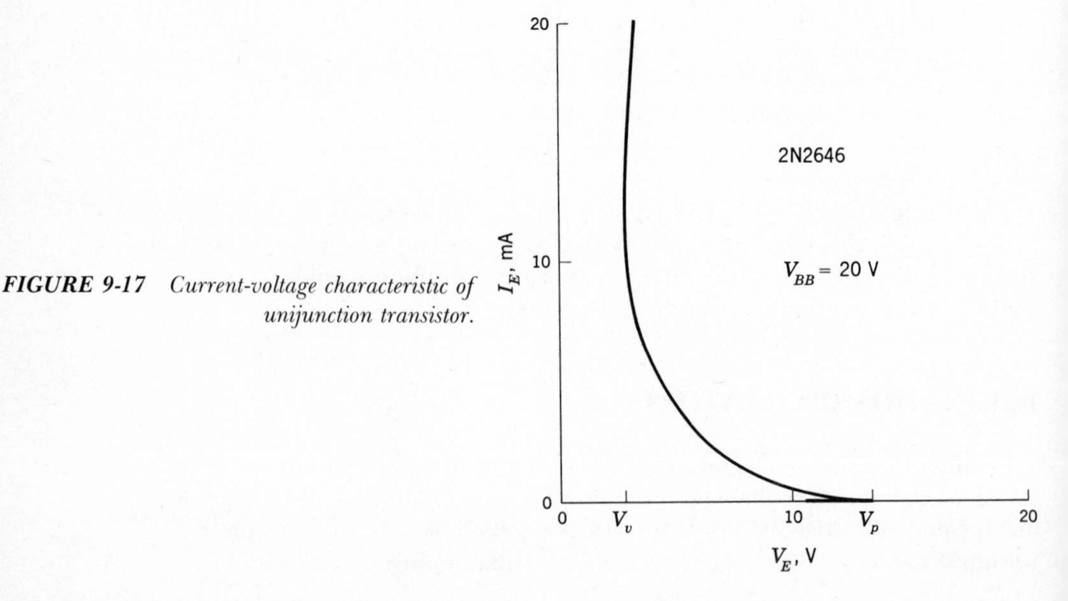

FIGURE 9-17 *Current-voltage characteristic of unijunction transistor.*

The unijunction oscillator operates as follows: Capacitor C charges through resistor R and the voltage across C increases exponentially until the potential V_p is attained. At this point the emitter junction becomes biased in the forward direction and the capacitor rapidly discharges through the emitter junction. When the capacitor potential drops to a low value (essentially equal to V_v) the emitter junction is again reverse-biased and the capacitor begins to recharge. The waveform at output terminal 1, Fig. 9-18*a*, is a series of RC charging curves with a peak-to-

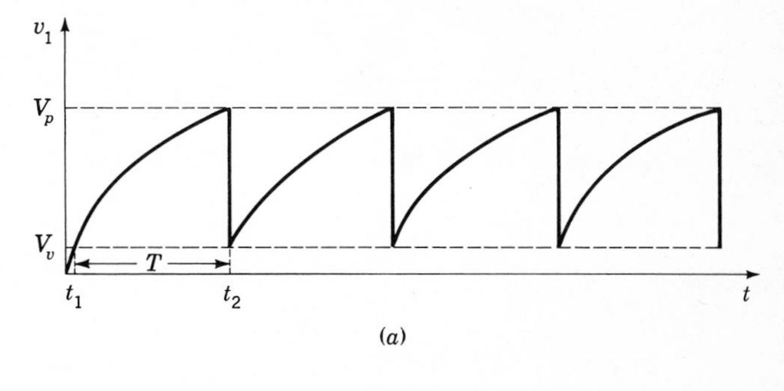

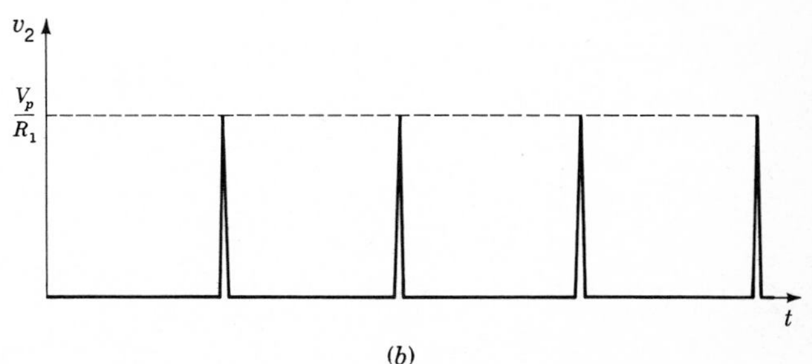

FIGURE 9-18 *Output waveforms of unijunction oscillator at (a) output terminal 1 and (b) output terminal 2.*

peak amplitude equal to $V_p - V_v$. Note that following every charging period the circuit "relaxes" back to the starting point. This is the origin of the terminology for this type of oscillator. The output waveform at terminal 2 is a series of sharp positive pulses, Fig. 9-18*b*.

The period of oscillation is found from the expression for the capacitor voltage in a simple RC circuit,

$$v = V(1 - e^{-t/RC})$$

The voltage reaches V_v at t_1,

$$V_v = V(1 - e^{-t_1/RC}) \tag{9-38}$$

Solving for t_1

$$t_1 = -RC \ln (1 - V_v/V) \tag{9-39}$$

Similarly, the voltage reaches V_p at t_2, so that

$$t_2 = -RC \ln (1 - V_p/V) \tag{9-40}$$

The period of the oscillation is just $t_2 - t_1$, or

$$T = RC \ln \frac{V - V_v}{V - V_p} \tag{9-41}$$

According to Eq. (9-41) the oscillation frequency depends upon the properties of the uninjunction transistor and magnitude of the supply voltage, as well as upon the circuit time constant. This equation assumes that the discharge time is zero, which is not true in practice. Because of the finite discharge time, uninjunction relaxation oscillators are limited to frequencies less than about 100 kHz.

The frequency stability of relaxation oscillators is inherently, and characteristically, very poor. The initiation of the discharge is a probabilistic phenomenon, which means that the discharge does not always start exactly at the potential V_p. Furthermore, the rate of change of the capacitor voltage is comparatively slow. This means that the exact instant at which the discharge is initiated varies slightly from cycle to cycle. In many applications, however, the poor frequency stability is an advantage since the oscillations can be *triggered,* or *synchronized* by an external signal.

Suppose, for example, that a synchronizing signal comprising a series of sharp voltage pulses is applied across the capacitor, Fig. 9-19. The sharp voltage rise of one of the pulses causes the capacitor voltage to

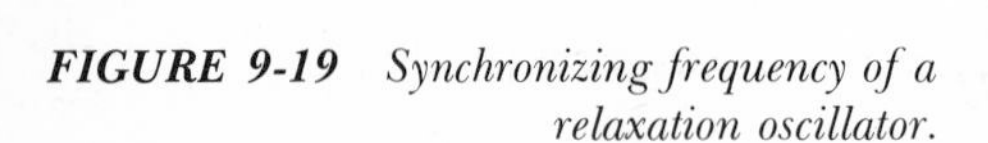

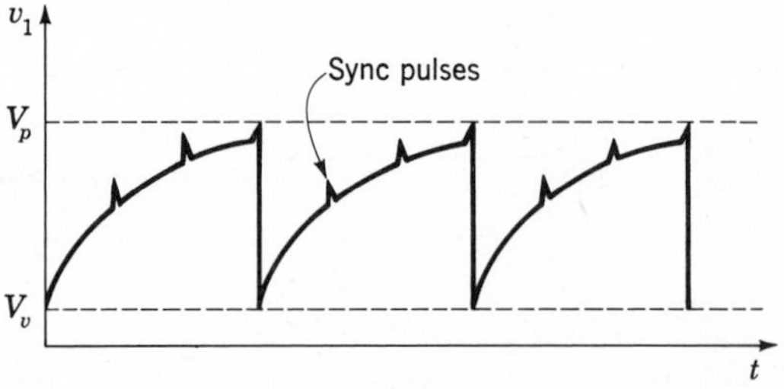

FIGURE 9-19 *Synchronizing frequency of a relaxation oscillator.*

exceed V_p at some point and the transistor is triggered into conduction. This repeats on successive cycles with the result that the frequency of oscillation is synchronized to the period of the pulse signal. Note that the period of the relaxation oscillator is a multiple or submultiple of the period of the pulses. In this way a relaxation oscillator acts as a frequency multiplier or divider. Frequency division by a factor as large as 10 or so is easily possible.

The free-running frequency of the relaxation oscillator must be reasonably close to a multiple or submultiple of the sync frequency. Since, however, the frequency stability of a relaxation oscillator is poor, the disparity may be as large as 20 percent or so without loss of synchronization. Triggering by a sinusoidal synchronizing voltage is also possible. Sharp pulses are more effective, however, since the time at which the breakdown voltage is attained is more definite.

The current-voltage characteristic of an SCR, Fig. 4-24, is similar in form to that of the unijunction transistor. However, the breakdown voltage is easily adjusted by varying the gate current. This means that the amplitude and frequency of oscillation can be changed in a relaxation oscillator. Furthermore, synchronizing signals can be introduced into the gate circuit to take advantage of the inherent amplification of the SCR.

A useful relaxation oscillator employing an SCR is shown in Fig. 9-20.

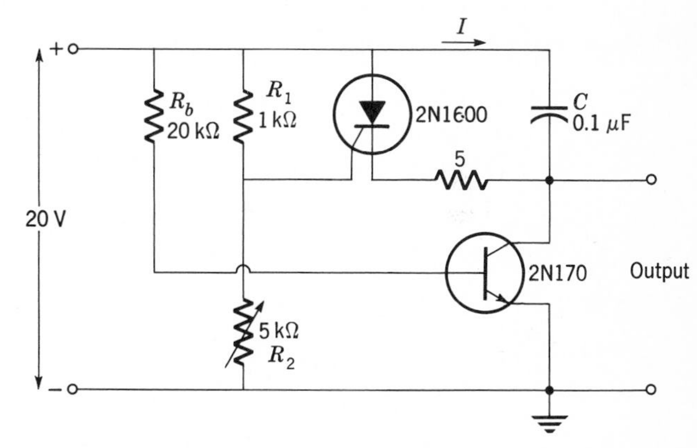

FIGURE 9-20 *Relaxation sawtooth generator.*

The gate potential is determined by the variable voltage divider comprising R_1 and R_2. Initially, when C is uncharged, the cathode of the SCR is at a potential of 20 V with respect to ground, so the gate is biased in the reverse direction. As the capacitor voltage increases, the SCR remains switched off until the cathode potential becomes slightly less positive than the gate. At this point the SCR switches on, the capacitor is discharged, and the cycle repeats. The purpose of the 5-Ω resistor in series with the cathode is to limit the discharge current to a safe value for the SCR. This is necessary because the internal resistance of the SCR in the on state is so small that the peak current can be destructive.

A linear sawtooth waveform is produced by the constant-current transistor. Since the charging current is constant, the capacitor voltage increases linearly with time according to

$$v = \frac{q}{C} = \frac{I}{C}t \tag{9-42}$$

where I is the current value and t is the time measured from when the capacitor voltage is zero.

A transistor is an excellent constant-current source since the collector current is essentially equal to the emitter current quite independent of the collector potential (refer to typical collector characteristics, Fig. 5-28). The emitter current is determined by the base bias resistor R_b. It can be made adjustable in order to vary the charging current and consequently the period of oscillation. The peak amplitude of the sawtooth is nearly as large as the dc supply voltage, which is advantageous in most applications.

As discussed in the previous section, relaxation oscillations are equally possible in the case of tunnel diodes. Because, however, the current-voltage characteristic of a tunnel diode exhibits an N-type negative resistance whereas both a unijunction transistor and an SCR have S-type negative resistances (refer to Chap. 5.), the operation of the oscillator is significantly different. Consider the tunnel-diode relaxation oscillator circuit in Fig. 9-21. Note that the reactive element in this case is an inductance whereas capacitors are used in both the unijunction and SCR oscillators. A large value of inductance and small capacitance (strays in the circuit of Fig. 9-21) lead to relaxation oscillations, as shown in the previous analysis.

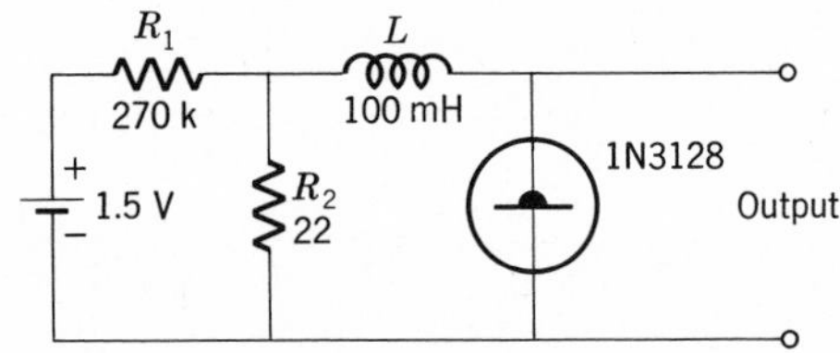

FIGURE 9-21 *Tunnel-diode relaxation oscillator.*

The operation of the tunnel-diode relaxation oscillator is as follows. Resistors R_1 and R_2 bias the diode to the negative-resistance region between the peak and valley potentials. The current initially increases from zero along segment 1, Fig. 9-22, at a rate determined by the series inductance L. When the current reaches I_p it cannot increase further and the diode potential suddenly changes to V_p'. Now the diode potential is greater

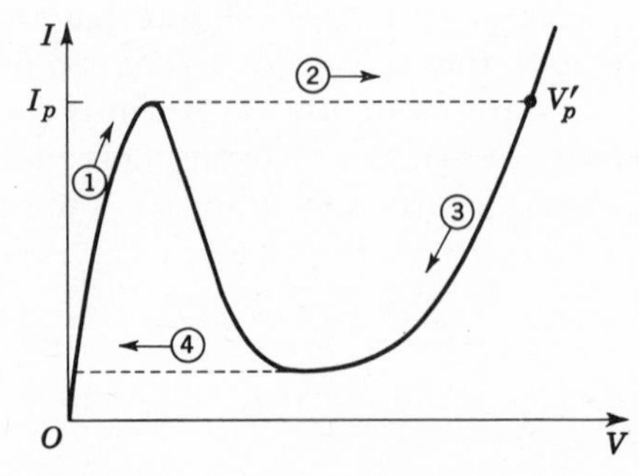

FIGURE 9-22 *Path of diode voltage in tunnel-diode relaxation oscillator.*

than the bias voltage, so the current decreases along segment 3, again at a rate fixed by the inductance. When the valley point is reached there is no way for the current to increase back up the curve, so the potential immediately drops back to nearly zero and the cycle repeats. Oscillation in the circuit depends upon the negative-resistance properties of the tunnel diode and the fact that current in an inductance cannot change rapidly, in close analogy with the capacitor and unijunction oscillator.

Blocking Oscillator A widely used circuit to generate pulse waveforms is the *blocking oscillator,* so called because the circuit cuts itself off, or *blocks,* periodically. The circuit of a blocking oscillator, Fig. 9-23*a*, looks similar

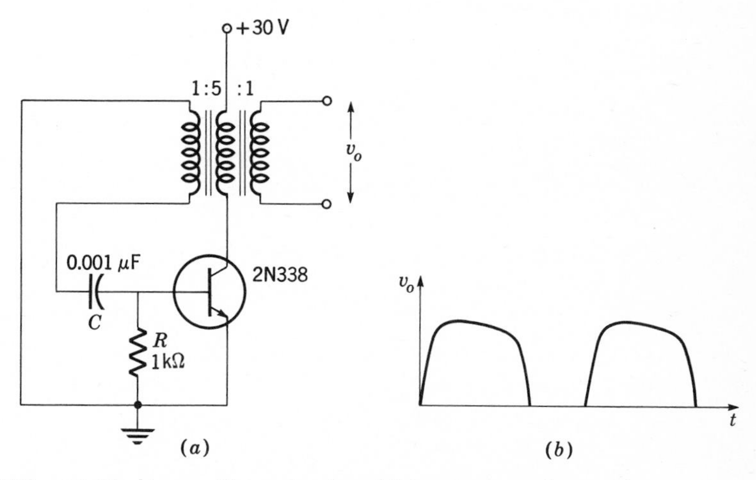

FIGURE 9-23 *(a) Blocking-oscillator circuit and (b) output waveform.*

to a feedback oscillator, except that the inductances are not tuned. In addition, the feedback ratio is very large, which means the feedback signal overdrives the transistor. Suppose the collector current is increasing; the positive feedback voltage further increases the collector current until the transistor is driven into saturation and the collector current no longer changes. The feedback is through a transformer so that when the rate of change of current stops, the feedback signal becomes zero. Simultaneously, the base capacitor, which was charged during the current pulse, discharges and gives a negative signal to the base. The collector current begins to decrease, again producing a feedback signal which rapidly decreases the collector current. The capacitor is charged and biases the transistor to cutoff. When the capacitor charge has drained away, the cycle repeats. The net result is a pulselike waveform, Fig. 9-23*b*.

In effect, the blocking-oscillator circuit tries to oscillate at its natural resonant frequency, but the strong feedback signal alternately drives the transistor into saturation and cutoff. The rise and decay times of the pulses are given essentially by the natural resonant frequency. This is purposely made high by minimizing stray capacitances so the rise and fall is rapid. The pulse length is determined by the effective inductance in the collector circuit, and the pulse-repetition frequency depends upon the RC time constant in the base circuit.

In some blocking oscillators dc base bias is used such that the transistor is cut off under quiescent conditions. A trigger voltage introduced across the base resistor starts the collector current and one output pulse is produced as described above. At the end of the output pulse the transistor remains cut off until another trigger signal arrives. Thus, the circuit produces an output pulse of specified amplitude and waveform whenever it is triggered into action. This form of blocking oscillator is often used as a pulse generator in digital-computer circuits.

Converters and Inverters It is often necessary to convert from a low dc supply voltage to one of a higher voltage (for example, in portable operation of vacuum-tube circuits from a low-voltage storage battery). Typically, vacuum tubes perform best at plate potentials above 100 V while a commercial storage battery, as in an automobile, may have a terminal voltage of only 12 V. There is no simple device analogous to a transformer for increasing dc potentials; instead *converters* perform this function. A converter is basically an oscillator that changes the dc input voltage to an ac signal which can be increased by transformer action and subsequently rectified to provide a higher dc potential. Most converters are designed to handle appreciable power; ratings from ten to several hundred watts are common.

Converter oscillators are designed to produce square waves since in this way the conversion efficiency is maximized. A square waveform means the oscillator transistors are alternately cut off or saturated and internal-power dissipation is a minimum. Furthermore, full-wave rectification of a square wave results in a dc output having minimum ripple, and this simplifies the filtering required. Oscillation frequencies are usually of the order of a few hundred to a thousand hertz; high frequencies are desirable to further simplify filtering, but transformer losses become significant if the frequency is too great.

Most often, converters are push-pull relaxation oscillators, Fig. 9-24, employing a special magnetic-core transformer to develop square-wave output. The magnetic core has a so-called *square-loop* hysteresis characteristic, Fig. 9-25. The essential feature is that magnetic flux in the core saturates suddenly as the current in the winding increases. In particular, the inductance of windings on the core is small when the core is saturated and large when the flux is changing along the vertical portion of the hys-

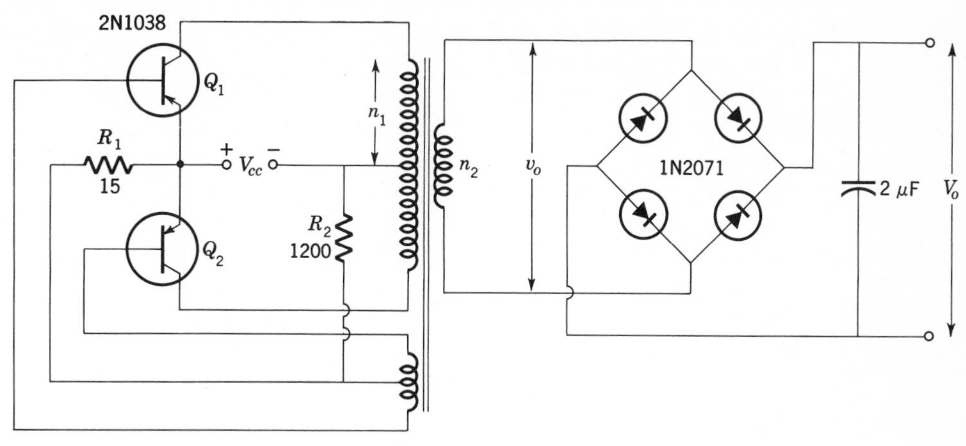

FIGURE 9-24 *15-W converter.*

teresis loop. This can be confirmed by referring to the definition of inductance, Eq. (2-52).

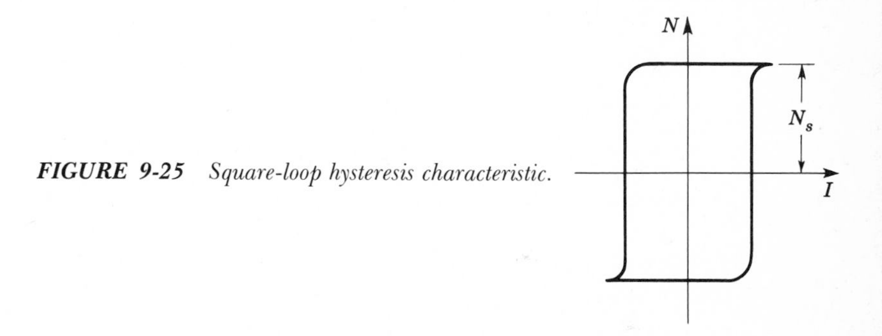

FIGURE 9-25 *Square-loop hysteresis characteristic.*

The performance of such a *saturated-core oscillator* can be described in the following way. Start by assuming that, say, transistor Q_1 in Fig. 9-24 is conducting. The feedback signal is very strong, so Q_1 is driven into saturation, while transistor Q_2 is cut off since the polarity of the feedback voltage applied to Q_2 is in the opposite direction. Since Q_1 is saturated, the emitter-collector voltage is small and the entire collector supply voltage V_{cc} is equal to the induced voltage across the transformer winding. Therefore, from Eqs. (2-52) and (2-53),

$$V_{cc} = L\frac{di}{dt} = n_1\frac{dN}{dt} \tag{9-43}$$

where L is the inductance of one-half of the primary winding consisting of n_1 turns, i is the current in the winding, and N is the magnetic flux in the core. According to Eq. (9-43) the rate of change of current and of magnetic flux are constant during this portion of the cycle, which corresponds to the vertical segment of the hysteresis loop in Fig. 9-25.

The current in Q_1 continues to increase at a constant rate until the core saturates. At this point the magnetic flux ceases to change and the feedback signal becomes zero. Since Q_1 no longer has an input signal, its collector current begins to decrease. This means that the core comes out of saturation and a positive feedback signal is generated which drives Q_1 into cutoff and at the same time forces Q_2 into saturation. The current in Q_2 now increases linearly according to Eq. (9-43) until the core is magnetically saturated in the reverse direction. At this point the cycle repeats. Thus, through the combined action of strong positive feedback and saturation of the transformer, each transistor is alternately fully conducting or cut off.

The current and flux waveform in the transformer primary is a triangular wave, Fig. 9-26*a*, since the current changes linearly on each half-

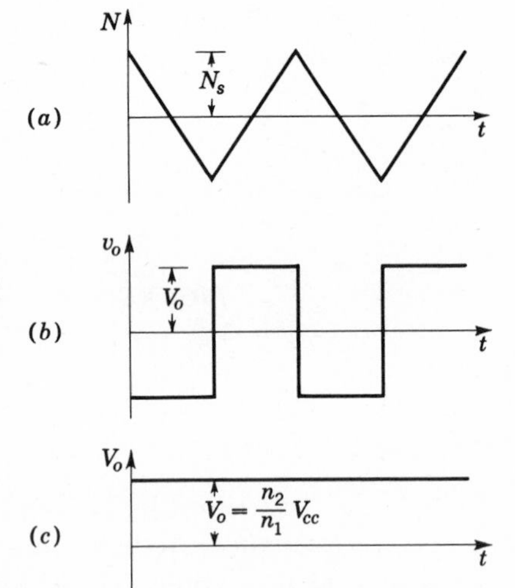

FIGURE 9-26 *Waveforms in push-pull converter. (a) Flux waveform in transformer, (b) secondary voltage, (c) rectified output voltage.*

cycle according to Eq. (9-43). The output voltage of the secondary is constant during each half-cycle, since

$$v_o = n_2 \frac{dN}{dt} = \frac{n_2}{n_1} V_{cc} \tag{9-44}$$

where Eq. (9-43) has been used, and n_2 is the number of turns on the secondary winding. The output waveform, Fig. 9-26*b*, is a square wave since the flux alternately increases and decreases. Full-wave rectification of the square wave results in a dc output voltage, Fig. 9-26*c*. Note that the magnitude of the output voltage is much larger than V_{cc} if $n_2 \gg n_1$, according to Eq. (9-44).

The frequency of oscillation is determined from Eq. (9-43) and the flux waveform, Fig. 9-26*a*,

$$V_{cc} = n_1 \frac{dN}{dt} = n_1 \frac{2N_s}{T/2} = 4n_1 N_s f \tag{9-45}$$

$$f = \frac{V_{cc}}{4n_1 N_s} \tag{9-46}$$

where N_s is the saturation flux density of the core. According to Eq. (9-46), the frequency is controlled primarily by the transformer characteristics.

It is usually desirable to include a filter capacitor, Fig. 9-24, in order to remove spurious switching transients associated with the transition of conduction from one transistor to the other. Additionally, it is common practice to connect a junction diode in parallel with each transistor to protect the transistors against voltage transients that might exceed the collector breakdown potential. Bias resistors R_1 and R_2 are useful in providing a small base bias to improve starting characteristics when the circuit is first turned on. Once oscillations begin, bias is unnecessary since the transistors are either fully conducting or cut off.

A saturated-core oscillator followed by a conventional power amplifier increases the power capability of the converter. Also, SCRs are used in place of transistors with some improvement in power efficiency. This is so because of the very low internal resistance characteristic of SCRs. Since the main function of the transistors is to provide a switching action, the use of SCRs in this connection is quite appropriate.

The unrectified output of a converter can be used as a source of alternating current to power ac devices such as motors or electronic circuits designed to be operated from the power mains. In this application the unit is called an *inverter* since it provides an ac output from a dc source, which is just the inverse of a conventional rectifier circuit. Inverters are also used to generate 400-Hz power suitable for use with servo systems because of the improved performance of servo systems at 400 Hz compared with the standard 60-Hz power-line frequency. Here, the 60-Hz power-main voltage is rectified and filtered to power the inverter, which supplies 400-Hz power. Thus the inverter may be looked upon as a power-frequency converter.

Usually it is desirable for the inverter frequency to be reasonably close to a specified value, generally either 60 Hz or 400 Hz, depending upon the application. A saturated-core oscillator is sufficiently stable for this use except for the change in frequency associated with any drop in the dc supply voltage under load, as given by Eq. (9-46). To compensate for the change in frequency from this effect, a voltage-regulator circuit is commonly included, as in the block diagram of a typical inverter, Fig.

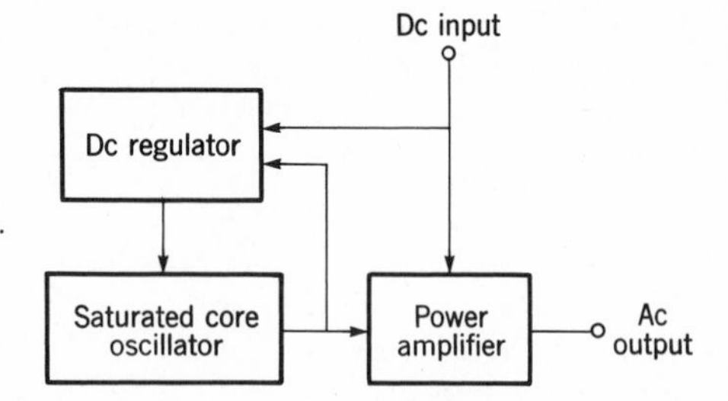

FIGURE 9-27 *Block diagram of dc-to-ac inverter.*

9-27. Most often, the regulator senses the output signal of the oscillator and adjusts the dc supply voltage to maintain the oscillator output at a constant level. This assures that the frequency of the oscillator remains substantially constant.

A conventional class B push-pull power amplifier is used to provide the necessary power output. If necessary, the power input to the amplifier can also be regulated to hold the output voltage constant. In many applications this refinement is not required, however. The inverter output is a square wave rather than sinusoidal and this may interfere with the proper operation of some devices. A low-pass *LC* output filter can be used to deliver a sinusoidal output waveform in this case.

SUGGESTIONS FOR FURTHER READING

Jacob Millman: "Vacuum-tube and Semiconductor Electronics," McGraw-Hill Book Company, New York, 1958.

Joseph A. Walston and John R. Miller (eds.): "Transistor Circuit Design," McGraw-Hill Book Company, New York, 1963.

"The Radio Amateur's Handbook" (published annually by the American Radio Relay League, West Hartford, Conn.).

EXERCISES

9-1 Derive Eq. (9-3) for the feedback network of a phase-shift oscillator. *Hint:* Use the technique of loop currents and solve for the current in the last resistor by determinants.

9-2 Draw the h-parameter equivalent circuit of the phase-shift oscillator, Fig. 9-1, and derive a relation analogous to Eq. (9-9) for the vacuum-tube oscillator.

Ans.: $\beta = h_{re} - [(\mathbf{Z}_i + h_{ie})/h_{fe}](h_{oe} + 1/R_L + 1/\mathbf{Z}_i)$

9-3 Using the results of Exercise 9-2, show that the circuit of Fig. 9-1 does oscillate. The h parameters of the 2N1414 are $h_{ie} = 1260\ \Omega$, $h_{re} = 3 \times 10^{-4}$, $h_{fe} = 60$, and $h_{oe} = 2.7 \times 10^{-5}$ mho. What is the frequency of oscillation?

Ans.: 1.28×10^{8} Hz

9-4 What is the value of R_1 in the Wien-bridge oscillator, Fig. 9-3, if the amplifier gain is 10? Repeat for gains of 100 and 1000.

Ans.: 1520 Ω, 2390 Ω, 2490 Ω

9-5 What factors are important in determining the low-frequency limit of a Wien-bridge oscillator? The high-frequency limit? Suggest improvements in the circuit of Fig. 9-3 that can increase the upper frequency limit.

9-6 By inspection, determine the voltage output of the power-oscillator circuit in Fig. 9-8.

Ans.: 19.8 V

9-7 Develop an expression analogous to Eq. (9-13) for the phase of the feedback voltage as a function of frequency in a tickler *LC* oscillator. Show that a resonant circuit with a large Q is desirable for maximum frequency stability.

9-8 Obtain an expression for the phase angle of the impedance of a quartz crystal near the parallel resonant frequency. Plot the phase angle as a function of frequency and compare with a plot of an *LC* resonant circuit having a Q of 200. Use the results of Exercise 9-7.

9-9 Find the period and amplitude of the output signal from the SCR relaxation oscillator, Fig. 9-20, if $h_{fe} = 20$ and the gate voltage required for turn-on is 3 V. Plot the output voltage waveform.

Ans.: 0.3 V, 17 V, 1.18×10^4 Hz, 6.67×10^5 Hz

9-10 Develop an expression analogous to Eq. (9-41) for the period of the tunnel-diode relaxation oscillator in Fig. 9-21. *Hint:* Approximate the positive resistance portions of the current-voltage characteristic by linear resistances R_a and R_b.

LABORATORY EXERCISES

9-A Wien Bridge Oscillator Resistance-capacitance oscillators are convenient generators of sinusoidal signals up to frequencies where tuned circuits must be used to reduce stray capacitance effects. The Wien bridge oscillator is widely used because of its simplicity and good performance. This experiment examines the properties of an integrated circuit Wien bridge oscillator, Fig. 9-28.

This oscillator uses the integrated-circuit operational amplifier diagramed in Fig. 8-18*a*. Compare the circuit with the more conventional form in Fig. 9-3. Verify that the phase shift is proper for oscillation. Explain the purpose of the frequency rolloff capacitor connected between terminals 9 and 10.

Determine the frequency and phase characteristic of the Wien bridge portion, ignoring the diode-resistor combination in parallel with R_2 for the moment. What is the frequency of oscillation and the signal loss that must be made up by the amplifier gain?

Isolate and analyze the amplitude-limiting circuit in parallel with R_2 by deter-

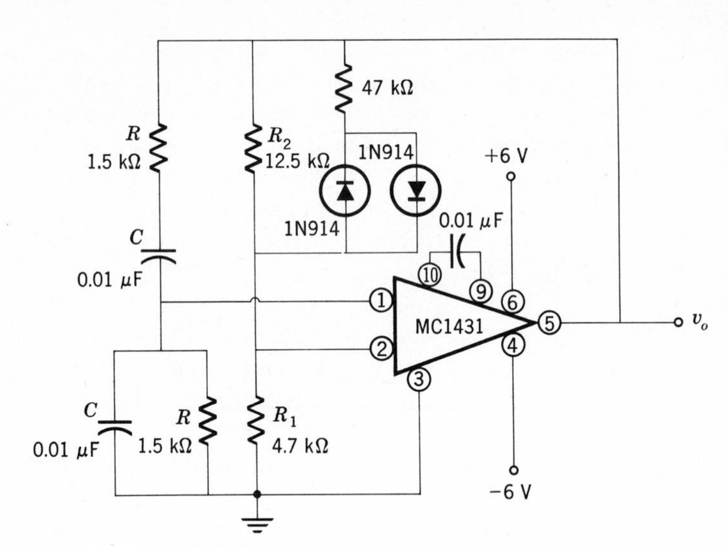

FIGURE 9-28 *Integrated-circuit Wien bridge oscillator.*

mining and plotting the signal voltage at terminal 2 and the reciprocal of the feedback ratio as a function of the output signal. What is the amplitude of the output signal?

9-B Unijunction Oscillator Relaxation oscillators can generate a wide variety of waveforms. One of the simplest and most versatile relaxation oscillators uses the unijunction transistor. The performance of the unijunction oscillator as a sawtooth generator, pulse generator, and frequency divider is examined in this experiment.

Consider the circuit illustrated in Fig. 9-16. Plot the load line on the current-voltage characteristic of the 2N2646 given in Fig. 9-17 and verify that the load line intersects the curve in the negative resistance region. Plot the waveform outputs at both terminal 1 and terminal 2. What is the frequency of oscillation and the peak amplitudes of the output signals? Be sure to observe the actual pulse waveform at terminal 2. What is the time constant associated with the decay of the pulse?

Introduce synchronizing pulses at terminal 1 that have an amplitude 10 percent as great as the peak sawtooth voltage and a frequency approximately five times that of the free-running oscillator frequency. Using the waveform plot prepared above, estimate the frequency range of the synchronizing pulses over which the oscillator will synchronize at a frequency one-fifth of the pulse frequency.

The relaxation oscillator is also an interval timer if the period of oscillation is made very long and if a switch is included in series with R to initiate the timing interval. The timing accuracy depends primarily upon the reverse current at the emitter junction because of its temperature dependance. Assuming that the capacitor charging current at breakdown should be 10 times the junction reverse current, what is the largest value of R possible and correspondingly, the longest interval that the circuit can indicate reliably? Assume the reverse current is 10^{-8}A.

Describe qualitatively the mode of operation if the voltage supply is 10 V, that is, less than V_p, so that one intersection is at a very small current, another is on the negative resistance portion, and the third is beyond the valley point. Note the analogy with the tunnel-diode oscillator.

chapter ten

ANALOG MEASUREMENTS

Electrical techniques are used to measure a wide variety of physical phenomena in both laboratory research and industrial control applications. Indeed, it is modern measurement practice to develop an electrical signal analogous to the phenomena of interest at the earliest opportunity so that subsequent processing can be carried out electronically. The great versatility of electronic circuits to amplify, modify, record, and detect electric signals is the principal reason they enjoy such widespread use. Circuits used to measure and process such analog signals and circuits for control purposes are investigated in this chapter. In addition, techniques for transporting signals from one point to another and recording them for subsequent analysis are considered.

VOLTAGE REGULATORS

Special circuits are widely used to stabilize the output voltage of a rectifier power supply to a greater extent than is possible with the simple zener diode circuit described in Chap. 4. Such a *voltage regulator* is an important and useful example of measurement and control with electronic circuits. The circuit measures the output voltage, compares it with the desired value, and adjusts conditions so that the difference between the two is zero. In this way control over the dc output voltage is maintained in the face of changes in load and also in ac line voltage.

Series Regulator The output voltage of a power supply is conveniently controlled by introducing a power transistor in series with the rectifier-filter and the load, Fig. 10-1. It is useful to analyze this circuit as a feedback

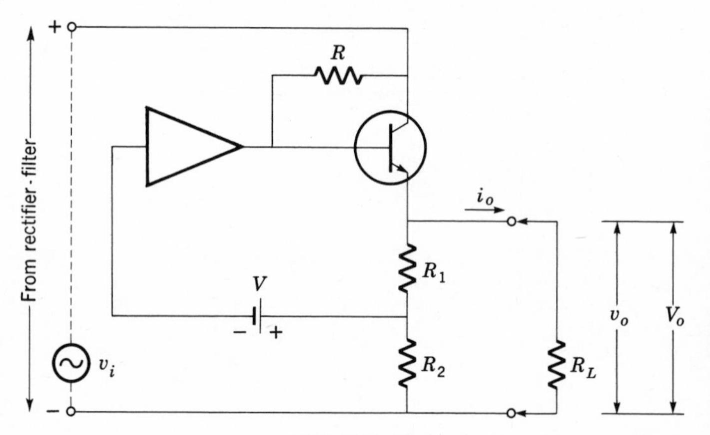

FIGURE 10-1 *Series voltage regulator.*

amplifier in which the reference voltage V is the input voltage and V_o is the output. Since the feedback ratio is $-R_2/(R_1 + R_2)$, the output voltage is controlled by the amplifier and power transistor to be, from Eq. (8-5),

$$V_o = -\frac{R_1 + R_2}{R_2} V$$

In effect, the voltage regulator compares a fraction of the dc output voltage with the reference voltage and adjusts the control transistor to maintain V_o constant.

The stabilizing effect of the regulator is best determined by analyzing the control transistor separately from the dc amplifier. Variations in the supply voltage are assumed to arise from an ac generator v_i in Fig. 10-1 and result in a variation v_o in the output voltage. The control transistor is essentially an emitter-follower amplifier and the appropriate equivalent

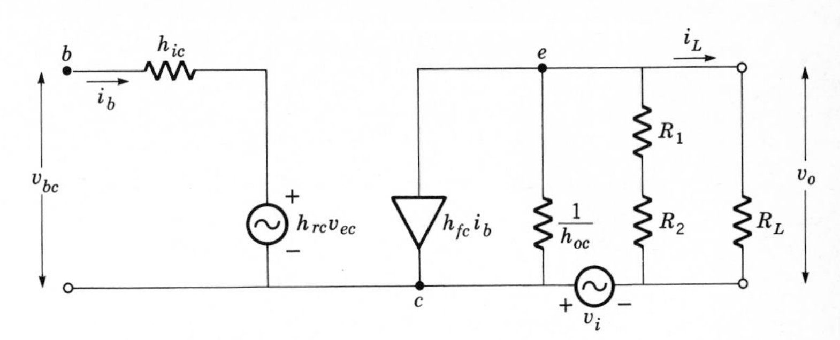

***FIGURE 10-2** Equivalent circuit of control transistor in Fig. 10-1.*

circuit is shown in Fig. 10-2. Analysis begins by writing an expression for the output current. Considering the currents at point e and neglecting the minor current drain in the feedback divider R_1R_2,

$$h_{fc}i_b + i_L + \frac{v_o - v_i}{1/h_{oc}} = 0 \tag{10-1}$$

where the third term is the current in $1/h_{oc}$. The base current is found by considering the voltages in the input circuit,

$$i_b = \frac{v_{bc} - h_{rc}v_{ec}}{h_{ic}} = -\frac{a\beta v_o + v_i + h_{rc}v_{ec}}{h_{ic}} \tag{10-2}$$

where the input voltage v_{bc} is v_i less the amplified feedback voltage $-a\beta v_o$, as may be best seen from Fig. 10-1. Finally, v_{ec} is determined by writing Kirchhoff's rule around the output circuit,

$$v_{ec} - v_o + v_i = 0 \tag{10-3}$$

Equations (10-2) and (10-3) are substituted into Eq. (10-1),

$$i_L = v_i\left[h_{oc} + \frac{h_{fc}(1 - h_{rc})}{h_{ic}}\right] - v_o\left[h_{oc} - \frac{h_{fc}(h_{rc} + a)}{h_{ic}}\right]$$

Solving for the output voltage,

$$v_o = v_i\frac{h_{ic}h_{oc} + h_{fc}(1 - h_{rc})}{h_{ic}h_{oc} - h_{fc}(h_{rc} + a\beta)} - i_L\frac{h_{ic}}{h_{ic}h_{oc} - h_{fc}(h_{rc} + a\beta)} \tag{10-4}$$

This expression has the form of the Thévenin equivalent of the regulated power supply in that the first term is the internal voltage generator and the second term is the voltage drop resulting from the load current in the equivalent internal resistance.

Before considering Eq. (10-4) further, it is useful to simplify the expression by considering the relative magnitudes of the h parameters. When this is done, Eq. (10-4) reduces to

$$v_o = v_i \frac{h_{ic}}{r_c(1 + a\beta)} - i_L \frac{h_{ic}(1 - \alpha)}{1 + a\beta} \tag{10-5}$$

which is sufficiently accurate for all practical purposes. According to Eq. (10.5) input voltage variations are reduced by a large factor through the feedback action of the regulator. Typical values of the power-transistor parameters in Eg. (10-5) are $h_{ic} = 1000\ \Omega$, $r_c = 10{,}000\ \Omega$, $a = 50$, and $\beta = 0.3$, so that input voltage changes are reduced by a factor of 160 at the output terminals.

Additionally, the effective internal impedance of the regulated power supply, which is the coefficient of i_L in Eq. (10-5), is very small. Taking $1 - \alpha = 0.05$, the internal resistance is only $3.1\ \Omega$. This means that changes in load current cause only minor changes in the regulated supply voltage. It is apparent that this simple regulator is effective against both input-voltage and output-current variations.

The control transistor in this regulator circuit must be capable of withstanding the full supply voltage without collector breakdown. Higher voltage regulators conventionally employ vacuum tubes as the control element because of their higher operating voltages. The circuit is identical to Fig. 10-1 with the transistor replaced by a triode or pentode. Analysis of the vacuum-tube regulator proceeds the same as that for the transistor version and the expression corresponding to Eq. (10-5) for the variation in output voltage is

$$v_o = \frac{v_i}{1 + \mu(1 + a\beta)} - i_L \frac{r_p}{1 + \mu(1 + a\beta)} \tag{10-6}$$

Typical values for the control tube in a regulator are $\mu = 10$ and $r_p = 2000\ \Omega$. Therefore, input voltage variations are reduced by a factor of 160 and the equivalent internal resistance is $12.5\ \Omega$, if the amplifier gain and feedback ratio are the same as in the transistor circuit.

Note that the factor by which input voltage variations are reduced at the output terminals applies to ripple voltages as well as to other voltage changes. Therefore output voltage ripple is very small in a regulated power supply. Both the reduction factor and the effective internal resistance can be improved by increasing the gain of the dc amplifier, according to Eqs. (10-5) and (10-6). In many critical applications the gain is made large enough that residual variations in the regulated voltage reflect the stability of the reference potential.

The series control transistor or tube must be capable of dissipating the heat generated by the entire load current without overheating. Power tubes and transistors are often used. It is feasible to employ two or more identical units connected in parallel, if this is necessary to carry the maximum load current.

Shunt Regulator Although the series regulator is most popular, it is also possible to connect the control transistor in parallel with the load, much as in the case of a simple zener-diode regulator, Fig. 10-3. The current

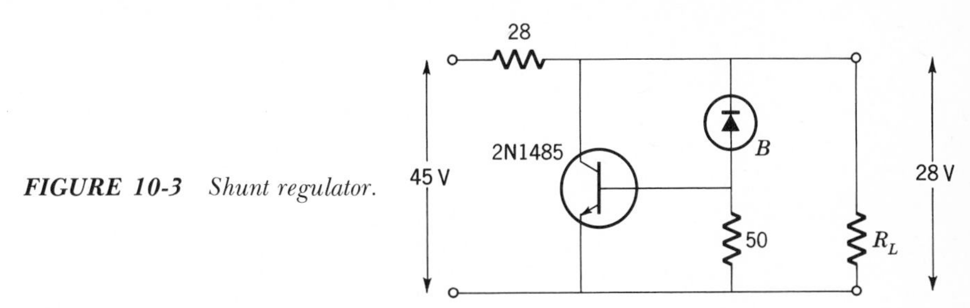

FIGURE 10-3 *Shunt regulator.*

through the transistor is controlled so that the voltage drop across the series resistor is the same independent of load-current changes. Effective voltage stabilization is obtained by such a *shunt regulator,* but the circuit suffers from the additional power loss in the series resistor. On the other hand, it is only necessary for the control transistor to carry a fraction of the full load current, which is a considerable advantage in high-current power supplies.

Practical Circuit A complete regulated power supply is illustrated in Fig. 10-4. The dc amplifier is a single grounded-emitter transistor stage

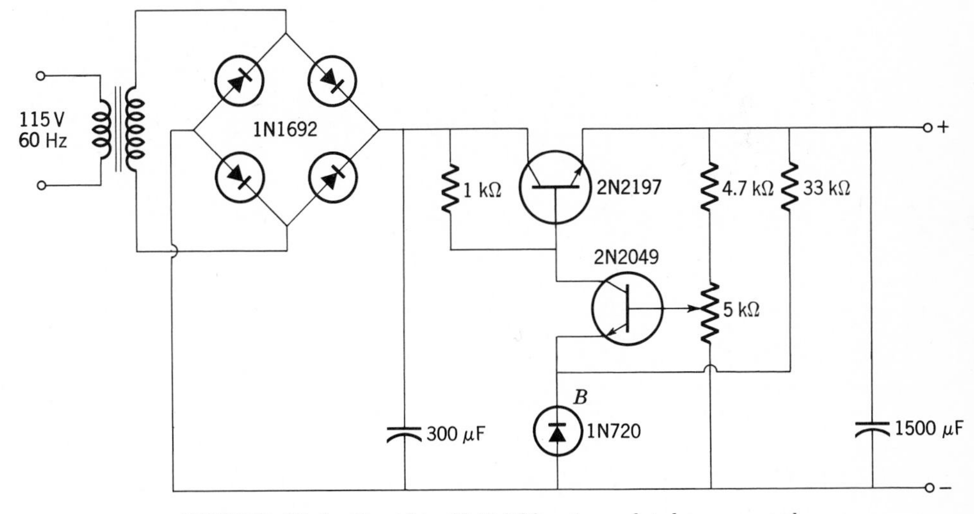

FIGURE 10-4 *Complete 40-V 500-mA regulated power supply.*

and the reference voltage is provided by an 18-V zener diode. The zener diode is placed in the emitter circuit of the amplifier, rather than in series with the feedback signal, so that one terminal of the diode can be grounded. The 33-kΩ resistor provides reverse potential for the diode. Note that the feedback ratio is adjustable. This enables the output voltage to be set at any desired value, within limits dictated by the operating conditions of the transistors. In this circuit, the output voltage can be adjusted over the range 40 to 50 V, while maintaining good regulation. The 1500-μF capacitor across the output terminals provides a very low effective internal impedance for ac signals.

Very precisely regulated voltages are obtained with more elaborate versions of this circuit. The gain is increased by using a multistage dc amplifier. In particular, operational amplifiers prove convenient because of their large gain and dual inputs, which facilitate comparing the output voltage with the reference voltage. In addition, several regulator transistors or tubes may be connected in parallel to enhance the maximum current capabilities of the regulator. In such circuits it is not difficult to control the output voltage within 0.1 percent over the entire range of load current. In fact, stabilities as great as one part in 10^5 can be achieved in special applications.

SERVOS

It is often useful to control mechanical position or motion in accordance with an electric signal. To assure that the desired motion takes place, a voltage feedback signal corresponding to the mechanical motion is developed. This voltage is compared with the input signal. When the two are equal, mechanical motion ceases since the mechanism has responded to the actuating signal. Feedback systems incorporating a mechanical link in the feedback network are called *servomechanisms*, or *servos*, from the Latin word for "slave," since the mechanical motion is the slave of the input signal. Servos are widely used in control applications, from laboratory chart recorders to automatic navigation systems.

Mechanical Feedback Consider the servo system sketched in block-diagram form in Fig. 10-5. The amplifier drives a mechanical actuator as, for example, a dc motor. Suppose the motor moves a mechanical arm to a position x and that the feedback voltage is proportional to the position of the arm and equal to kx. So long as the *error signal*, which is the difference between the input and feedback voltages, is not equal to zero, the amplified error signal continues to drive the motor. When the feedback signal equals the actuating signal, the error is zero, and motion ceases. Thus

$$v_i - v_f = v_i - kx_0 = 0 \qquad (10\text{-}7)$$

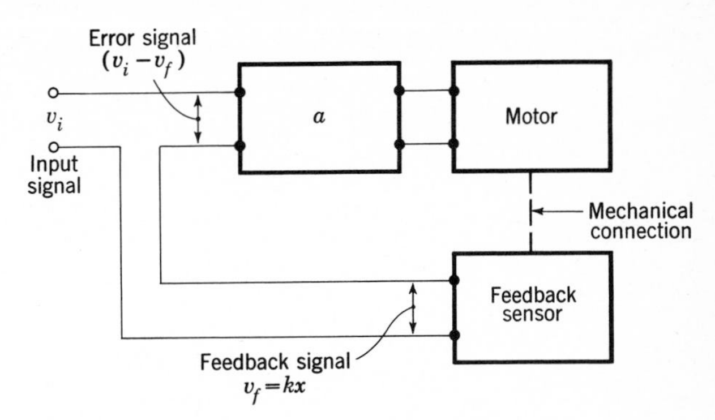

FIGURE 10-5 *Block diagram of servo system.*

so that

$$x_0 = \frac{v_i}{k} \tag{10-8}$$

According to Eq. (10-8), the equilibrium position of the mechanical motion x_0 is directly proportional to the input signal.

The servo cannot respond instantaneously to rapid changes in the input signal because of mechanical inertia and friction. According to the above discussion, the motive force from the motor is proportional to the error signal. This force is opposed by the inertial and frictional forces (which are assumed to be proportional to the velocity), so that

$$Ka(v_i - v_f) = m\frac{d^2x}{dt^2} + b\frac{dx}{dt} \tag{10-9}$$

where K is the motor constant, a is the amplifier gain, m is the mass of the moving parts, and b is a friction constant. Inserting Eq. (10-8), the differential equation for the response of the system is

$$m\frac{d^2x}{dt^2} + b\frac{dx}{dt} + (Kak)x = (Kak)x_0 \tag{10-10}$$

This equation can be solved by the methods used in Chap. 2 or by analog-computer methods (compare Eq. 8-45). In fact, the equation is identical in form to the one for an *RLC* circuit, Eq. (2-93). Based on the discussion in Chap. 2, a ringing effect is anticipated and the solutions are of two general kinds. If the frictional force is small compared with the driving force, a damped oscillation results, Fig. 10-6. On the other hand, if the frictional force is large, the system is *overdamped* and requires an excessively long time to reach the equilibrium position. Most servo systems are designed so they are slightly underdamped because this minimizes the response time. This results in a small *overshoot*, Fig. 10-6. If desirable, additional damping

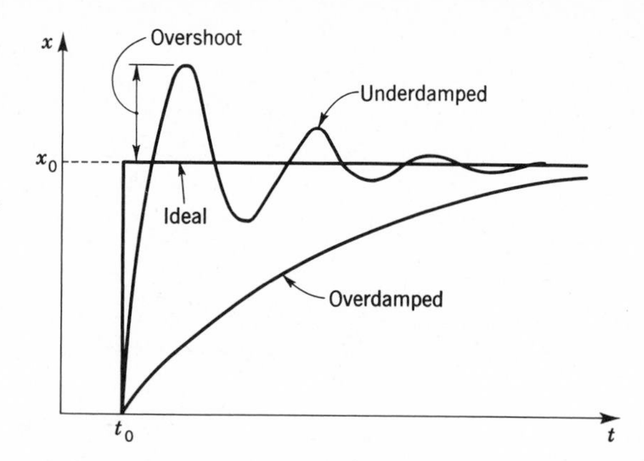

FIGURE 10-6 *Characteristic responses of servo to sudden input signal.*

may be introduced by including an RC filter in the input circuit. In this way the servo is never subjected to a signal change faster than the time constant of the filter.

The Recording Potentiometer A common laboratory instrument based on servo principles is the *recording potentiometer.* This is a self-balancing dc potentiometer in which the balance position is marked on a moving paper strip or chart producing a continuous record of the input voltage. The heart of the system, Fig. 10-7, is the slide wire R_s, in which there is a cali-

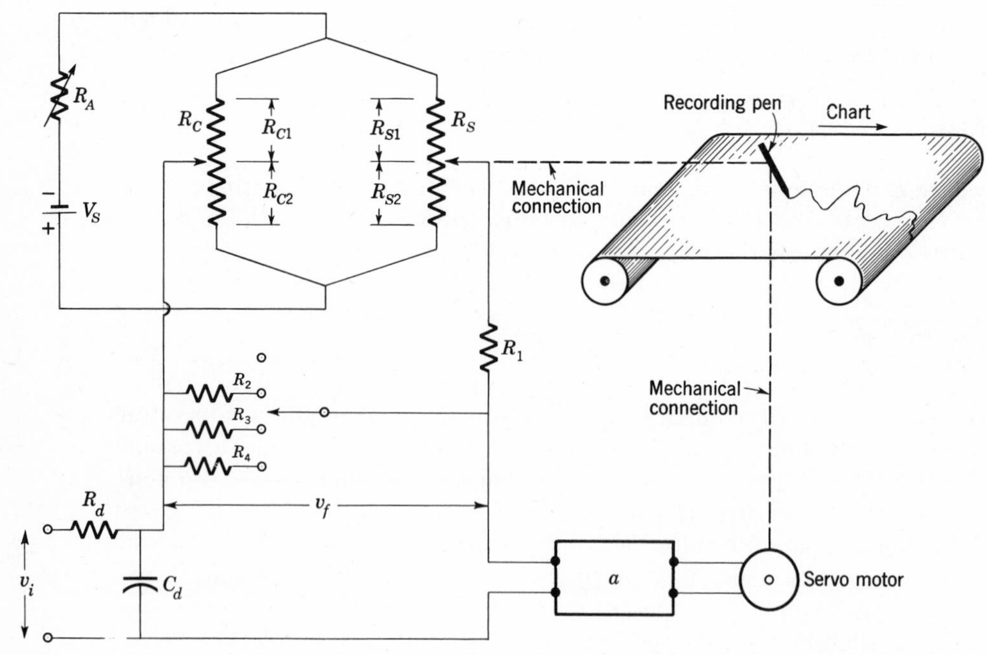

FIGURE 10-7 *Recording-potentiometer servo.*

brated current supplied by the battery V_s and variable resistor R_A. The current is standardized by comparison with a standard battery, as described in Chap. 1. If less accuracy is satisfactory, V_s may be a mercury battery which supplies a relatively constant current.

The potentiometer in parallel with the slide-wire resistance forms a bridge circuit which is balanced when, from Eq. (1-69),

$$\frac{R_{C1}}{R_{C2}} = \frac{R_{S1}}{R_{S2}} \tag{10-11}$$

Under this condition the feedback voltage v_f is zero. In the absence of an input signal, the adjustable contact on the slide wire is moved by the servomotor until balance is achieved. The recording pen mechanically attached to the slide-wire contact continuously records the balance position on the chart. This means that the zero position ($v_i = 0$) may be placed anywhere on the chart by adjusting R_C.

When an input voltage is applied, an initial signal $v_i - v_f$ is fed to the amplifier. The servomotor drives the slide-wire contact to a new equilibrium position where the unbalanced bridge voltage equals the input voltage, and $v_i - v_f = 0$. The displacement of the sliding contact is a linear function of the input voltage, as can be shown in the following way. Using Eq. (1-81), the unbalance voltage from the bridge is

$$v_f = V\left(\frac{R_{C1}}{R_{C1} + R_{C2}} - \frac{R_{S1}}{R_{S1} + R_{S2}}\right) \tag{10-12}$$

where V is the voltage across the bridge. If the slide wire is uniform, and the zero position, $x = 0$, is taken to be at the center,

$$\begin{aligned} R_{S1} &= \frac{R_S}{2} - mx \\ R_{S2} &= \frac{R_S}{2} + mx \end{aligned} \tag{10-13}$$

where m is a constant of the slide-wire. Introducing Eq. (10-13) into Eq. (10-12) and noting that $R_{C1} = R_{C2}$ since the zero position is at the center,

$$\begin{aligned} v_f &= V\left(\frac{1}{2} - \frac{R_S/2 - mx}{R_S/2 - mx + R_S/2 + mx}\right) \\ &= V\left(\frac{1}{2} - \frac{1}{2} + \frac{mx}{R_S}\right) = \frac{V}{R_S}mx \end{aligned} \tag{10-14}$$

Therefore, according to Eq. (10-8),

$$x_0 = \frac{1}{mI_S} v_i \tag{10-15}$$

where $I_S = V/R_S$ is the standardized current in the slide wire. Equation (10-15) shows that the deflection of the slider, and hence of the recording pen, is proportional to the input signal. It is common practice to select m and the standard current I_S such that the full chart width is an integral voltage such as 100 mV.

Resistors R_1 through R_4 in Fig. 10-7 constitute a voltage divider for changing the sensitivity of the potentiometer. When the divider is in the circuit, only a fraction of the unbalance voltage is available as the feedback signal and the sensitivity is increased correspondingly. The combination of R_d and C_d is a damping filter at the input to optimize the response time of the recorder.

The recording potentiometer is a very versatile instrument in the laboratory. It combines high sensitivity for dc signals and the inherently high input impedance of a potentiometer circuit with automatic operation and a permanent record. It is widely used as part of other instruments, such as optical spectrometers, x-ray diffractometers, and gas chromatographs, or as a separate unit.

Servo Amplifiers Amplifiers used in servo systems are of several types, depending upon the application and the kind of servomotor used. Simple systems may use a dc motor, so that a dc amplifier and power amplifier are required. This approach is subject to the drift problems of dc amplifiers and requires that the entire power of the servomotor come from the amplifier. Most often, two-phase ac motors are used in which one phase is supplied by the ac line voltage. An ac power amplifier controlled by the servo amplifier feeds the second-phase winding of the motor. Thus only a fraction of the total power of the motor is required from the amplifier.

In this case a chopper amplifier develops an ac voltage corresponding to the dc input signal. This is followed by a conventional ac amplifier and a push-pull class B power amplifier to drive the motor. Chopping frequencies of 60 and 400 Hz are standard, although the latter is preferred as it permits a faster response.

Since the amplifier need only handle signals of the chopping frequency, it is permissible to employ a rectified but unfiltered collector supply voltage of the chopping frequency, Fig. 10-8. The efficiency of the power amplifier is greater than that of a conventional class B amplifier using a dc collector supply voltage. The reason for this is that the pulsating collector supply reduces the average power dissipation in each transistor. It is conventional to drive the servomotor directly without an output transformer, although this requires a center-tapped motor winding. The other phase winding of the motor is fed directly from the ac line through a phase-shifting capacitor C_2. This provides a 90° phase shift between the currents in the two windings, as required by a two-phase motor. Capacitor C_1 resonates the

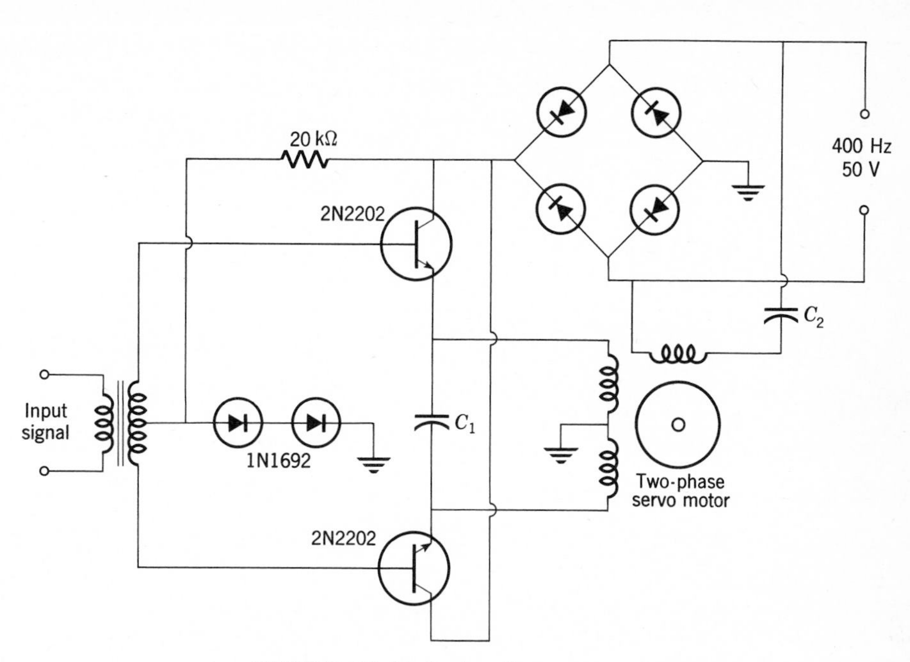

FIGURE 10-8 *Push-pull servo power amplifier.*

motor winding at the chopping frequency so that the motor presents a resistive load to the amplifier. The circuit of Fig. 10-8 is a grounded-collector push-pull amplifier using diodes for temperature compensation of the base bias, as described in Chap. 7. Grounded-emitter and common-base circuits are also used for servo amplifiers.

The increased efficiency accompanying an unfiltered collector supply is analyzed in the following way. Each transistor is considered a switch with internal resistance R_T which depends upon the magnitude of the input signal. The internal resistance is very large for a small input signal and reaches a lower limit called the collector saturation resistance for large values of input signal. The internal resistance of the transistors determines the magnitude of the supply voltage that is delivered to the load. This analysis is epitomized by an equivalent circuit consisting of R_T in series with the load resistance R_L and a voltage generator $V_p/\sqrt{2}$ for each transistor. V_p is the peak value of the rectified collector supply potential, Fig. 10-9.

FIGURE 10-9

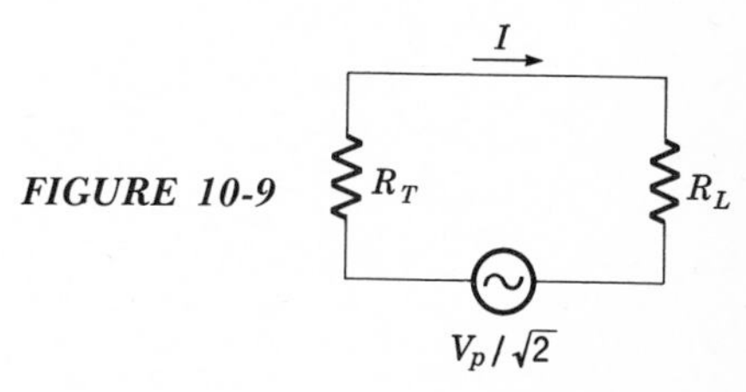

Using this equivalent circuit, the average power delivered to the load by each transistor is

$$P_L = \frac{1}{2} I^2 R_L = \frac{V_p{}^2 R_L}{4(R_L + R_T)^2} \tag{10-16}$$

where the factor $\frac{1}{2}$ results from the fact that each transistor is active for only one-half of the input cycle because of the class B operation of the power amplifier. Similarly, the power dissipated in each transistor is

$$P_T = \frac{1}{2} I^2 R_T = \frac{V_p{}^2 R_T}{4(R_L + R_T)^2} = \frac{V_p{}^2}{4R_L} \frac{R_T/R_L}{(1 + R_T/R_L)^2} \tag{10-17}$$

The power efficiency is, from Eqs. (10-16) and (10-17),

$$\eta = \frac{P_L}{P_L + P_T} = \frac{R_L}{R_L + R_T} = \frac{1}{1 + R_T/R_L} \tag{10-18}$$

which indicates that the efficiency approaches 100 percent when $R_T \ll R_L$. Although this ideal is not achieved in practical amplifiers, efficiencies greater than those of conventional class B stages can be attained.

Note that according to Eq. (10-17) the power dissipated in each transistor is small when R_T is large, that is, when the input signal is zero. Transistor dissipation is also small when maximum power is delivered to the load ($R_T \ll R_L$) corresponding to a large input signal. This means that the power lost in the power amplifier is smallest not only when the servo is quiescent, but also when the servo is striving to attain a new stable position corresponding to a change in the input signal. These two conditions are predominant in normal servo applications. This analysis shows that deleterious heating effects caused by transistor dissipation are minimized by the use of unfiltered collector supply voltage.

TRANSMISSION LINES AND WAVEGUIDES

Usually the components of a measuring system are physically separate from one another. For example, several distinct electronic instruments may be connected together in cascade to make a measurement. The signal must then be transported from the output of one unit to the input of the next. This is done by connecting the various circuits with *transmission lines*. In the simplest case a transmission line is just two wires, as has been implicitly assumed in previous chapters. A transmission line must, however, faithfully transmit the signal between instruments with a minimum of waveform or amplitude distortion. If this is not the case, the ultimate output of the system may be erroneous.

In low-frequency and dc systems it is only necessary to consider the resistivity of the conductors in the transmission line, as well as, perhaps, shunt resistance between conductors. These effects are examined by straightforward circuit analysis as studied in Chap. 1. Even a straight piece of wire has associated with it a small inductance however, and there is also a small capacitance between the two conductors. At high frequencies the reactances of these unavoidable inductances and capacitances become significant and influence signal propagation along the transmission line. This is particularly important in pulse circuits because of the high frequencies associated with pulse waveforms. In many practical high-frequency circuits transmission lines only a few inches long introduce objectionable waveform distortions.

Characteristic Impedance It is useful to represent the incremental impedances of a short section of transmission line by a series resistance r_1 and inductance l per unit length together with a shunt resistance r_2 and capacitance c per unit length, as in Fig. 10-10. Thus the series and shunt

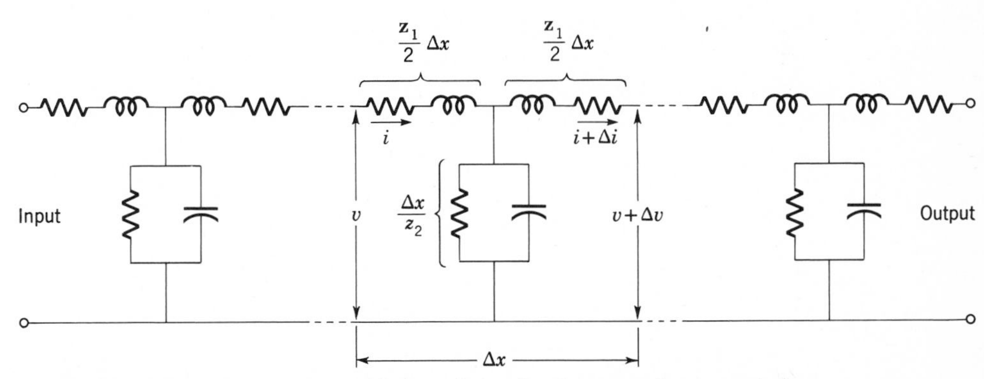

FIGURE 10-10 *Representation of transmission line in terms of series and shunt impedances per incremental length.*

complex impedances of a small section Δx of the line are

$$\mathbf{z}_1 \Delta x = (r_1 + j\omega l)\ \Delta x$$

$$\frac{1}{\mathbf{z}_2} \Delta x = \left(\frac{1}{r_2} + j\omega c\right) \Delta x \tag{10-19}$$

Variations of voltage and current along the transmission line are produced in response to an input signal. Consider one section Δx of the line for which the input voltage is v and the output voltage is incrementally different, $v + \Delta v$. Similarly, the input and output currents of the line element are i and $i + \Delta i$ respectively. The voltage equation around the outside loop of this circuit is then

$$v - i\frac{\mathbf{z}_1}{2}\Delta x - (i + \Delta i)\ \frac{\mathbf{z}_1}{2}\ \Delta x = v + \Delta v \tag{10-20}$$

Dividing through by Δx,

$$\frac{\Delta v}{\Delta x} = -\mathbf{z}_1 i - \frac{\mathbf{z}_1}{2} \Delta_i \tag{10-21}$$

In the limit as $\Delta x \to 0$, $\Delta v/\Delta x$ becomes the derivative dv/dx and the second term on the right side of Eq. (10-21) vanishes,

$$\frac{dv}{dx} = -\mathbf{z}_1 i \tag{10-22}$$

Similarly, the voltage equation including the shunt path is

$$v - i\frac{\mathbf{z}_1}{2} \Delta x - [i - (i + \Delta i)] \frac{\mathbf{z}_2}{\Delta x} = 0 \tag{10-23}$$

$$v = i\frac{\mathbf{z}_1}{2} \Delta x + \frac{\Delta i}{\Delta x}\mathbf{z}_2 = 0 \tag{10-24}$$

In the limit, $\Delta x \to 0$, Eq. (10-24) reduces to

$$\frac{di}{dx} = -\frac{1}{\mathbf{z}_2} v \tag{10-25}$$

Differentiating Eq. (10-22) with respect to x and using Eq. (10-25), the result is

$$\frac{d^2v}{dx^2} = -\mathbf{z}_1 \frac{di}{dx}$$

$$\frac{d^2v}{dx_2} = \frac{\mathbf{z}_1}{\mathbf{z}_2} v \tag{10-26}$$

Similarly, differentiating Eq. (10-25) with respect to x and using Eq. (10-22) yields an equation in i alone,

$$\frac{d^2i}{dx^2} = \frac{\mathbf{z}_1}{\mathbf{z}_2} i \tag{10-27}$$

Equations (10-26) and (10-27) are called the transmission-line equations. Their solutions yield the current and voltage at any point along the line.

The general solution of Eq. (10-26) is

$$v = Ae^{-\gamma x} + Be^{\gamma x} \tag{10-28}$$

where A and B are constants that depend upon the input and output conditions of the transmission line, and γ is called the *propagation constant.* The propagation constant is a complex number with a real part α and an imaginary part β given by

$$\gamma = \alpha + j\beta = \sqrt{\frac{\mathbf{z}_1}{\mathbf{z}_2}} \tag{10-29}$$

as can be verified by substituting Eqs. (10-28) and (10-29) into the transmission-line equation, Eq. (10-26).

Since the differential equation for i, (20-27), is identical in form to the voltage equation, (10-26), the solution also has the same form. It is possible to evaluate the arbitrary constants in terms of A and B by using Eq. (10-25). The result is

$$i = \frac{A}{\mathbf{Z}_c} e^{-\gamma x} - \frac{B}{\mathbf{Z}_c} e^{\gamma x} \tag{10-30}$$

where

$$\mathbf{Z}_c = \sqrt{\mathbf{z}_1 \mathbf{z}_2} \tag{10-31}$$

is called the *characteristic impedance* of the line. Equations (10-28) and (10-30) give the variation of voltage and current with distance along the line. We postpone examination of their significance to consider the meaning of the characteristic impedance.

Suppose a transmission line d meters long is terminated in its characteristic impedance, as in Fig. 10-11. Using Eqs. (10-28) and (10-30), the output

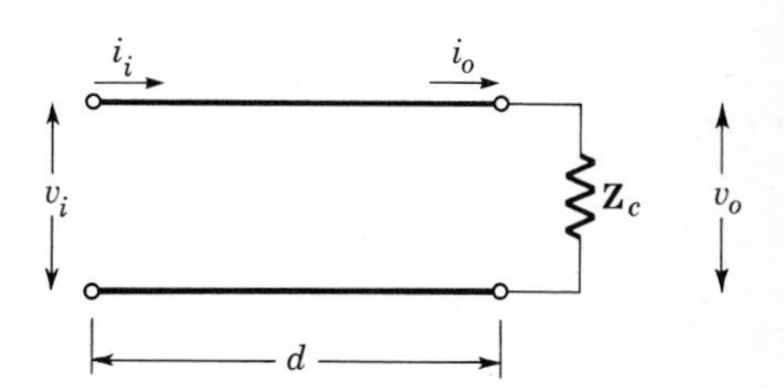

FIGURE 10-11 Transmission line terminated in its characteristic impedance.

voltage v_o can be written

$$v_o = i_o \mathbf{Z}_c$$

$$Ae^{-\gamma d} + Be^{\gamma d} = \mathbf{Z}_c \left(\frac{A}{\mathbf{Z}_c} e^{-\gamma d} - \frac{B}{\mathbf{Z}_c} e^{\gamma d} \right) \tag{10-32}$$

which means

$$B = -B = 0$$

At the input end $x = 0$, so that

$$v_i = A$$

Therefore, the voltage and current equations become

$$v = v_i e^{-\gamma x}$$

$$i = \frac{v_i}{\mathbf{Z}_c} e^{-\gamma x} \tag{10-33}$$

In particular, the input impedance of the line is

$$\mathbf{Z}_i = \frac{v_i}{i_i} = \frac{v_i}{v_i/\mathbf{Z}_c} = \mathbf{Z}_c \tag{10-34}$$

This means that the input impedance of a transmission line terminated in its characteristic impedance is simply equal to the characteristic impedance, independent of the length of the line. For this reason, as well as an equally significant one discussed subsequently, a transmission line is most often terminated in its characteristic impedance.

In many applications the series and shunt resistances of the transmission line can be neglected in comparison with the reactances. Thus, if $r_1 \ll \omega l$ and $1/r_2 \ll \omega c$, the characteristic impedance, Eq. (10-31), reduces to

$$\mathbf{Z}_c = \sqrt{\mathbf{z}_1\mathbf{z}_2} = \sqrt{\frac{r_1 + j\omega l}{1/r_2 + j\omega c}} = \sqrt{\frac{l}{c}} \tag{10-35}$$

which is a pure resistance.

The magnitude of $\mathbf{Z}_c$ depends upon the geometry of the line since both l and c are functions of the wire size, shape, separation, etc. In particular, the characteristic impedance of a transmission line comprising two parallel wires is given by

$$\mathbf{Z}_c = 276 \log \frac{D}{a} \tag{10-36}$$

where D is the separation between the wires and a is the radius of each conductor. A transmission line commonly used to connect television antennas to television receivers has $D/a = 10$, so that $\mathbf{Z}_c = 276\ \Omega$. Actually, such "twinlead" has a characteristic impedance closer to 300 Ω. The difference is attributable to the dielectric constant of the insulation separating the two conductors. Equation (10-36) applies to the case of air (dielectric constant of unity) between the wires.

Coaxial transmission lines are also commonly employed. The characteristic impedance of a coaxial line is given by

$$\mathbf{Z}_c = 138 \log \frac{b}{a} \tag{10-37}$$

where b is the radius of the outer conductor and a is the radius of the central wire. A common value for the characteristic impedance of a coaxial line is 72 Ω, although other values are also commercially available.

Delay Time According to Eq. (10-33) the voltage along a transmission line terminated in its characteristic impedance is given by

$$v = v_i e^{-j\beta x} \tag{10-38}$$

if the series and shunt resistances can be neglected. Suppose that the input signal is sinusoidal. Then, recognizing that

$$e^{-j\beta x} = \cos \beta x - j \sin \beta x \qquad (10\text{-}39)$$

Eq. (10-38) becomes

$$v = V_i \sin \omega t(\cos \beta x - j \sin \beta x) \qquad (10\text{-}40)$$

This can be written equivalently as

$$v = V_i \sin (\omega t - \beta x) \qquad (10\text{-}41)$$

Equation (10-41) represents a wave that is periodic both in time and in space.

A plot of Eq. (10-41) for one cycle when $t = 0$ is illustrated in Fig. (10-12).

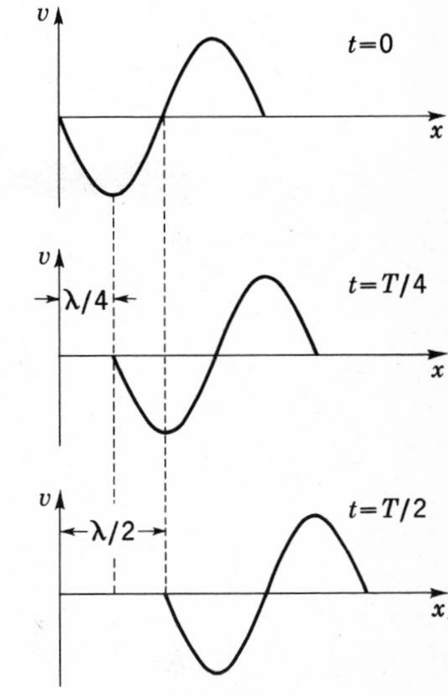

FIGURE 10-12 *Motion of sinusoidal signal along transmission line.*

Similarly, plots of Eq. (10-41) when $t = T/4$ and $t = T/2$ (where T is the period of the input signal) shows that the input signal moves away from the input end of the line. The *wavelength* λ of the signal at any given time is the distance along the line at which the voltage has gone through one complete cycle of 2π radians. Therefore, from Eq. (10-41)

$$\lambda = \frac{2\pi}{\beta} \qquad (10\text{-}42)$$

Note that, according to Fig. 10-12, the input signal moves a distance $\lambda/4$ in a time $T/4$, a distance $\lambda/2$ in a time $T/2$, etc. Therefore, the speed with which the input signal is transported along the line is

$$v = \frac{\lambda/4}{T/4} = \frac{\lambda}{T}$$

$$v = \lambda f \qquad (10\text{-}43)$$

where f is the frequency of the input signal. Equation (10-43) is a fundamental relation between the wavelength, frequency, and velocity of any propagating phenomenon.

The time required for an input signal to travel from the input of a transmission line to the output end is given by, using (10-43),

$$t_d = \frac{d}{v} = \frac{d}{\lambda f}$$

$$t_d = \frac{d}{\lambda} T \tag{10-44}$$

where d is the length of the line. According to (10-44), the time delay in terms of the period of the input signal depends upon the length of the line in terms of the wavelength. The time delay is a function of the parameters of the line since $\lambda = 2\pi/\beta$ and β depend upon the line geometry.

The propagation characteristics of transmission lines are often used to introduce a given time delay between input and output signals. Such *delay lines* are also constructed with actual inductances and capacitances connected according to Fig. 10-10. This is necessary if long delay times are required, since the delay time increases when l and c are large.

Reflections and Resonance We now consider a transmission line that is not terminated by an impedance equal to its characteristic impedance. As a useful illustration, consider the case when the output end is short-circuited so that $v_o = 0$. Thus at $x = d$, where d is the length of the line,

$$0 = Ae^{-j\beta d} + Be^{j\beta d} \tag{10-45}$$

Similarly, from Eq. (10-30),

$$i_0 = \frac{A}{\mathbf{Z}_c} e^{-j\beta d} - \frac{B}{\mathbf{Z}_c} e^{j\beta d} \tag{10-46}$$

Solving (10-45) and (10-46) for A and B yields

$$A = \frac{i_o \mathbf{Z}_c}{2} e^{j\beta d} \tag{10-47}$$

and

$$B = -\frac{i_o \mathbf{Z}_c}{2} e^{-j\beta d} \tag{10-48}$$

Using Eq. (10-28) the voltage on the line is

$$v = \frac{i_o \mathbf{Z}_c}{2} e^{j\beta d} e^{-j\beta x} - \frac{i_o \mathbf{Z}_c}{2} e^{-j\beta d} e^{j\beta x}$$

$$v = \frac{i_o \mathbf{Z}_c}{2} (e^{-j\beta(x-d)} - e^{j\beta(x-d)}) \tag{10-49}$$

Similarly, the current is, from Eq. (10-30),

$$i = \frac{i_o}{2} (e^{-j\beta(x-d)} + e^{j\beta(x-d)}) \tag{10-50}$$

Comparing Eqs. (10-49) and (10-50) with Eq. (10-33) we interpret the second term in (10-49) and (10-50) as a wave traveling from the receiving end to the input end. That is, the incident wave is *reflected* at the output of the line because of the improper termination at the receiving end. Reflections are avoided in pulse applications particularly since pulses that are reflected back and forth between the input and output ends of the line completely obscure the true input signal. This is the second major reason why a transmission line is normally terminated in its characteristic impedance.

A short-circuited transmission line does have useful properties, however, as can be illustrated by computing the input impedance. According to Eqs. (10-49) and (10-50) the complex input impedance is

$$\mathbf{Z}_i = \frac{v_i}{i_i} \mathbf{Z}_c \frac{e^{j\beta d} - e^{-j\beta d}}{e^{j\beta d} + e^{-j\beta d}} \tag{10-51}$$

Using Eq. (10-39), this becomes

$$\begin{aligned} \mathbf{Z}_i &= \mathbf{Z}_c \frac{\cos \beta d + j \sin \beta d - \cos \beta d + j \sin \beta d}{\cos \beta d + j \sin \beta d + \cos \beta d - j \sin \beta d} \\ &= j\mathbf{Z}_c \tan \beta d \\ &= j\mathbf{Z}_c \tan 2\pi \frac{d}{\lambda} \end{aligned} \tag{10-52}$$

According to Eq. (10-52) the input impedance of a short-circuited transmission line is equal to zero when the length of the line is such that $d = \lambda/2$, because $\tan \pi = 0$. Of particular importance is the case when $d = \lambda/4$, for which $\mathbf{Z}_i \to \infty$. That is, the line acts as a parallel resonant circuit at the frequency corresponding to $d = \lambda/4$. Such quarter-wavelength lines are often used as resonant circuits at very high frequencies. This is useful since normal parallel resonant circuits consisting of a single inductance and capacitance are not possible at extreme frequencies because of the very small inductance and capacitance values required. Analogous results can be achieved with transmission lines that are open-circuited at the receiving end.

Waveguides At very high frequencies, where the signal wavelength is of the order of the spacing between conductors, parallel-wire and coaxial transmission lines do not efficiently transport signals from the input to the output end. Such signal frequencies are propagated in hollow conductors, called *waveguides.* Analysis of the transmission properties of waveguides is accomplished by focusing attention on the electric and magnetic fields in the interior of the hollow conductor, rather than on the currents in the conductor itself. In effect, the signal is carried by waves of electric and magnetic fields and the function of the hollow conductor is to guide these waves from the transmitting to the receiving end.

Some appreciation of the electric-field configuration in a circular waveguide is attained by considering the situation in a coaxial line as the central conductor is made smaller and smaller until it vanishes altogether, Fig. 10-13. When the central conductor is large, the electric field in the space

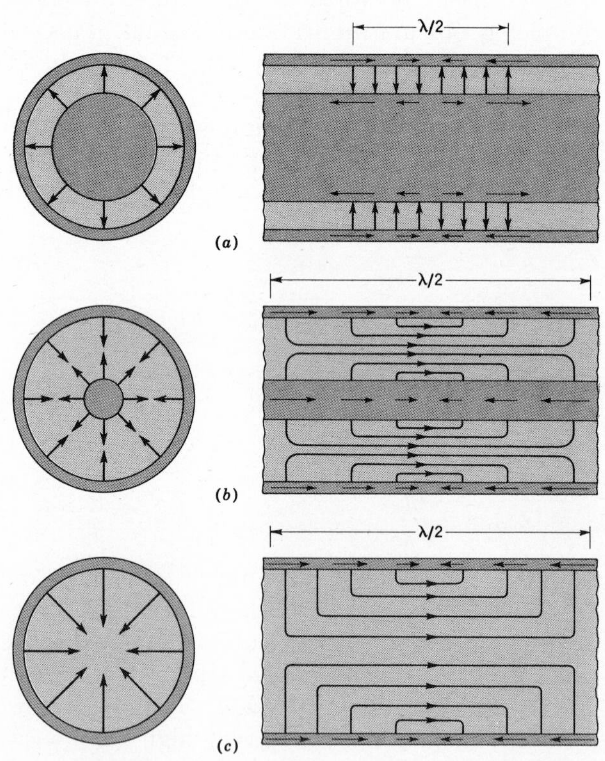

FIGURE 10-13 *Electric-field configurations (a and b) in coaxial transmission lines and (c) in hollow circular waveguide.*

between conductors is essentially radial, Fig. 10-13*a*, and currents are present in both conductors corresponding to the electric-field pattern. A small central conductor results in an electric-field configuration that originates and terminates on the same conductor, Fig. 10-13*b*. This suggests that the central wire may be eliminated altogether; the electric-field pattern that results is shown in Fig. 10-13*c*. One reason for the increased efficiency of the waveguide over that of the two-conductor line is the fact that resistance losses in the center wire are eliminated.

Waveguides with a rectangular cross section are more commonly employed than are circular waveguides because the electric-field configurations are somewhat less complicated. This means that input and output terminations are simpler. In both types the electric-field configuration is such that only signals of frequencies higher than a certain *cutoff frequency* can be propagated; that is, a waveguide acts as a high-pass filter. For example, the longest wavelength that can be transmitted through a rectangular waveguide having the longer cross-sectional dimension a is given by

$$\lambda_c = 2a \tag{10-53}$$

Minimum losses are achieved if the waveguide dimension is only just sufficient in cross section to pass the lowest signal frequency of interest. For this reason, the size of a waveguide is a good indication of the signal frequency it is designed to transmit. A relation analogous to Eq. (10-53) applies to circular waveguides also.

If the receiving end of a waveguide is short-circuited, a resonance effect identical in principle to that of a short-circuited transmission line is observed. Most often, both ends of the waveguide are closed, which produces a *resonant cavity* useful as a resonant circuit at microwave frequencies.

ANALOG INSTRUMENTS

Oscilloscopes We have touched several times previously upon the circuits and applications of that most versatile instrument, the cathode-ray oscilloscope. The complete block diagram of a typical unit, Fig. 10-14, includes vertical and horizontal amplifiers coupled to the vertical and horizontal deflection plates of the cathode-ray tube, a triggered sawtooth, or ramp generator to provide a linear horizontal sweep, and suitable power-supply circuits. In the most common mode of operation, the waveform of interest is applied to the vertical input terminals. The vertical amplifier increases the amplitude sufficiently to cause an appreciable vertical deflection of the electron beam. A portion of the amplified signal is used to derive a trigger pulse for the sweep generator in order to synchronize the sweep signal with the input waveform. The output of the sweep generator is amplified and applied to the horizontal deflection plates. The sweep generator also provides a pulse to the electron gun of the cathode-ray tube (CRT) which blanks off the electron beam during retrace so that the spot can return to the starting point without leaving a spurious trace.

Sometimes it is advantageous to trigger the sweep generator with an ex-

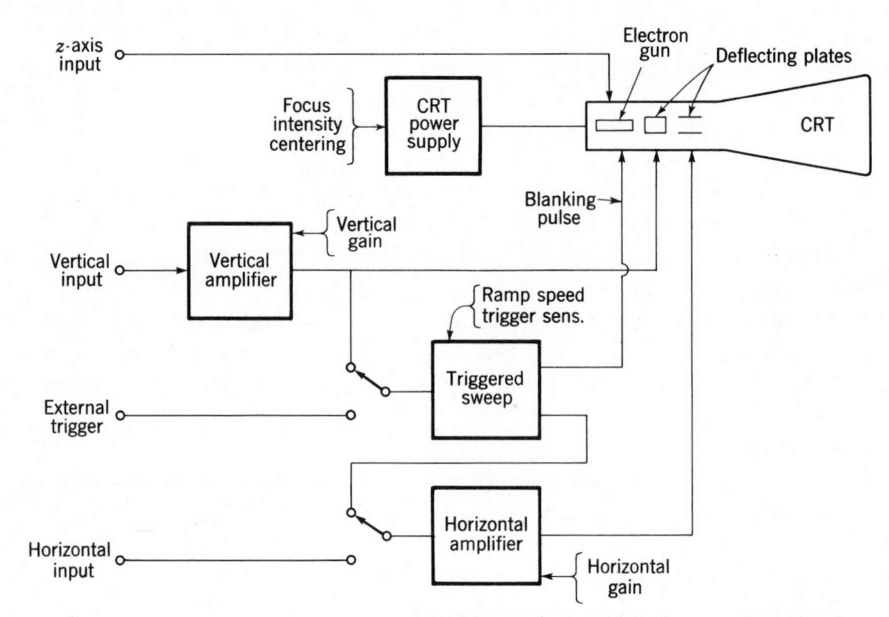

FIGURE 10-14 *Block diagram of oscilloscope.*

ternal signal as, for example, if it is necessary to start the sweep before the vertical signal amplitude is large. For this purpose, an external trigger input is provided to which the generator input can be connected. When the linear sweep is not desired, as in the case of Lissajous figures (Chap. 2), the input of the horizontal amplifier is connected to the horizontal input terminals and external waveforms can be placed on both the horizontal and vertical deflection plates. It is also possible to modulate the intensity of the electron beam by altering the grid potential in the electron gun; this external terminal is conventionally termed the z-axis input to distinguish it from the horizontal (x-axis) and vertical (y-axis) inputs.

The vertical and horizontal amplifiers are quite often of the dc-coupled differential type. Voltage-divider gain controls are included to adjust the amplitude of the deflection on the screen. Many different sweep circuits are in common use; a typical version is described in Chap. 11. Sweep-speed and trigger-sensitivity controls are necessary here in order to set the sweep for best waveform display. The power supply for the CRT is designed so that dc potentials applied to the electron-gun elements can be adjusted to achieve the optimum focus and intensity of the trace. Similarly, the quiescent position of the spot on the tube face is adjusted by altering the dc voltages on both the horizontal and vertical deflection plates.

The ability of a conventional oscilloscope to display very-high-frequency signals is limited by the high-frequency cutoff of the vertical amplifier and by the luminous efficiency of the cathode-ray-tube screen at high spot speeds. As one example of the versatility of CRT instruments, consider the so-called *sampling oscilloscope,* which can display repetitive waveforms at frequencies far in excess of a conventional instrument. The input waveform amplitude is sampled at successively later points in its cycle and then reassembled into a replica of the input waveform on the CRT screen. Since the input signal is repetitive, the sampling can be done every tenth cycle, as illustrated in Fig. 10-15, which means the sampling and display circuits operate at only one-tenth of the input frequency.

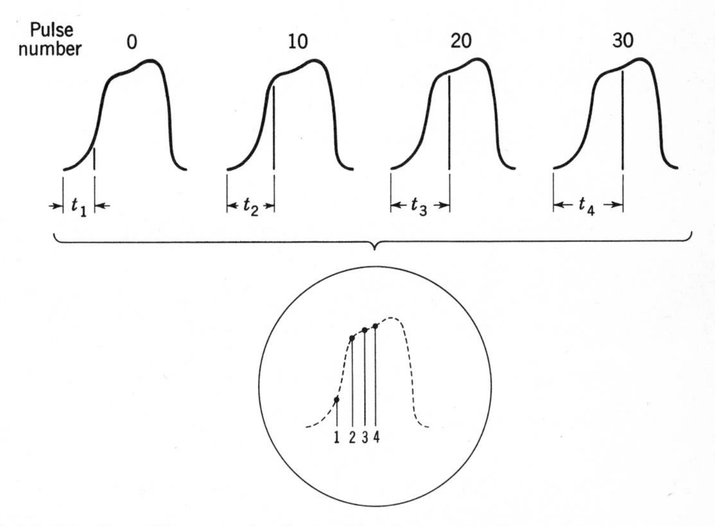

FIGURE 10-15 *By taking sample of repetitive input waveform every tenth cycle, waveshape can be reconstructed by circuits with maximum frequency response only one-tenth of the input frequency.*

The basic circuit diagram of a sampling oscilloscope, Fig. 10-16, employs a pulse generator to produce a series of successively delayed pulses and a

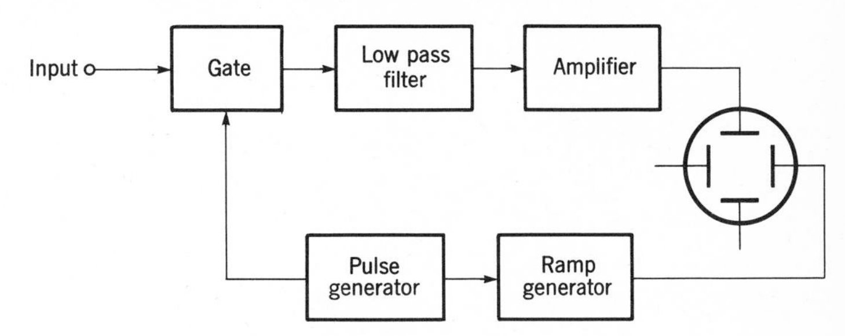

FIGURE 10-16 *Block diagram of sampling oscilloscope.*

gate circuit which is active only during the pulse duration. The gate output is passed through a low-pass filter and applied to the CRT. Note that although most of the circuit need not be capable of handling the signal frequency, the pulse generator must produce very sharp pulses. Several different circuits such as the blocking oscillator are available to perform this function. It proves easier to generate extremely narrow pulses than to design an amplifier with significant gain at corresponding frequencies.

Waveform Analyzer Complex waveforms may be thought of as combinations of harmonically related sine waves, according to the method of Fourier analysis discussed in Chap. 2. Instruments capable of determining the amplitude of the frequency components in a waveform, called *waveform analyzers,* in effect carry out Fourier analysis experimentally. As such, they are frequency-measuring instruments.

A waveform analyzer is basically a sharply tuned ac amplifier-voltmeter which has a significant response at only one frequency. The input signal is heterodyned with the output of a variable-frequency sinusoidal oscillator to produce a signal at the response frequency of the amplifier, Fig. 10-17.

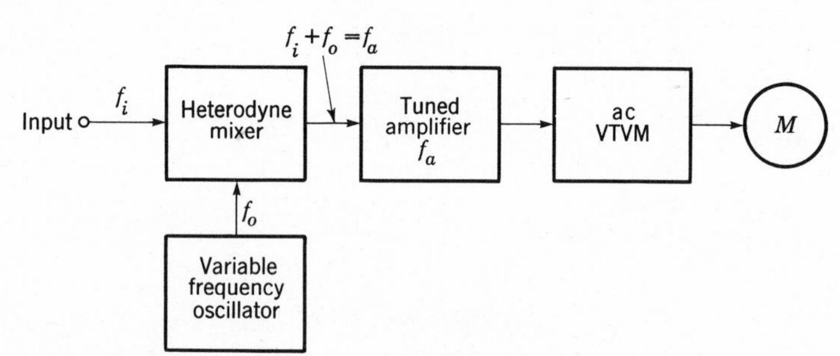

FIGURE 10-17 *Block diagram of waveform analyzer for determining frequency components in complex waveform.*

That is, according to Eq. (4-38), two sine waves suitably mixed produce a sine wave at the sum (and difference) frequency. If the oscillator frequency is f_o while the input waveform has a frequency component f_i, the combination will result in an output signal if

$$f_i + f_o = f_a \tag{10-54}$$

where f_a is the center frequency of the tuned amplifier. Thus, by tuning the oscillator frequency over a given interval, the output meter deflects each time Eq. (10-54) is satisfied for a frequency component of the input waveform. The magnitude of the meter deflection is proportional to the amplitude of each input-signal component. In this way, the entire frequency complement of the input waveform is determined.

The heterodyne technique is used in preference to simply tuning the amplifier itself in order to obtain the greatest possible frequency coverage. An oscillator circuit capable of output frequencies over a wide band is simpler than a tunable amplifier covering the same frequency interval. Furthermore a fixed-tuned amplifier has a constant peak response and bandpass characteristic, both difficult to achieve in a tunable system. The frequency dial of the waveform analyzer is calibrated in terms of $f_a - f_o$, so that the unknown frequency is indicated directly.

Magnetic Recorder One of the most useful signal recording schemes is based on the *magnetic-recorder* principle in which an electric waveform is recorded in the form of magnetization on a ferromagnetic chart, or tape. After recording, the tape is passed back through the instrument and a voltage signal corresponding to the original input is obtained. This makes it possible to observe the signal again and again, if necessary, and to analyze the waveform by as many different techniques as may be required.

The magnetic tape, most often in the form of a ferromagnetic oxide powder coated onto a plastic tape, is magnetized in accordance with the signal by the magnetic field of the recording head, Fig. 10-18*a*. Signal

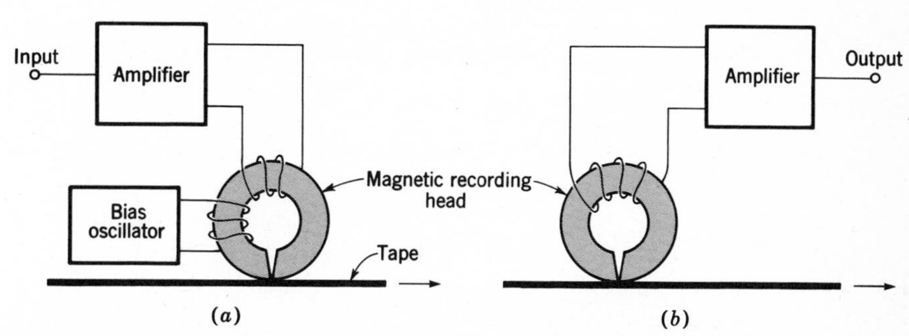

FIGURE 10-18 *(a) Recording and (b) playback of electric signals using magnetic recording techniques.*

current produces a magnetic field at a sharply defined gap in the core of the recording head and this field permanently magnetizes the tape as it passes the gap. A high-frequency ac bias signal is also applied to the recording head to improve the linearity of the recording process. In playback, Fig. 10-18*b*, the tape is moved past the head again and the magnetic field of the tape induces a voltage in the coil of the head. Since the magnetization of the tape corresponds to variations in the original signal, voltages induced in the head replicate the input waveform. Playback does not change the tape magnetization, so that the signal can be reproduced as many times as desired.

The action of the ac bias frequency can be understood by considering the typical nonlinear properties of the magnetic tape as illustrated by the hysteresis loop in Fig. 10-19. In the absence of ac bias, the recorded magnetization is not a linear function of the signal field and the playback signal

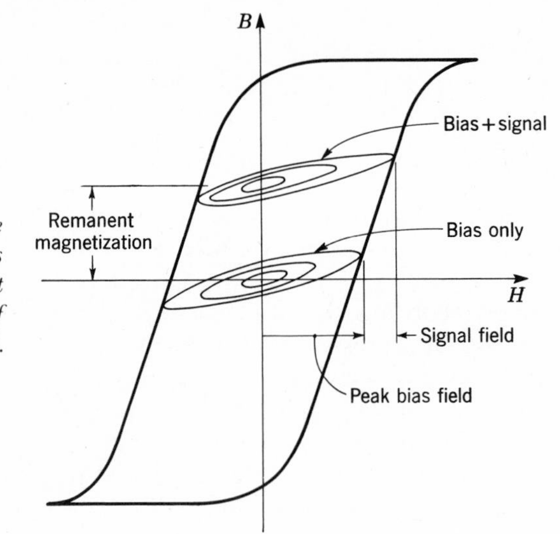

FIGURE 10-19 Action of ac bias is to cause magnetic tape material to traverse minor hysteresis loops as it passes under recording head. Remanent magnetization is then a linear function of signal field.

is strongly distorted. The ac bias causes the magnetization of the material to traverse minor hysteresis loops as shown. As an element of tape moves away from the recording gap, the size of the loops decreases to zero. Thus, in the absence of signal the resulting magnetization is equal to zero. The signal field displaces the minor loop so that the remanent magnetization is a finite value. Because of the straight sides of the major hysteresis loop, the relation between the remanent magnetization and the signal field is linear. The ac bias signal is of the order of five times the maximum signal frequency and the peak amplitude is approximately equal to the coercive force of the tape.

The highest frequency that can be recorded depends upon the tape speed and width of the gap in the record-playback head. Consider the enlarged view of the playback head shown in Fig. 10-20 and assume that the tape has been recorded with a sinusoidal signal. The magnetic flux of the tape is then given by

$$\phi_t = Mdw \sin \frac{2\pi x}{\lambda}$$

where M is the peak magnetization, d and w are the thickness and width of the tape, respectively, and λ is the wavelength of the sinusoidal magnetic signal on the tape. The flux induced in the playback head due to an element of tape dx depends upon the reluctance of the air gap, so that

$$d\phi = \phi_t \frac{dx}{l} \tag{10-55}$$

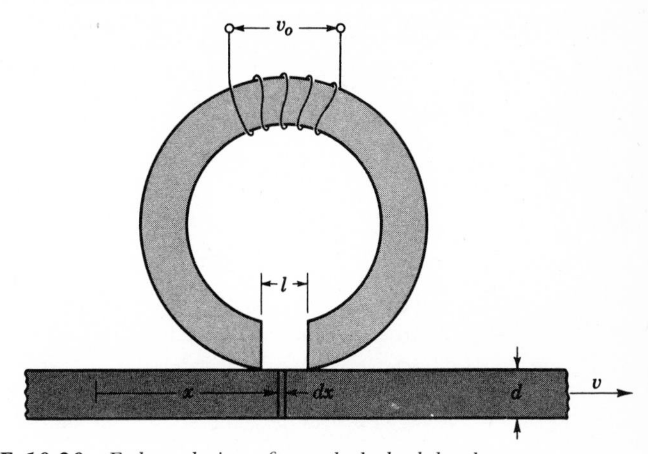

FIGURE 10-20 *Enlarged view of record-playback head showing resolution limit determined by width of gap.*

The total flux in the playback head is found by integrating Eq. (10-55) over the length of the gap,

$$\phi = \int_{x-l/2}^{x+l/2} d\phi = \frac{Mdw}{l} \int_{x-l/2}^{x+l/2} \sin 2\pi \frac{x}{l} dx \tag{10-56}$$

Equation (10-56) can be integrated directly and the result simplified by means of a trigonometric identity to yield

$$\phi = -Mdw \sin \frac{2\pi x}{l} \frac{\sin (\pi l/\lambda)}{\pi l/\lambda} \tag{10-57}$$

The output voltage fron the N-turn coil on the head depends upon the rate of change of flux caused by tape motion.

$$v = -N \frac{d\phi}{dt} = MdwN \frac{2\pi}{l} \frac{dx}{dt} \cos \frac{2\pi x}{l} \frac{\sin (\pi l/\lambda)}{\pi l/\lambda} \tag{10-58}$$

But $dx/dt = v$, the tape velocity, and $v = \lambda f$, where f is the playback signal frequency. After introducing $x = vt$, Eq. (10-58) can be put in the form

$$v = MdwN\omega \cos \omega t \frac{\sin (\pi l/\lambda)}{\pi l/\lambda} \tag{10-59}$$

Finally, the rms playback voltage is simply

$$V = M'dwN\omega \frac{\sin (\pi l/\lambda)}{\pi l/\lambda} \tag{10-60}$$

where M' is the rms magnetization on the tape. Putting Eq. (10-60) in terms of the playback frequency,

$$\frac{\pi l}{\lambda}=\frac{\pi l}{v/f}=\frac{\omega l}{2v} \tag{10-61}$$

the result is

$$V=M'dwN\omega\frac{\sin(\omega l/2v)}{\omega l/2v} \tag{10-62}$$

According to Eq. (10-62), the playback voltage is small at low frequencies, rises to a maximum, and falls sharply to zero where

$$\frac{\omega_m l}{2v}=\frac{2\pi f_m l}{2v}=\frac{f_m l}{v}=\pi \tag{10-63}$$

or

$$f_m=\frac{v}{l} \tag{10-64}$$

Since $f=v/\lambda$, Eq. (10-64) can also be written

$$\lambda_m=l \tag{10-65}$$

The meaning of Eq. (10-65) is that the minimum wavelength that can be played back is equal to the gap length. That is, the maximum frequency depends upon the tape speed, according to Eq. (10-64). Experimental results, Fig. (10-21), are in general agreement with Eq. (10-62).

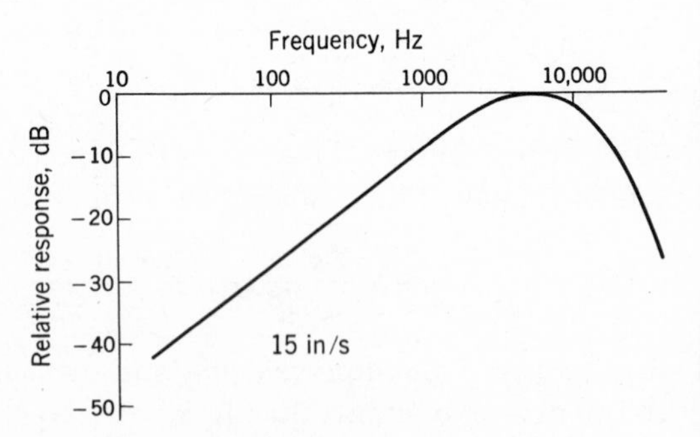

FIGURE 10-21 *Experimental frequency-response characteristic of magnetic recorder playback head. Fall at high frequencies is determined by size of the gap in head.*

The falloff in playback signal at low frequencies is a result of the small rate of change of flux at long tape wavelengths. Uniform frequency response is achieved by a *compensated* playback amplifier that has large gain at low frequencies and decreasing gain at high frequencies. The high-frequency response depends upon the recording-head gap length and tape speed. With suitable choice of parameters, frequencies as high as 10 MHz can be accommodated.

Although major use of magnetic recorders is in reproducing the amplitude and frequency characteristics of the input signal exactly, many other possibilities also exist. For example, according to Eq. (10-64) the playback frequency depends upon the tape speed. Therefore a very-long-duration, low-frequency recording can be played back at high tape speed. The resulting high-frequency signal can be analyzed in a much shorter time than required for the original recording. Conversely, a high-frequency recording can be played back at a slow tape speed with a corresponding reduction in the frequency components in the signal. In this way, the signal can be analyzed with instruments having only a low-frequency capability. Time compression or expansion ratios greater than factors of 1000 are possible.

Active Filter The signal waveforms analogous to physical quantities can be processed readily by means of electronic circuits. In particular, operational amplifiers can add and subtract, multiply and divide, differentiate and integrate, etc., signal waveforms, as well as simply increase their amplitudes. This by no means exhausts the possibilities, however, and as a further example, the properties of a so-called *active filter* are analyzed in the following fashion.

Consider the operational amplifier in Fig. 10-22 in which both the inverting and noninverting inputs have the operational feedback connection. Because of the large input impedance of the amplifier, the current into the amplifier input terminals is negligible so that

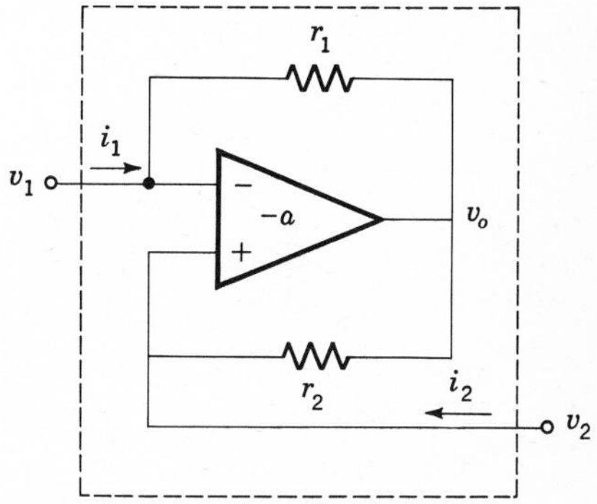

FIGURE 10-22 *Operational amplifier connected as a negative impedance converter.*

$$i_1 = -\frac{v_o}{r_1} \quad \text{and} \quad i_2 = -\frac{v_o}{r_2} \tag{10-66}$$

which means

$$i_1 = \frac{r_2}{r_2} i_2 \tag{10-67}$$

Writing Kirchhoff's voltage equation around the outer loop,

$$v_1 - i_1 r_1 + i_2 r_2 - v_2 = 0 \tag{10-68}$$

Inserting Eq. (10-67), the relationship between the input and output voltages is simply

$$v_1 = v_2 \tag{10-69}$$

Consider now the apparent input impedance of the circuit when an impedance $\mathbf{Z}_2$ is connected to the output terminals. To focus on the impedance transformation, use the block diagram, Fig. 10-23, in which the

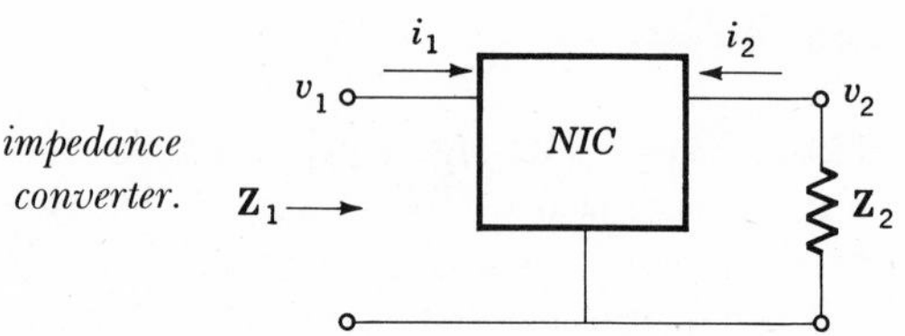

FIGURE 10-23 *Block diagram of a negative impedance converter.*

portion in the dashed rectangle in Fig. 10-22 is labeled *NIC,* for reasons justified below. According to Fig. 10-23, the impedance $\mathbf{Z}_2$ is given by

$$\mathbf{Z}_2 = \frac{-i_2}{v_2} \tag{10-70}$$

The input impedance is

$$\begin{aligned} \mathbf{Z}_1 &= \frac{v_1}{i_1} = \frac{v_2}{r_2 i_2 / r_1} \\ \mathbf{Z}_1 &= -\frac{r_1}{r_2}\mathbf{Z}_2 \end{aligned} \tag{10-71}$$

where Eqs. (10-67) and (10-69) have been used. According to Eq. (10-71) the circuit has the interesting property that the apparent input impedance is the negative of the impedance connected to the output terminals. That is, the circuit is a *negative impedance converter,* or *NIC.* If, for example, the output impedance is a capacitor, the input impedance is an inductance. Furthermore, the magnitude of the inductance depends upon the scale factor r_1/r_2.

A simple application of this useful property is the *active bandpass filter* in Fig. 10-24. The input-output characteristic of this network is just

$$\frac{v_o}{v_i} = \frac{v_2}{i_1 \mathbf{Z}_1 + v_1} = \frac{1}{1 + (i_1/v_1)\mathbf{Z}_1} \tag{10-72}$$

where Eq. (10-69) has been used. Finally, inserting Eq. (10-71)

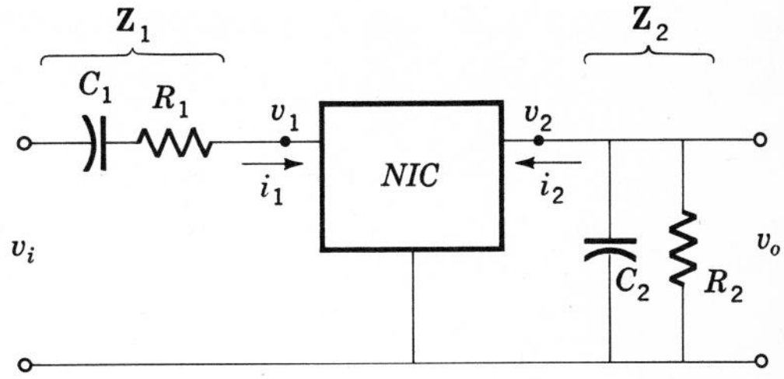

FIGURE 10-24 *Active bandpass filter.*

$$\left(\frac{v_o}{v_i}\right)^{-1} = 1 - \frac{r_2}{r_1}\frac{\mathbf{Z}_1}{\mathbf{Z}_2} \tag{10-73}$$

The ratio $\mathbf{Z}_1/\mathbf{Z}_2$ is

$$\frac{\mathbf{Z}_1}{\mathbf{Z}_2} = \left(R_1 - \frac{j}{\omega C_1}\right)\frac{R_2 - j/\omega C_2}{-jR_2/\omega C_2} = \left(\frac{R_1}{R_2} + \frac{C_2}{C_1}\right) + j\omega\left(R_1 C_2 - \frac{1}{\omega^2 C_1 R_2}\right) \tag{10-74}$$

The circuit is in resonance when the imaginary part vanishes,

$$\omega_0{}^2 = \frac{1}{R_1 C_1 R_2 C_2} \tag{10-75}$$

Inserting Eq. (10-74) and Eq. (10-75) into Eq. (10-73), the result is

$$\left(\frac{v_o}{v_1}\right)^{-1} = \left[1 - \frac{r_2}{r_1}\left(\frac{R_1}{R_2} + \frac{C_1}{C_2}\right)\right] - j\frac{r_2}{r_1}\sqrt{\frac{R_1 C_2}{R_2 C_1}}\left(\frac{\omega}{\omega_0} - \frac{\omega_0}{\omega}\right) \tag{10-76}$$

This expression should be compared with Eq. (3-59) for an *RLC* circuit. In the simple case where $R_1 = R_2$ and $C_1 = C_2$, the input-output characteristic is

$$\left(\frac{v_o}{v_i}\right)^{-1} = \left(1 - 2\frac{r_2}{r_1}\right) - j\frac{r_2}{r_1}\left(\frac{\omega}{\omega_0} - \frac{\omega_0}{\omega}\right) \tag{10-77}$$

Note that the equivalent Q of the circuit is just r_2/r_1, which means that the Q can be adjusted independently of the resonant frequency. Alternatively, it is possible to adjust the resonant frequency by changing either R_1 (and R_2) or C_1 (and C_2) without changing Q. The output signal may exceed the input voltage at resonance if the real term in Eq. (10-77) is less than unity, so that an active filter may also have gain.

SUGGESTIONS FOR FURTHER READING

S. D. Prensky: "Electronic Instrumentation," Prentice-Hall, Inc., Englewood Cliffs, N.J., 1963.

M. B. Stout: "Basic Electrical Measurement," 2d ed., Prentice-Hall, Inc., Englewood Cliffs, N.J., 1960.
J. J. Studer: "Electronic Circuits and Instrumentation Systems," John Wiley & Sons, Inc., New York, 1963.
F. E. Terman and J. M. Pettit: "Electronic Measurements," 2d ed., McGraw-Hill Book Company, New York, 1952.

EXERCISES

10-1 Determine the regulation factor and internal resistance of the transistor regulated power supply, Fig. 10-4. Assume properties of the control transistor given in the text and that the h parameters of the 2N2049 are the same as those given in Exercise 6-11.

Ans.: 4.5×10^{-4}, 2.3 Ω

10-2 Analyze the simple shunt regulator in Fig. 10-3 to determine the effective internal resistance and regulation factor. The h parameters are $h_{ie} = 100\ \Omega$, $h_{re} = 10^{-4}$, $h_{fe} = 60$, $h_{oe} = 175 \times 10^{-6}$ mho.

Ans.: 5.9×10^{-2}; 1.6 Ω

10-3 Calculate the time delay of a 100-MHz signal transmitted along a transmission line 10 m long. Assume the properties of the line are such that the wavelength on the line is 1 m.

Ans.: 10^{-7} s

10-4 What is the input impedance of a transmission line if the output end is an open circuit? Compare the condition for parallel resonance with that corresponding to the short-circuited line, as given by Eq. (10-52).

Ans.: $-j\mathbf{Z}_c \cot (2\pi d/\lambda)$; $\lambda/2$

10-5 What is the minimum tape speed necessary if a magnetic recorder is used to record signals up to 1 MHz? Assume that the recording gap is 1 μm long.

Ans.: 1m/s

LABORATORY EXERCISES

10-A Voltage Regulator The voltage regulator circuit is a useful example of measurement and control using feedback techniques. Fairly elaborate circuits are required to produce precisely stabilized voltages, but these are all based on the same principles as those discussed in the text. The output voltage is measured, compared with a standard, and then adjusted so that the error signal is zero. The purpose of this experiment is to examine a practical voltage regulator and establish its operating performance.

Consider the voltage regulator shown in Fig. 10-25. Note that the input stage of the dc amplifier is a difference amplifier, which facilitates comparison of the output voltage with the standard voltage, 12 V, across the zener diode. Three control transistors in parallel permit greater load current than is possible with a single

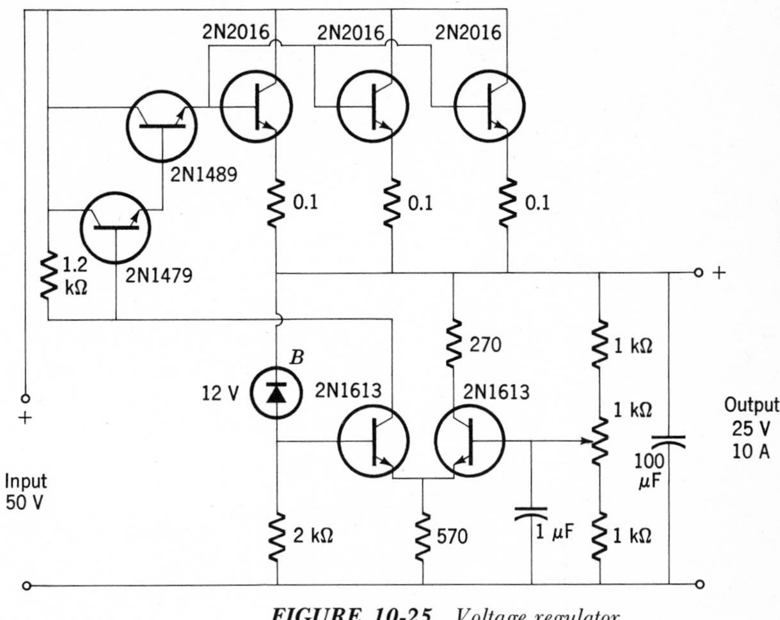

FIGURE 10-25 *Voltage regulator.*

transistor without exceeding the maximum ratings of each transistor. Evidently, the output voltage is adjustable by means of the potentiometer in the input circuit.

Focus attention on the input difference amplifier. Describe how the reference voltage and the output voltage are combined to develop an error signal. Do this by finding the voltage at the 1.2-kΩ collector load resistor as a function of the output voltage of the regulator assuming that the 2N1479 transistor is disconnected for the moment so that the regulator is inoperative. Appropriate values of small-signal parameters for the 2N1613's can be obtained from the collector characteristics of type 2N1415 in Appendix 1, since these two transistor types are similar.

Next consider the two emitter-follower Darlington stages together with the control transistors. Determine the load current as a function of input current to the 2N1479 assuming the characteristics of the 2N1479 are the same as those of the type 2N1715 given in Appendix 1 and that the 2N1489 is similar to the 2N2016 given in Fig. 7-18. Do this graphically for load resistances of 2.5, 5.0, and 10 Ω.

What is the range over which the output voltage may be adjusted with the 1-kΩ potentiometer? Plot the output voltage as a function of current from no-load to a load current of 10 A with the potentiometer set at the middle of the range using the Thévenin expression, Eq. (10-5). Does the circuit regulate over the entire range of output current? What is the percentage change in output voltage from no load to full load?

Now plot the output voltage as a function of input voltage from 0 to 60 V for a load current of 5A. What is the percentage change in output voltage for a change in input voltage from 40 to 50 V? Estimate the range of input voltages over which the regulator is effective by noting where one or more of the transistors is cut off or saturated.

Finally, replace the difference amplifier with a type MC1431 integrated-circuit operational amplifier. Using the Thévenin equivalent circuit of the regulator and the properties of the MC1431 given in Table 8-1, calculate the input voltage variation reduction factor and the effective internal resistance of the regulator. Is this an improvement over the original circuit? What unfavorable features might develop with this modification?

10-B Furnace Temperature Controller Feedback systems are used to control the value of many physical variables. A typical application is the regulation of the temperature of an electrically heated furnace. In this case there is a thermal link in the feedback network. The purpose of this experiment is to study the operation of a practical temperature-control circuit.

Consider the furnace temperature-control system illustrated in Fig. 10-26. The

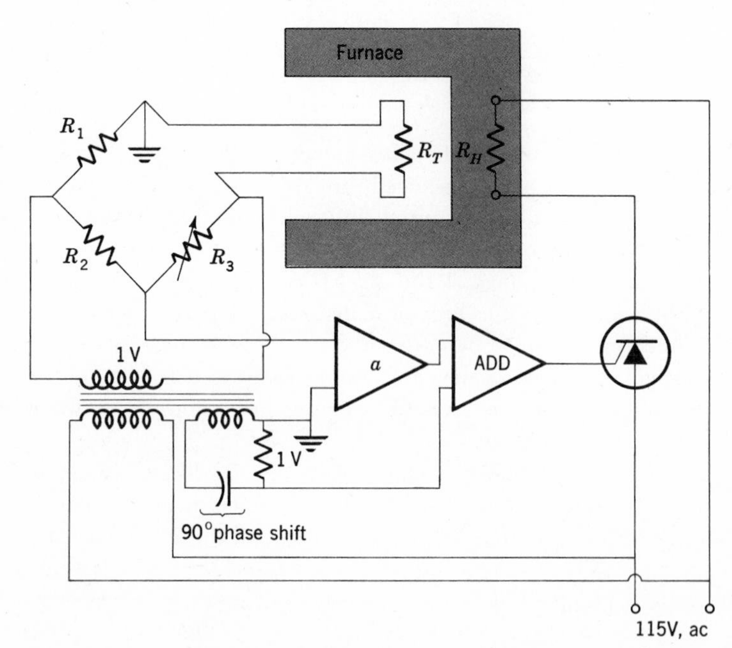

FIGURE 10-26 *Furnace temperature-control system.*

furnace is heated by current in R_H which is controlled by the SCR. The temperature of the furnace is measured by a platinum resistance thermometer, R_T. Changes in furnace temperature resulting in variations in R_T unbalance the bridge and produce an ac signal. This signal is amplified and added to a reference signal which has been shifted in phase by 90° to yield the gate voltage for the SCR. In this way the furnace temperature is regulated to ±1°C at any temperature between 100 and 1000°C.

Investigate the operation of this circuit as a feedback network and verify that the system uses negative feedback. To do this, first choose appropriate values for the bridge resistors if the resistance thermometer is 50 Ω at room temperature and the temperature coefficient of resistance is 0.004/°C. Assume that the furnace temperature above room temperature is proportional to the power in R_H and that 500 W is required to achieve the highest temperature. How is the desired temperature set initially?

Sketch gate-control waveforms for furnace temperatures above, below, and near 500°C and determine the furnace-heater-current waveforms. Determine the relationship between furnace temperature and the output signal from the amplifier. Describe why the 90° phase shift is present in the reference signal.

Plot the temperature of the furnace, rms furnace-heater current, and the bridge output signal as a function of R_3. Choose an appropriate value of amplifier gain so that the furnace temperature remains within ±1°C.

chapter eleven

PULSE CIRCUITS

Most circuits discussed in preceding chapters are designed for dc or sinusoidal signal waveforms. Signals which consist of a series of rapid transitions from one voltage magnitude to another, such as, for example, a square wave, are of equal importance. Although such pulse waveforms may be resolved into their sinusoidal harmonic components using the method of Fourier analysis, it is usually more convenient to consider the signal pulses themselves.

Electric signals from many physical phenomena, most notably reactions in nuclear physics, are inherently pulses. In addition, modern communication techniques, such as radar, television, and data telemetry, involve the transmission of pulses from the transmitter to the receiver. In these applications the size and relative timing of the pulses represent an encoding of the information transmitted. Circuits specifically designed to be used with pulses are discussed in this chapter.

PULSE AMPLIFIERS

Rise Time The pulse size and shape representing a physical effect carry the information that describes the phenomenon. For example, the voltage pulse from a Geiger counter which has detected a nuclear disintegration may be characterized by the exact time of onset of the pulse, its magnitude, and its duration. Since the output of many detectors is small, it is often necessary to amplify the pulse signal before measurement. Vacuum-tube and transistor amplifiers are used for this purpose. An ideal pulse amplifier introduces no change in the pulse shape other than increasing its amplitude, and practical circuits can come close to this performance.

The important characteristics of pulse amplifiers, gain, frequency response, stability, and distortion are similar to those already discussed in connection with conventional amplifiers. To illustrate, in particular, the significance of frequency response, consider the ideal square pulse shown in Fig. 11-1 and the nonideal pulse shape resulting from passing

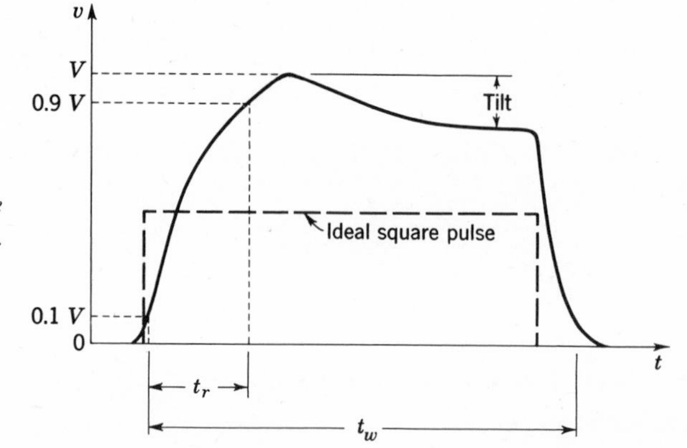

FIGURE 11-1 *Comparison of ideal square voltage pulse with amplified, nonideal pulse.*

this waveform through a practical amplifier. Note that the amplified pulse does not rise instantaneously to the maximum value. This is a result of the high-frequency cutoff of the amplifier caused by shunt capacitance. Consider the case of a single-stage transistor or vacuum-tube amplifier, which can be represented for present purposes by the input circuit of Fig. 11-2, in which C_s represents the total shunt capacitance (compare Fig.

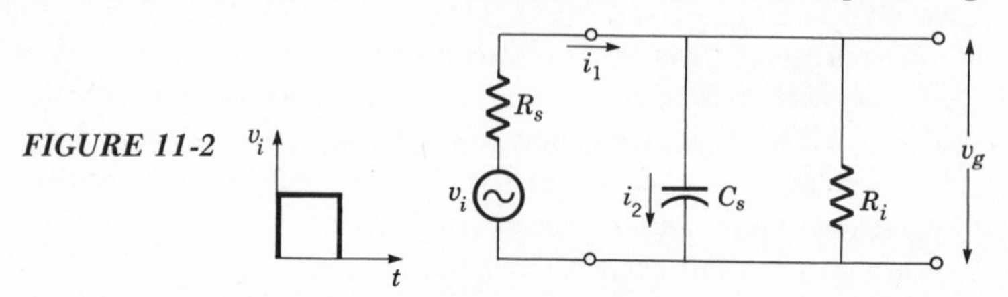

FIGURE 11-2

7-11), R_i is the input impedance, and R_s is the output impedance of the preceding stage. The waveform of the output pulse v_g due to a square voltage pulse v_i starting at $t=0$ is found by analyzing the circuit using techniques described in Chap. 2. The voltage and current equations of the circuit are

$$v_i = R_s i_1 + v_g$$

$$0 = \frac{1}{C_s}\int i_2\,dt - v_g \tag{11-1}$$

$$i_1 = i_2 + \frac{v_g}{R_i}$$

Upon differentiating the second equation and substituting for i_1 and i_2 from the remaining expressions, the differential equation of the circuit is just

$$C_s \frac{dv_g}{dt} + \left(\frac{1}{R_s} + \frac{1}{R_i}\right) v_g = \frac{v_i}{R_s} \tag{11-2}$$

The solution to Eq. (11-2) is

$$v_g = \frac{v_i}{1 + R_s/R_i}\left(1 - e^{-t/\tau}\right) \tag{11-3}$$

where

$$\tau = \frac{C_s}{1/R_s + 1/R_i} \tag{11-4}$$

is the time constant of the circuit.

According to Eq. (11-3) the output voltage rises exponentially to its final value at a rate determined by the circuit time constant. It is desirable to make the time constant as small as possible for minimum distortion of the leading edge of the pulse.

At the end of the input pulse, the output voltage decays back to zero at a rate again determined by the circuit time constant. Note, however, that the preceding stage may be completely cut off at the end of the pulse, so that $R_s \cong \infty$ and $\tau = R_i C_s$ at the trailing edge. In vacuum-tube and FET amplifiers $R_s \ll R_g$ during conduction and the time constant at the onset of the pulse is $\tau = R_s C_s$. This means that the rise and fall of the amplified pulse are likely to have different time constants. The rise time is usually the faster. By contrast, $R_i \ll R_s$ in transistor amplifiers and the time constant for both rise and fall is simply $R_i C_s$.

It is convenient to define the *rise time* t_r of a pulse as the time required for the voltage to go from 10 to 90 percent of its final value, as indicated in Fig. 11-1. A useful approximate expression for t_r may be developed by introducing

$$v_g = 0.9 \frac{v_i}{1 + R_s/R_i} \tag{11-5}$$

into Eq. (11-3). The result is solved for t_r,

$$0.9 \frac{v_i}{1 + R_s/R_i} = \frac{v_i}{1 + R_s/R_i} (1 - e^{-t_r/\tau}) \tag{11-6}$$

$$0.1 = e^{-t_r/\tau}$$

$$t_r = \tau \ln 10 \tag{11-7}$$

Introducing the upper cutoff frequency $2\pi f_0 = 1/\tau$ of the amplifier from Eq. (7-12),

$$t_r = \frac{\ln 10}{2\pi f_0} \cong \frac{1}{3f_0} \tag{11-8}$$

According to Eq. (11-8) an amplifier with a good high-frequency response is necessary if the pulse rise time is short. For example, if the pulse rise time is 1 μs, the amplifier bandwidth must extend to 370 kHz. In this connection it should be remembered that the bandwidth of a multiple-stage amplifier is always poorer than that of any individual stage. Inverse feedback is commonly used in pulse amplifiers to improve high-frequency characteristics.

Tilt A second form of pulse distortion is departure of the top of the pulse from a constant value, called *tilt,* in Fig. 11-1. This is a result of low-frequency cutoff of the amplifier. In effect, the pulse length represents a dc signal of short duration. Therefore, the appropriate single-stage equivalent circuit, Fig. 11-3, includes the interstage coupling capacitor

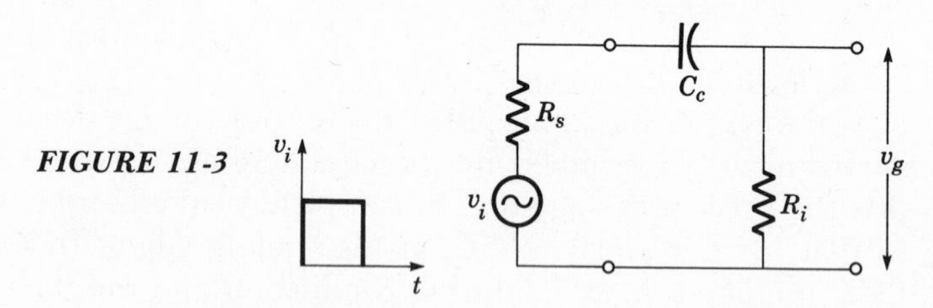

FIGURE 11-3

C_c (compare Fig. 7-5). The output voltage accompanying a square-wave input v_i beginning at $t = 0$ is found immediately from Eq. (2-82),

$$v_g = v_i e^{-t/\tau} \tag{11-9}$$

where

$$\tau = (R_i + R_s)C_c \tag{11-10}$$

Clearly, the circuit time constant τ must be long for minimum decay of output voltage during the pulse. This means that a large value of C_c is required.

The tilt is small in most practical situations, so Eq. (11-9) can be approximated by the first two terms of a series expansion,

$$v_g = v_i \left(1 - \frac{t}{\tau}\right) \tag{11-11}$$

The percentage tilt $P = \Delta v_g / v_i$ at the end of the pulse duration t_w is then

$$P = \frac{\Delta v_g}{v_i} = \frac{t_w}{\tau} \tag{11-12}$$

Introducing the low-frequency cutoff $2\pi f_0 = 1/\tau$ from Eq. (7-4),

$$t_w = \frac{P}{2\pi f_0} \tag{11-13}$$

Notice that a long pulse requires an amplifier with an excellent low-frequency response. For example, suppose that tilt distortion must be less than 1 percent. Then the low-frequency cutoff of an amplifier required to handle 1-ms pulses must extend down to 1.6 Hz.

The results of these two sections provide a convenient technique for determining response characteristics of amplifiers. A square wave is applied to the input, and the output waveform is displayed on an oscilloscope. The frequency of the square wave is reduced until tilt is measurable in the output waveform and the low-frequency cutoff determined using Eq. (11-13). Similarly, the square-wave frequency is increased until the rise time of the amplified pulse is observed on the oscilloscope. Then Eq. (11-8) gives the high-frequency cutoff of the amplifier. Of course, it is assumed that the square-wave signal source and the response characteristics of the oscilloscope are sufficiently good so that they introduce negligible pulse distortion.

Such *square-wave testing* is particularly advantageous in adjusting amplifiers for optimum response because the results of any changes are immediately apparent in the output pulse waveform. This makes repeated laborious determinations of the sine-wave frequency-response characteristic unnecessary. Waveforms similar to those of Figs. 2-34 and 2-36 are often seen, depending upon the relative values of the square-wave frequency and amplifier cutoff frequency.

MULTIVIBRATORS

Astable Multivibrator Two-transistor or two-tube feedback circuits called *multivibrators* find extensive application in pulse circuits. In one or another of its many modifications a multivibrator can be used to generate nonsinusoidal waveforms, to discriminate between pulses of different amplitudes, and to count the number of pulses in a signal. The multivibrator is a basic building block in pulse and digital circuits.

It is easiest to begin with a study of the so-called astable multivibrator, Fig. 11-4, the terminology for which will become clear as we proceed.

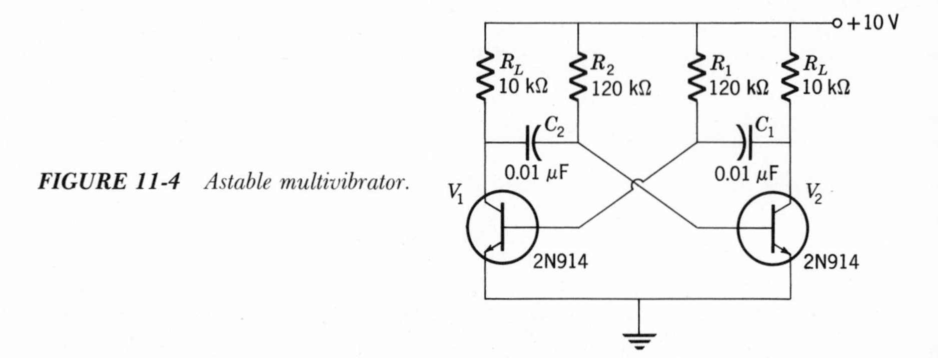

FIGURE 11-4 *Astable multivibrator.*

This circuit can be recognized as a two-stage *RC* amplifier with the output coupled back to the input. The feedback ratio is unity and positive because of the 180° phase shift in each stage; thus, the circuit oscillates. Because of the very strong feedback signals, the transistors are driven into either cutoff or saturation, and nonsinusoidal oscillations are generated.

Suppose that at some instant the feedback voltage drives V_1 into cutoff; this implies that V_2 is conducting because of the relative phase shift between the two stages. The voltage drop across the collector resistor of V_2 puts the collector at nearly ground potential, and C_1 charges through R_1 toward the collector supply potential. When the voltage across C_1 increases sufficiently to bias the emitter junction of V_1 in the forward direction, V_1 begins to conduct. The collector voltage of V_1 drops and V_2 is driven into cutoff through coupling capacitor C_2. Now, C_2 charges through R_2 until V_2 becomes forward-biased and the cycle repeats. These actions will be illustrated graphically when voltage waveforms in the circuit are considered.

Notice that the circuit alternates between a state in which V_1 is conducting and simultaneously V_2 is cut off and a state in which V_1 is cut off and V_2 is conducting. The transistion between these two states is rapid because of the strong feedback signals. The time in each state depends upon the coupling capacitor and bias resistor time constant. Since each transistor is driven alternately into cutoff and saturation, the voltage waveform at either collector is essentially a square wave with a peak amplitude equal to the collector supply voltage.

This picture is confirmed by actual collector voltage waveforms, Fig. 11-5. The triangular base voltage waveforms illustrate the alternate

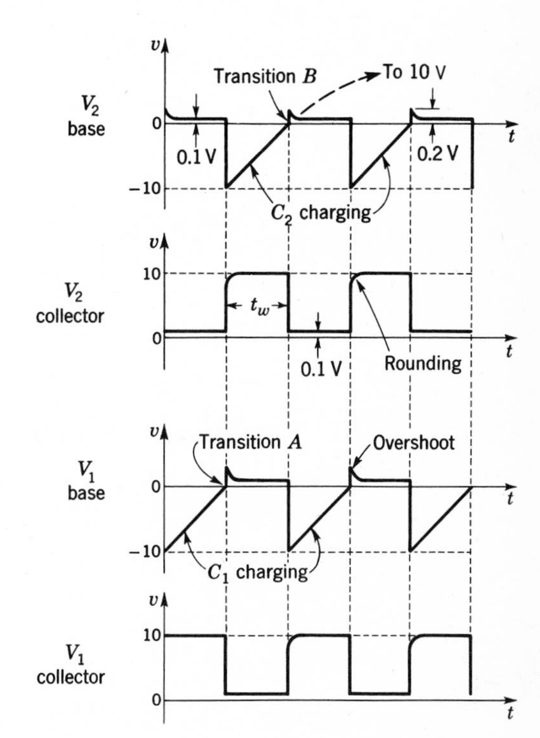

FIGURE 11-5 *Collector and base voltage waveforms in transistor multivibrator.*

charging and discharging of the coupling capacitors. Of particular interest is the very low voltage drop across the transistors when saturated ($\approx$ 0.1 V) and the fact that the transition from one state to another is initiated when the base voltage of the cutoff transistor just slightly exceeds zero. Note, too, that there is a small voltage spike, or *overshoot*, in the base voltage waveform at the transition from one state to the other. The reason for the overshoot may be seen by focusing attention on capacitor C_2 when V_1 goes from conducting to cutoff. Consider that portion of the circuit shown in Fig. 11-6*a* where the on-to-off transition of V_1 is represented by a switch. Further simplification in the circuit is effected by ignoring R_2 since $R_L \ll R_2$; in addition, the forward-biased emitter junction of V_2 is represented approximately by a simple resistor r, Fig. 11-6*b*. According to the waveforms of Fig. 11-5, V_1 switches from saturation to cutoff at transition B when the base voltage of V_2 (and hence the voltage across C_2) is essentially zero. Thus, when the switch in the circuit of Fig. 11-6*b* opens, C_2 begins to charge through R_L and r and the charging current results in an overshoot voltage drop across the emitter junction resistance. From

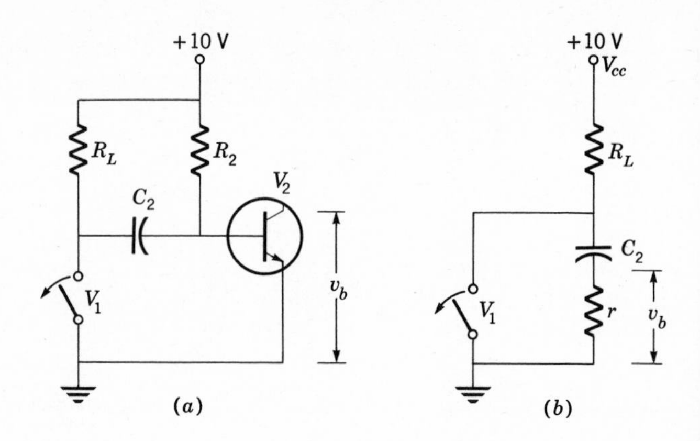

FIGURE 11-6 *Equivalent circuits of the V_2 portion of Fig. 10-4 showing the origin of overshoot.*

Fig. 11-6*b* the peak value of this voltage is

$$v_{\text{overshoot}} = \frac{V_{cc}}{R_L + r} r \tag{11-14}$$

while the time constant of the overshoot spike is

$$\tau_{\text{overshoot}} = (R_L + r)C_2 \tag{11-15}$$

The charging current through R_L also prevents the collector voltage of V_1 from rising immediately to V_{cc}. This is the source of the rounding of the collector waveform pulse edge shown in Fig. 11-5. A typical value for r is 200 Ω, so that the peak overshoot in the multivibrator of Fig. 10-4 is $10(200/10{,}200) = 0.2$ V. This small value, plus the relatively short time constant, Eq. (11-15), means the overshoot introduces only a minor departure from a square collector waveform.

The pulse width in a multivibrator depends upon the time constant of C_1 (or C_2) charging through R_1 (or R_2). A value for t_w can be written down immediately from the expression for the period of a relaxation oscillator derived in the previous chapter, Eq. (9-41). Comparing the waveforms in Fig. 11-5 with those of Fig. 9-18,

$$t_w = R_1C_1 \ln \frac{V_{cc} - (-V_{cc})}{V_{cc} - 0}$$

$$t_w = R_1C_1 \ln 2 \tag{11-16}$$

When $R_1 = R_2$ and $C_1 = C_2$ the waveform is a square wave of frequency $f = 1/2t_w$. It is equally possible to generate asymmetrical waveforms by choosing nonequal values for, say, the coupling capacitors. The result is alternating pulses of widths given by Eq. (11-16) for each time constant.

Astable multivibrators are relaxation oscillators and can therefore be synchronized by application of external signals. Positive synchronizing pulses can be applied to either base to trigger the other transistor into conduction. Multivibrators are often used in this way as frequency dividers. The circuit is also a useful square-wave generator, and, in conjunction with a differentiating circuit, yields sharp, spike waveforms.

The Binary A dc-coupled multivibrator can exist in either of its two states indefinitely. It can be caused to make a transition from one state to the other by an external trigger pulse. Because of its two stable states, a multivibrator with dc coupling is called a *binary*. A binary is also commonly referred to as a *flip-flop*.

Consider the typical transistor binary in Fig. 11-7. Note that conven-

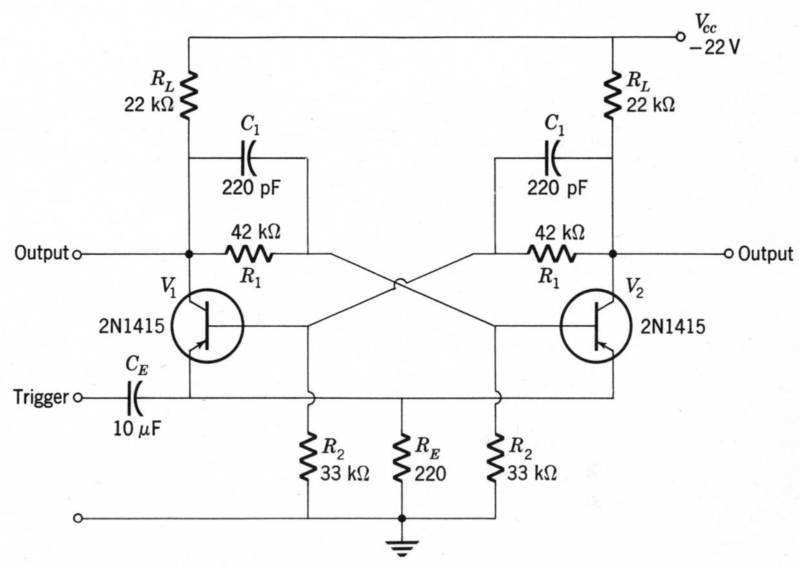

FIGURE 11-7 *Binary circuit is dc-coupled multivibrator.*

tional dc coupling using resistors R_1 and R_2 is used. In other respects the circuit is similar to the ac-coupled, astable multivibrator. Trigger pulses are introduced across the emitter resistor R_E. The purpose of capacitors C_1 is to improve triggering action, as will be discussed subsequently. Output is conventionally taken from either, or both, collectors.

It is important to verify that a stable state of the binary exists when one transistor is cut off and the other is saturated. To begin, assume that V_1 in Fig. 11-7 is cut off so that its collector current is negligible. Collector current in the saturated transistor V_2 is found by drawing a load line corresponding to the total load resistance $R_L + R_E$ on the collector characteristics. The intersection of this load line with the current axis gives the collector current, 9.3 mA, and also shows that a base current of 125 μA is needed to reach saturation. The bias on V_1 may now be calculated using the portion of the circuit shown in Fig. 11-8*a*, where the

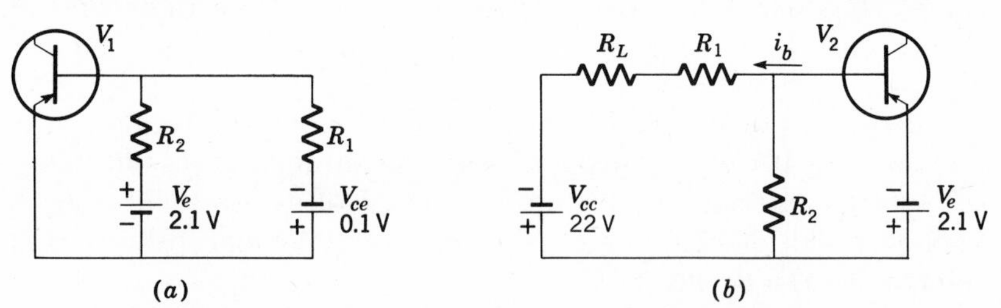

FIGURE 11-8 *Equivalent circuits of transistor binary used to show that binary has only two stable states.*

emitter voltage $V_e = i_c R_E = 9.3 \times 10^{-3} \times 220 = 2.1$ V. The emitter-collector voltage of V_2 is taken to be 0.1 V when saturated. Actually, a more accurate value for the emitter-collector voltage may be determined from transistor manufacturer's data, but it is usually small enough to be ignored in any event. Using Fig. 11-8*a*, the emitter base voltage of V_1 is

$$V_{eb} = V_e \frac{R_1}{R_1 + R_2} = 2.1 \frac{42}{42 + 33} = 1.2 \text{ V}$$

This value of reverse bias at the emitter junction is much more than required to keep V_1 cut off, as assumed.

To show that V_2 is biased to saturation, focus attention on that portion of the circuit shown in Fig. 11-8*b*. The resistance R_2 may be ignored in comparison with the resistance of the emitter junction, and the current through V_1 is negligible because it is cut off. Therefore,

$$i_b = \frac{V_{cc} - V_e}{R_L + R_1} = \frac{22.5 - 2.1}{2.2 + 42} 10^{-3} = 460 \ \mu\text{A}$$

This is much greater than the 125 μA needed to keep V_2 in saturation. Thus, it is established that V_1 is cut off and V_2 is saturated. Because of the symmetry of the circuit, it is obvious that an equally stable state is the one in which V_1 is saturated and V_2 cut off. When the binary switches from one configuration to the other the collector voltage changes from essentially the emitter potential in the saturated condition to a voltage equal to V_{cc} less the drop across R_L resulting from the base bias current. Therefore the output signal is

$$v_o = (22.5 - 2.2 \times 10^3 \times 460 \times 10^{-6}) - 2.1 = 19.4 \text{ V}$$

The circuit is triggered into switching from one state to the other by a positive voltage pulse applied across the emitter resistor. This drives the saturated transistor out of saturation and also tends to turn on the cutoff transistor. As current in the saturated transistor decreases, the collector signal is coupled to the cutoff transistor, further turning it on. Once initiated, the circuit flips itself over because of the strong regenerative feedback. Each successive trigger pulse causes the binary to assume its alternate state. The trigger-pulse amplitude must be somewhat greater than the quiescent voltage across R_E in order to initiate the transfer; in the circuit of Fig. 11-7, a positive pulse of 2.5 V is sufficient.

It is of interest to inquire if the binary has any other stable states than the two considered so far. According to the above analysis, it is not possible for both transistors to be cut off simultaneously, for then the base bias conditions on one transistor would be inappropriate. Similarly, both transistors cannot be saturated at the same time. A state in which equal, but not saturation, currents are in each transistor is possible in principle, but this is a state of unstable equilibrium. Any slight departure from exact balance upsets the equilibrium and the circuit reverts to one of the stable states. Suppose, for example, that a random-noise voltage decreases the current in V_1 slightly. The signal at the collector of V_1 tends to increase the current in V_2, and the collector signal of V_2 further reduces the current in V_1. This regenerative action continues until V_1 is cut off and V_2 is saturated. We conclude that the only two stable states of a binary are with one transistor cut off and the other saturated, and vice versa. Note that minor noise voltages cannot cause a transition out of one of the stable states because the transistors are cut off and saturated. Therefore, the amplifier gain is very small and regeneration is not effective.

The speed with which a binary makes the transition from one state to the other limits the trigger pulse rate for reliable switching. Stray shunt capacitances reduce the transition speed, and the same considerations that apply to the high-frequency response of amplifiers are important in binary circuits. This is not surprising since during the transition the circuit really operates as a feedback amplifier. In particular, the total shunt capacitance from base to ground is the limiting factor. This is so because this capacitance must be charged through the relatively high resistance of R_1 in order to change the base voltage corresponding to one state to that of the other state. The time required to charge the input capacitance is reduced by shunting both R_1 resistors with small capacitances C_1, called *commutating capacitors.* When commutating capacitors are used, the time between states is reduced to essentially that required for C_1 to charge through the parallel combination R_1 and R_2. Thus, the resolving time of the binary in Fig. 11-7 is

$$t = \frac{C_1}{1/R_1 + 1/R_2} = 220 \times 10^{-12} \frac{42 \times 33}{75} \times 10^3 = 4.1\ \mu\text{s}$$

This means that the binary can respond reliably to sequential trigger pulses only if they are separated by about 5 μs. With special design and additional circuit complications binaries having resolving times of the order of 0.01 μs are possible.

The versatility of a binary is enhanced by the variety of inputs which may cause a change of state. A particularly flexible and popular arrangement, Fig. 11-9, employs three separate inputs *R*, *S*, and *T*, and four

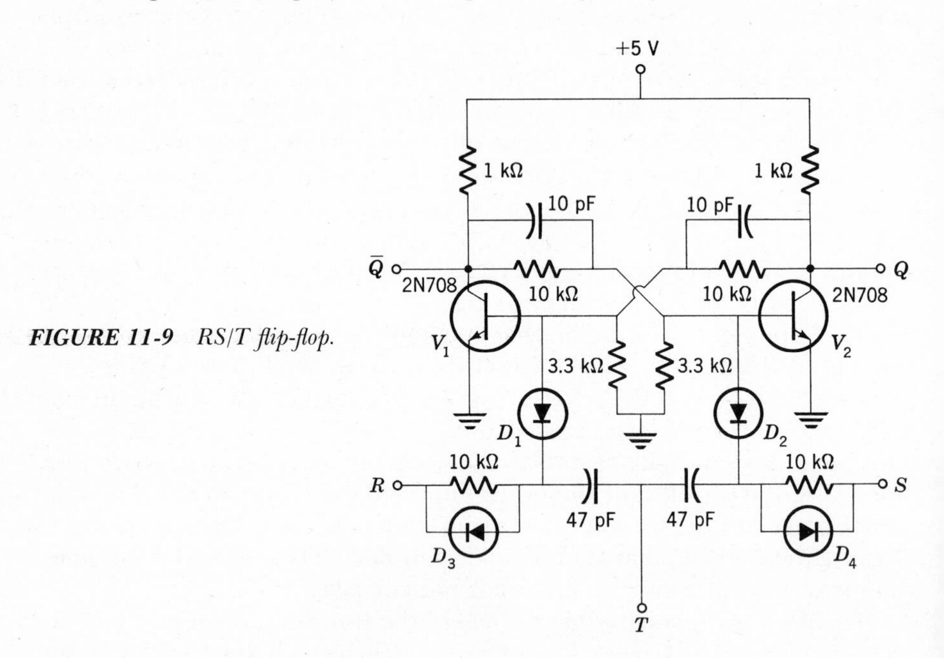

FIGURE 11-9 *RS/T flip-flop.*

diodes to steer input pulses to the desired transistor. Suppose, for example, input *S* is at ground potential, input *R* is at +5 V, and a positive pulse is applied to terminal *T*. The leading and trailing edges of the pulse are differentiated by the 47-pF capacitors. The positive leading edge is blocked by both D_1 and D_2 and shunted to ground by D_3; thus, nothing happens. The negative trailing edge is passed by D_2 and turns transistor V_2 off. The negative pulse is not transmitted through D_1 because of the positive potential at *R*. The result is that V_2 is cutoff and V_1 is conducting and the binary is said to be *set*.

If the *R* and *S* inputs are reversed so that *R* is at ground potential and *S* is at +5 V, a positive pulse applied to terminal *T* results in the state in which V_1 is cutoff and V_2 is conducting. The binary is then said to be *reset*, or *cleared*. Note that a positive potential at *S* causes the binary to be set by an input pulse while a positive potential at *R* causes it to be reset. In effect these control potentials steer input pulses at *T* to the corresponding transistor and turn it off. The transitions between states are timed or "clocked" by the signal at *T*. Therefore, the *T* input is sometimes referred to as the *clock* connection; the entire configuration in Fig. 11-9 is known as an *RS/T* flip-flop.

For many applications it is desired that the binary change state upon successive input pulses at terminal T. This is easily accomplished by connecting S to output terminal Q and R to output terminal $\overline{Q}$, Fig. 11-9. Suppose, for example that V_2 is conducting and V_1 is cut off so that S is essentially at ground potential while R is essentially at +5 V. According to the preceding discussion, the negative trailing edge of a positive pulse applied to T will be steered to V_2, turning it off so that the binary changes state. Now S is at +5 V and R is grounded so that the next pulse is steered to V_1 and the original state is restored. This action is much like that of a mechanical toggle switch and for this reason T is also called the *toggle* input.

A useful binary circuit called the *Schmitt trigger* (after its inventor) results if one collector-to-base coupling is omitted and replaced by feedback through a common-emitter resistor, Fig. 11-10. This circuit has two stable

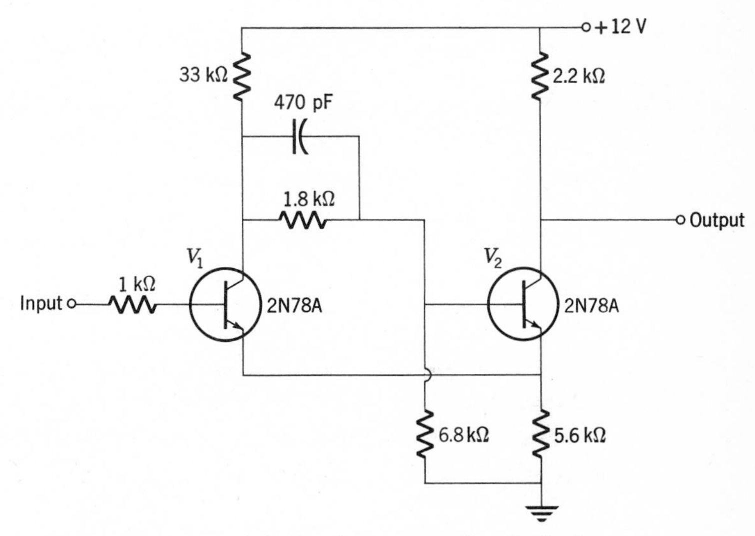

FIGURE 11-10 *Schmitt trigger uses emitter feedback.*

states and the magnitude of the input voltage determines which of the two is possible. Suppose the input voltage is zero and assume that V_1 is cut off while V_2 is conducting but not saturated. Since V_1 is cut off, the quiescent current of V_2 is found by determining the operating point. The voltage drop across the emitter resistor is large enough to cut off V_1, as initially assumed. In the circuit of Fig. 11-10, for example, the emitter bias is 6.6 V. Thus, the normal state of the Schmitt trigger has V_1 cut off and V_2 conducting.

Suppose now the input voltage in Fig. 11-10 is increased from zero. Nothing happens until the input voltage just exceeds the base bias, 6.6 V, at which point V_1 begins to conduct. The bias on V_2 is reduced and the emitter bias voltage decreases; this, in turn, increases current in V_1. Regenerative action continues until V_1 is conducting and V_2 is cut off. This state is maintained so long as the input voltage is greater than 6.6 V. As the input voltage is decreased, the circuit does not regain its original state until an input voltage significantly smaller than 6.6 V is attained. The reason for this is that the base bias on V_2 is much lower in this state than previously because the collector of V_1 is at a lower potential. Therefore the current in V_1 must be considerably reduced before the smaller emitter voltage and increased collector potential of V_1 combine to increase the base current of V_2 and it begins to turn on. In the circuit of Fig. 11-10 it is necessary to reduce the input voltage to 5.2 V before regeneration returns the circuit to the original state.

An interesting application of the Schmitt trigger is as a pulse regenerator or *squaring circuit,* illustrated in Fig. 11-11. The input signal, a series of

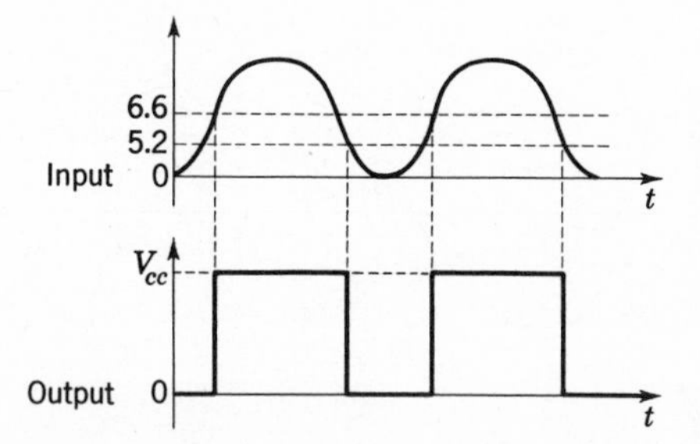

FIGURE 11-11 *Waveforms in Schmitt trigger used as squaring circuit.*

degraded pulse waveforms, is converted to a square pulse output as the circuit is triggered back and forth between its two states by the input waveform. Note that even a sinusoidal signal may be converted to a square wave in this way.

The Schmitt trigger is also useful as a *pulse height discriminator* to measure the voltage pulse amplitudes of an input signal. Each time an input pulse exceeds the trigger threshold the circuit generates an output pulse. Thus, by varying, say, a dc voltage in series with the input signal, the range of pulse sizes in a signal can be determined. A precision of the order of 0.1 V is possible with this circuit. Note that the output pulses are all of uniform amplitude, independent of the input trigger pulse. This is useful in that subsequent circuits need not be capable of handling a wide range of pulse sizes.

Monostable Multivibrator A final form of multivibrator is the partial combination of a binary and an astable multivibrator. Both dc and ac coupling are employed, Fig. 11-12. As might be anticipated, this circuit has only one stable state: with V_1 cut off and V_2 saturated. When triggered out of this state by a negative pulse applied to the grid of V_2, V_1 becomes conducting and V_2 is cut off. This state lasts for a time determined by the

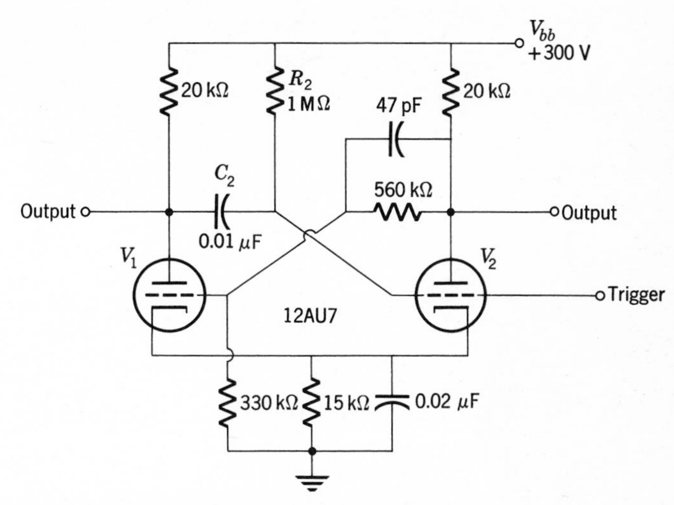

FIGURE 11-12 *Monostable multivibrator uses both ac and dc coupling between stages.*

R_2C_2 time constant, after which the circuit reverts spontaneously to the stable configuration. Because of this action the circuit is called a *one-shot* or *monostable* multivibrator. This completes the roster of multivibrator types: the binary, which has two stable states and is therefore bistable; the monostable multivibrator with its one stable state; and the astable circuit which is a free-running oscillator having no stable states.

A useful version of the monostable multivibrator uses cathode feedback for dc coupling as in the Schmitt trigger. In this circuit, Fig. 11-13, the

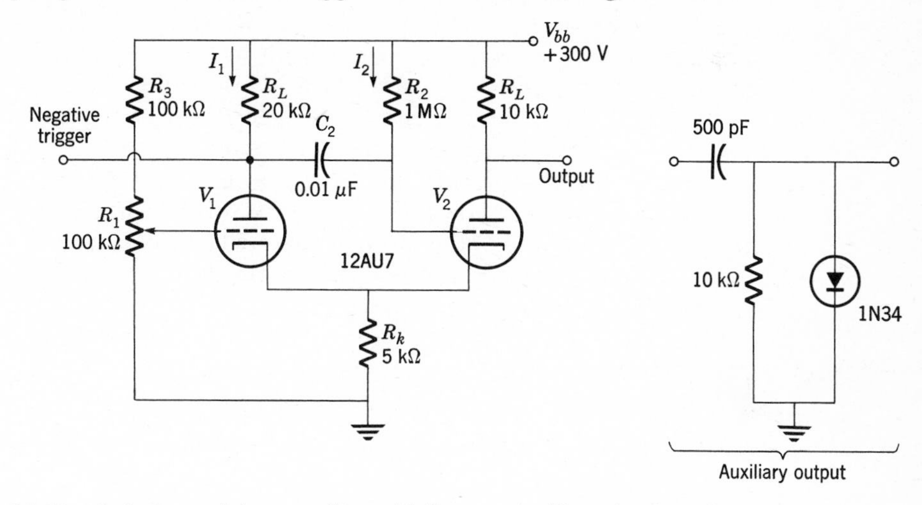

FIGURE 11-13 *Cathode-coupled monostable multivibrator. Auxiliary circuit produces negative output pulses similar to input trigger pulses but delayed by multivibrator.*

time that V_1 spends in the conducting state is easily varied by changing the bias using potentiometer R_1. This action can be understood by focusing attention on the voltage waveform at the grid of V_2, Fig. 11-14. In the

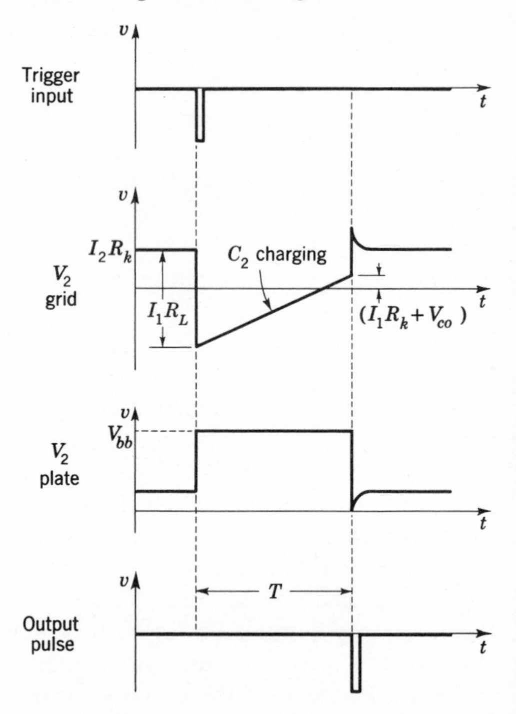

FIGURE 11-14 *Waveforms in monostable multivibrator used as variable time delay. Delay time T is adjusted by changing I_1.*

stable state with V_2 conducting, the grid voltage is essentially at cathode potential I_2R_k because of the positive grid voltage applied through R_2. This bias causes a small forward current in the grid-cathode diode and the resulting drop across R_2 places the grid just slightly positive with respect to the cathode. When the circuit is triggered, V_1 conducts and the grid voltage drops momentarily by the amount I_1R_L because of coupling through C_2. Now C_2 begins charging through R_2 until the grid-cathode potential of V_2 reaches cutoff V_{co}. The grid voltage with respect to ground at this time is the sum of the cathode potential I_1R_k plus the cutoff potential V_{co} at the given plate voltage. When the grid reaches this potential, V_2 begins to conduct and the circuit returns to its stable state.

The time T that the circuit remains in its second state can be determined from the expression for the period of a relaxation oscillator, Eq. (9-41). Comparing the waveforms of Fig. 9-18 with those of Fig. 11-14,

$$T = R_2C_2 \ln \frac{V_{bb} - (I_1R_k - I_1R_L)}{V_{bb} - (I_1R_k + V_{co})}$$

$$T = R_2C_2 \ln \frac{1 - (I_2R_k - I_1R_L)/V_{bb}}{1 - (I_1R_k + V_{co})/V_{bb}} \tag{11-17}$$

Expanding the log term in a series and retaining only the first term is equivalent to assuming that the portion of the exponential charging characteristic is linear. Thus, Eq. (11-17) becomes

$$T = R_2C_2\left(\frac{I_1R_L - I_2R_k}{V_{bb}} + \frac{I_1R_k + V_{co}}{V_{bb}}\right)$$

$$T = R_2C_2 \frac{R_L + R_k}{V_{bb}} I_1 + \frac{R_2C_2}{V_{bb}} (V_{co} - I_2R_k) \tag{11-18}$$

The first term in Eq. (11-18) increases with I_1, while the second term is a constant set by the circuit design. The current I_1 in V_1 during conduction is determined by the grid bias of V_1, which can be adjusted by the potentiometer. Actually, because of the cathode-follower action of V_1, the current is very nearly a linear function of the dc grid bias. Thus, the conduction period can be altered simply by changing the grid bias on V_1 and, in practice, T varies linearly with bias to an accuracy of about 1 percent.

The output voltage waveform, Fig. 11-14, is a square pulse of duration T. If this waveform is differentiated and the positive peaks clipped (refer to Fig. 11-13), the result is a negative output pulse delayed by a time T from the negative input trigger pulse. Therefore, this cathode-coupled monostable multivibrator is a convenient circuit for introducing a time delay in pulse circuits. The duration of the delay is adjusted by varying the grid bias on the input tube.

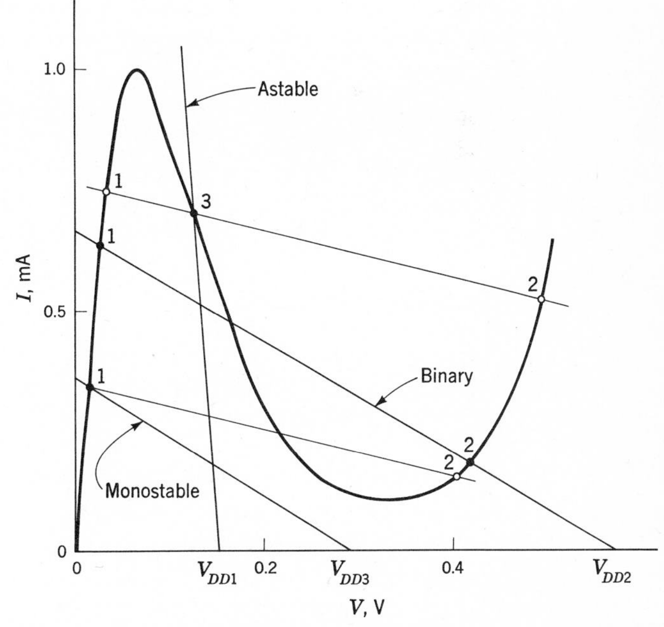

FIGURE 11-15 *The three types of tunnel diode multivibrators depend upon circuit parameters.*

It is interesting to observe that all three types of multivibrator operation, astable, binary, and monostable, are possible with negative resistance devices, depending upon circuit parameters. Consider the current-voltage characteristic of the tunnel diode in Fig. 11-15. According to the analysis in Chap. 9, relaxation oscillations are obtained when the dc load line and V_{DD1} are such that the operating point is on the negative resistance portion of the current-voltage characteristic, and the ac load line intersects the curve at two other points.

When the dc load line intersects the curve at three places, there are two stable states, one at point 1 and the other at point 2. The circuit may be triggered from one binary state to the other by pulses of suitable polarity. Finally, if V_{DD3} is small enough for the operating point to be on the low current portion of the characteristic, the circuit is monostable. An input pulse can cause a momentary transition to point 2, which is followed by a spontaneous transition back to point 1.

COUNTERS

Binary and Decade Scalers A single binary circuit produces one output pulse for every two input trigger pulses. This is so because two trigger pulses cause the binary to shift from one stable state to the other and then back to the first state again. If the output pulse of one binary is used to trigger the input of a second binary, a total of four input pulses produces one output pulse, as will be examined in greater detail presently. Thus, a series of cascaded binary circuits divides, or scales down, the input pulse rate by a factor depending upon the number of cascaded stages. One major application of such a *scaler* is to reduce the pulse rate to the point where the output pulses can be recorded by an electromechanical register. A scaler may also be looked upon as an electronic *counter*. The state of each binary tells the total number of input pulses at any time.

Consider the four-binary scaler of Fig. 11-16. Each binary is similar to Fig. 11-7 except that a separate base bias voltage is used rather than an emitter bias resistor, and triggering is accomplished with steering diodes. Note the similarity of this trigger circuit to Fig. 11-9 with S and R connected to Q and $\overline{Q}$, respectively. The diode connected to the collector of the conducting transistor is reverse biased and blocks the negative trigger pulse. Correspondingly, the other diode is unbiased and steers the pulse to the base of the conducting transistor, thus turning it off. Each binary thus toggles with successive input pulses.

A pulse amplifier is used at the input to provide trigger pulses of sufficient amplitude to cause a transition in binary B_1. Similar trigger amplifiers are often placed between the binaries in order to isolate each stage and also to assure reliable triggering action. Suppose that, initially, the

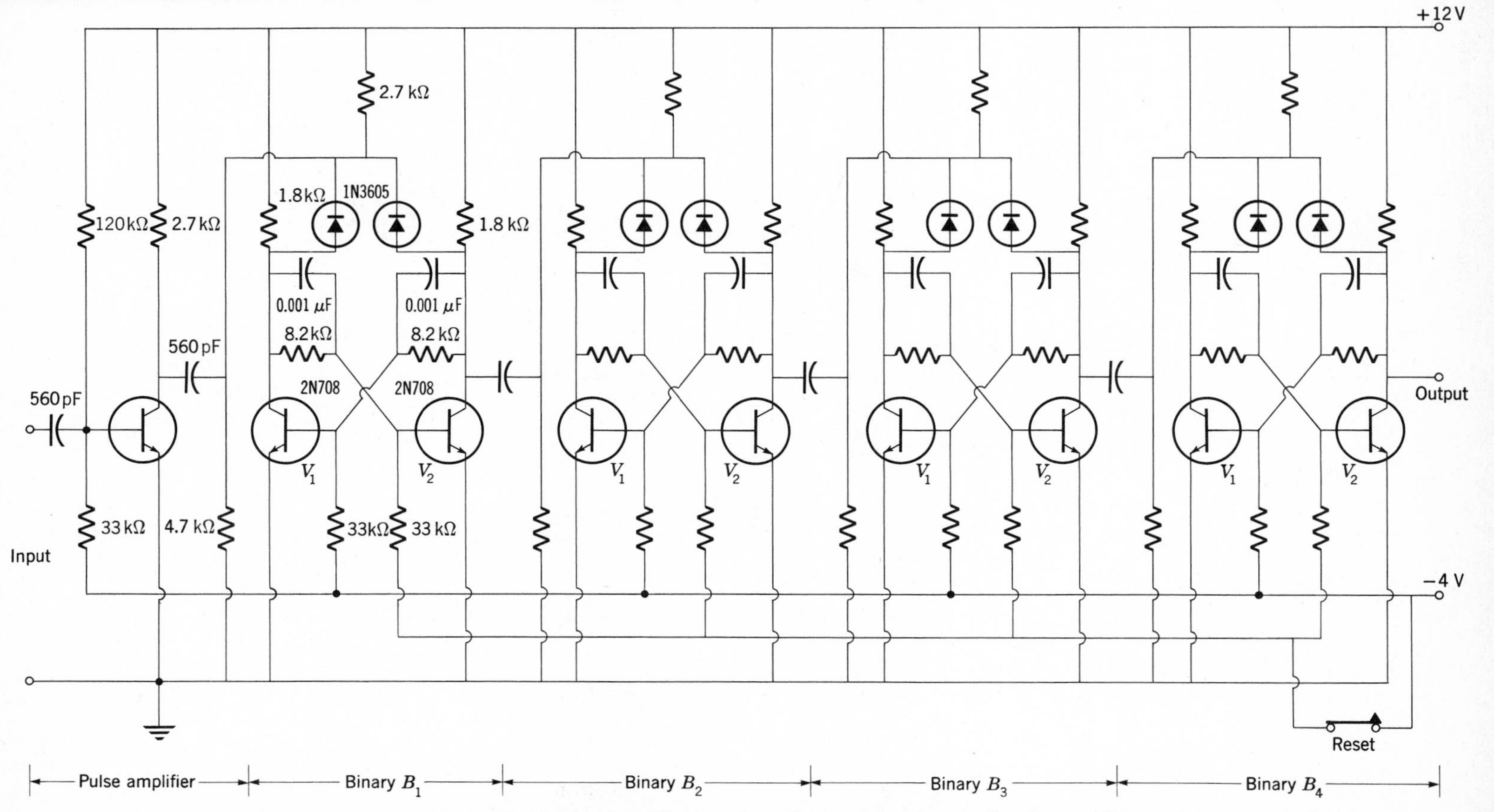

FIGURE 11-16 *Four-binary scaler. Component values in B_2, B_3, and B_4 are the same as in first stage.*

V_2 transistors are all conducting and the V_1 transistors are cut off. This condition is achieved using the *reset* switch. When the switch is opened momentarily a strong forward bias is placed on all V_2 transistors, which assures that they conduct. With the reset switch closed, the binaries perform normally.

For convenience, we designate the state of any binary as state 0 if the output transistor V_2 is saturated. Similarly, when the output transistor is cut off, the binary is said to be in state 1. Starting from the condition in which all four binaries are in state 0, consider the waveforms at the output transistors of each binary as a result of a regular sequence of input trigger pulses, Fig. 11-17. The first input pulse causes binary B_1 to make

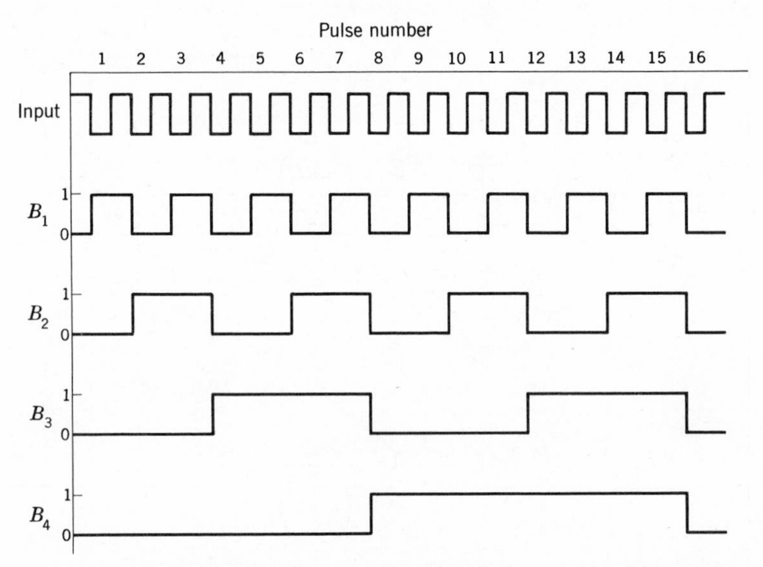

FIGURE 11-17 *Waveforms in scale-of-16 counter.*

a transition from state 0 to state 1. This means that the output transistor goes from saturation to cutoff and the output voltage increases. This positive step is blocked from triggering the second binary by the steering diodes coupling B_1 to B_2. Therefore the result of one input pulse is to put B_1 in state 1 while the other binaries remain in state 0.

At the second input pulse, B_1 returns from state 1 back to state 0. The resulting negative output voltage step triggers B_2 from state 0 to 1. Binary B_3 remains in state 0 because the output voltage from B_2 is a positive signal. The net result of two input pulses is that B_2 is in state 1, while the other three are in state 0. This process continues with further input pulses as illustrated in Fig. 11-17 until at the sixteenth pulse a negative output pulse is produced. The cascade combination of four binaries is, accordingly, called a *scale-of-16* counter. Greater scaling factors are achieved simply by adding additional binary stages.

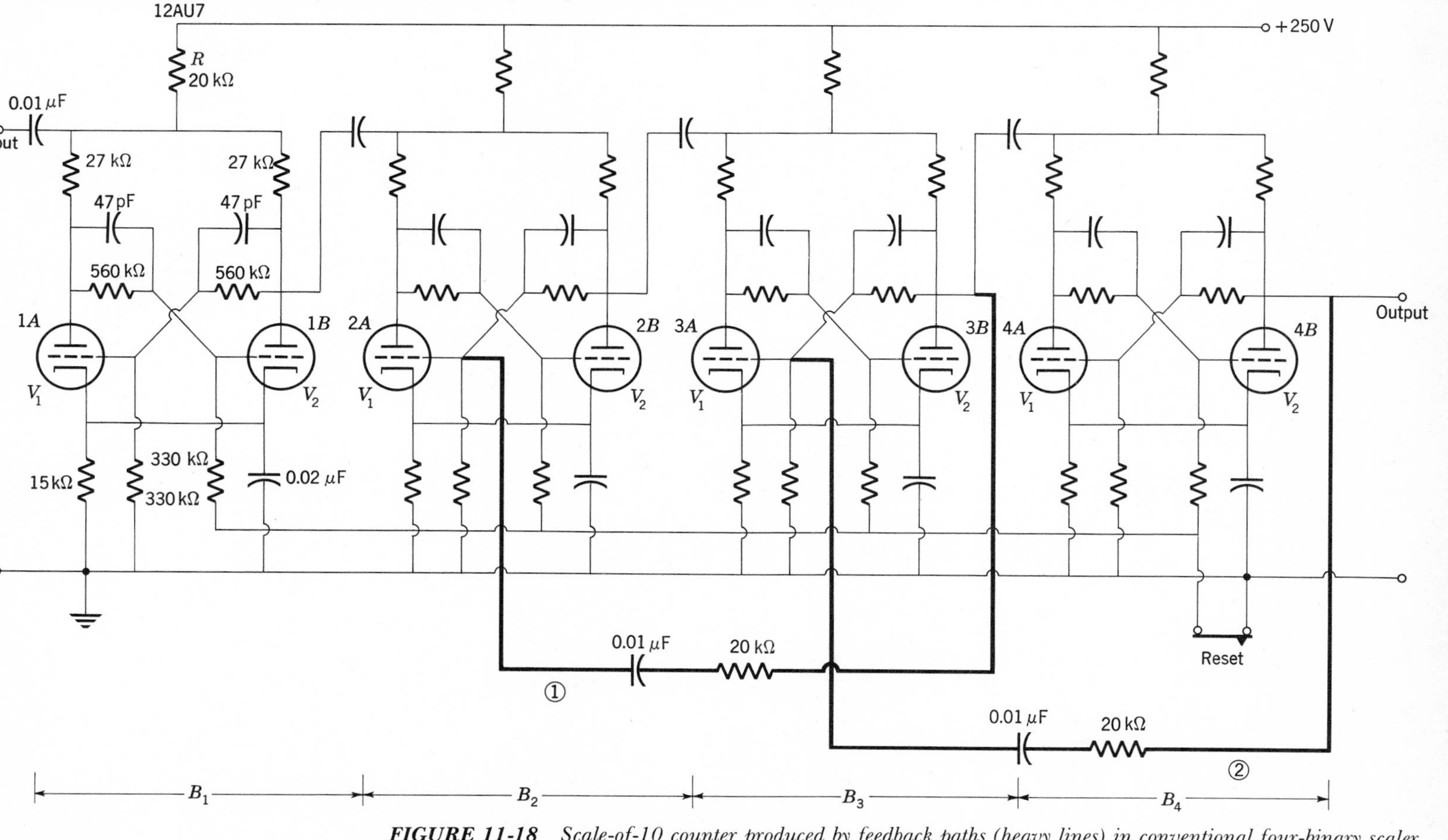

FIGURE 11-18 Scale-of-10 counter produced by feedback paths (heavy lines) in conventional four-binary scaler.

Because of the two stable states of each stage, a binary scaler basically counts by 2s, or rather, by powers of 2, a process which is explored in greater depth in the next chapter. It is somewhat inconvenient to determine the total number of counts corresponding to any given combination of 0 and 1 states for each binary in a scaler because we are not familiar with the binary number system. Rather, we prefer to count in the decimal system, and for this purpose *decade scalers* are used. The scale of a binary counter may be converted to a decade counter by several feedback arrangements which return pulses from certain stages to preceding binaries in the cascade.

For example, a decade counter based on a scale-of-16 binary is illustrated in Fig. 11-18. In this vacuum-tube version the method of coupling trigger pulses from one binary to the next does not use steering diodes. Trigger pulses are developed across a resistor R in series with both plates of the succeeding binary. Thus, when the preceding binary changes states, a positive pulse or a negative pulse is applied to both plates. A binary is triggered best by turning off the on tube, so that only the negative trigger pulses are effective. To convert the scale-of-16 counter into a decade counter, two feedback circuits are added: one from the output of B_3 to the grid of V_1 in binary B_2; the other from the output of B_4 to the grid of V_1 in binary B_3. These feedback paths are shown by the heavy lines in Fig. 11-18.

The operation of the decade counter is best understood by examining the plate voltage waveforms at each output tube, Fig. 11-19. Through the third count the circuit operates as a normal scaler. At the fourth input pulse binary B_1 makes a transition from state 1 to 0; this triggers binary B_2 into a similar transition, and the output of B_2 causes B_3 to make a

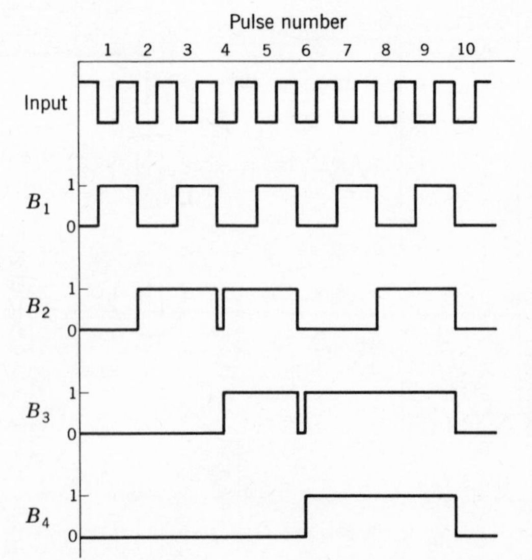

FIGURE 11-19 *Waveforms in decade counter.*

transition from state 0 to state 1. At this time the positive pulse from the output of B_3 is coupled back through the first feedback path to the grid of V_1 in binary B_2. Since B_2 is in the 0 state, V_1 is cut off and the positive feedback pulse induces a transition from state 0 to 1 in B_2. Similarly, at the sixth input pulse binary B_3 first makes a transition from state 1 to 0 and subsequently reverses in response to the feedback pulse from B_4 through the second feedback path. The net result is that a negative output pulse is delivered by the last binary at the tenth count. That is, the circuit is a decade scaler.

Note that at the fourth count B_2 must settle in state 0 before the feedback pulse arrives to reverse the state. The delay between these two events is provided by the inherent transition time of the B_3 stage. A similar comment applies to the double transition in B_3 at the sixth input pulse. The necessary delay limits the rate at which the counter can reliably follow rapid input pulses.

Visual Displays Most often, the output of a scaler is recorded by an electromechanical ratcheting register. A power output stage driven by the last binary is usually necessary to provide sufficient energy to actuate the register. Since a mechanical register typically cannot operate faster than about 10 pulses per second, a scaling factor at least sufficient to reduce the input pulse rate to this magnitude is required. Conversely, the maximum useful scaling factor is that which reduces the maximum input pulse rate to the limiting rate of the mechanical register. The maximum input pulse rate is set by the resolving time of the first binary stage.

The mechanical register totals the number of output pulses. The total number of input pulses is the product of the register total times the scaling factor plus the number of counts represented by the state of each binary in the scaler. In vacuum tube circuits the state of each binary is indicated by a small neon lamp. The lamp is in series with a current-limiting resistor connected between each output tube plate and ground, Fig. 11-20.

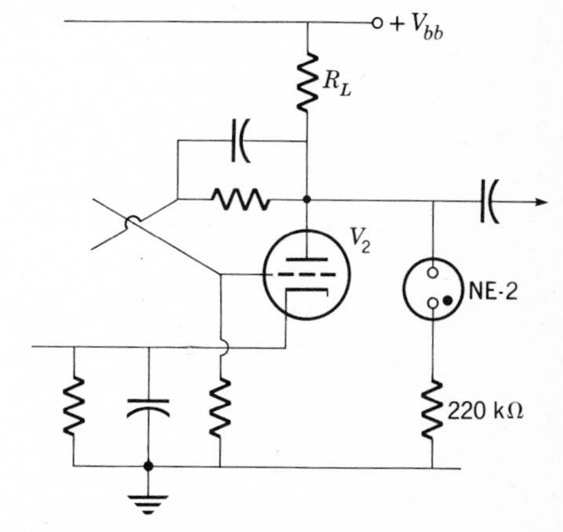

FIGURE 11-20 *Neon-lamp indicator lights when binary is in state 1 with V_2 cut off.*

With the binary in state 0, V_2 is conducting and the plate voltage is insufficient to light the lamp. When the binary makes a transition to state 1, V_2 is cut off and the plate voltage rises sufficiently to light the lamp. The neon bulbs connected to the four binaries in a scale-of-16 counter are labeled 1, 2, 4, and 8, respectively. The count is determined by adding the numbers associated with those neon bulbs which are lit.

In decade scalers a neon bulb is associated with each digit from 0 to 9 and the count is indicated directly in the decimal system. A method for interconnecting the 10 neon bulbs to a decade counter to provide this display is illustrated in Fig. 11-21. The voltages in parentheses correspond

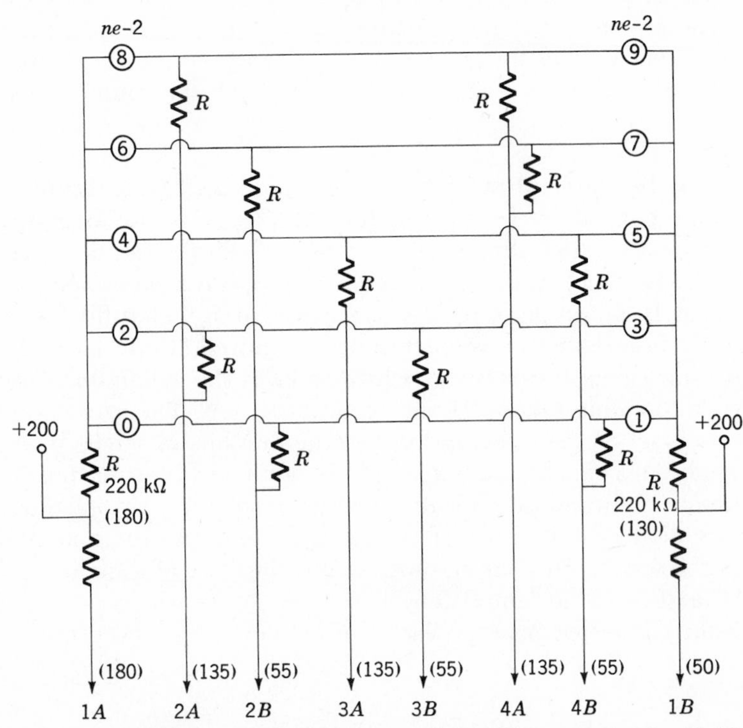

FIGURE 11-21 *Decade neon display for counter in Fig. 11-19.*

to the reset condition when the count is zero and the 0 light is lit. The performance of this display circuit is considered in Exercise 11-5. When three or four decade counters are cascaded, the neon indicator lights for a given decade are conventionally arranged in a vertical column parallel to the lights of the preceding decade. Thus the number of counts in the system is indicated directly in terms of the units, tens, hundreds, thousands, etc., corresponding to each decade column.

Several indicator schemes have been developed that display the actual number corresponding to the count in a decade scaler. One of these, called a *Nixie* tube, is basically a 10-cathode neon tube in which each cathode is a wire shaped into the form of one of the digits from 0 to 9. These are connected to the scaler in a fashion similar to that used for the 10 individual neon bulbs, so that only one cathode is lit at a time. The characteristic neon glow covers the entire cathode wire and this numeral appears at the face of the tube, Fig. 11-22. Thus, one Nixie tube is associ-

FIGURE 11-22 *Nixie tube numerical indicator.* (Burroughs Corporation.)

ated with each decade and when the tubes corresponding to cascaded decades are arranged in a row, the count is indicated directly as a decimal number. Several other schemes which also present a numerical digit display are in use.

The potentials in transistor circuits are normally low compared with the voltages needed to light a neon bulb, and an auxiliary switching circuit is used to turn on the neon bulb or Nixie. A typical circuit, Fig. 11-23, uses a transistor switch activated by the collector potential of the

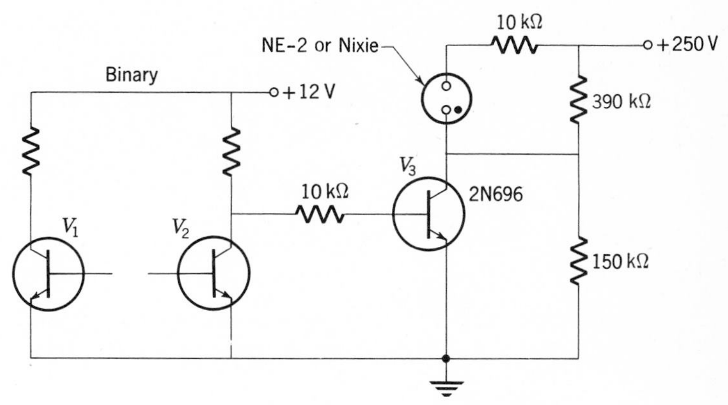

FIGURE 11-23 *Switching transistor used to light neon or Nixie indicator in transistor counters.*

binary output transistor. When V_2 is conducting, its collector potential is low and the base bias on V_3 is so small that it is cut off. Therefore, the Nixie does not light. V_3 is turned on when the binary is in state 1 and the collector potential of V_2 rises sufficiently to provide base bias for V_3. With V_3 saturated the full 250-V power supply potential is applied across the Nixie, which causes it to light. Note that the voltage swing at the collector of the switching transistor need only be somewhat greater than the difference between the breakdown and maintaining voltage of the neon tube. That is, it is not necessary for the transistor to be capable of withstanding the full power-supply potential. Note also that a switching circuit similar to Fig. 11-23 is required for each of the separate cathodes of a Nixie tube.

Dekatron Counter A very useful decade counter that requires a minimum of components is based on the Dekatron tube. This is a gas-filled discharge tube containing a central anode and 30 cathodes arranged around its periphery. Successive input pulses cause a glowing spot to move circumferentially around the face of the tube, thus indicating the number of counts. At the tenth input pulse the Dekatron delivers an output pulse which may be used to trigger a succeeding stage.

The movement of the glowing spot may be understood by referring to Fig. 11-24. The Dekatron contains a single anode, 10 glow cathodes (K_0,

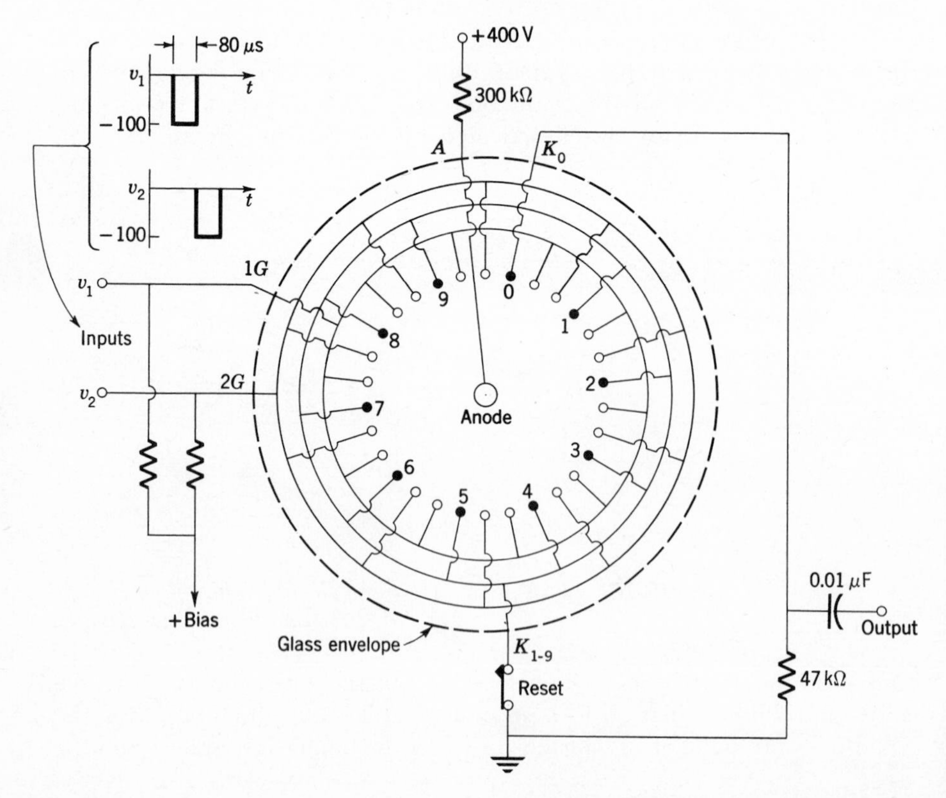

FIGURE 11-24 *Dekatron counter tube.*

$K_1, K_2, \ldots, K_9$), 10 $1G$ guide cathodes ($1G_0, 1G_1, 1G_2, \ldots, 1G_9$), and 10 $2G$ guide cathodes ($2G_0, 2G_1, 2G_2, \ldots, 2G_9$). K_0 is the output terminal, and cathodes K_1 to K_9 are tied together and brought to a single base pin. Similarly, all $1G$ electrodes are connected together, as are all $2G$ cathodes. A small positive bias is applied to the $1G$ and $2G$ cathodes and a normally closed reset switch is included in the K_1-K_9 lead.

Opening the reset switch momentarily causes a discharge between K_0 and the anode, since K_0 is then the most negative cathode. With the reset switch closed, the circuit is set to count pulses. A negative input pulse of about 100 V applied to $1G$ causes the discharge to transfer from K_0 to $1G_0$. Although all $1G$ cathodes experience the same large anode-cathode potential resulting from the negative pulse, cathode $1G_0$ is nearest to the discharge previously existing at K_0. Gas ions remaining in the space near K_0 preferentially continue the discharge at $1G_0$. Just at the termination of the pulse on $1G_0$ a second negative pulse is applied to the $2G$ cathodes. The discharge now transfers to $2G_0$, since it is closest to $1G_0$. Finally, at the end of the second pulse the discharge jumps to K_1 because it is then at a lower potential than $2G_0$. The net result of the duality of negative pulses is that the glowing cathode spot has moved from K_0 to K_1. The spot remains in this position until a second pair of trigger pulses is applied to $1G$ and $2G$, at which time the spot moves to K_2. The sequence continues until at the tenth input pulse pair the discharge returns to K_0. The discharge current through the cathode resistor in series with K_0 results in a positive output pulse.

The circuit diagram of a three-stage, scale-of-1000 decade counter based on the Dekatron is shown in Fig. 11-25. Each counter stage is driven by a dual triode pulse-forming network which generates two properly timed negative pulses from a single positive input pulse. It is instructive to examine the way in which these pulses are produced. The output signal of each Dekatron (and the input to the scaler) is a 30-V positive step. This is differentiated by the combination R_1C_1 to yield an exponential pulse with a time constant of $(2.2 \times 10^5)(680 \times 10^{-12}) = 150$ μs. The positive peak is clipped by the grid-cathode diode of V_1 together with R_4 to give an approximately rectangular pulse 150 μs long at the plate of V_1. This 130-V negative pulse is applied to the 1G cathodes. The output of V_1 is differentiated by R_2C_2, producing first a negative pulse followed by a positive pulse at the leading and trailing edges of the rectangular signal. The negative pulse cuts off V_2, but since this tube is biased nearly to the cutoff position anyway, the output signal is negligible. The positive pulse, $R_2C_2 = (2.2 \times 10^5) \times (100 \times 10^{-12}) = 22$ μs long, is clipped by the grid-cathode diode of V_2 to yield a 130-V negative rectangular pulse at the plate of V_2. The combination R_3C_3 integrates this output, making the pulse length $(1 \times 10^6)(100 \times 10^{-12}) = 100$ μs, which is suitable to drive the $2G$ electrodes. Note that the $2G$ pulse starts at the trailing edge of the $1G$ pulse, as required for proper operation of the Dekatron.

The Dekatron combines the functions of decade counting and display

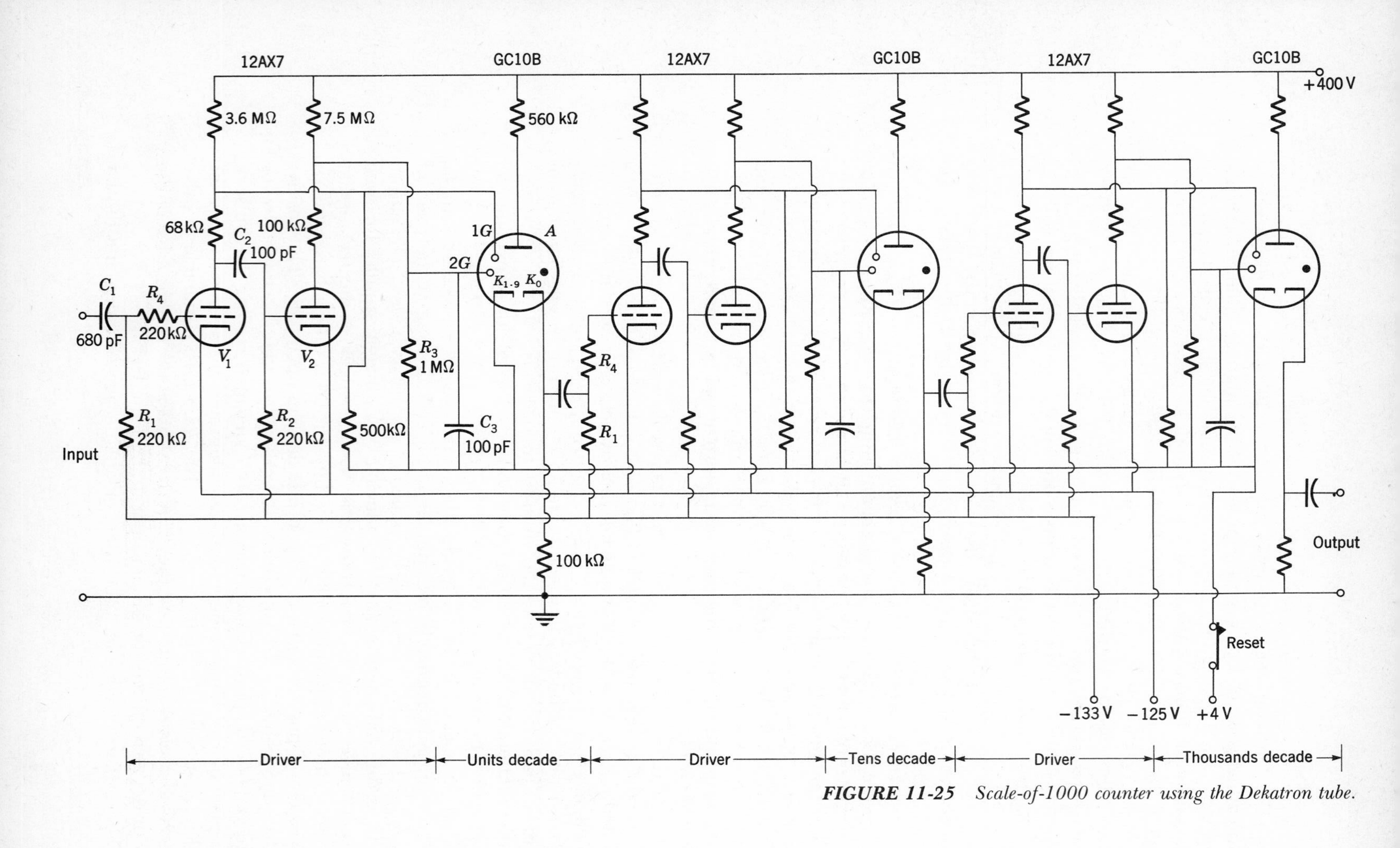

FIGURE 11-25 *Scale-of-1000 counter using the Dekatron tube.*

all in one unit, and comparatively simple scaling circuits are possible. The scale-of-1000 counter in Fig. 11-25, for example, employs a total of only six tubes, compared with at least twelve tubes for a decade counter based on binary circuitry. The major drawback of Dekatron scalers is the maximum counting rate, which is limited by the time required to transfer the discharge from one cathode to another. Note that a time interval of $150 + 100 = 250$ μs is necessary to register one count. The input pulse interval cannot be a shorter time than this, so that the maximum counting rate is $1/250$ μs $= 4000$ Hz. Other Dekatrons are capable of proper performance up to 20 kHz but even this rate is much lower than that possible with transistor and vacuum-tube binaries.

WAVEFORM GENERATORS

Several techniques for generating nonsinusoidal waveforms have been discussed. These include blocking oscillators, relaxation oscillators, and multivibrator circuits. The output signals from these generators can be further modified through the use of diode clippers and clamps to select a portion or set the dc level of a waveform. Also, integrating and differentiating circuits can be used to modify waveforms in specific ways. A number of additional waveform generators which further illustrate the variety of waveshapes of interest in pulse circuits are described in this section.

Diode Pump Consider the action of the so-called *diode-pump* circuit, Fig. 11-26, in response to negative-going signal pulses. On each negative

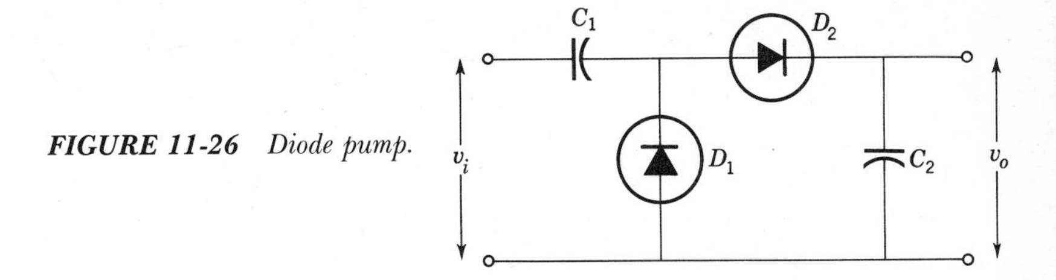

FIGURE 11-26 *Diode pump.*

pulse, diode D_1 conducts and C_1 charges to the peak value of the input pulse. Since the input pulse biases D_2 in reverse, no charge reaches C_2. Thus the charge on C_1 is

$$Q_1 = C_1 v_i \tag{11-19}$$

During the time that the input voltage is zero, the voltage on C_1 biases D_2 in the forward direction and C_2 becomes charged. Since the two capacitors are now effectively connected in parallel, the charge on C_2 is

$$Q_2 = C_2 v_o = C_2 \frac{Q_1}{C_1 + C_2}$$

or

$$Q_2 = \frac{Q_1}{1 + C_1/C_2} \tag{11-20}$$

If $C_2 \gg C_1$, $Q_2 = Q_1$ and the charge has effectively been pumped from the source to C_1 and then to C_2. Note that under this condition C_1 is essentially discharged and the process can be repeated on the next cycle.

The output voltage after the first cycle is

$$v_o = \frac{Q_2}{C_2} = \frac{v_i}{1 + C_2/C_1} \cong \frac{C_1}{C_2} v_i \tag{11-21}$$

where Eqs. (11-19) and (11-20) have been used. On each subsequent cycle a voltage increment given by Eq. (11-21) appears across C_2 as long as the total voltage remains small compared with the input pulse amplitude. The result is a *staircase* waveform, Fig. 11-27. In most applications the voltage across C_2 is rapidly discharged by an auxiliary circuit after a finite number of steps and the staircase waveform then repeats.

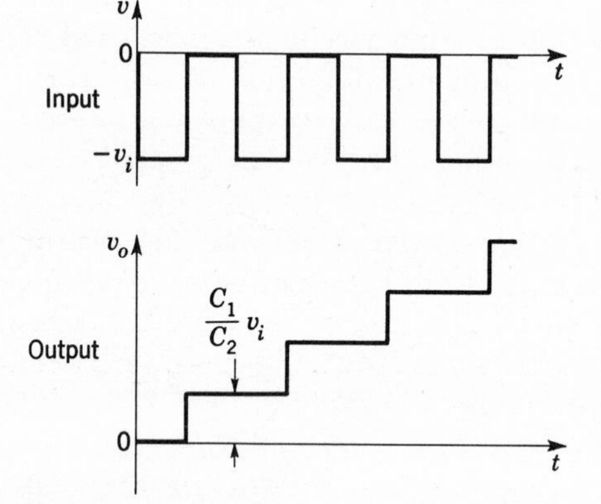

FIGURE 11-27 *Staircase waveform produced by diode pump when input is negative square wave.*

Ramp Generator A voltage that increases linearly with time, called a *ramp*, is widely used for the horizontal sweep voltage in oscilloscopes. Recurring ramps such as those produced by relaxation oscillators discussed in Chap. 9 are referred to as *sawtooth* waves. Other ramp generators are designed to produce a single ramp waveform when triggered by an external signal.

Figure 11-28 is the circuit diagram of a ramp generator called the *Miller sweep*, which yields a linear ramp with a peak amplitude nearly equal to the dc supply potential. Actually the circuit is basically an operational-

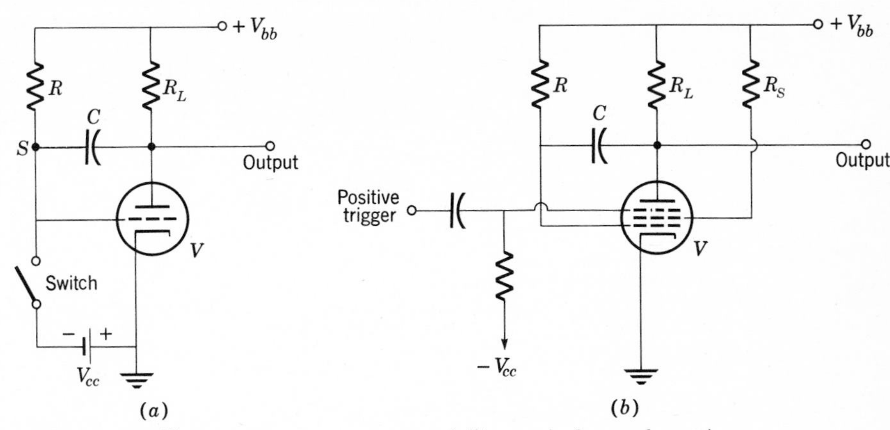

FIGURE 11-28 *(a) Miller sweep ramp generator and (b) practical pentode version.*

amplifier integrator, Fig. 8-15, in which the pentode is the amplifier and the input signal, applied through resistor R, is the constant voltage V_{bb}. Since the integral of a constant increases linearly, the output voltage is a ramp.

It is convenient to start and stop the integration by means of suppressor grid control. Normally the suppressor grid is biased to plate current cutoff and the control grid is at cathode potential. A positive pulse applied to the suppressor cancels the bias, and the tube operates as an amplifier yielding a negative output ramp. At the end of the input pulse the tube is cut off and C discharges through the grid-cathode diode. The circuit is then ready for another actuating pulse.

Circuits similar to the Miller sweep are often used in *triggered-sweep* oscilloscopes, which are particularly useful for examing waveforms of transient signals. A typical block diagram of the sweep circuit for such an oscilloscope, Fig. 11-29, starts with a Schmitt trigger activated by the

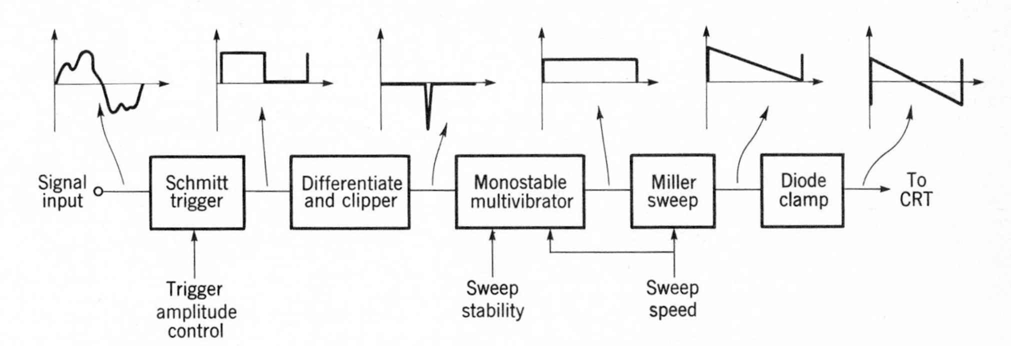

FIGURE 11-29 *Block diagram and typical waveforms of triggered-sweep circuit used in oscilloscopes.*

input signal applied to the vertical deflection amplifier. The output of the Schmitt trigger is differentiated and clipped to produce a sharp negative pulse suitable for triggering a monostable multivibrator. The multivibrator generates a square pulse input to the Miller integrator. The output ramp is clamped at a voltage level such that the sweep starts at the desired place on the oscilloscope screen.

A number of adjustable controls are included to increase the flexibility of the circuit. For example, bias on the Schmitt trigger can be adjusted to permit the circuit to trigger on input signals of different amplitudes (*trigger amplitude,* in Fig. 11-29). Similarly, the bias of the multivibrator is variable (*sweep stability*) so that it may also be operated as a synchronized astable multivibrator to produce recurring sweeps. The sweep speed is altered by changing the charging capacitor (and/or resistor) in the Miller sweep circuit. At the same time, it is necessary to change the time constant of the multivibrator (by switching capacitors) so that the sweep duration is commensurate with the sweep speed.

Pulse Generators Pulse waveforms of endless variety can be generated using combinations of multivibrator circuits. Consider the two cascaded monostable multivibrators in Fig. 11-30. A negative trigger pulse causes

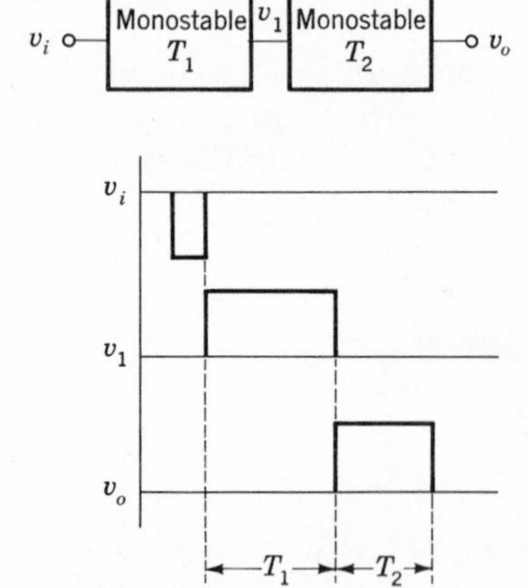

FIGURE 11-30 *Cascaded monostable multivibrators.*

a positive output pulse with a duration T_1 (compare Fig. 11-14). The negative trailing edge of this pulse triggers a second monostable multivibrator and the output is a square pulse of duration T_2. Thus, this combination generates one standard-shaped output pulse delayed by a specified interval following each input trigger pulse.

A simple binary triggered by a repeating waveform produces an output square wave with half-cycles as precisely regular as the frequency of the input signal. This is so because the binary changes state on every negative swing of the input wave and the precision of the square wave output depends only upon the precision of the successive negative swings of the input signal.

Binary triggering may be accomplished by pulses fed directly to only one or the other transistor base. In this case a negative pulse induces a transition only when the transistor is conducting. Such *unsymmetrical* triggering is useful when a binary is triggered from two separate sources, one applied to each base. The output of this binary is a square pulse with a duration equal to the time interval between input pulses.

SUGGESTIONS FOR FURTHER READING

W. C. Elmore and M. Sands: "Electronics," McGraw-Hill Book Company, New York, 1949.

F. J. M. Farley: "Elements of Pulse Circuits," John Wiley & Sons, Inc., New York, 1963.

J. Millman and H. Taub: "Pulse and Digital Circuits," McGraw-Hill Book Company, New York, 1956.

Samuel Weber (ed.): "Digital Circuits," McGraw-Hill Book Company, New York, 1964.

EXERCISES

11-1 Select appropriate square-wave test frequencies to confirm that an amplifier has a passband extending from 20 to 20,000 Hz. Sketch the expected output waveforms at both frequencies.

Ans.: 314 Hz, 3 kHz

11-2 Sketch the voltage waveforms at the collector and base of both transistors of the astable multivibrator, Fig. 11-4, if $C_2 = 0.05\ \mu F$. Compare with the waveforms in Fig. 11-5.

11-3 Astable multivibrators are often used as light flashers to mark road barricades and construction work, etc. Qualitatively describe the operation of a typical circuit, Fig. 11-31, starting from the time when V_1 and V_2 are off and V_1 is starting to turn on.

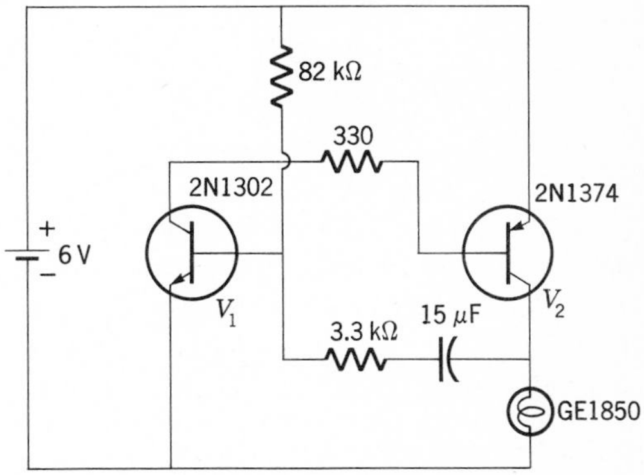

FIGURE 11-31 *Light-flasher circuit analyzed in Exercise 11-3.*

11-4 What scale factor is needed to count pulses having a repetition rate of 1000 pulses per second if the maximum rate of the mechanical register is 10 counts per second? Is there any advantage in a greater scale factor?

Ans.: 100; no

11-5 The neon-tube display for a decade counter uses the circuit of Fig. 11-21 in conjunction with the counter in Fig. 11-18. With the aid of the waveform chart, Fig. 11-19, show that with this circuit the proper neon tube lights corresponding to the indicated count. Assume that the breakdown voltage of the neon tubes is 65 V and the maintaining voltage is 55 V.

11-6 Analyze the Dekatron driver tube circuit, Fig. 11-24, semiquantitatively by using an equivalent circuit in which each tube is represented by a switch in series with a battery representing the cathode-plate voltage in the saturated state. Plot the output waveforms.

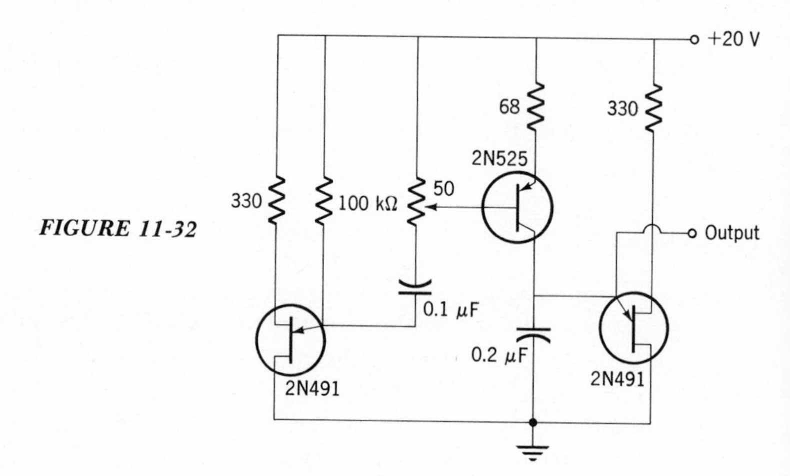

FIGURE 11-32

11-7 Sketch the waveform at the output of the circuit in Fig. 11-32 with the 50-Ω potentiometer at the center. Repeat for the potentiometer set one-fourth of the way from the +20-V terminal. Assume the 2N491 characteristics are the same as those of the 2N2646 in Fig. 9-17.

11-8 Sketch the output waveforms at both plates of the monostable multivibrator circuit in Fig. 11-12 following a single negative trigger pulse.

11-9 Draw a block diagram of a simple *pulse-height analyzer* using a Schmitt trigger and counter to measure the peak pulse voltages in a signal containing a distribution of pulse heights. Describe how you would use this to determine the pulse-height distribution of pulses.

11-10 Show how you would use the Dekatron counter in Fig. 11-25 and a scaler to count pulses at rates up to 250 kHz. What is the maximum count accuracy possible with this arrangement.

Ans.: 0.1%

LABORATORY EXERCISES

11-A Pulse Generator Multivibrator circuits can generate a rich variety of waveforms and are widely used as pulse generators. The purpose of this experiment is to examine the operation of a practical pulse generator. This is accomplished primarily by determining the waveforms in various portions of the circuit.

Examine the circuit illustrated in Fig. 11-33. Is the multivibrator astable, bistable, or monostable? Estimate the duration of the output pulse. Also estimate the frequency of the relaxation oscillator. Now describe qualitatively the operation of the circuit.

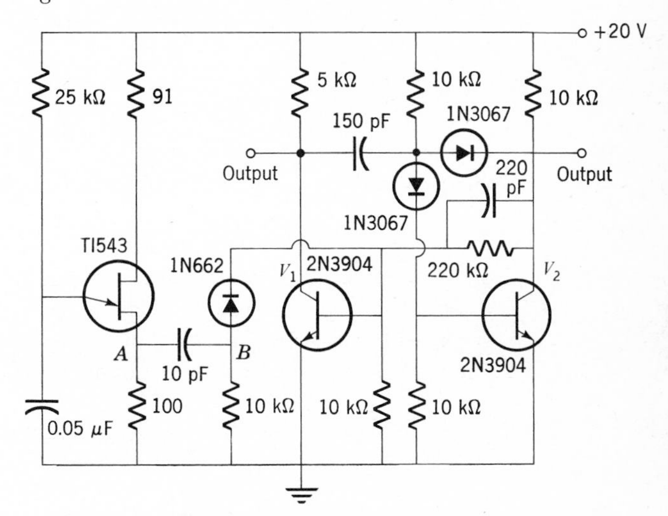

FIGURE 11-33 *Pulse generator.*

Plot the waveform at point A assuming the characteristics of the T1543 unijunction transistor are the same as those of type 2N2646 in Fig. 9-17. Does the pulse repetition rate agree with your preliminary estimate? What is the function of the 1N662 coupling diode?

Quantitatively sketch the waveforms at the bases and collectors of both 2N3904 transistors. Do this first of all ignoring the 1N3067 diodes. Suggest the effect of the 1N3067 diodes upon the saturation condition of V_2. Now plot the output waveform of a single pulse. What is the pulse width? Polarity? What is the ratio of the interval between pulses to the pulse width? It may help you to know that in the actual instrument this ratio is about 1000.

11-B The Binary Counter The binary is a unique circuit that has memory, can count, and is also useful as a waveform generator. Cascaded flip-flops exhibit extremely versatile properties depending upon how their input terminals are interconnected. In this experiment the performance of a simple binary counter is examined quantitatively.

Consider the binary counter in Fig. 11-16. First determine the voltages at the base and collector of each transistor in binary B_1. Assume that the collector characteristics of the transistors are similar to those of type 2N930 given in Appendix 1. Verify that one transistor is conducting while the other is cut off. Based on these voltages, which steering diode is under reverse bias after the reset switch is opened momentarily? What magnitude and polarity trigger pulse must be applied to the steering diodes in order to cause a transition in B_1? Now determine the magnitude and polarity of the input pulse required to activate the counter. Finally, determine the magnitude of the output pulse of binary B_1 and verify that it is adequate to induce a transition in binary B_2.

Determine a value for the maximum input-pulse frequency for reliable counting. Do this by drawing the ac equivalent circuit associated with one base circuit. Since during the transition each transistor acts as an amplifier, it is appropriate to ascertain the high-frequency cutoff point of this circuit. Remember to use Eq. (7-12) and assume $C_s = 200$ pF. Is the high-frequency limit determined predominately by C_s or by the commutating capacitors? Quantitatively sketch the waveform at the output of B_1 if the interval between trigger pulses is equal to the period of the high-frequency cutoff. Is reliable operation obtained at this counting rate? Which binary stage in a counter must have the highest counting-rate capability?

Cascade four RS/T flip-flops as shown in Fig. 11-34 and apply a square wave

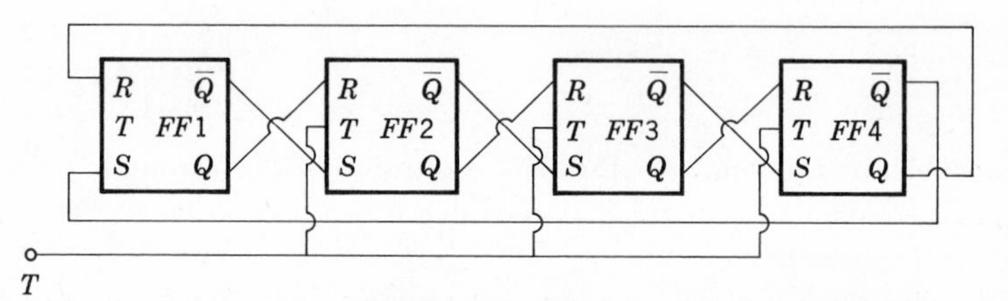

FIGURE 11-34 *Cascaded clocked flip-flops.*

to terminal T. Put $FF1$ in state 1 and all the others in state 0. Plot the waveforms at Q for all flip-flops for the first 10 pulses of the square wave. It is helpful to arrange the waveforms as in Fig. 11-17. Make a simple statement of what the circuit is doing.

chapter twelve

DIGITAL ELECTRONICS

Numerical digits can be represented by electrical signals which have only two possible magnitudes, say, zero and some finite value. In this situation only the existance of one signal state or the other is significant and the actual magnitude is relatively unimportant. Therefore, digital circuits need have only two stable conditions represented by a transistor fully conducting, or completely cut off. Such circuits are inherently more reliable than conventional analog circuits, which must handle a continuous range of signal levels. Any given number can be represented by a digital waveform, so accuracy is not limited by the stability of circuit parameters. Digital signals are manipulated by circuits according to specific logic statements which make possible exceedingly flexible and powerful processing of information. The advantages of reliability, accuracy, and flexibility were first attained in digital computers but now find application in a wide range of instrumentation and measuring circuits.

DIGITAL LOGIC

Binary Numbers Voltage signals are used to represent the digits of numbers in order that mathematical logic operations such as addition and subtraction can be carried out by electronic circuits. The stability advantages of digital circuits are best realized if the signal waveforms have only two amplitudes, called state 0 and state 1, or *off* and *on*. This is so because a transistor, for example, need then only be either completely cut off or fully conducting. Since only two digits are available, the *binary* number system is used, rather than the more familiar *decimal* system based on the use of 10 digits. It is, of course, not particularly surprising that a counting system based on the number of fingers we possess is not the most efficient one for electronic circuits. Although the binary system is less familiar, there is no difference in principle between the two number systems.

Consider, for example, the meaning of the decimal number 528. This array of decimal digits is a shorthand notation for the increasing powers of ten in the number. Thus,

$$
\begin{aligned}
528 &= (5 \times 10^2) + (2 \times 10^1) + (8 \times 10^0) \\
&= \quad 500 \quad + \quad 20 \quad + \quad 8 \quad = 528
\end{aligned}
$$

Similarly, the two digits in the binary system, 0 and 1, tell the number of increasing powers of two in the number. The binary number 10110, for example, means

$$
\begin{aligned}
10110 &= (1 \times 2^4) + (0 \times 2^3) + (1 \times 2^2) + (1 \times 2^1) + (0 \times 2^0) \\
&= \quad 16 \quad + \quad 0 \quad + \quad 4 \quad + \quad 2 \quad + \quad 0 \\
&= \quad 22
\end{aligned}
$$

That is, the binary number 10110 represents the same quantity as the decimal number 22.

Table 12-1 contains the decimal-binary equivalents for the numbers from 0 through 16. The binary-number representations in this table are identical to the states of the four-binary counter of Fig. 11-16. The correspondence can be established by noting the state of each binary at every input count using the waveforms in Fig. 11-17. This is the basis for the earlier statement that a binary scaler counts by powers of 2 and is one illustration of the convenience attendant on using binary-number representation in electronic digital circuits.

Arithmetic manipulations with binary numbers involve mathematical logic quite familiar from decimal numbers. Addition and subtraction operations involve digits that are carried and borrowed in an analogous

TABLE 12-1 BINARY AND DECIMAL NUMBERS

Decimal number	*Binary number*
0	0000
1	0001
2	0010
3	0011
4	0100
5	0101
6	0110
7	0111
8	1000
9	1001
10	1010
11	1011
12	1100
13	1101
14	1110
15	1111
16	10000

TABLE 12-2 TRUTH TABLE FOR ADDITION

A	*B*	*Sum*	*Carry*
0	0	0	0
0	1	1	0
1	0	1	0
1	1	0	1

fashion. It is useful to state the mathematical logic contained in the addition of two binary numbers A and B by means of a *truth table,* Table 12-2, which accounts for all of the possible combinations of A and B. A digit is carried as in the fourth entry in Table 12-2 by shifting it to the next higher position, that is, to the left. For example, the sum of 0110 and 0101 is written

$$\begin{array}{r} 0110 \\ +0101 \\ \hline 1011 \end{array} \qquad \begin{array}{r} 6 \\ +\ 5 \\ \hline 11 \end{array}$$

Binary numbers can be represented by voltage waveforms with regularly spaced pulses of uniform amplitude. Conventionally, the pulses corresponding to increasing powers of 2 appear in time sequence beginning with 2^0. The waveforms of the binary numbers 1011 and 0011 are illustrated in Fig. 12-1, together with their sum, 1110. Note that the pulse

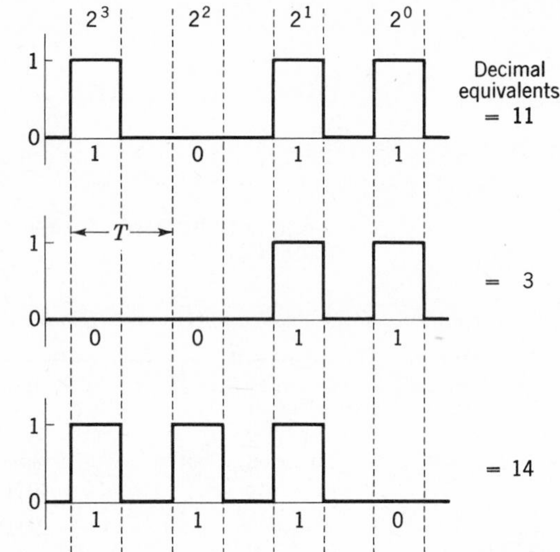

FIGURE 12-1 *Waveforms of the binary numbers 1011 and 0011 and their sum, 1110.*

lengths of both 0 and 1 digits are equal. One digit, be it a 0 or a 1, is termed a *bit*, which is a contraction for *binary digit*.

It should be noted in passing that the waveform of the sum bears no direct relationship to the waveforms of the two numbers. That is, the waveform of the sum is not obtained simply by adding the voltages of the waveforms representing the individual binary numbers. This points up the fact that waveforms in digital circuits are a coded representation of numbers and the voltages of the waveforms have no significance in themselves.

Logic Gates The various operations performed on digital waveforms are accomplished using circuits termed *logic gates* which have two or more inputs and one output. The output waveform of a logic gate depends upon the input waveforms and the input-output characteristic of the circuit described in terms of mathematical logic.

Consider the logic statement "if A is true AND B is true, then T is true," which is written symbolically, as

$$A \cdot B = T \tag{12-1}$$

This is called the AND concept and is signified by the dot between the two quantities A and B in Eq. (12-1). Conventionally the condition true is identified with state 1 and the opposite condition false is identified with state 0 in digital circuits. In this fashion a truth table for the AND operation

TABLE 12-3 TRUTH TABLE FOR AND

A	B	T
0	0	0
0	1	0
1	0	0
1	1	1

may be devised as in Table 12-3. Extending the truth table for more than two input quantities is easily accomplished by applying the logic in Table 12-3 several times to each pair of inputs. According to Table 12-3 an output is obtained (that is, T is true) only when A *and* B are true.

A simple three-input diode AND gate that accomplishes the logic in Table 12-3 is illustrated in Fig. 12-2*a*. The diodes are biased in the low-resist-

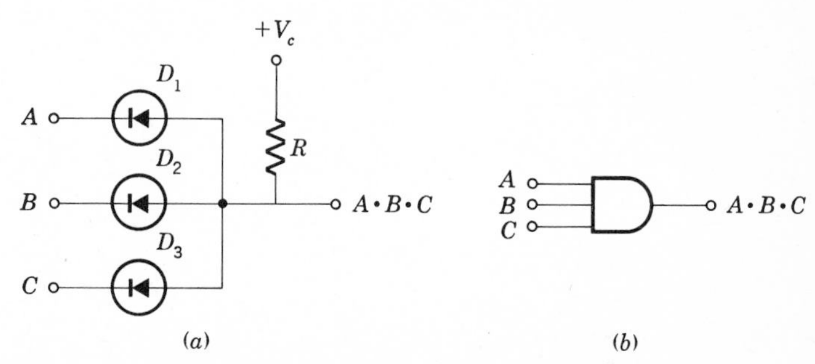

FIGURE 12-2 *(a) Diode* AND *gate and (b) circuit symbol.*

ance forward direction and the output voltage is zero, which means that the output condition is state 0. If positive input voltage signals somewhat greater than V_c are applied simultaneously to all three inputs, the diodes become reverse-biased and the output voltage rises to V_c, or state 1. Note, however, that if even one input remains at state 0, the corresponding diode is under forward bias and the output signal remains at state 0. This action is logically described by saying that an output signal is obtained only when there are A AND B AND C input signals.

The AND gate is also called a *coincidence circuit* in nuclear radiation measurements since, in effect, it detects the coincidence of pulses presented to the inputs. The operation of an AND gate should not be confused with the mathematical operation of addition. That is, the output of an AND gate is not the sum of the input signals, as may be verified by noting the differences between the corresponding truth tables. AND gates are so extensively used in digital electronic circuits that it is convenient to use the special symbol in Fig. 12-2*b*.

The logical-OR statement is "if A OR B is true, then T is true," written as

$$A + B = T \tag{12-2}$$

where the + symbol indicates the OR concept. The corresponding truth table, Table 12-4, shows that an output is obtained whenever any input is present.

TABLE 12-4 TRUTH TABLE FOR OR

A	B	T
0	0	0
0	1	1
1	0	1
1	1	1

A diode OR gate and its circuit symbol are shown in Fig. 12-3*a* and *b*, respectively. Note that a positive signal at any input biases the correspond-

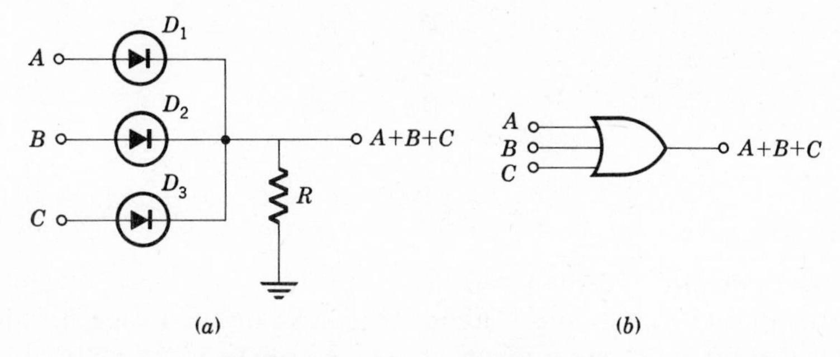

***FIGURE 12-3** (a) Diode OR gate and (b) circuit symbol.*

ing diode in the forward direction and appears at the output. At the same time the other diodes are reverse-biased so that signals fed back to the other inputs are negligible. In effect the OR gate is a simple mixing circuit that brings several input signals to a common output with minimum interaction between signal sources.

In binary logic the NOT operation inverts the polarity of an input signal, as indicated schematically in Fig. 12-4. Since a digital waveform has only two states, 0 and 1, if at any given instant the signal is in state 1 it may

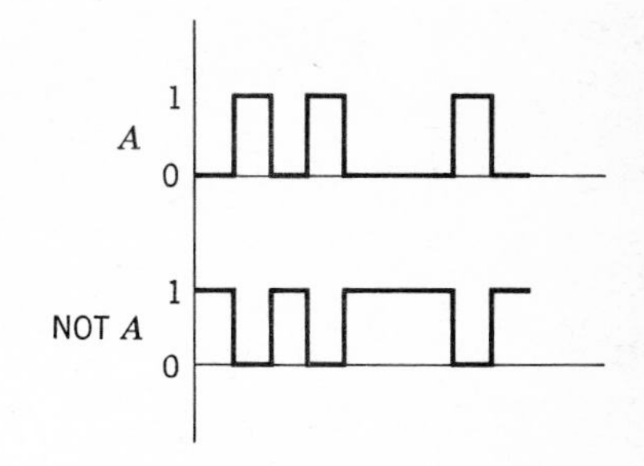

FIGURE 12-4 *Waveforms of binary numbers* A *and* NOT A.

equally be said to be in state NOT 0. The symbol for NOT is a bar over the quantity,

$$\text{NOT } A = \bar{A} \tag{12-3}$$

and the truth table is quite simple, Table 12-5. A simple grounded-emitter amplifier produces the logic in Table 12-5 as a result of the 180° phase shift between input and output. Accordingly, the appropriate circuit symbol for a NOT or *inverter* gate is similar to that previously used for an amplifier, Fig. 12-5. The small circle at the apex of the triangle (sometimes placed at the input) specifically indicates inversion.

TABLE 12-5 TRUTH TABLE FOR NOT

A	$\bar{A}$
0	1
1	0

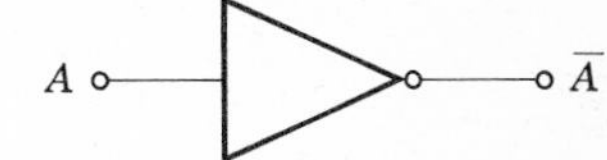

FIGURE 12-5 *Inverter circuit symbol for* NOT *operation.*

It is common practice to combine the NOT inversion with other logic-gate functions. This results in amplification of the signals in the logic gate itself and consequently preserves the signal amplitude throughout a circuit of multiple logic gates. Consider the single-transistor amplifier in Fig. 12-6*a* with three equivalent inputs so that the waveform at each input may appear at the output terminal. This is the same logic as in Table 12-4 and therefore this circuit is a NOT-OR, or NOR, gate. The circuit symbol, Fig. 12-6*b*, includes the small circle to indicate inversion. A NOT-AND, or NAND, gate is equally possible and a typical circuit is il-

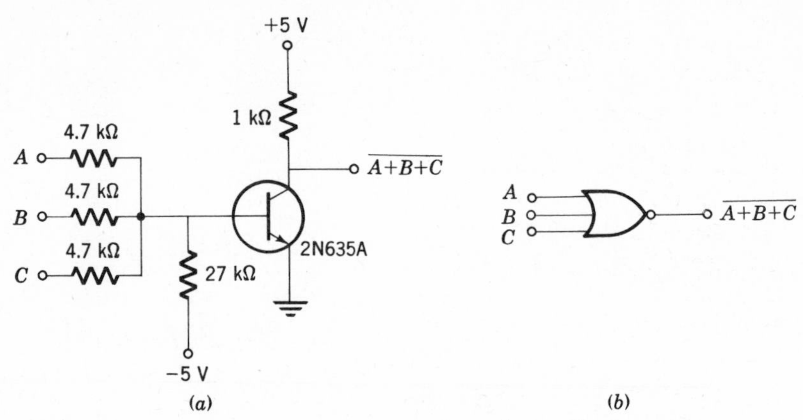

FIGURE 12-6 *(a) Simple* NOR *gate performs the combined operation* NOT-OR *and (b) circuit symbol.*

lustrated in a later section. The circuit symbol for a NAND gate is shown in Fig. 12-7. Truth tables for NOR and NAND gates are simply obtained by applying the NOT operation to the last columns of the truth tables for OR, Table 12-4, and for AND, Table 12-3.

FIGURE 12-7 NAND *gate.* A B C $\overline{A \cdot B \cdot C}$

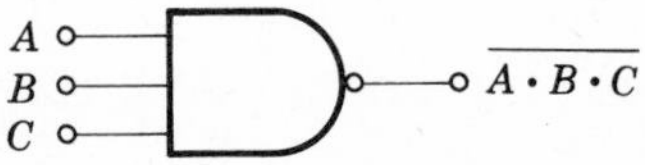

A logic gate related to the NAND gate is a combination of an AND gate and a NOR gate, Fig. 12-8. The operation of this circuit is best examined

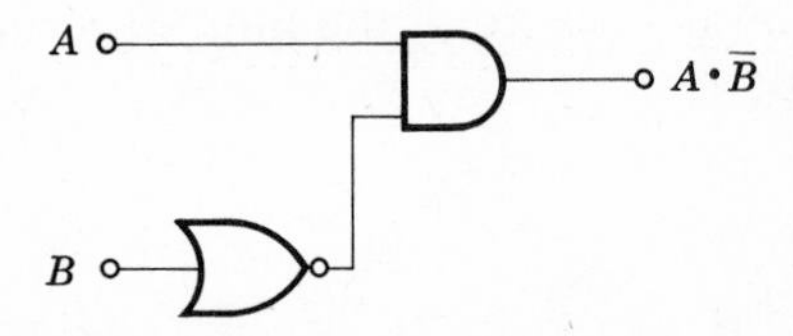

FIGURE 12-8 *Combination of an* AND *gate and a* NOR *gate to produce an anticoincidence circuit.*

by developing the appropriate truth table, as in Table 12-6. The output of the NOR gate $\overline{B}$ is found with the aid of Table 12-5 and the output $A \cdot \overline{B}$ uses columns A and $\overline{B}$ in connection with Table 12-3. According to the truth table, no output is ever achieved when signal B is present. That is, the B input inhibits the A output and in pulse applications the circuit is termed an *anticoincidence* circuit by analogy with the coincidence circuit.

One further simple logic statement is the exclusive-OR function. The everyday usage of *or* is ambiguous for it can mean "one or the other or both" or "one or the other and not both." The logical-OR function pre-

TABLE 12-6 TRUTH TABLE FOR ANTICOINCIDENCE CIRCUIT

A	B	$\bar{B}$	$A \cdot \bar{B}$
0	0	1	0
0	1	0	0
1	0	1	1
1	1	0	0

viously defined is the "one or the other or both" situation according to the truth table, Table 12-4. It could correctly be called the inclusive-OR function. The exclusive-OR statement is

$$A \oplus B = A \cdot \bar{B} + B \cdot \bar{A} \qquad (12\text{-}4)$$

which can be implemented by the array of logic gates in Fig. 12-9. Here

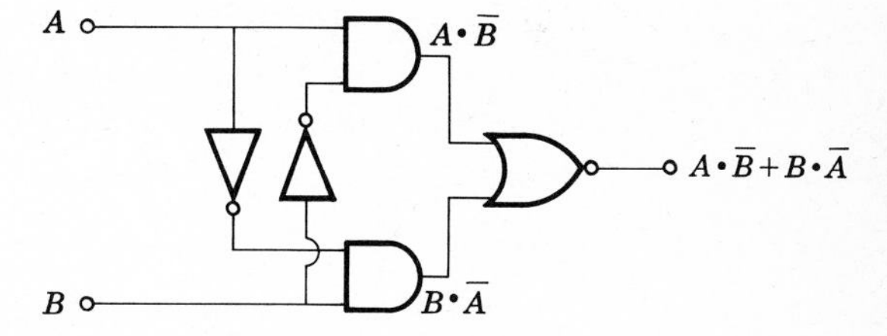

FIGURE 12-9 Exclusive-OR logic.

again, it is useful to examine the truth table, Table 12-7, which is constructed with the aid of Tables 12-3, 12-4, and 12-5.

TABLE 12-7 TRUTH TABLE FOR EXCLUSIVE-OR

A	B	$\bar{A}$	$\bar{B}$	$A \cdot \bar{B}$	$B \cdot \bar{A}$	$A \cdot \bar{B} + B \cdot \bar{A}$
0	0	1	1	0	0	0
0	1	1	0	0	1	1
1	0	0	1	1	0	1
1	1	0	0	0	0	0

According to the last column in Table 12-7, an output from the exclusive-OR circuit is produced only when either A or B is present, but not both, as expected.

Boolean Algebra The design of logical networks is greatly aided by the logical algebra developed in the last century by George Boole, an English mathematician. The theorems of Boolean algebra are used to simplify digital logic networks in much the same fashion mathematical logic is used to manipulate ordinary algebraic expressions. One major difference, of course, is that the variables in Boolean expressions can assume only one of two possible values.

A list of theorems in Boolean algebra is given in Table 12-8. Theorems 1 through 4 may be recognized as OR logic and are proved using the OR truth table, Table 12-4. Similarly, Theorems 5 through 8 are based on the AND concept in Table 12-3. Theorem 9 is a formal definition of the NOT function. The commutation, association, distribution, and absorption theorems are relatively straightforward and may be proved by means of truth tables.

TABLE 12-8 BOOLEAN ALGEBRA THEOREMS

OR function	1	$0 + A = A$
	2	$1 + A = 1$
	3	$A + A = A$
	4	$A + \overline{A} = 1$
AND function	5	$0 \cdot A = 0$
	6	$1 \cdot A = A$
	7	$A \cdot A = A$
	8	$A \cdot \overline{A} = 0$
NOT function	9	$(\overline{\overline{A}}) = A$
Commutation	10	$A + B = B + A$
	11	$A \cdot B = B \cdot A$
Association	12	$A + (B + C) = (A + B) + C$
	13	$A \cdot (B \cdot C) = (A \cdot B) \cdot C$
Distribution	14	$A \cdot (B + C) = A \cdot B + A \cdot C$
	15	$(A + B) \cdot (A + C) = A + B \cdot C$
Absorption	16	$A + A \cdot B = A$
	17	$A \cdot (A + B) = A$
De Morgan's Theorems	18	$\overline{(A + B)} = \overline{A} \cdot \overline{B}$
	19	$\overline{A \cdot B} = \overline{A} + \overline{B}$

De Morgan's theorems, numbers 18 and 19, are particularly interesting as they show a useful relationship between and AND and OR functions. Theorem 18 is proved in Table 12-9 by first finding $(\overline{A + B})$ and then separately finding $\overline{A} \cdot \overline{B}$. The result is that the two truth tables are equal and the theorem is proved. A similar process may be used to establish Theorem 19. The basic duality of Boolean algebra is expressed by these

TABLE 12-9 PROOF OF DE MORGAN THEOREM

A	B	$A+B$	$\overline{(A+B)}$	$\overline{A}$	$\overline{B}$	$\overline{A}\cdot\overline{B}$
0	0	0	1	1	1	1
0	1	1	0	1	0	0
1	0	1	0	0	1	0
1	1	1	0	0	0	0

theorems and can also be noted in pairs of other expressions in Table 12-8. Compare, for example the two association theorems, and also Theorem 1 with Theorem 6. In each case the OR and AND functions are dually related.

An elementary example of the use of Boolean algebra to simplify logic networks is provided by the circuit in Fig. 12-10 which contains three

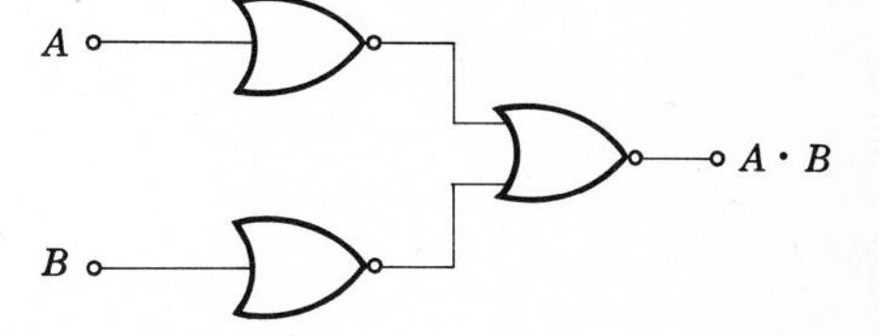

FIGURE 12-10 *Three* NOR *gates produce an* AND *gate.*

NOR gates. The outputs of the two input gates are $\overline{A}$ and $\overline{B}$, respectively, so that the output of the total circuit may be written down and subsequently simplified by means of De Morgan's theorem 18

$$\overline{(\overline{A}+\overline{B})} = \overline{\overline{A}} \cdot \overline{\overline{B}} = A \cdot B \tag{12-5}$$

In arriving at Eq. (12-5), Theorem 9 is also used. According to this result, the combination of three NOR gates may be simplified to one simple AND gate.

A second example of logic-network reduction is the exclusive-OR logic network illustrated in Fig. 12-9 and repeated in Fig. 12-11*a*. Observe that the output logic may be written in the form

$$A \cdot \overline{B} + B \cdot \overline{A} = \overline{\overline{(A \cdot \overline{B}) + (B \cdot \overline{A})}} = \overline{\overline{(A \cdot \overline{B})} \cdot \overline{(B \cdot \overline{A})}} \tag{12-6}$$

where Theorems 9 and 18 have been used. The last logic statement may be accomplished by replacing the output OR gate with a NAND gate, if the inputs are $\overline{(A \cdot \overline{B})}$ and $\overline{(B \cdot \overline{A})}$. These logic functions are easily produced by replacing the two AND gates with NAND gates. The result is the logic network in Fig. 12-11*b*. This network may be further simplified to Fig. 12-11*c* by noting that a NAND gate can replace the two inverters. This is proved by determining the output of the upper NAND gate,

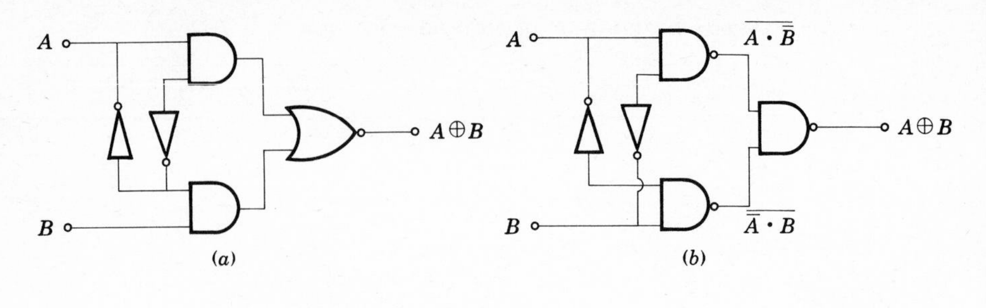

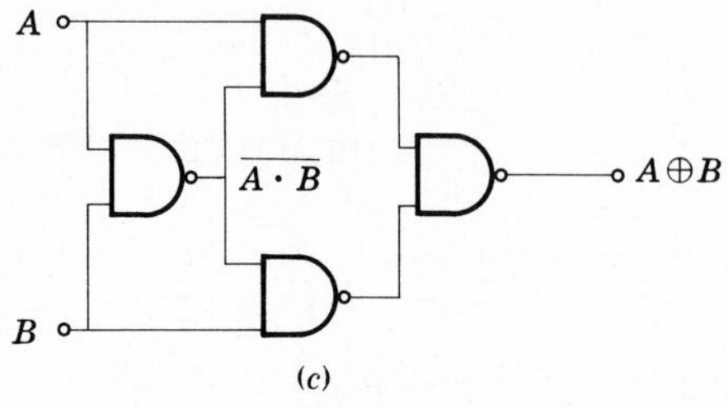

FIGURE 12-11 *Reduction of exclusive-*OR *logic.*

$$\overline{A \cdot (A \cdot B)} = A \cdot (\overline{\overline{A} + \overline{B}}) = \overline{(A \cdot \overline{A}) + (A \cdot \overline{B})} \qquad \text{Theorems 19 and 14}$$

$$= \overline{0 + (A \cdot \overline{B})} = \overline{0} \cdot \overline{(A \cdot \overline{B})} \qquad \text{Theorems 8 and 18}$$

$$= 1 \cdot \overline{(A \cdot \overline{B})} = \overline{(A \cdot \overline{B})} \qquad \text{Theorem 6} \qquad (12\text{-}7)$$

A similar analysis shows that the output of the lower NAND gate is $\overline{(B \cdot \overline{A})}$. Therefore the network in Fig. 12-11*c* is exactly logically equivalent to the exclusive-OR logic in Fig. 12-11*a*.

The results of the two previous paragraphs illustrate an important feature of Boolean logic networks. This is that any logic function may be accomplished with NOR gates alone, or with NAND gates alone. Although it may seem wasteful to employ three NOR gates to achieve AND logic as in Fig. 12-10, it is an appreciable convenience to be able to assemble large logic networks of several thousand gates from the same simple subunit. This greatly simplifies design, production, and maintenance considerations. Furthermore, it turns out that, as discussed in a later section, the operating characteristics of either the NAND or NOR gate of a particular design are more favorable than the other type of gate. Therefore, considerable advantage is realized in using only the more favorable form. The choice of which to use is entirely a matter of convenience and the logic statements may be converted from one form to the other by means of the theorems of Boolean algebra in Table 12-8.

LOGIC CIRCUITS

Logic Signals In the preceding discussions it has been assumed that signal waveforms corresponding to state 1 are at a small positive potential while state 0 is represented by ground potential. This is in keeping with current practice in which the voltage of a state 1 bit is approximately +5 V and that of a state 0 bit is near 0 V. This choice is completely arbitrary, of course, since Boolean algebra does not depend on the specific choice of signals used but only upon the presence of two distinct states.

It is interesting to note the effect of interchanging the designation of logic levels upon actual gate circuits. That is, suppose state 1 is represented by 0 V and state 0 by +5 V. This is called *negative logic* because state 1 is more negative than state 0. Consider the action of the diode AND gate in Fig. 12-2*a* in negative logic. The diodes are biased in the forward direction so that the output is state 1 when any one of the inputs is at ground potential, or state 1. The output is positive, state 0, only when all inputs are positive, or state 0. The truth table for this logic is written in Table 12-10 for the case of only two inputs. Table 12-10 is identical to Table 12-4 for the OR function. According to this analysis, an AND gate in positive logic becomes an OR gate in negative logic. This also means that a positive logic OR gate is a negative logic AND gate.

TABLE 12-10 TRUTH TABLE FOR NEGATIVE LOGIC AND GATE

A	B	T
0	0	0
0	1	1
1	0	1
1	1	1

It is useful to think of an inverter gate as a logic level exchanger since a state 1 input yields a state 0 output, according to Table 12-5. Thus, for example, a logical AND function can be produced using a positive logic OR gate by first inverting signals to negative logic. The negative logic signals are acted upon by the positive logic OR gate to give negative AND logic. Subsequently, the signals may be converted back to positive logic by means of another inverter gate. The steps in this process can be symbolized in the following fashion

1. Positive logic signals A, B
2. Negative logic signals $\bar{A}, \bar{B}$

3. Positive OR gate $\overline{A} + \overline{B}$
4. Theorem 19 $\overline{A \cdot B}$
5. Positive logic signal $A \cdot B$ (12-8)

This illustrates how the combination of positive and negative logic may be analyzed by means of the standard theorems of Boolean algebra.

There are two fundamentally different ways of transmitting signals representing the several bits of a complete binary number. The first, already considered in connection with Fig. 12-1, employs a time sequence of pulses corresponding to increasing powers of 2 in the number. This is called *serial* representation and is indicated schematically in Fig. 12-12*a*.

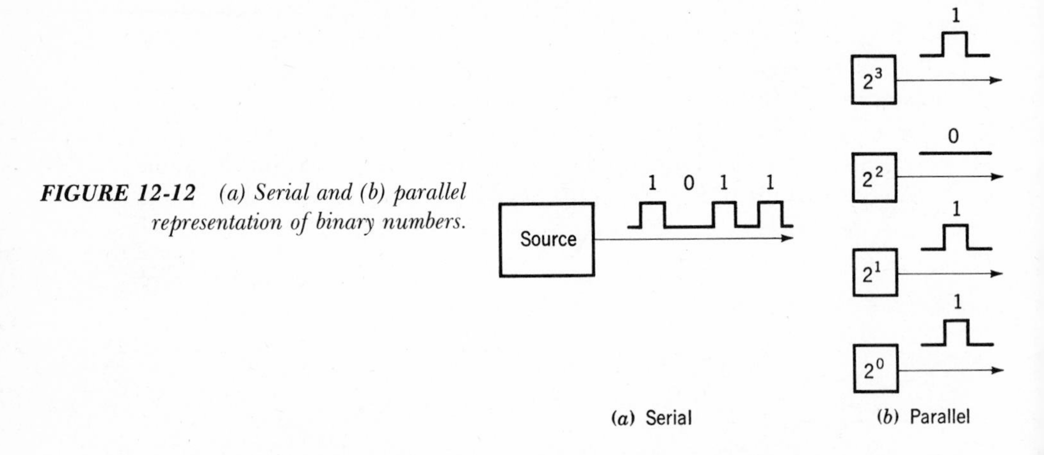

FIGURE 12-12 *(a) Serial and (b) parallel representation of binary numbers.*

An alternate method, *parallel* representation, illustrated in Fig. 12-12*b*, is equally possible. In parallel transmission, pulses representing the increasing powers of 2 in the number all appear simultaneously.

Both serial and parallel representation of numbers are commonly used in the same digital circuitry, as is examined in subsequent sections. Clearly, serial transmission has some advantages since only one signal path is required while a path for each bit is necessary in the case of parallel transmission. On the other hand, a time interval of only one bit length is needed to transmit an entire digital number when the parallel representation is used, whereas the time required is much longer for the serial representation and the time interval also depends upon the number of digits in the number. The method selected in any given instance depends upon which feature is more advantageous.

DTL and TTL Logic Several different designs of electronic circuits are used as logic gates. The simplest, illustrated in Figs. 12-2*a* and 12-3*a*, use diodes and, accordingly, are known as *diode logic*, or DL, gates. These simple circuits perform the desired logic function satisfactorily but are not easily cascaded to carry out complicated logical functions. The reason for this is the inevitable signal loss through the circuit and the impedance mismatch between the output of one gate and the input of the next.

Much more satisfactory in these respects is *resistor-transistor logic,* RTL, already shown in Fig. 12-6*a* as a NOR gate. Actually this design is used only in the form of a NOR gate, but this is sufficient to carry out all logic functions, as pointed out previously. The transistor provides gain so that logic signals are not degraded in traversing the gate, and the transistor also significantly reduces the effect of load impedance upon the input circuit. One useful measure of this effect is the *fan-out* characteristic. Fan-out refers to the number of identical logic gates that can be connected to the output without seriously impairing circuit operation. Clearly a large fan-out is advantageous since complex logic networks are then possible. A fan-out of about 5 is characteristic of RTL gates. Better fan-out and somewhat faster circuit operation is achieved by shunting the summing resistors by capacitors, very much like the commutating capacitors in a binary. This configuration is called RCTL, *resistor-capacitor-transistor logic.*

The combination of a diode AND gate and an inverter, Fig. 12-13, yields

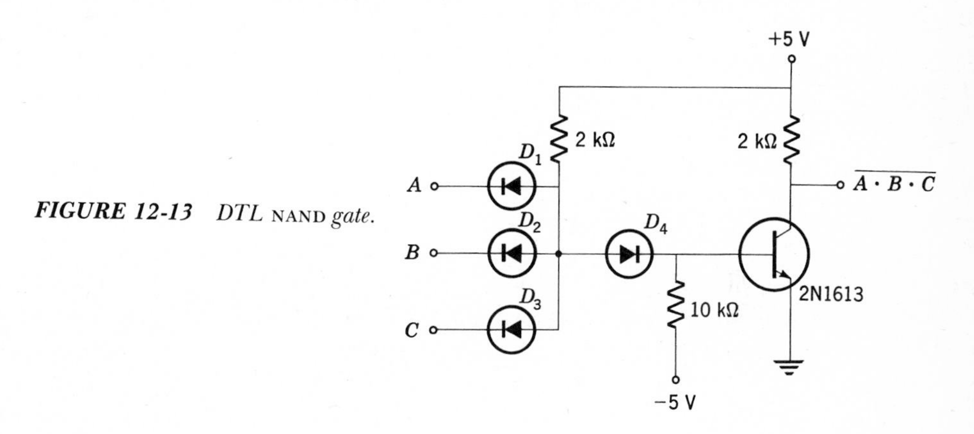

FIGURE 12-13 *DTL* NAND *gate.*

a very useful NAND gate. Diodes D_1, D_2, and D_3 perform the logic function, and the transistor inverts and amplifies the signal. This circuit is capable of higher operating speeds than the RTL gate and has even further reduced output-input interaction because of coupling diode D_4 (which, however, can be omitted at some degradation in circuit performance). Fan-out in DTL gates can be as large as 10. Although an analogous NOR gate is possible, DTL logic normally employs only NAND gates.

All the foregoing logic gates are available as integrated circuits, although, obviously, discrete component circuits can also be used. The flexibility of integrated-circuit fabrication permits *transistor-transistor logic,* TTL, in which a multiple-emitter transistor is the logic element, Fig. 12-14.

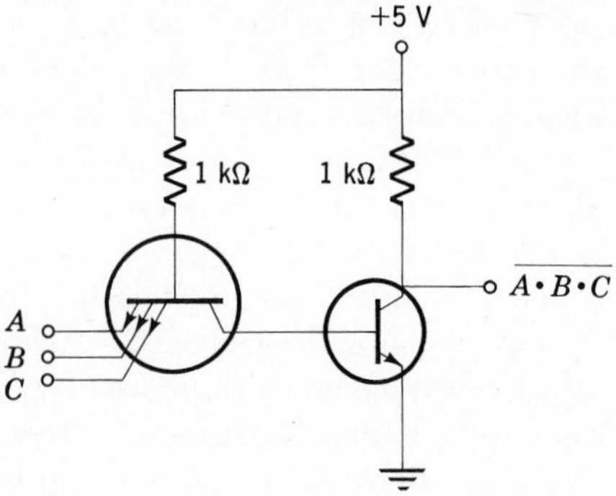

FIGURE 12-14 *Integrated-circuit TTL* NAND *gate.*

The operation of the circuit may be looked upon as a DTL gate with the diode anodes in common. When any one of the emitters is at ground potential, state 0, the collector is essentially at ground and the output transistor is cut off since no current flows in its emitter-base junction. With the output transistor cut off the output is at state 1. If all three emitters are at state 1, both transistors conduct and the output goes to state 0. This is logically equivalent to a NAND gate.

TTL logic has a fan-out as great as 15, and eight or more input emitters are not unusual. Furthermore operating speed is faster than the other gate types and the so-called noise-immunity properties are better. Logic speed is usually stated in terms of the *propagation delay* of the gate, that is, the time between introducing a change of logic state at the input and achieving a change at the output. The propagation delay is about 10×10^{-9} s (called nanoseconds, or ns) for TTL gates compared to values of 20 ns for DTL and 25 ns in the case of RTL gates.

The noise immunity of logic gates is an important parameter that specifies how well a gate remains in a given logic state in the face of spurious signals due, for example, to changes in supply voltages, etc. Noise immunity is the difference between the minimum effective input signal in state 1 and the maximum state 0 output signal. A large value implies the gate is most likely to remain in a given logic state. Noise immunity exceeds 1 V for TTL gates, is about 0.75 V for DTL gates and may be as small as 0.2 V for RTL gates.

A wide variety of other gate circuits has been devised, some based on FET circuits and others using grounded-collector transistors rather than the grounded-emitter configuration, etc. Also, integrated-circuit gates of considerable complexity, accompanied by operational advantages, are readily available. In every case, however, the basic principles are similar to those illustrated in the foregoing discussion.

ADD gates It is most pertinent to examine the addition of binary numbers since addition is an arithmatic operation basic to numerical computation.

Multiplication, for example, can then be accomplished simply by repeated additions. Similarly, division can be carried out by the inverse of addition, that is, subtraction. Logic gates are used to achieve binary addition in the following two-step way. First, the digits in each column are added and then the carry digits are added to the column representing the next higher power of 2. This is identical to conventional addition as illustrated in a previous section.

The truth table for the addition of bits, Table 12-2, shows that the sum S of two bits is just exclusive-OR logic, Table 12-7, while the carry C is AND logic, Table 12-3. Therefore, the addition of bit A to bit B is stated by the following logic

$$S = A \oplus B \tag{12-9}$$

$$C = A \cdot B \tag{12-10}$$

A network to accomplish this logic is called a *half-adder,* and one form is illustrated in Fig. 12-15. Note that Eq. (12-10) is immediately evident,

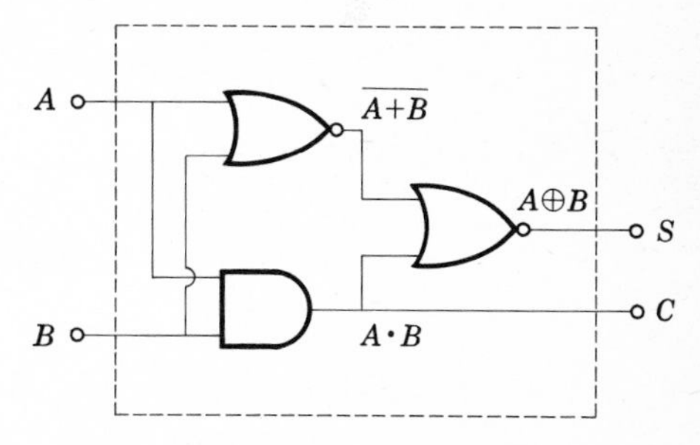

FIGURE 12-15 *Half-adder circuit.*

while the sum output is

$$S = \overline{(\overline{A+B}) + (A \cdot B)} = (\overline{\overline{A+B}}) \cdot (\overline{A \cdot B}) \qquad \text{Theorem 18}$$

$$= (A + B) \cdot (\overline{A} + \overline{B}) \qquad \text{Theorems 9 and 19}$$

$$= A \cdot \overline{A} + A \cdot \overline{B} + B \cdot \overline{A} + B \cdot \overline{B} \qquad \text{Theorem 14}$$

$$= A \cdot \overline{B} + B \cdot \overline{A} \qquad \text{Theorem 8}$$

which establishes Eq. (12-9).

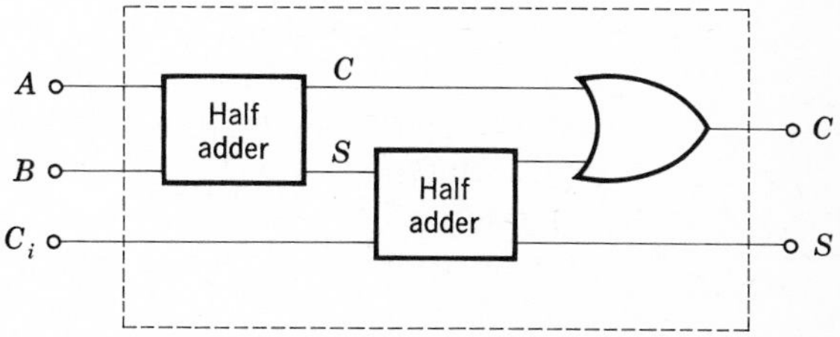

FIGURE 12-16 *Full-adder.*

The half-adder accomplishes the first step in the addition of numbers, that is, the addition of bits. Two half-adders are combined into a *full adder,* Fig. 12-16, in order to add carry bits to the bit sum. The OR gate is needed to include the possibility that adding the carry bit to the sum generates a new carry bit.

The way the full adder is used to add binary numbers depends upon whether the numbers to be added are available in serial or parallel representation. In serial representation, Fig. 12-17*a*, the waveforms of the two

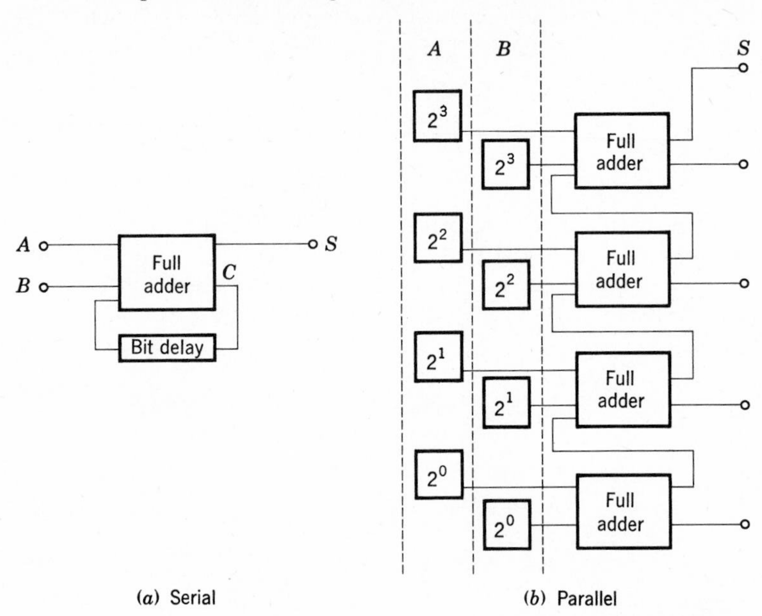

FIGURE 12-17 *(a) Serial and (b) parallel addition.*

numbers are introduced to a full-adder circuit, and each bit sum is added in sequence. Any carry bit produced is returned to the carry input delayed by a time interval equal to the time between successive bits. This delay has the effect of adding the carry bit to the next column of bits in the binary number. A convenient 1-bit delay might be the combination of two cascaded binaries discussed in the previous chapter.

Addition in parallel representation is similar but a full adder is required for each digit, Fig. 12-17*b*. In this case the carry bit from the least significant bit addition is introduced into the full adder of the next most significant bit, etc. Although parallel addition may appear to require very elaborate circuitry since some 28 logic gates are required to add the two 4-digit numbers in Fig. 12-17*b*, it should be recalled that fabrication processes in integrated circuits are readily capable of producing elaborate networks. Thus a complete integrated-circuit full-adder gate is physically no larger than the container needed to provide the necessary terminals.

INFORMATION REGISTERS

Flip-flops If two inverting gates are cross coupled, as in the case of the two NAND gates in Fig. 12-18, the combination has two stable states. Sup-

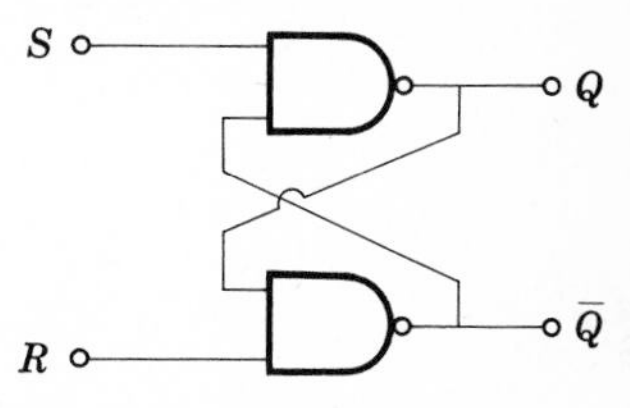

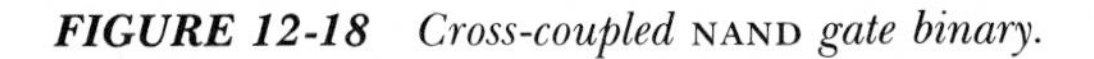

FIGURE 12-18 *Cross-coupled* NAND *gate binary.*

pose, for example, the S input (for *set*) is state 1 and the R input (for *reset*, or *clear*) is also state 1. If the upper NAND gate has a state 1 at the other input, its output, Q, must be state 0. The state 0 and state 1 input at the lower NAND gate makes the $\bar{Q}$ output state 1, as originally assumed at the upper gate input. That is, Q and $\bar{Q}$ always have opposite logic states. Note that, if S is state 1 and R state 0, the output Q becomes state 0 and $\bar{Q}$ becomes state 1. Actually, this cross-coupled logic gate is quite equivalent to the binary discussed in the previous chapter, as may be verified by comparing the logic diagram, Fig. 12-18, with the circuit of a binary, Fig. 11-7, in the case of RTL logic, Fig. 12-6*b*.

The truth table for a NAND binary is given in Table 12-11. When S is state zero the flip-flop is said to be *set* since $Q = 1$. Similarly, a state 0 input to R *clears* the flip-flop by putting Q at state 0. Note that the binary is, in effect, a memory circuit since it indicates which input was at state 0 last. Simultaneous state 1 inputs to both R and S leave the circuit in its original state while two state 0 inputs gives an indeterminate result since the logic of the circuit is not satisfied and the final state after these input signals are removed is only a matter of chance. A similar binary results from cross-coupled NOR gates; the corresponding truth table is related but not identical to Table 12-11.

TABLE 12-11 TRUTH TABLE FOR NAND BINARY

R	S	Q
0	0	Indeterminate
0	1	0
1	0	1
1	1	Q

If input gates are added to the NAND flip-flop, Fig. 12-19, the circuit responds to input logic signals only when the T, or *clock* input is in state 1. The reason for this is that when the clock input, T, is at state 0, the input

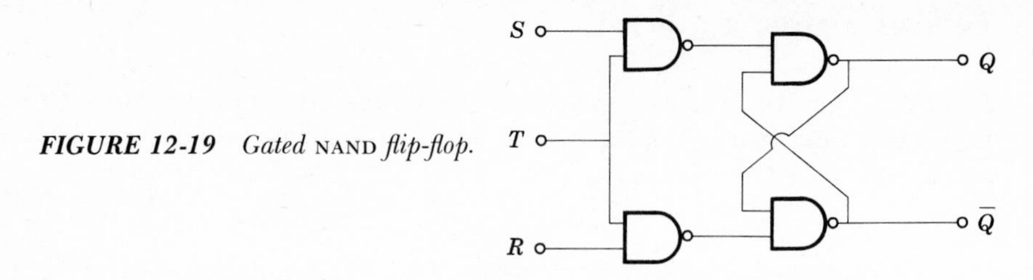

FIGURE 12-19 *Gated* NAND *flip-flop.*

gates must both have state 1 outputs (independent of the signals at R and S). According to Table 12-11, the binary stays in its original state under this condition. When the T input is in state 1, logic signals at R and S can cause transitions in the circuit, as described by Table 12-12.

TABLE 12-12 TRUTH TABLE FOR GATED NAND **FLIP-FLOP**

R	S	Q
0	0	Q
0	1	1
1	0	0
1	1	Indeterminate

Note that this logic is reversed compared to the simple NAND flip-flop because of inversion in the input gates. In effect, this circuit is controlled, or gated, by signals at the T input since logic signals at R and S are only effective when T is at state 1. Note the similarity of this configuration with the RS/T binary discussed in Chap. 11, Fig. 11-9. The present gated flip-flop is completely dc coupled, however, and responds to logic signals even if they are not pulses.

It is not possible to connect the output terminals of a gated flip-flop to the input terminals as in the RS/T binary in order to make the circuit toggle. The reason for this is that the output cannot change state while connected to the input signals. Two cascaded flip-flops can be used, however, to achieve toggling while retaining the advantages of dc coupling. The logic network of a *RS master-slave* flip-flop in Fig. 12-20 illustrates this technique.

Note that the circuit is two gated flip-flops controlled by the same clock input. The slave flip-flop is gated by an inverted clock signal, however, and because of the 220-Ω resistor in the clock line is activated at slightly a different time than the master flip-flop. When the clock is at state zero, logic signals at the R and S inputs are ineffective, as in a regular gated flip-flop. The inverted clock signal is state 1 at the slave flip-flop, however, and the slave assumes the same state as the master. As the clock signal goes towards state 1, the input of the clock inverter gate reaches state 0

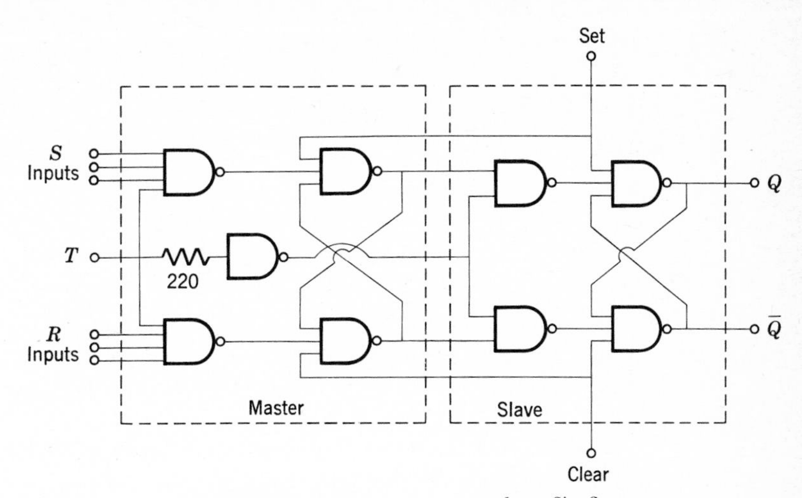

FIGURE 12-20 *RS master-slave flip-flop.*

before the master input gates reach state 1. Therefore, the slave input gates isolate the slave flip-flop from the master, and the state of the slave flip-flop stores that of the master.

Next, the master input gates receive the state 1 clock signal and the logic signals at R and S determine the state of the master flip-flop according to Table 12-12. At the end of the clock pulse, the master input gates isolate the master flip-flop from the R and S inputs. With the clock back at state 0, the inverted clock signal at the slave flip-flop causes the slave to take up the state of the master flip-flop. Thus the output state reflects the logic signals at R and S, but the output state is not attained until the end of the clock pulse. Therefore, output changes cannot appear at the input terminals during the clock pulse.

Note that multiple terminals are available at both the R and S inputs. According to Table 12-12 a state 0 at any S input puts Q at state 0, and a state 0 at any R input puts Q at state 1. Direct set and clear inputs are also provided. A state 0 signal at the set input overrides all other signals and produces $Q = 1$. Similarly, a state 0 signal at the reset input makes $Q = 0$ and the flip-flop is cleared.

Toggle action is achieved by connecting $\bar{Q}$ in Fig. 12-20 to one of the S inputs and Q to one of the R inputs. The remaining S terminals are renamed J inputs and the free R terminals are called K inputs. The configuration is known as a *JK master-slave* flip-flop. The new truth table, given in Table 12-13, is similar to that for a gated NAND flip-flop except that the circuit toggles, as indicated by the last entry. Stated in words, the logic of a JK flip-flop is: a state 0 at any J input prevents a state 1 at Q, a state 0 at any K input prevents a state 0 at Q, and the circuit toggles unless inhibited by a state 0 at either J or K inputs. Note also that the tog-

TABLE 12-13 TRUTH TABLE FOR NAND *JK* MASTER-SLAVE FLIP-FLOP

$J_1 \cdot J_2$	$K_1 \cdot K_2$	Q
0	0	Q
0	1	0
1	0	1
1	1	$\bar{Q}$

gling action is in response to the signals at T whether or not they are pulses.

Such elaborate flip-flops are only practical in integrated-circuit form. A typical *JK* master-slave flip-flop in TTL logic uses a total 18 transistors yet is contained with a conveniently sized package, Fig. 12-21.

FIGURE 12-21 *Integrated circuit JK master-slave flip-flop.* (Motorola Semiconductor Products, Inc.)

A useful modification of a gated flip-flop has a single input terminal connected to the S input and uses an inverter to connect $\bar{S}$ to the R input, Fig. 12-22. The appropriate truth table, Table 12-14, shows that the D

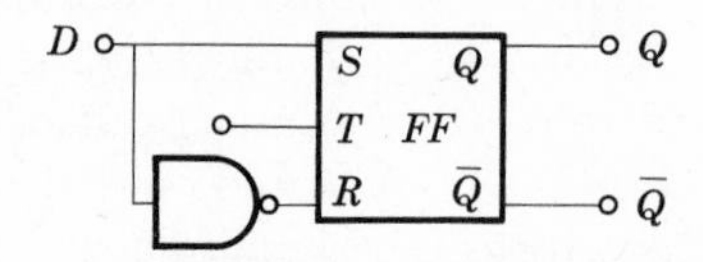

FIGURE 12-22 *Data latch flip-flop.*

TABLE 12-14 TRUTH TABLE FOR *D* FLIP-FLOP

D	Q
0	0
1	1

flip-flop (for *data latch*) stores the logic signal at the input terminal during the last interval that the clock was in state 1. An application of this circuit is considered in the next section.

Counters Cascaded flip-flops are similar to binary counters discussed in the previous chapter. The multiple inputs and dc coupling of a flip-flop compared to a simple binary result in much more flexible and useful operation, however. Cascaded flip-flops, called *registers,* are widely used to count, store, and manipulate logic signals and signals representing binary numbers.

Consider the so-called synchronous binary counter in Fig. 12-23, which

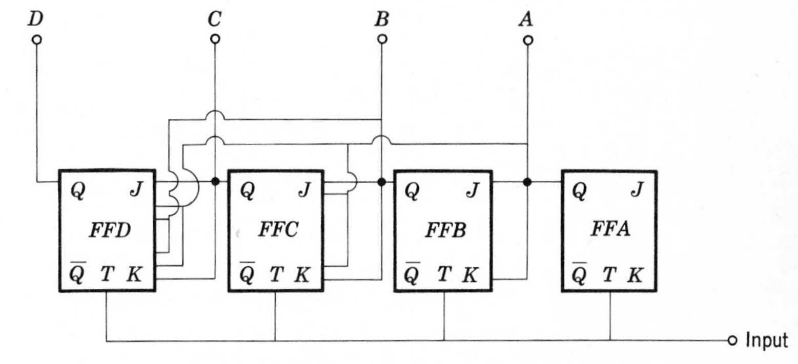

FIGURE 12-23 *Synchronous binary counter.*

is composed of four cascaded *JK* flip-flops. Note that the input terminal is on the right so that a digital number stored in the register is presented with its most significant digit on the left as it is normally read. Each flip-flop is toggled by input pulses applied to all clock inputs. Interconnections between the flip-flops inhibit transitions until the appropriate count is reached. Note, for example, that *FFA* toggles on every input pulse, while *FFD* is inhibited until the eighth (and sixteenth) input pulse when all its *J* and *K* inputs are at state 1. The waveforms are identical to those for the nonsynchronous binary counter, Fig. 11-17, since the final state of each *JK* flip-flop is attained on the negative transition of the clock pulse. This waveform pattern may be verified by using the truth table for the *JK* flip-flop, Table 12-13.

An asynchronous binary counter using flip-flops is achieved by connecting the *Q* output from the first flip-flop to the *T* input of the second, and so on. This is identical to the circuit connections of the cascaded binaries of Chap. 11. The main advantage of the synchronous counter over an asynchronous counter is reduction of propagation delays on, for example, the eighth and sixteenth counts when all four flip-flops change state. In the asynchronous counter each flip-flop is triggered sequentially, so that a total time equal to the propagation delay of each flip-flop times the number of cascaded flip-flops is required for the last transition. At high count rates the input flip-flop may already have responded to succeeding input pulses before the output flip-flop reaches its final state. Thus, the state of the counter at a given instant does not represent any exact number.

This situation is eliminated in the synchronous counter since all flip-flops make transitions at the same time in response to the signal at the T inputs.

It is usually convenient to display information in decimal form, and therefore conversion from the binary number system to the decimal system is necessary. Decoding logic to accomplish this change becomes very complicated when more than 4 or 5 bits are involved. Instead, it is often more convenient to store each digit of a decimal number in binary form separately. Thus each digit of the decimal number is represented by a four-flip-flop register counting from 0 to 9, and, for example, five such registers can store decimal numbers from 0 to 99,999. The decoding logic need therefore handle only decimal numbers from 0 to 9 and large numbers are readily accommodated by adding an additional register for each decade.

A four-flip-flop register is capable of counting to 16, but a total count of 10 is desired in such a *binary-coded decimal,* or BCD, counter. This can easily be accomplished by interconnections between flip-flops, as in Fig. 12-24. The resulting decade counter has a different counting pattern than

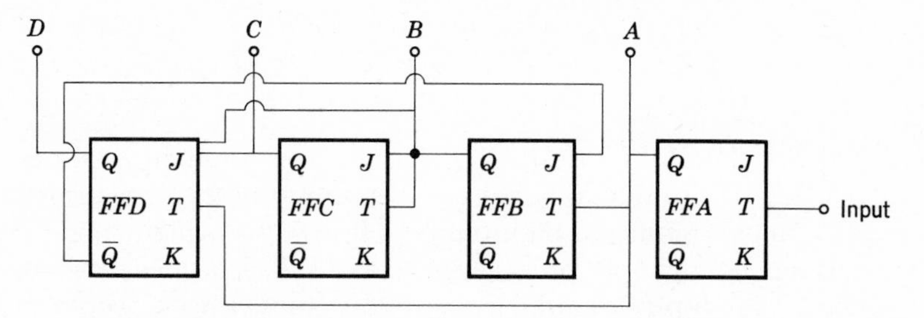

FIGURE 12-24 *Asynchronous binary-coded decimal counter.*

the decade counter discussed in Chap. 11. Input pulses cause *FFA* to toggle and *FFA*, in turn, toggles *FFB* until the eighth count when the Q output from *FFD* is inhibiting. Similarly, *FFD* is inhibited by the Q outputs of *FFB* and *FFC* at its J inputs, one or the other of which is at state 0 until the seventh count. *FFD* flips on the eighth input pulse. The ninth pulse toggles *FFA* only and on the tenth count both *FFA* and *FFD* flip so that all Q's are at state zero and the cycle is complete. This is best illustrated by a table showing the state of each flip-flop as a function of the input count, Table 12-15. Note that this pattern is identical to the waveforms of the binary counter, Fig. 11-17. That is, the counting pattern of this BCD counter is the same as the first nine states of a scale-of-16 counter. This is called the natural 8421 code and is the one most often used. The decade counter in Chap. 11 is an example of a different pattern.

The cycle, or *modulus,* of a BCD counter is 10 since all flip-flops are cleared every 10 counts. Similarly, the modulus of a four-flip-flop binary counter is 16. Through appropriate logic connections, counters of any modulus are possible. Note, for example, that the combination of *FFB, FFC,* and *FFD* of the BCD counter is a *modulo*-5 counter if the input is at *T* of *FFB*. Alternatively, a simple binary counter may be converted to a counter of any desired modulus by using the direct clear inputs and a logic circuit to clear all flip-flops at the proper count. This technique is illustrated in Fig. 12-25 for a direct-cleared modulo-7 counter. The cir-

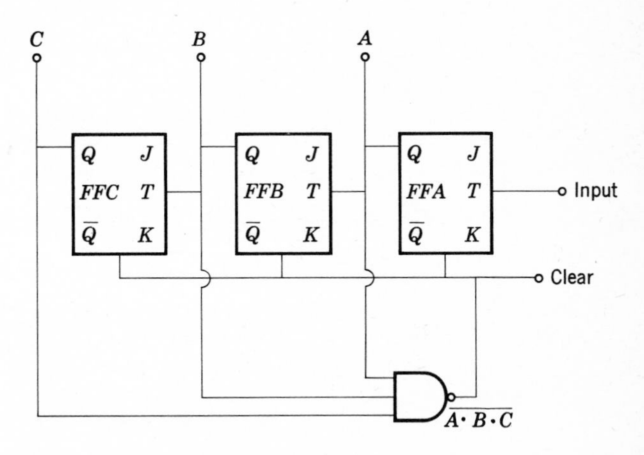

FIGURE 12-25 *Modulo-7 counter using direct clearing.*

cuit counts as a normal binary until the seventh input pulse when the *Q* outputs are all state 1 (refer to Table 12-15) for the first time. The output of the NAND gate goes to state 0, all the flip-flops are cleared, and the count cycle repeats.

TABLE 12-15 BCD COUNTER STATES

Count	*FFD*	*FFC*	*FFB*	*FFA*
0	0	0	0	0
1	0	0	0	1
2	0	0	1	0
3	0	0	1	1
4	0	1	0	0
5	0	1	0	1
6	0	1	1	0
7	0	1	1	1
8	1	0	0	0
9	1	0	0	1

It is often desirable to store the total count in a register, as for display purposes, while the counter proceeds to count a new signal waveform.

One way to accomplish this is by means of a *D* flip-flop connected to each counter flip-flop, Fig. 12-26. A state 1 pulse at the memory transfer termi-

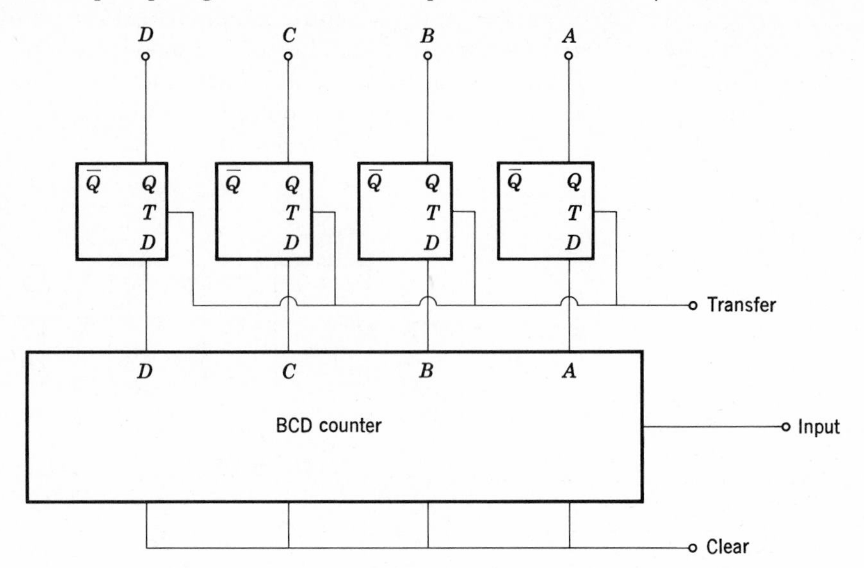

FIGURE 12-26 *BCD counter and storage resistor.*

nal causes each *D* flip-flop to assume the state of its respective counter flip-flop, according to Table 12-14. Therefore, the number in the BCD counter at the time of the transfer pulse is stored in the memory register while the counter is cleared to count again. The output of the memory register may be used to activate a numerical display that remains constant until the next transfer pulse. Note that this configuration also converts a serial representation into the parallel representation of the binary number.

As pointed out previously, a numerical decimal display of, say, the num-

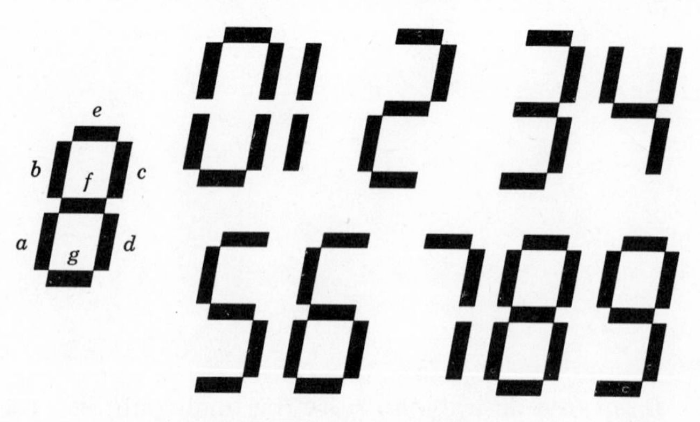

FIGURE 12-27 *Seven-segment decimal display.*

ber stored in the memory register of Fig. 12-26 is advantageous. Indeed, nearly all present-day digital instruments have this feature. One method of decoding binary signals suitable for decimal display by means of neon lamps or Nixie tubes was described in Chap. 11. A similar logic code can be developed to drive a Nixie tube from a BCD counter.

Another popular technique uses a seven-segment display as illustrated in Fig. 12-27. Each segment is lit by a small lamp, and the appropriate combination of lit segments displays the numbers from 0 to 9. A truth table specifying the lit segments for each numeral is easily established, Table 12-16.

TABLE 12-16 TRUTH TABLE FOR SEVEN-SEGMENT DECIMAL DISPLAY

	Segment						
Decimal number	*a*	*b*	*c*	*d*	*e*	*f*	*g*
0	1	1	1	1	1	0	1
1	1	1	0	0	0	0	0
2	1	0	1	0	1	1	1
3	0	0	1	1	1	1	1
4	0	1	1	1	0	1	0
5	0	1	0	1	1	1	1
6	1	1	0	1	1	1	1
7	0	0	1	1	1	0	0
8	1	1	1	1	1	1	1
9	0	1	1	1	1	1	1

Decoding the binary output signals from a BCD counter, Table 12-15, to light the proper segments as specified by Table 12-16 is accomplished with an array of OR and AND gates. For example, according to the two truth tables the logic necessary to light the lamp in segment a, which is activated for decimal numbers 0, 1, 2, 6, 8, is

$$a = (\bar{A} \cdot \bar{B} \cdot \bar{C} \cdot \bar{D}) + (A \cdot \bar{B} \cdot \bar{C} \cdot \bar{D}) + (\bar{A} \cdot B \cdot \bar{C} \cdot \bar{D}) + (\bar{A} \cdot B \cdot C \cdot \bar{D}) + (\bar{A} \cdot \bar{B} \cdot \bar{C} \cdot D) \quad (12\text{-}11)$$

A similar logic expression can be written for the other six lamps. The expressions can be simplified by noting that if *FFD* is in state 1 the number must be either 8 or 9 and the state of *FFA* distinguishes between them. Also, for certain segments such as d and e the logic statement for NOT lit is shorter since these segments are activated on eight of the ten numerals.

In addition to these simplifications, note that both Q and $\bar{Q}$ are available from each flip-flop. The decoding logic according to Eq. (12-11) and the six other similar relations simplified by these observations may be accomplished by an array of 13 NOR gates and 5 inverter gates. Actually, complete integrated-circuit decoder gates are commercially available which accept the four input signals and provide the proper outputs at seven output terminals.

Memory Registers A combination of flip-flops arranged to store logic signals is called a register, and the four D flip-flops so grouped in Fig. 12-26 is a typical example. Actually, of course, a set of cascaded binaries is also a register since it stores information. It is often useful to shift the information stored in a register along the chain of flip-flops without changing the relative waveform of the logic signal and such a device is called a *shift register*.

Consider the 4-bit shift register of JK flip-flops in Fig. 12-28*a* together with typical corresponding waveforms in Fig. 12-28*b*. The input signal, I, and $\overline{I}$ are applied to the J and K inputs, respectively. Whenever I is state

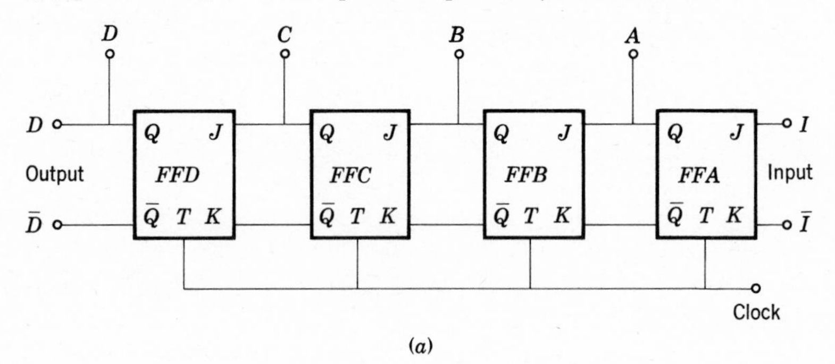

(*a*)

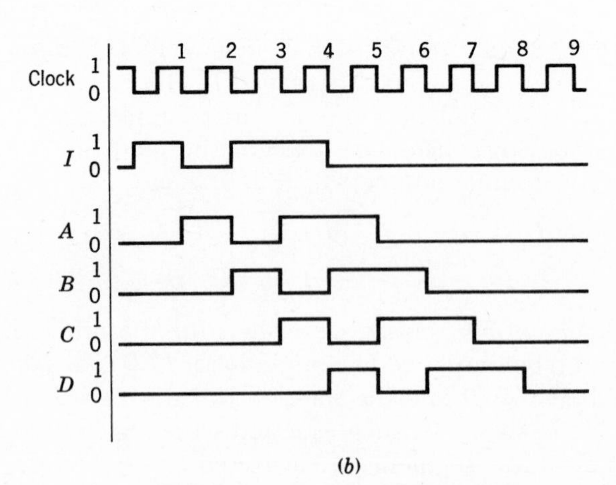

(*b*)

FIGURE 12-28 *(a) Four-bit shift register and (b) typical waveforms.*

1, the same logic appears at Q of *FFA* at the trailing edge of the clock pulse. This, in turn results in state 1 at Q of *FFB* at the cessation of the next clock pulse, and so on. That is, information presented to the input appears at each succeeding flip-flop output delayed by one additional clock pulse. Therefore, the information pattern is shifted intact along the register, as indicated by the typical waveforms in Fig. 12-28*b*.

If the output of a shift register is returned to the input, the logic signals circulate continuously. A useful arrangement of such a *circulating register* is shown in Fig. 12-29. With the control input at state 1, logic signals at the

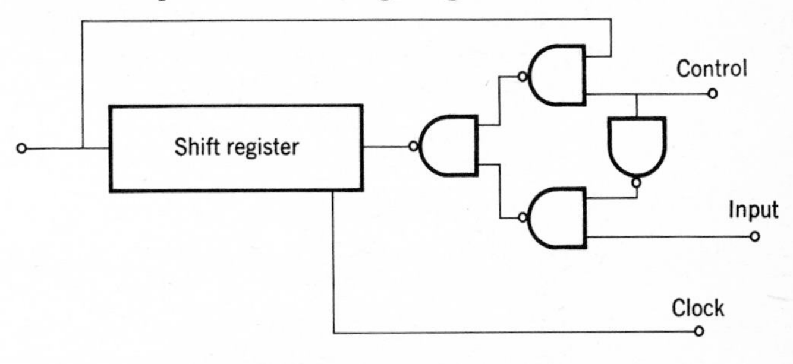

FIGURE 12-29 *Circulating register.*

output of the register are reintroduced at the input. The signals reappear at the output every m pulses, where m is the number of flip-flops in the register. New logic signals can be introduced by a state 0 signal at the control terminal. Integrated-circuit circulating registers several hundred bits long are available.

A related register with useful properties is obtained by connecting the outputs of the shift register in Fig. 12-28*a* to the input terminals and regarding the clock terminal as the input signal. First, a state 1 signal is placed in *FFA* and state 0 in all other flip-flops using the direct-set and direct-clear inputs on each flip-flop. According to the waveforms in Fig. 12-28*b* this single 1 circulates around the register in response to input pulses at the clock terminal. The successive states of this 4-bit *ring counter* are shown in Table 12-17.

TABLE 12-17 STATES OF 4-BIT RING COUNTER

Input Pulse	*FFD*	*FFC*	*FFB*	*FFA*
1	0	0	0	1
2	0	0	1	0
3	0	1	0	0
4	1	0	0	0

Note that this operation is, in effect, a modulo-4 counter, and that the total decimal count is readily discernible by, say, connecting an indicator to each flip-flop output. That is, the ring counter is self-decoding. This useful feature compensates for the fact that only one bit per flip-flop may be accommodated.

DIGITAL INSTRUMENTS

Digital instruments use the logic circuits and techniques described in the foregoing to carry out measurements and process data. A significant advantage of digital over analog instruments is that digital data is decisive. The least significant bit must be either a 1 or a 0 so that precision is increased simply by employing additional digits. Furthermore, digital signals can be amplified indefinitely and stored accurately. Lastly, because of the off-on character of digital circuits, drift and stability problems are inherently negligible.

On the other hand, digital instruments tend to be more complex than their analog counterparts. Mathematical operations such as integration, differentiation, and addition require rather elaborate logic circuits, whereas simple RC circuits suffice for analog signals. Despite this inherent complexity and attendent higher cost the advantages of digital instruments are sufficient to assure their ever-expanding use in measurement and control applications.

Time and Frequency Meters The time interval between two events can be accurately measured by counting the number of cycles of a stable oscil-

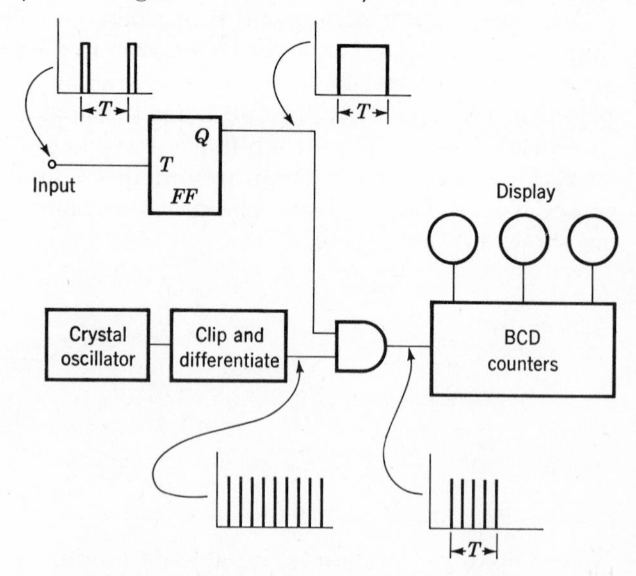

FIGURE 12-30 *Block diagram and waveforms of time-interval meter.*

lator during the time interval between pulses signaling the two events. This is accomplished, Fig. 12-30, by triggering a binary by the input pulses to provide a gate pulse with a duration equal to the interval between pulses. If the oscillator frequency is 1000 Hz, the cascaded decade counters display the time interval directly in milliseconds.

The timing accuracy depends upon oscillator stability and is subject to a one-cycle uncertainty, depending upon when the gate pulse is initiated relative to the oscillator pulses. One extra count can be registered if the gate pulse starts during an oscillator pulse compared with the count recorded if the gate opens between pulses, as illustrated in Fig. 12-31.

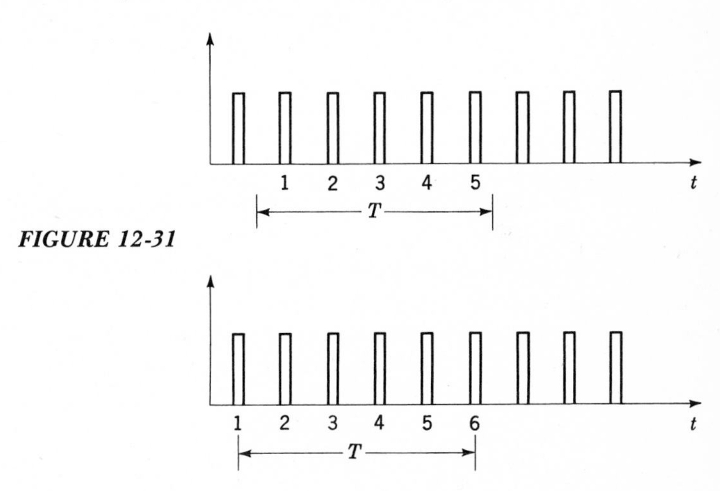

FIGURE 12-31

Because of this 1-bit uncertainty it is desirable to have the oscillator frequency high enough to ensure a large total count.

Similar techniques are used to measure the frequency of an unknown oscillator. The number of cycles in the unknown signal is counted during an accurately known time interval. For this purpose the input signal is first converted into a series of pulses, one per cycle, by means of a Schmitt trigger followed by a differentiator and clipper, as shown in Fig. 12-32. The number of input pulses is counted during a given time interval. If, for example, the interval is 1 s, the frequency of the unknown is displayed directly in hertz.

It is convenient to derive the time interval by scaling down the output of a stable crystal oscillator. The output pulse from the last flip-flop of the scaler is used directly to control the gate. Note that the frequency range of the instrument may be changed easily by changing the scale factor of the oscillator counter. In effect, the instrument compares the unknown frequency of the input signal with the known frequency of the crystal oscillator.

The frequency meter is started by closing the start switch and subsequently opening it after one counting interval. In practical instruments,

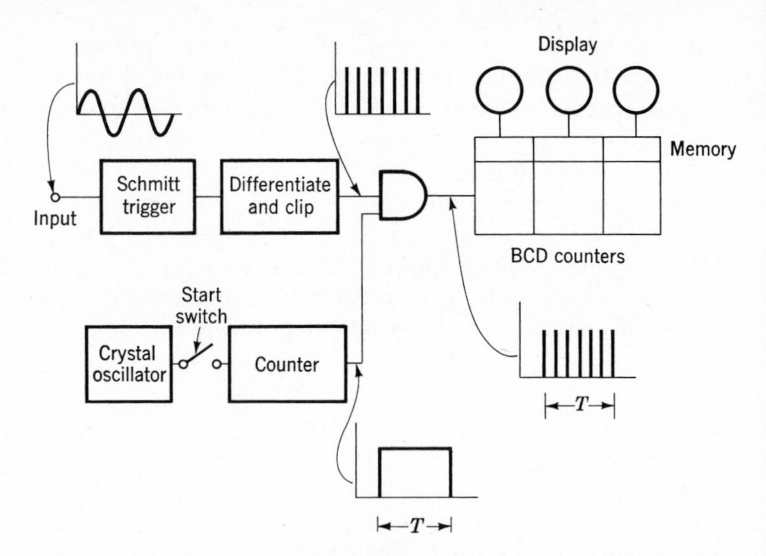

FIGURE 12-32 *Digital frequency meter.*

control circuits are included to perform this operation, as well as to transfer the count to memory registers, and to clear the frequency-count register. In this mode the instrument repetitively samples the input frequency and can, for example, detect and display slow frequency changes of the input signal. Control circuits employ the same logic gates and networks as used in the signal-handling portions of the instrument.

The frequency measurement is subject to the same 1-bit uncertainty as in the case of the time-interval meter. The effect is minimized by choosing a scaling factor for the oscillator counter that maximizes the number of input signal pulses during the gate interval. Alternatively, low-frequency input signals can be measured by deriving the gate pulse from one period of the input signal and counting the number of oscillator pulses during this time, just as in the time interval meter.

Digital Voltmeter A basic feature of digital instruments is conversion of input signals into digital data. The instruments discussed in the previous section employ quite simple pulse circuits to accomplish this *analog-to-digital* conversion. A number of different techniques have been devised to convert other analog signals into digital signals. A typical example of interest is used in one type of *digital voltmeter,* an instrument that provides a direct-reading numerical display of the voltage being measured.

Consider the simplified block diagram of a digital voltmeter sketched in Fig. 12-33. The unknown dc voltage applied to the input terminals is compared with a staircase waveform produced by a diode pump circuit fed from a 10-kHz oscillator. When the staircase signal is equal to

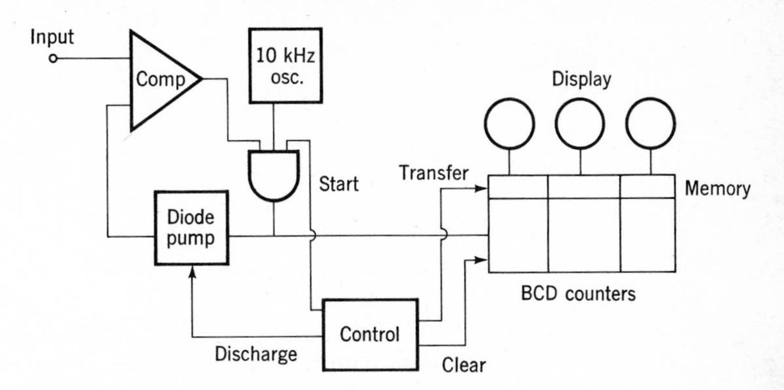

FIGURE 12-33 *Simplified block diagram of digital voltmeter.* (Princeton Applied Research Corporation.)

the input voltage, the comparator interrupts transmission of further oscillator pulses. The parameters of the diode pump are selected so that the height of each step of the staircase is equal to 1 mV. Therefore, the number of steps is directly equal to the unknown voltage in millivolts, and the steps are counted and displayed by a three-decade counter, memory register, and decimal display.

The control circuits subsequently execute a memory transfer for display and discharge the diode pump capacitor so that the cycle repeats. A new reading is attained in less than 1 s. Actually, in practical instruments the control circuit has other functions as well. It is possible, for example, to change input-multiplier resistors automatically so that voltages from 100 mV to 999 V can be measured with the same 1-digit precision. In addition, the circuit can automatically adjust to either input-voltage polarity.

In control applications of digital circuits it is common for digital data to be reconverted to analog form after the desired processing steps have been completed. The analog output signal is then used to adjust the system being controlled. Circuits for *digital-to-analog* conversion tend to be less elaborate than the A to D conversion and also depend upon whether the digital signals are presented in serial or parallel representation.

Digital Filter Digital electronic logic circuits permit signal processing of an entirely different nature than is possible using analog instruments. In particular, signal-averaging instruments capable of discerning weak signals in the presence of random noise are very effective. These types of instruments are, in effect, small digital computers which are designed to carry out certain highly specialized calculations. It is useful to examine the operation of the so-called *digital filter* in this connection.

The digital filter performs much the same task as, for example, an inductance-capacitance filter network, but in a fashion that can provide filter parameters not possible with conventional component values. The operation of a digital filter rests upon the same duality between frequency and time already noted in connection with the Fourier series representation of complex waveforms. That is, the spectrum of frequency components in a signal is an equally valid description of the signal as is the time variation of the signal given by the waveform. Correspondingly, while most often a filter network is described by means of its frequency response characteristic, an equally useful description is possible using the transient properties of the circuit. Note that the frequency-response characteristic is given by the ratio of the output signal, $v_o(f)$, to the input signal, $v_i(f)$ at a given frequency,

$$\frac{v_o(f)}{v_i(f)} = F(f) \tag{12-12}$$

A similar input-output relation exists in the time domain, except that the input and output signals may be noted at separate times

$$\frac{v_o(t)}{v_i(t')} = T(t,t') \tag{12-13}$$

Both $F(f)$ and $T(t,t')$ are satisfactory descriptions of filter network properties.

It is possible to develop an expression for $T(t,t')$ if the frequency-response characteristic is known in much the same fashion as the Fourier components of a complex waveform are found. The details of this transformation are not of direct interest here. It is important to observe, however, that the time-like representation permits a direct calculation of filter performance by digital techniques in the following way. The output of the digital filter at any instant is written in terms of the input signal at several discrete times earlier,

$$v_o(t) = a_1 v_i(t - \delta) + a_2 v_i(t - 2\delta) + a_3 v_i(t - 3\delta) + \cdots \tag{12-14}$$

Here, the a coefficients are the representation of the $T(t,t')$ characteristic of the filter. The calculations indicated by Eq. (12-14) are accomplished by the circuit shown in Fig. 12-34. First, the input signal is sampled at discrete intervals and digitized. Each sample is delayed an appropriate time interval, multiplied by a constant characteristic of the filter performance desired, and the result is summed to yield the output signal. The output may be converted to analog form, if desirable.

The digital filter, in effect, continuously calculates the filter output arising from input signals. Since filter performance is described by the a coefficients which are easily adjustable, the same digital filter may be used to synthesize a variety of frequency-response characteristics. No inductances are needed, which is often advantageous in low-frequency

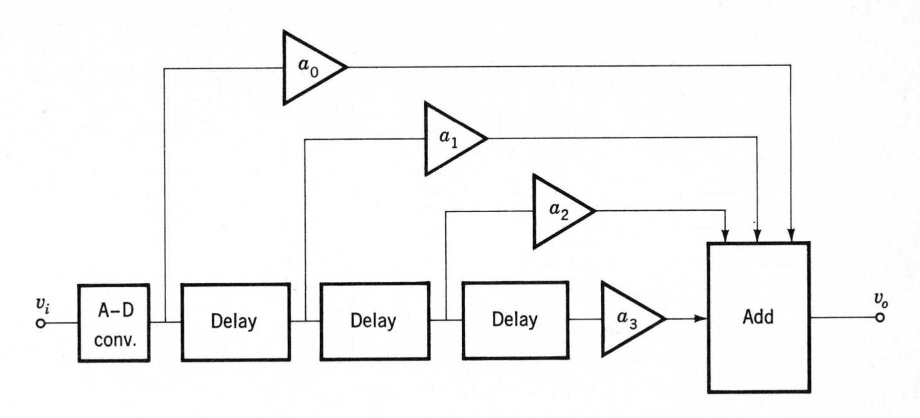

FIGURE 12-34 *Block diagram of digital filter.*

and integrated-circuit applications. Furthermore, because of the discreteness of digital data, precise performance, such as very large attenuation at specific frequencies, is achievable.

THE DIGITAL COMPUTER

Organization A *digital computer* is a complex array of logic gates and associated circuitry organized to perform logic computations by manipulating waveforms representing digital numbers. The great power of electronic digital processing stems from the variety of phenomena that can be represented and analyzed logically. Typical examples range from scientific computations which obey some physical law to bookkeeping activities which follow the principles of accounting. Because the speed of electronic circuits is so great, digital computers can complete extremely complex and extensive calculations in practical lengths of time.

The circuits of digital computers are designed to carry out numerical calculations of all kinds. Therefore, the machine is furnished specific instructions pertaining to any desired computation, in addition to all the digital numbers involved. These instructions are represented by pulse waveforms similar to those representing numbers. A complete set of instructions, together with the numbers associated with a given problem, is called the *program.*

A digital computer comprises five major parts: input, output, memory, control, and arithmetic units, Fig. 12-35. The input and output units present the program to the machine and retrieve the final results. Most often these are magnetic tape recorders which are capable of recording and reproducing the digital waveforms representing the instructions and numbers of the program as well as the waveforms representing the results.

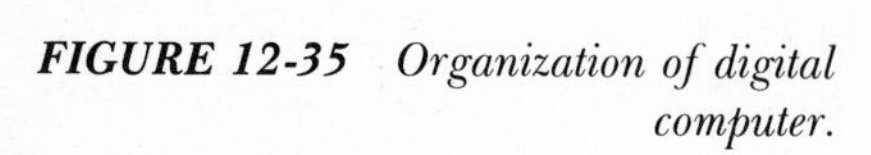

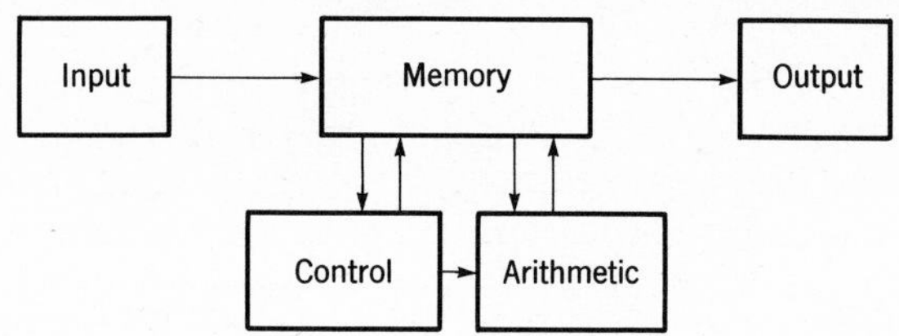

FIGURE 12-35 *Organization of digital computer.*

The recorded output signals are subsequently converted into printed form by passing the output magnetic tapes through a separate unit, called a *printer*. In effect, the input and output units are the means by which human operators communicate with the machine.

The computer *memory* stores each instruction and number of the program until needed during the course of the computation. In one type of memory, individual bits are stored by magnetizing a small ring, or core, of magnetic material in either the clockwise or counterclockwise direction. Information is retrieved by sensing the direction of magnetization in each core and generating a corresponding signal waveform. The instructions of the program are stored separately from numbers since the two are used in fundamentally different ways in the calculation. The location of each number is specified by an *address* which is used to tell the machine where a particular number is stored in the memory.

The *arithmetic* unit contains logic circuits such as ADD gates which actually perform the calculations specified by the program. The arithmetic unit also includes *registers* for temporarily storing a digital number. The reason for this is that, ordinarily, only one number at a time is recalled from the memory. If two numbers are to be added, the first one must be stored until the second one is available before they can be presented simultaneously to the ADD gate.

The function of the *control* unit is to interpret each instruction and set the circuits of the computer accordingly. It is the most heterogeneous part of the computer and is composed of logic gates, registers, and a *clock* circuit which regulates the basic speed at which the machine operates. The clock provides a continuous train of pulses which are used in connection with the various logic gates to direct the digital waveforms through the various parts of the computer. Clock pulses are used, for example, to sense the magnetization of the magnetic cores in the memory; this means that the time interval between bits of a number read from the memory is the same as that of the clock pulses.

A digital computer is operated by first inserting the program into the memory. When the machine is started, the control unit reads the first instruction, prepares the circuits of the computer accordingly, and causes the proper number to be read from the memory as specified by the address in the instruction. At the completion of the indicated operation, the result is returned to the memory at the address also specified in the program. The control unit then passes on to the next instruction. The machine proceeds sequentially through the instructions, placing final results in the output unit until the final instruction *stop* is reached. At this point the computation is finished and the desired result is located in the output unit.

It is useful to illustrate by means of a rudimentary example how the control unit prepares the circuits of the machine in accordance with the instructions. Consider the partial block diagram, Fig. 12-36, in which

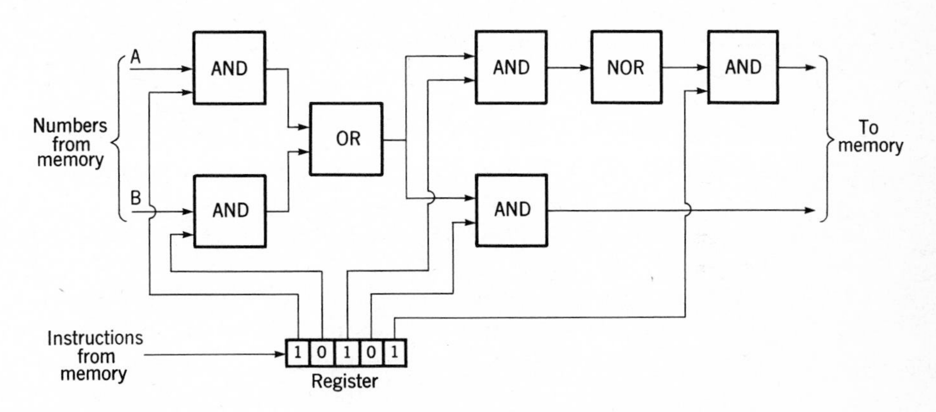

FIGURE 12-36 *Illustrating how an instruction determines path of number through circuit as in digital computer.*

either of two numbers can be operated upon and returned to the memory, depending upon the instruction. Suppose the instruction 10101, as shown, is abstracted from the memory and stored in a register connected to the several AND gates. This means that the number A passes through the circuit, emerges as NOT A, and is returned to the memory. The number B is not used since it does not pass the first AND gate. Other instructions cause different operations to be performed, as may be determined by following the path through the circuit in each case:

Instruction	*Interpretation*
10101	Take A and return NOT A to memory
10010	Take A and return A to memory
01101	Take B and return NOT B to memory
01010	Take B and return B to memory

In effect, instructions set up signal paths through the computer by activating logic gates of the control unit.

Programming It is not feasible, or necessary, to trace the entire signal path for each instruction through the complete circuit when preparing a program. In the design of the machine, a tabulation of the instructions to carry out various operations is developed, corresponding to the control logic circuits. This *dictionary of machine language* is used to write programs. It obviates the need to refer to the circuits of the computer itself. Nevertheless, because each minute step must be detailed in the instructions, the preparation of a program is a long and arduous task and programming errors are frequent. For example, transposing a 0 for a 1 in any one instruction may result in a "nonsense" instruction, or, more seriously, in a false instruction, and such errors are difficult to locate. To help this situation, *assembler* programs have been prepared which can translate instructions written in a more natural, alphabetic code into machine language.

Large-scale digital machines are provided with previously prepared *compiler* programs which allow programs to be written in a stylized algebraic language that can be read and understood by the person preparing the program. As this program is typed on a special electric typewriter, an encoded punched-paper tape is prepared simultaneously. When this tape is passed through the computer input the compiler program in the memory operates to produce a magnetic-tape output which is a computer program in machine language. In effect, the computer compiles its own program by translating the stylized language program into machine language. Subsequently, the machine-language program is run on the computer in standard fashion. This technique is feasible because most programs comprise a large number of similar, repetitive steps. The net result is that programs can be prepared by relatively unskilled users with relatively little effort than would be required if machine language were used directly.

SUGGESTIONS FOR FURTHER READING

Thomas C. Bartee: "Digital Computer Fundamentals," McGraw-Hill Book Company, New York, 1966.

I. H. Gould and F. S. Ellis: "Digital Computer Technology," Reinhold Publishing Corporation, New York, 1963.

H. V. Malmstadt and C. G. Enke: "Digital Electronics for Scientists," W. A. Benjamin, Inc., New York, 1969.

EXERCISES

12-1 Carry out the binary additions for 20 + 14, 15 + 34, and 56 + 25. Check by forming the binary numbers of the decimal sums. Sketch the waveforms of the individual numbers and their sums.

Ans.: 100010; 110001; 1010001

12-2 Carry out the binary subtractions for 25 − 14, 35 − 16, and 12 − 3. Check by forming the binary numbers of the decimal subtractions. Sketch the waveforms of the individual numbers and their differences.

Ans.: 1011; 10011; 1001

12-3 Sketch the waveforms of the binary numbers 1101 and 1001 and their sum. Compare the waveform of the sum with the output waveform of an AND gate with 1101 and 1001 as input signals.

12-4 Using De Morgan's theorem simplify the logic expression for NOT exclusive-OR. Why is this sometimes called an *equality* comparator?

Ans.: $A \cdot B + \overline{A} \cdot \overline{B}$

12-5 Develop the truth table for the full adder in Fig. 12-16.

12-6 Sketch the waveforms at the output of each gate of a serial full adder using Figs. 12-17*a*, 12-16, and 12-15, if the input waveforms are the binary numbers 1101 and 1001. Verify that the output waveform represents the sum of the inputs.

12-7 Develop the truth tables for a NOR binary and a gated NOR binary and contrast with Tables 12-11 and 12-12.

12-8 Specify the states of the 4-bit ring counter similar to Table 12-17 if the Q output of FFD is connected to J of FFA and $\overline{Q}$ of FFD to K of FFA. Assume the initial state of this so-called *switch-tail* counter is all flip-flops cleared.

12-9 Devise a control logic circuit for the digital frequency meter in Fig. 12-32 which causes the instrument to repetitively display the frequency of the input signal.

12-10 Devise a block diagram of a digital voltmeter in which the output of a ramp generator is compared with the unknown voltage and the time from beginning of the ramp to the equality point is measured by a time-interval meter.

LABORATORY EXERCISES

12-A The Digital Voltmeter The digital voltmeter exemplifies many important features of digital instruments such as analog-to-digital conversion, logic circuitry, and digital counting and display. Furthermore, it is an important and convenient laboratory instrument in its own right. The purpose of this experiment is to become familiar with the performance of a typical digital voltmeter.

Examine the block diagram of the digital voltmeter in Fig. 12-33 until you really understand the important facets of circuit operation. You should confirm your understanding by devising control circuit logic to cause the instrument to operate automatically. One convenient form for the control circuit is based on a multivibrator having a 1-s period to establish the timing of the capacitor discharge, memory transfer, clearing the counter, etc. Be sure your instrument works.

Quantitatively sketch voltage waveforms at the diode pump output, input to the BCD counter, at each flip-flop in the counter for the least significant decimal digit, and at all the memory flip-flops for three cycles of operation if the input is

0.27 V on the 1.00-V scale. Take the oscillator frequency to be 10 kHz and one step in the staircase to be 1 mV. Now also sketch the control logic waveforms to show these operate the instrument.

Suppose the input signal is $\frac{1}{2}$ V dc together with a $\frac{1}{2}$-V (peak-to-peak) sawtooth waveform having a period of 2 s. Sketch the input signal, diode pump, and comparator output waveforms. Do you consider this situation a major deficiency of this simplified digital voltmeter? How would you recognize this situation from the decimal display?

Focus attention on the input comparator. Is an operational amplifier suitable in this application? What should the polarity of the staircase waveform be compared to the input signal (and what should the input voltage polarity be?), in order to provide the appropriate logic signal for the gate? Sketch the voltage signals at the input and output of the comparator for the last few steps just before and up to the equality point. Should the comparator gain be large or small for best performance? Determine the minimum value of gain appropriate if the gate signal required is 1 V.

12-B The Digital Filter The digital filter is a simple but useful example of signal processing by digital logic circuits which is used in several different kinds of digital instruments. It exemplifies the versatility inherent in digital electronic techniques. This experiment examines the properties of an elementary digital filter.

Refer to the block diagram of a simple digital filter, Fig. 12-34, which computes the filter output according to Eq. (12-14). A relation corresponding to Eq. (12-14) in the case of an elementary filter is

$$\begin{aligned} v_o(t) = {} & 0.0008v_i(t - 10^{-2}) - 0.0025v_i(t - 2 \times 10^{-2}) \\ & + 0.0094v_i(t - 3 \times 10^{-2}) - 0.035v_i(t - 4 \times 10^{-2}) + 0.189v_i(t - 5 \times 10^{-2}) \\ & + 0.189v_i(t - 6 \times 10^{-2}) - 0.035v_i(t - 7 \times 10^{-2}) + 0.0094v_i(t - 8 \times 10^{-2}) \\ & - 0.0025v_i(t - 9 \times 10^{-2}) + 0.0008v_i(t - 10^{-1}) \qquad (12\text{-}15) \end{aligned}$$

Note that, according to Eq. (12-15) the sampling rate is 100 Hz.

Determine the frequency-response characteristic of the filter. Do this by introducing sine waves of different frequencies at the input. Sample the input signal at the 100 Hz rate for one full cycle and compute the digital output signals according to Eq. (12-15). Use these values to determine the rms output signal at each frequency. It is easiest to undertake these calculations with a simple Fortran program. This, in effect, uses your general-purpose digital computer to simulate the digital filter. Finally, plot the frequency-response characteristic of the filter over a frequency interval from 0.1 to 50 Hz. Based on the frequency response characteristic, what is the name of this filter?

Determine the output waveform resulting from a 2-V peak-to-peak 1-Hz square-wave input signal. Does the result conform to what you expected?

Most often digital filters are used to achieve more complicated input-output functions. By the same technique used above, determine the frequency response characteristic for a digital filter represented by

$$\begin{aligned} v_o(t) = {} & 0.12v_i(t - 10^{-2}) - 0.044v_i(t - 2 \times 10^{-2}) + 0.127v_i(t - 3 \times 10^{-2}) \\ & - 0.327v_i(t - 4 \times 10^{-2}) + 0.90v_i(t - 5 \times 10^{-2}) + 0.90v_i(t - 6 \times 10^{-2}) \\ & - 0.327v_i(t - 7 \times 10^{-2}) + 0.127v_i(t - 8 \times 10^{-2}) - 0.044v_i(t - 9 \times 10^{-2}) \\ & + 0.012v_i(t - 10^{-1}) \end{aligned} \qquad (12\text{-}16)$$

Plot the characteristic and contrast its properties with those for conventional filters. Note how easy it is to implement different properties with the same basic instrument.

appendix one

CHARACTERISTIC CURVES OF TUBES AND TRANSISTORS

The following pages contain characteristic curves of tubes and transistors discussed in the text and examined in the Exercises. These curves have been taken directly from manufacturers' publications and are typical of information available to circuit designers. Note the minor differences in terminology used in these curves compared with the text.

The vacuum-tube characteristics are reproduced with the permission of Radio Corporation of America, as are those for transistor types 2N35 and 2N175. Data for types 2N338, 2N930, 2N1719, and 2N2499 are reproduced with the permission of Texas Instruments, Inc. The General Electric Co. supplied the characteristic curves for the type 2N1415.

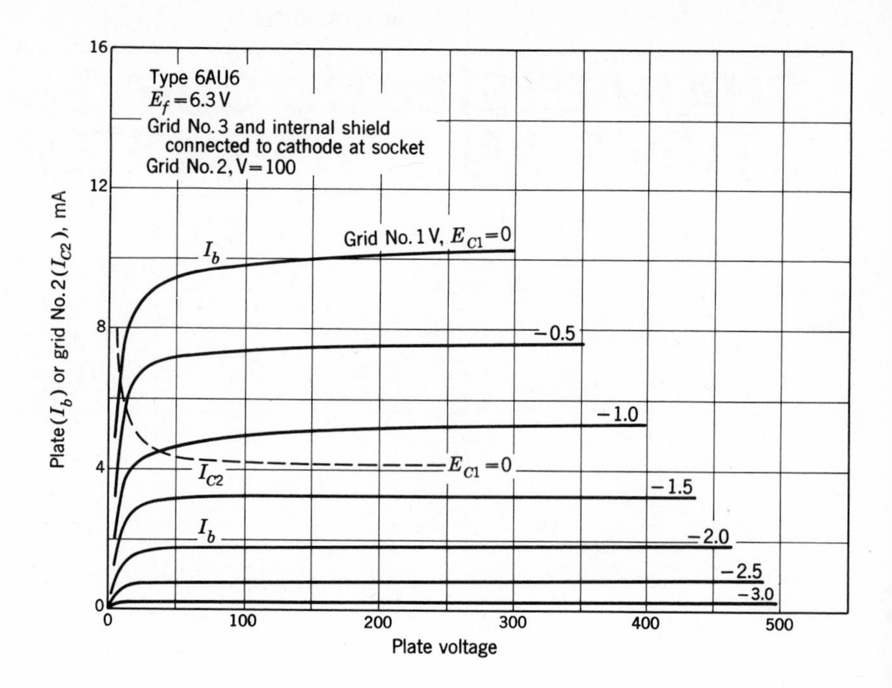

Type 6AU6
$E_f = 6.3$ V
Grid No. 3 and internal shield connected to cathode at socket
Grid No. 2, V=100
Grid No. 1 V, $E_{C1}=0$
I_b
I_{C2}
$E_{C1}=0$
−0.5
−1.0
−1.5
−2.0
−2.5
−3.0
0
4
8
12
16
100
200
300
400
500
Plate voltage
Plate (I_b) or grid No. 2 (I_{C2}), mA

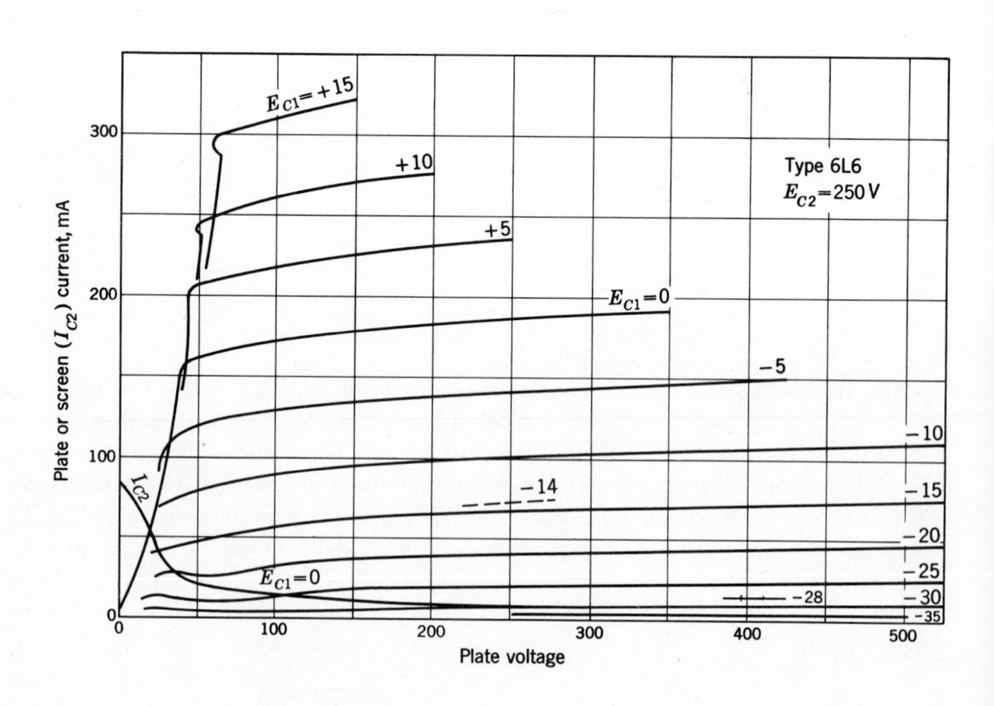

Type 6L6
E_{C2}=250 V
E_{C1}= +15
+10
+5
E_{C1}=0
−5
−10
−14
−15
−20
−25
−28
−30
−35
I_{C2}
E_{C1}=0
0
100
200
300
100
200
300
400
500
Plate voltage
Plate or screen (I_{C2}) current, mA

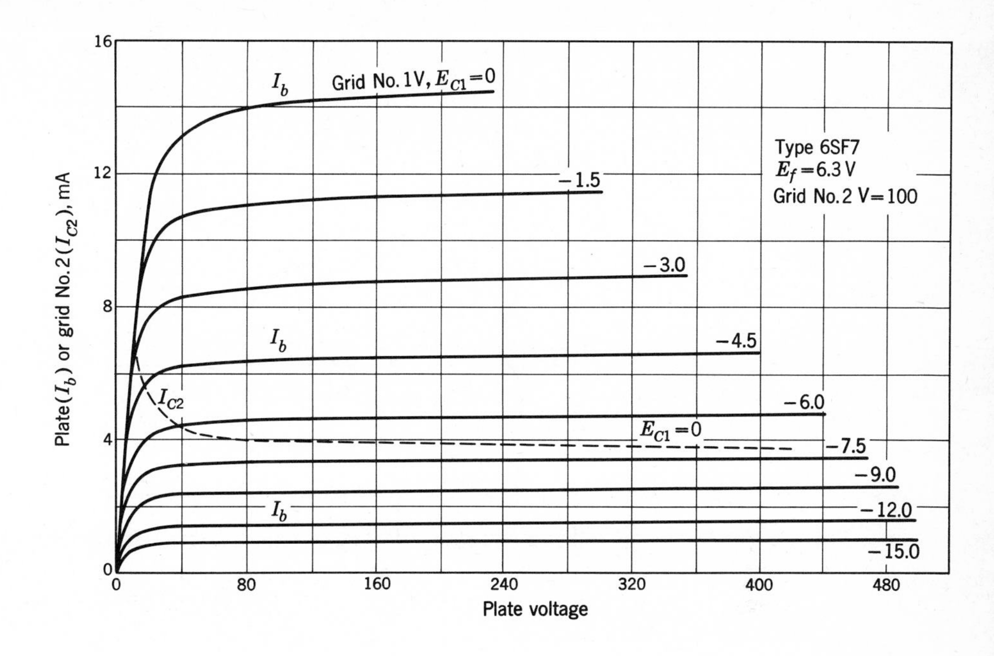

I_b
Grid No. 1V, $E_{C1}=0$
Type 6SF7
$E_f=6.3$ V
Grid No. 2 V=100
−1.5
−3.0
I_b
−4.5
I_{C2}
−6.0
$E_{C1}=0$
−7.5
−9.0
I_b
−12.0
−15.0
Plate (I_b) or grid No. 2 (I_{C2}), mA
16
12
8
4
0
0
80
160
240
320
400
480
Plate voltage

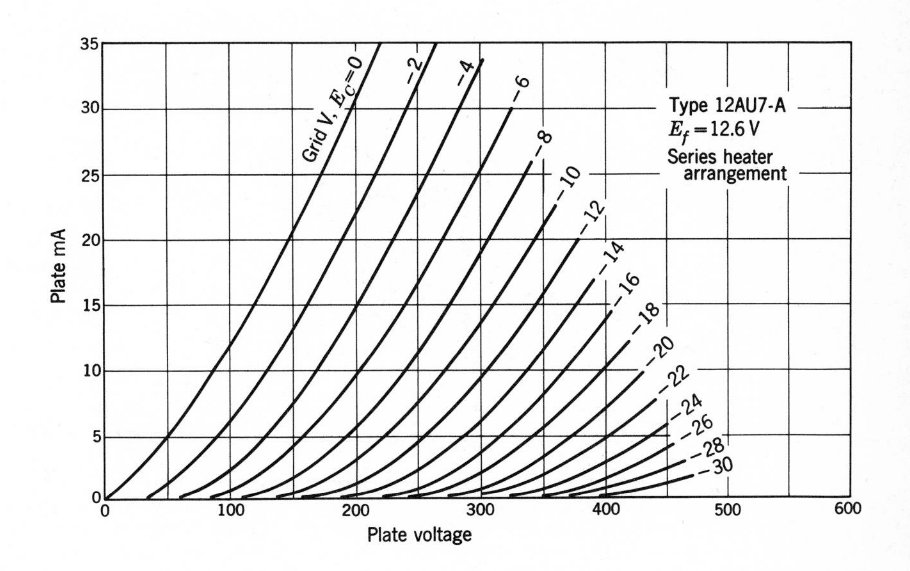

Grid V, $E_C=0$
−2
−4
−6
−8
−10
−12
−14
−16
−18
−20
−22
−24
−26
−28
−30
Type 12AU7-A
$E_f=12.6$ V
Series heater
arrangement
Plate mA
35
30
25
20
15
10
5
0
0
100
200
300
400
500
600
Plate voltage

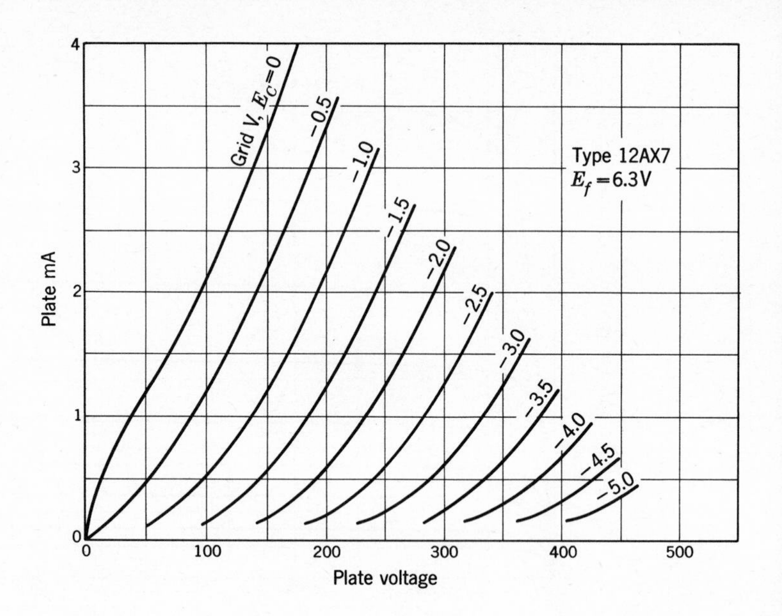
Plate mA
Plate voltage
Grid V, E_C=0
−0.5
−1.0
−1.5
−2.0
−2.5
−3.0
−3.5
−4.0
−4.5
−5.0
Type 12AX7
E_f = 6.3V
0
1
2
3
4
100
200
300
400
500

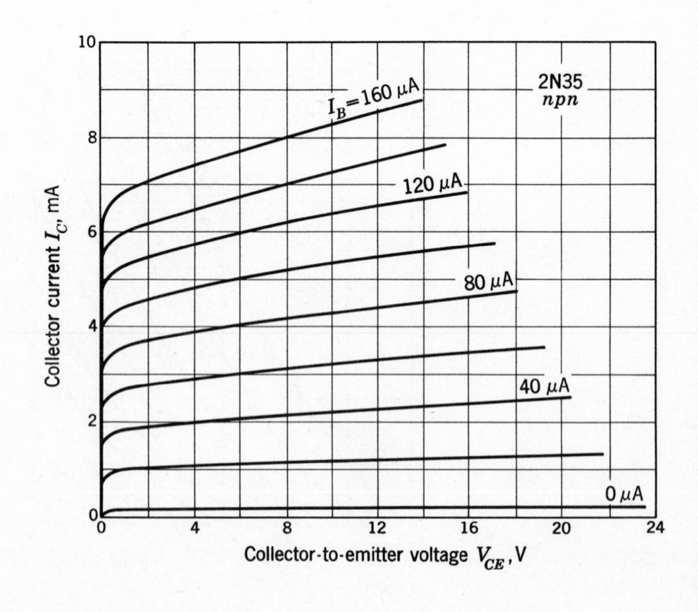
Collector current I_C, mA
Collector-to-emitter voltage V_{CE}, V
I_B=160 μA
120 μA
80 μA
40 μA
0 μA
2N35
npn
0
2
4
6
8
10
4
8
12
16
20
24

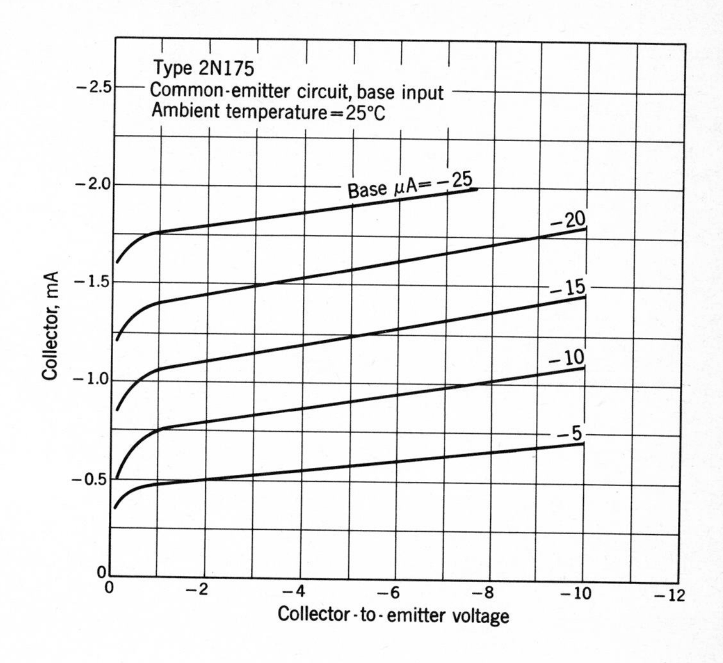
Type 2N175
Common-emitter circuit, base input
Ambient temperature=25°C
Base μA=−25
−20
−15
−10
−5
Collector, mA
−2.5
−2.0
−1.5
−1.0
−0.5
0
0
−2
−4
−6
−8
−10
−12
Collector-to-emitter voltage

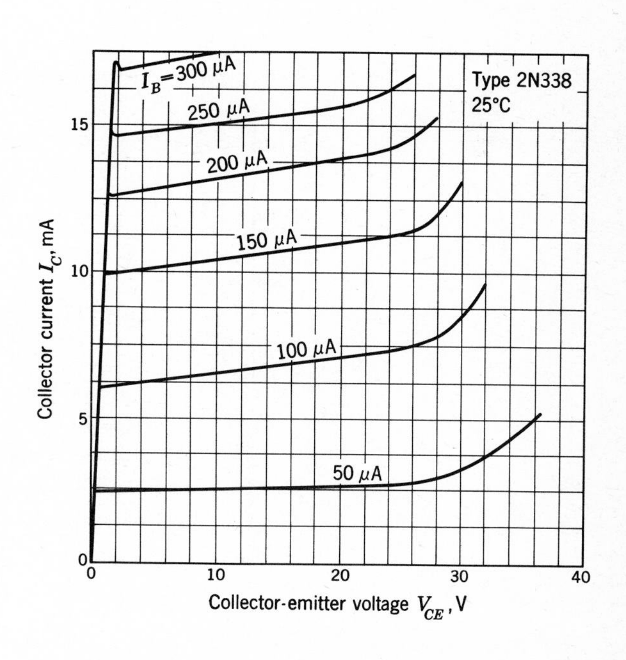
I_B=300 μA
250 μA
200 μA
150 μA
100 μA
50 μA
Type 2N338
25°C
Collector current I_C, mA
15
10
5
0
0
10
20
30
40
Collector-emitter voltage V_{CE}, V

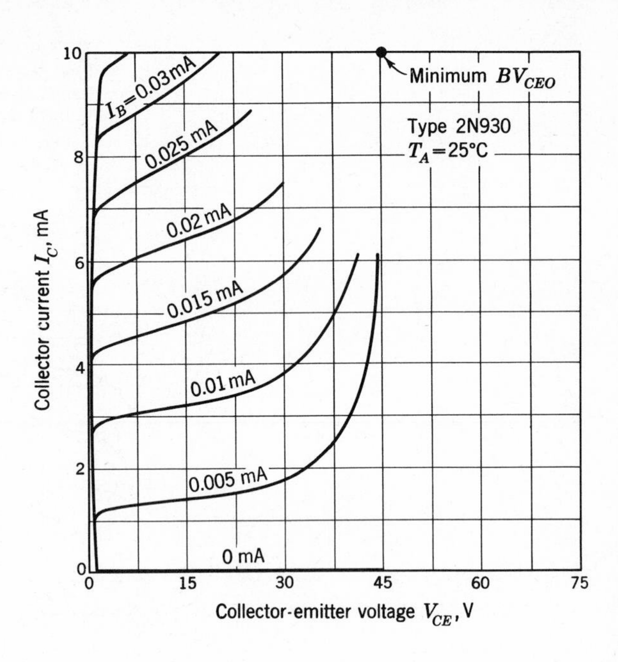
10
8
6
4
2
0
Collector current I_C, mA
I_B=0.03mA
0.025 mA
0.02 mA
0.015 mA
0.01 mA
0.005 mA
0 mA
Minimum BV_{CEO}
Type 2N930
T_A=25°C
0
15
30
45
60
75
Collector-emitter voltage V_{CE}, V

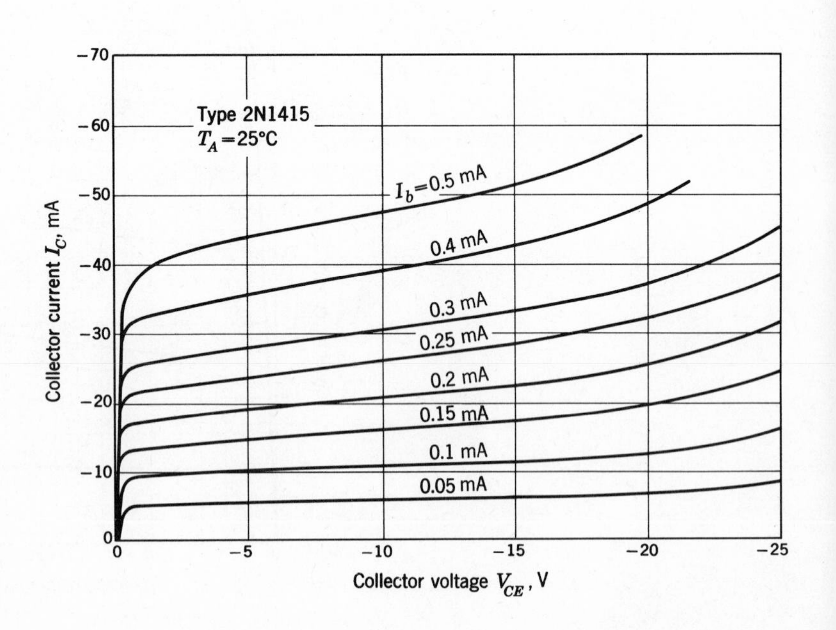
−70
−60
−50
−40
−30
−20
−10
0
Collector current I_C, mA
Type 2N1415
T_A=25°C
I_b=0.5 mA
0.4 mA
0.3 mA
0.25 mA
0.2 mA
0.15 mA
0.1 mA
0.05 mA
0
−5
−10
−15
−20
−25
Collector voltage V_{CE}, V

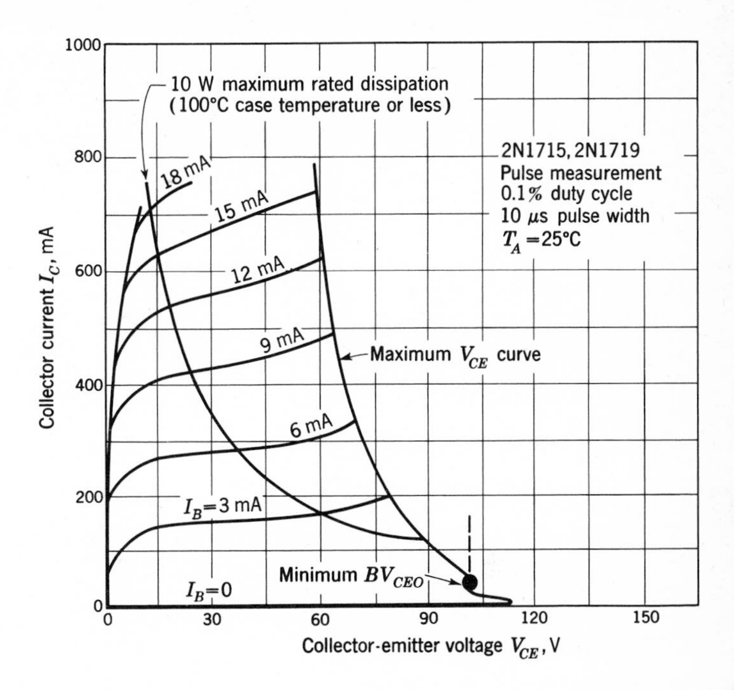

10 W maximum rated dissipation
(100°C case temperature or less)
2N1715, 2N1719
Pulse measurement
0.1% duty cycle
10 μs pulse width
T_A = 25°C
18 mA
15 mA
12 mA
9 mA
6 mA
I_B = 3 mA
I_B = 0
Maximum V_{CE} curve
Minimum BV_{CEO}
Collector current I_C, mA
1000
800
600
400
200
0
30
60
90
120
150
Collector-emitter voltage V_{CE}, V

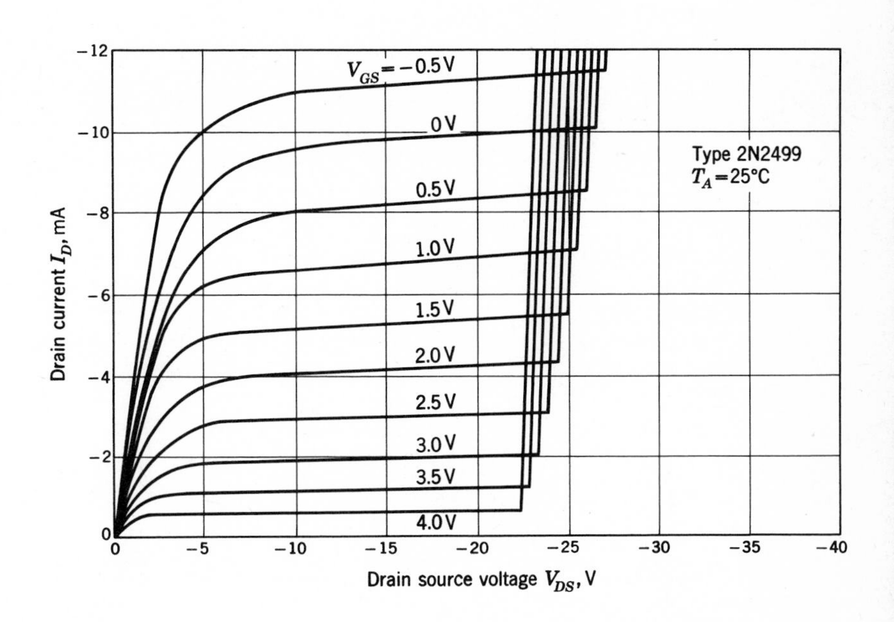

V_{GS} = −0.5 V
0 V
0.5 V
1.0 V
1.5 V
2.0 V
2.5 V
3.0 V
3.5 V
4.0 V
Type 2N2499
T_A = 25°C
Drain current I_D, mA
−12
−10
−8
−6
−4
−2
0
−5
−10
−15
−20
−25
−30
−35
−40
Drain source voltage V_{DS}, V

appendix two

FORTRAN PROGRAMS FOR LABORATORY EXERCISES

The simulated laboratory experience provided by Laboratory Exercises at the end of each chapter may be enhanced in several instances by the use of simple Fortran programs. Typical Fortran IV programs for these Exercises are illustrated in this Appendix.

The programs for Laboratory Exercise 1-A, The Wheatstone Bridge, 3-B, Bridged-T and Twin-T Filters, and 12-B, The Digital Filter, in effect simulate data collection in the laboratory. That is, the programs generate an array of numbers which are equivalent to the results of measurements on a given experimental circuit. This data is used by the student to prepare graphs or calculate results much as an experimenter reduces laboratory data. The student may alter experimental conditions in the program and note corresponding changes in the data.

Similarly, programs for Laboratory Exercises 2-B, Square Waves, and 12-B, The Digital Filter, yield waveforms which simulate waveform analysis using an oscilloscope. This is particularly effective if the programs are written to produce a graphical output directly. Those included here suggest graphical-output display using the conventional line printer because of its universal availability. More elaborate automatic plotting devices can also be used to improve the waveform display. In either case, here again the student can alter experimental conditions and note the ensuing changes in the waveform much as if he were using an oscilloscope in the laboratory.

The Fortran program for Laboratory Experiment 7-B, Random Noise Signals, is used primarily to reduce the labor of calculation. It also may be viewed as simulating noise measurements at different instrumental bandwidths.

In each instance the Fortran programs illustrated are known to operate satisfactorily. Other approaches are possible, of course, and students may wish to prepare their own programs. The control-card statements peculiar to particular computer centers are not indicated.

```
C        WHEATSTONE            BRIDGE
      DIMENSION R3(200),VR5(200)
      DATA R1,R2,R4,R5/1.0E2,1.0E2,1.1E2,1.0E2/
   11 FORMAT(20X,F20.3,20X,F20.3)
      DO 10 I=1,200,1
      R3(I)=(200*I+1790)/199
      VTH=12.0*((R1/(R1+R3(I)))-(R2/(R2+R4)))
      RTH=R1*R3(I)/(R1+R3(I))+R2*R4/(R2+R4)
      VR5(I)=VTH*(R5/(RTH+R5))
   10 WRITE(6,11) R3(I),VR5(I)
      STOP
      END
```

FIGURE A2-1 *Fortran program for Laboratory Exercise 1-A, The Wheatstone Bridge, for the case of an equal-arm bridge with a matched detector.*

```
C                               S Q U A R E   W A V E S
C
      REAL L
      LOGICAL OPENED
      COMMON / FLAG / OPENED
      OPENED = .FALSE.
      INTEGER FILTER, LO PASS, HI PASS
      COMMON / GLOBAL / OMEGA, TUNE, FILTER, LO PASS, HI PASS
      LO PASS = 0
      HI PASS = 1
      EXTERNAL SIGNAL
C
C         THIS PROGRAM USES FOURIER ANALYSES TO STUDY BOTH LOW PASS
C     AND HIGH PASS (RLC) FILTERS. THE RESULTS ARE GRAPHED.
C
C     TUNE           -->   IS THE FUNDAMENTAL FREQUENCY TO WHICH THE
C                          SQUARE WAVE OSCILLATOR IS TUNED. 'TUNE' MUST BE
C                          GIVEN A VALUE WHENEVER THE INPUT CALLS FOR IT.
C                          THAT IS, THE PREVIOUS SETTING OF 'TUNE' IS NOT
C                          RETAINED.
C     OMEGA          -->   IS THE HALF-POWER FREQUENCY OF THE RLC FILTER.
C                          IF AND ONLY IF THE INPUT FIELD FOR 'OMEGA' IS
C                          LEFT BLANK, NON BLANK FIELDS ARE EXPECTED FOR:
C                             R   -->   THE RESISTANCE IN THE NETWORK;
C                             L   -->   THE INDUCTANCE; AND
C                             C   -->   THE CAPACITANCE.
C                          IN THIS CASE 'R', 'L', AND 'C' ARE USED TO COMPUTE
C                          'OMEGA'. IF A NEGATIVE 'R' IS READ, THE PROGRAM
C                          TERMINATES NORMALLY.
C
C     THE SUBROUTINE 'ONPLOT' ALSO EXPECTS CARD INPUT. SEE THAT ROUTINE
C     FOR FURTHER DOCUMENTATION. NO OTHER ROUTINE REQUIRES CARD INPUT.
C     THE ROUTINE 'ONPLOT' ALSO REQUIRES A DIRECT ACCESS DEVICE ASSIGNED
C     TO LOGICAL I/O UNIT NUMBER 9. IT'S PURPOSE IS SCRATCH FILE FOR
C     REFORMATING IN E NOTATION.
C         THE LANGUAGE USED THROUGH-OUT IS STANDARD ANSI FORTRAN.
1     READ(5,2) TUNE, OMEGA,                        R, L, C
2     FORMAT (5(E9.2, 3X))
      IF (R .LT. 0.)  STOP
      CALL ON PLOT
      IF (OMEGA) 3,3,4
3     FILTER = LO PASS
      OMEGA = R/L
      CALL O SCOPE(SIGNAL)
      CALL ON PLOT
      FILTER = HI PASS
      OMEGA = 1./(R*C)
      CALL O SCOPE(SIGNAL)
      GO TO 1
4     FILTER = LO PASS
      CALL O SCOPE(SIGNAL)
      CALL ON PLOT
      FILTER = HI PASS
      CALL O SCOPE(SIGNAL)
      GO TO 1
C     THIS PROGRAM WRITTEN BY JAMES VANDENDORPE. IIT/1971
      END
```

FIGURE A2-2 *Main program for Laboratory Exercise 2-B, Square Waves.*

```
      FUNCTION SIGNAL (TIME)
C          THIS FUNCTION COMPUTES THE FOURIER SERIES OF A FILTER WHOSE
C     INPUT IS A SQUARE WAVE. THE FUNDAMENTAL FREQUENCY OF THIS SERIES
C     IS GIVEN BY 'TUNE'. THE TERMS OF THIS SERIES ARE COLLECTED UNTIL
C     EITHER NEW TERMS WOULD BE INSIGNIFICANT OR UNTIL TWENTY TERMS HAVE
C     OCCURED. THE FOUIER COEFFICIENTS AND PHASE ARE CHOOSEN FOR THE
C     FILTER THAT THEY REPRESENT. THE FINAL VALUE OF SIGNAL IS THE
C     FILTERS OUTPUT.
C
C ----'VP' IS THE AMPLITUDE CONSTANT.
      VP = 1.
      REAL TERM(20),N
      COMMON /GLOBAL/ OMEGA,TUNE
      THETA=TUNE*TIME
      CALL FILTR (TUNE,COEFF,PHASE)
      TERM(1) = COEFF*SIN(THETA + PHASE)
      SIGNAL = 0.
      CUT OFF = COEFF*5.E-4
C---- COLLECT ONLY THE FOURIER TERMS THAT WILL BE USED.
      DO 1 M=2,20
      N=2.*M-1.
      CALL FILTR (TUNE*N,COEFF,PHASE)
      TERM(M) = (1./N)*COEFF*SIN(N*THETA + PHASE)
      IF (COEFF .LT. CUT OFF)  GO TO 2
    1 CONTINUE
C---- SUM THESE TERMS.
    2 DO 3 I=1,M
    3 SIGNAL=SIGNAL+TERM(M-I+1)
C---- MULTIPLY BY THE NORMALIZING COEFICIENT.
      SIGNAL = VP*SIGNAL
      WRITE(6,4) TIME,SIGNAL,M
    4 FORMAT (1X,'TIME =',F6.4,'; SIGNAL =',E13.7,'; # OF FOURIER TERMS'
     A   ,' =',I3,'.')
      RETURN
      END
```

(*a*)

```
      SUBROUTINE FILTR(NTUNE,COEFF,PHASE)
C          THIS ROUTINE SIMULATES A FILTER BY ALTERING THE FOURIER
C     COEFFICIENTS, 'COEFF', AND 'PHASE' OF THE INPUT SIGNAL.
      REAL NTUNE
      INTEGER FILTER, LO PASS
      COMMON /GLOBAL/ OMEGA, TUNE, FILTER, LO PASS
      IF (FILTER .EQ. LO PASS)  GO TO 1
      COEFF = 1./SQRT(1. + (OMEGA/NTUNE)**2)
      PHASE = ATAN(OMEGA/NTUNE)
      RETURN
1     COEFF = 1./SQRT(1. + (NTUNE/OMEGA)**2)
      PHASE = ATAN(-NTUNE/OMEGA)
      RETURN
      END
```

(*b*)

FIGURE A2-3 *(a) and (b) subroutines for Laboratory Exercise 2-B.*

```
      SUBROUTINE OSCOPE (F)
C         THIS SUBROUTINE PLOTS THE FUNCTION SUPPLIED IN IT'S ARGUMENT.
C     IN THIS SENSE IT SIMULATES AN OSCILLOSCOPE. THE LANGUAGE USED HERE
C     IS STANDARD ANSI FORTRAN.
C         THE FOLLOWING IS A DICTIONARY OF LABELS USED IN THIS ROUTINE:
C
C                     -->     THE EXTERNAL FUNCTION WHICH IS TO BE PLOTTED.
C     DOMAIN(1)       -->     LOW BOUND OF THE DOMAIN (IE. X-AXIS).
C     DOMAIN(2)       -->     HIGH BOUND OF THE DOMAIN.
C     RANGE(1)        -->     THE LOW BOUND OF THE RANGE.
C     RANGE(2)        -->     THE HIGH BOUND OF THE RANGE.
C     SLATE(Y,X)      -->     THE IMAGE OF THE PAGE ON WHICH THE GRAPH IS
C                             TO BE WRITTEN (IE. THE PRINT BUFFER).
C     POINT           -->     THE CHARACTER THAT IS TO BE USED TO MARK THE
C                             GRAPH AT FUNCTION POINTS.
C     NUMPTS          -->     THE NUMBER OF POINTS WHICH ARE TO BE PLOTTED
C                             INSIDE THE DOMAIN.
C     Y UP            -->     THE X-AXIS IS LOCATED ('Y UP' + 1) LINES FROM
C                             THE PAGES BOTTOM. THIS QUANTITY IS ALSO USED AS
C                             A PAGE COORDINATE FOR LOCATING LABELS.
C     X RIGHT         -->     THE Y-AXIS IS LOCATED ('X RIGHT' + 1) SPACES
C                             FROM THE PAGES LEFT MARGIN. THE QUANTIY ALSO
C                             ACTS AS A COORDINATE FOR POSITIONING LABELS.
C     X AXIS          -->     A HOLLERITH FLAG WHICH EQUALS X; HENCE IT MEANS
C                             THE X AXIS IS INTENDED.
C     Y AXIS          -->     A HOLLERITH FLAG WHICH EQUALS Y; HENCE IT MEANS
C                             THE Y AXIS IS INTENDED.
C
C     : : : : : : : : : : : : : : : : : : : : : : : : : : : : : : : : : : :
C                             P O I N T S     P L O T T E D
      INTEGER SLATE(66,131),YUP,XRIGHT,XAXIS,YAXIS,POINT,START
      DIMENSION DOMAIN(3), RANGE(3), ARRAY(14)
      REAL LO BND
      LOGICAL SCOPE,OPENED
      SCOPE = .TRUE.
      IF (.NOT.OPENED) GO TO 12
   10 X = DOMAIN(1)
      Y UP = I SAVE Y
      X RIGHT = I SAVE X
      DOMAIN(3)=ABS(DOMAIN(2)-DOMAIN(1))
      RANGE(3) = ABS(RANGE(2)-RANGE(1))
      DELTA = DOMAIN(3)/NUMPTS
      X RIGHT = MAX0(XRIGHT,9)
      DO 11 I=1,NUMPTS
      X = X + DELTA
      Y = F(X)
C
      IF (RANGE(3)*DOMAIN(3) .LT. .5*RANGE(3)*DOMAIN(3))  STOP 9
      BASE = Y UP
      IF (RANGE(1)*RANGE(2) .LT. 0)
     A   BASE = ((66.-YUP)*ABS(RANGE(1)))/RANGE(3) + YUP
      IY = BASE + Y*(65-YUP)/RANGE(3)
      IF ((RANGE(1) .LT. 0) .AND. (RANGE(2) .LT. 0))
     A   IY = BASE + ABS(Y*(65.-YUP)/RANGE(3))
      IY = MIN0(MAX0(IY,YUP+1),66)
      IX = X RIGHT + ABS((130.-XRIGHT)*X/DOMAIN(3))
      IX = MIN0(MAX0(IX,XRIGHT+1),131)
C
   11 SLATE(IY,IX)=POINT
      WRITE (6,35)
      WRITE (6,36) ((SLATE(67-I,J),J=1,131),I=1,66)
      RETURN
```

FIGURE A2-4 *Subroutine used in connection with Laboratory Exercise 2-B which uses computer line printer to display waveforms graphically.*

```
C                                  L A Y O U T S    E D I T E D
      ENTRY ON PLOT
C....      THIS ENRTY POINT PREPARES A TWO DIMENSIONAL ARRAY ('SLATE')
C.... WITH THE APPROPRIATE HEADERS, LABELS AND AXIS SO THAT THE WRITTEN
C.... SLATE WILL REPRESENT A GRAPH.
C....      THE WORDING, AS IT WILL APPEAR ON THE PRINTED GRAPH, IS READ
C.... OFF DATA CARDS. EACH OF THESE CARDS ARE PRECEEDED BY A CARD WITH
C.... ONE NUMERIC CONTROL CHARACTER PUNCHED IN COLUMN 1. THIS CONTROL
C.... CHARACTER IS CALLED 'JOB'.
C....      THE VALUE 'JOB' TAKES DETERMINES THE COURSE OF ACTION THAT
C.... 'ON PLOT' PERFORMS WHEN IT READS THE NEXT DATA CARD. THE
C.... SUBROUTINE CONTINUES READING CARDS AND ARRANGING SLATE UNTIL
C.... 'JOB'=0 OR 'JOB'=1 . WHEN THESE VALUES OCCUR, RETURN IS MADE TO
C.... THE CALLING ROUTINE. THE VALUES THAT 'JOB' MAY TAKE, FOLLOW:
C....  JOB                          A C T I O N   T A K E N
C.... 1 OR 0          -->    RETURN TO CALLER.
C....    2            -->    SET UP ANOTHER X-AXIS.
C....    3            -->    INSERT A LABEL IN THE X DIRECTION.
C....    4            -->    SET UP ANOTHER Y-AXIS.
C....    5            -->    INSERT A LABEL ALONG THE Y-AXIS.
C....    6            -->    ERASE A ROW OF SLATE.
C....    7            -->    ERASE A COLUMN OF SLATE.
C....    8            -->    WIPE THE SLATE CLEAN.
C....    9            -->    CHOOSE A NEW CHARACTER TO PLOT WITH AND SPECIFY
C....                        THE NUMBER OF POINTS TO BE PLOTTED.
C....      AFTER A VALUE OF 'JOB' IS CHOOSEN CONSULT THE JUMP-TABLE
C.... (WHICH IS INDEXED BY 'JOB' VALUE) FOR A FORTRAN EXPLANATION OF THE
C.... DATA CARD WHICH FOLLOWS THE CARD WITH 'JOB'S VALUE ON IT.
C
      SCOPE=.FALSE.
      INTEGER TEMP(66),AXIS
      COMMON /FLAG/ OPENED
      IF (OPENED) GO TO 14
   12 DO 13 I=1,66
      DO 13 J=1,131
   13 SLATE(I,J) = 1H
      X AXIS = 1HX
      Y AXIS = 1HY
C     . . . . . . . . . . . . DEFAULT  PARAMETERS . . . . . .
      POINT = 1H.                                            .
      NUMPTS = 60                                            .
      I SAVE X = 9                                           .
      I SAVE Y = 5                                           .
      DOMAIN(1) = -10.                                       .
      DOMAIN(2) = +10.                                       .
      RANGE(1) = -10.                                        .
      RANGE(2) = +10.                                        .
C     . . . . . . . . . . . . . . . . . . . . . . . . . . . .
      OPENED=.TRUE.
      IF (SCOPE) GO TO 10
   14 READ (5,37) JOB
C     . . . . . . . . . . . . . JUMP TABLE . . . . . . . . . .
   15 JOB = MAX0(JOB,1)                                      .
      GO TO (1,2,3,4,5,6,7,8,9), JOB                         .
     :                                                       .
1     RETURN                                                 .
2     AXIS = X AXIS                                          .
      GO TO 24                                               .
     :                                                       .
3     READ (5,16) YUP,XRIGHT,(TEMP(I),I=1,66),JOB            .
   16 FORMAT (I2,3X,I3,3X,66A1/I1)                           .
      K=0                                                    .
      J = MIN0(XRIGHT+65,131)                                .
      DO 17 I=XRIGHT,J                                       .
      K=K+1                                                  .
   17 SLATE(YUP,I) = TEMP(K)                                 .
      GO TO 15                                               .
     :                                                       .
4     AXIS = Y AXIS                                          .
      GO TO 24                                               .
     :                                                       .
5     ND = 66 - ISAVEY                                       .
      READ (5,18) XRIGHT,(SLATE(67-I,XRIGHT),I=1,ND)         .
   18 FORMAT (I3,66A1)                                       .
      READ (5,37) JOB                                        .
      GO TO 15                                               .
     :                                                       .
6     READ (5,19) YUP,XRIGHT,JOB                             .
   19 FORMAT (I2,3X,I3/I1)                                   .
      DO 20 I=XRIGHT,131                                     .
   20 SLATE(YUP,I) = 1H                                      .
      GO TO 15                                               .
```

FIGURE A2-5 *Subroutine used in connection with program in Fig. A2-4.*

```
7        READ (5,19) YUP,XRIGHT,JOB
         DO 21 I=YUP,66
      21 SLATE(I,XRIGHT) = 1H
         GO TO 15
        :
8        READ (5,19) YUP,XRIGHT,JOB
         DO 22 I=YUP,66
         DO 22 J=XRIGHT,131
      22 SLATE(I,J) = 1H
         GO TO 15
        :
9        READ (5,23) POINT,NUMPTS,JOB
      23 FORMAT (A1,3X,I3/I1)
         GO TO 15
        :
C       . . . . . . . . . . . . . . . . . . . . . . . . . . . . . . . . . . .
C                        A X E S    I N S T A L L E D
C....        SET UP EITHER THE X AXIS OR THE Y AXIS, 'YUP' UNITS FROM THE
C....   PAGE'S BOTTOM AND 'XRIGHT' UNITS FROM THE PAGE'S LEFT EDGE. EITHER
C....   AXIS IS CALIBRATED OVER A DECIMAL SCALE WHOSE DOMAIN OR RANGE IS
C....   FROM THE LO-BOUND TO THE HI-BOUND.
C
      24 READ (5,25) YUP,XRIGHT,LOBND,HIBND,JOB
      25 FORMAT (I2,3X,I3,2(3X,F9.2)/I1)
         J = 0
         X RIGHT = MAX0(XRIGHT,9)
         Y UP = MAX0(YUP,2)
         IF (AXIS.NE.X AXIS)  GO TO 26
         LENGTH = 121
         I SAVE X = X RIGHT
         START = X RIGHT
         DOMAIN(1) = LO BND
         DOMAIN(2) = HI BND
      26 IF (AXIS.NE.Y AXIS)  GO TO 30
         LENGTH = 56
         I SAVE Y = Y UP
         START = Y UP
         RANGE(1) = LO BND
         RANGE(2) = HI BND
      30 UNIT10 = (HI BND - LO BND)/((IFIX(LENGTH - START)/10) + 1.)
         ARRAY(1) = LO BND
         DO 34 I=START,LENGTH,10
         J = J + 1
         ARRAY(J+1) = ARRAY(J) + UNIT10
C
C....        THIS SECTION DRAWS PARTS OF LINES ALONG THE X OR Y AXIS.
C....    WHEN CALLED ONLY TEN POINTS OF THE LINE ARE PRINTED AT A TIME.
C
         I8 = I + 8
         IF (AXIS.EQ.X AXIS)  GO TO 32
         SLATE(I,XRIGHT)=1H+
         DO 31 K=I,I8
      31 SLATE(K+1,XRIGHT) = 1H:
         SLATE(I8+2,XRIGHT) = 1H+
         GO TO 34
      32 SLATE(YUP,I) = 1H+
         DO 33 K=I,I8
      33 SLATE(YUP,K+1) = 1H-
         SLATE(YUP,I8+2) = 1H+
      34 CONTINUE
         J = J + 1
C
C.... THIS SECTION CALIBRATES X OR Y AXIS OVER THE PRESCRIBED INTERVAL.
C
         WRITE (9,38) (ARRAY(L),L=1,J)
         K1 = X RIGHT - 8
         ND = X RIGHT - 1
         REWIND 9
         IF (AXIS.EQ.XAXIS) READ (9,39) (SLATE(YUP-1,L-1),L=XRIGHT,129)
         IF (AXIS.EQ.YAXIS) READ (9,40) ((SLATE(I,K),K=K1,ND),I=YUP,66,10)
         REWIND 9
         GO TO 15
      35 FORMAT (1H1)
      36 FORMAT (66(1X,131A1/))
      37 FORMAT (I1)
      38 FORMAT (13(E8.3,2X))
      39 FORMAT (130A1)
      40 FORMAT (7(8A1,2X))
C        THIS PROGRAM WRITTEN BY JAMES VANDENDORPE, IIT/1971
         END
```

FIGURE A2-6 *Continuation of subroutine in Fig. A2-5.*

```
C        B R I D G E D -- T    A N D    T W I N -- T    F I L T E R S
C
C          TO PREPARE DATA, ALWAYS BEGIN IN COLUMN ONE USING ONE CARD PER
C        DATUM. THE FIRST DATUM, 'WO', IS THE CHARACTERISTIC FREQUENCY OF
C        THE FILTER. THE SECOND DATUM, 'N', IS THE INTEGER VALUE OF THE
C        NUMBER OF FREQUENCY DATA CARDS TO FOLLOW THIS CARD. THE NEXT 'N'
C        CARDS ARE THE FREQUENCIES 'W', WHICH ARE INPUT TO THIS FILTER.
C        AFTER THIS IS A CARD FOR THE RATIO R2 TO R1. THE FINAL DATA CARD
C        IS A CONTROL CARD (VALUE 1->3), SEE 'JMP' USAGE BELOW.
C
      DIMENSION W(100)
      DEGREE(RADIAN) = 57.2956*RADIAN
1     READ(5,2) WO
2     FORMAT (E15.3)
      READ (5,3) N
3     FORMAT (I2)
      READ(5,2) (W(I), I=1,N)
C
C                                      TWIN-T FILTERS
C                                     CHOOSE C1 = 2*C2
      WRITE(6,4)
4     FORMAT ('1TWIN-T:'///)
      WRITE(6,5) WO
5     FORMAT (1X, 'THE CHARACTERISTIC  FREQUENCY, WO, OF THIS FILTER =',
     A   E8.3, ' HZ.'/1X, 'LET W = THE FREQUENCY OF THE INPUT SIGNAL.'
     B   /1X, 'THE FREQUENCY CHARACTERISTC ARE REPRESENTED BY THE ',
     C   'RATIO, VO/VI, OF OUTPUT TO INPUT VOLTAGE;'/41X, 'W', 13X,
     D   'W/WO', 11X, 'VO/VI', 10X, 'PHASE'/39X, 'IN HZ.', 38X,
     E   'IN DEGREES')
      DO 7  I = 1,N
      FRATIO = W(I)/WO
      X = WO/W(I) - FRATIO
      VRATIO = 1./(1.+16./X**2)
      RADIAN = ATAN(-1./X)
      PHASE = DEGREE(RADIAN)
      WRITE(6,6) W(I), FRATIO, VRATIO, PHASE
6     FORMAT ('0', 30X, 3E15.3, F15.2)
7     CONTINUE
C
C                                      BRIDGED-T FILTERS
C                                       CHOOSE  C1 = C2
8     READ(5,2) RRATIO
      WRITE(6,9) RRATIO
9     FORMAT ('1BRIDGED-T: '///1X, 'THE RATIO OF R2 TO ',
     A   'R1 IS', E8.3, ' (IE. R2/R1).')
      WRITE(6,5) WO
      DO 10 I=1,N
      FRATIO = W(I)/WO
      X = FRATIO - WO/W(I)
      VRATIO = (1./RRATIO)*SQRT(4.*(1. + X) + (1. + RRATIO)*X**2)
      RADIAN = ATAN(SQRT(RRATIO)*X/(2. + X**2))
      PHASE = DEGREE(RADIAN)
10    WRITE(6,6)  W(I), FRATIO, VRATIO, PHASE
      READ(5,11) JMP
11    FORMAT (I1)
      GO TO (1,8,12), JMP
12    STOP
      END
```

FIGURE A2-7 *Fortran program for Laboratory Exercise 3-B, Bridged-T and Twin-T Filters.*

FIGURE A2-8 *Main program for Laboratory Exercise 7-B, Random Noise Signals.*

```
C                   R A N D O M    N O I S E    S I G N A L S
C
C         A SERIES OF VOLTAGE MEASURMENTS ARE READ, THESE ARE AVERAGED
C      ('VDC'), THE ROOT-MEAN-SQUARE IS CALCULATED ('VRMS'), THE VARIANCE
C      ('VR') AND THE RATIO ('R=VDC/VR') IS COMPUTED. 'VDC' IS RECOMPUTED
C      OVER INTERVALS, AND THE STATISTICS ARE LISTED.
C         THE DATA ARE ARRANGED ONE DATUM PER CARD. DECIMAL POINTS
C      SHOULD BE INSERTED FOR ALL DATA EXCEPT 'N' AND ONE SHOULD ALWAYS
C      BEGIN PUNCHING DATA IN COLUMN ONE.
C         THE FIRST DATUM, 'N' IS THE NUMBER OF VOLTAGE MEASUREMENTS TO
C      FOLLOW THIS CARD. NOW COMES THE 'N' MEASURENTS, 'X'. THE NEXT CARD
C      GIVES THE TIME IN SECONDS BETWEEN MEASUREMENTS. NEXT THE INTERVAL
C      OVER WHICH THE AVERAGE IS TO BE CALCULATED IS SPECIFIED IN
C      SECONDS. THE FINAL CARD IS A CONTROL CARD (SEE 'JMP' USAGE BELOW).
C         THE LANGUAGE USED IS STANDARD ANSI FORTRAN.
C
      DIMENSION X(500), SQUARE(500)
      LOGICAL GETRMS
      COMMON ANSWER(250), GET RMS
      RATIO (VDC, VR) = VDC / SQRT(ABS(VR))
      IDUMMY = 0
      DUMMY = 0.
      LBL = 6HVDC =
      GETRMS = .FALSE.
      READ(5,1) N, (X(I), I=1,N)
1     FORMAT (I3/(E12.3))
      READ(5,2) TIME
2     FORMAT (E12.3)
      M = N/3
      K = N - 3*M
      L = 3*M + 1
      WRITE(6,3) TIME, (X(I), X(I+M), X(I+M*2), I=1,M)
3     FORMAT (1H1, 30X, 'SUCCESSIVE VOLTAGE READINGS OF NOISY SIGNAL'
     A //36X, 'SAMPLE INTERVAL:', E12.3, ' SEC.'//// (3(10X, E12.3,
     B  10X)))
      IF (K .NE. 0)  WRITE(6,4) (X(I),I=L,K)
4     FORMAT (3(10X, E12.3, 10X))
      REAL INTRVL
      INTRVL = TIME*N
      CALL MEAN(N,X,TIME,DUMMY,INTRVL)
      VDC = ANSWER(1)
      CALL STAT(N,X,TIME,SQUARE,IDUMMY)
      VRMS = ANSWER(1)
      CALL MS(N,X,TIME,SQUARE,INTRVL)
      VR = ANSWER(1) - VDC**2
      R = RATIO(VDC, VR)
      WRITE(6,5)
5     FORMAT ('1 CONSIDER ENTIRE INTERVAL:'/1X, 25 (1H-))
      WRITE(6,6) LBL, VDC
6     FORMAT (10X,A6,E7.3,16X,A6,E7.3,16X,A6,E7.3)
      WRITE(6,7) VR, R, VRMS
7     FORMAT (5X, 'VARIANCE = ', E7.3/8X, 'RATIO = ', E7.3/9X, 'VRMS =',
     A  E8.3)
8     READ(5,2) INTRVL
      M = IFIX(N*TIME/INTRVL)/3
      K = IFIX(N*TIME/INTRVL + .5)
      L = K - 3
      CALL MEAN(N,X,TIME,DUMMY,INTRVL)
      WRITE(6,9) INTRVL
9     FORMAT ('0', E8.3 , ' SECONDS INTERVAL:' /1X, 25 (1H-))
      IF (M .GT. 0)  WRITE(6,6) (LBL,ANSWER(I),LBL,ANSWER(I+1),LBL,
     A  ANSWER(I+2), I=1,L,3)
      IF (K .EQ. 3*M + 2)  WRITE(6,6) LBL,ANSWER(K-1),LBL,ANSWER(K)
      IF (K .EQ. 3*M + 1) WRITE(6,6) LBL,ANSWER(K)
      INTRVL = TIME*K
      CALL MS(K,ANSWER,TIME,SQUARE,INTRVL)
      VR = ANSWER(1) - VDC**2
      R = RATIO(VDC, VR)
      VRMS = SQRT(ANSWER(1))
      WRITE(6,7) VR, R, VRMS
      READ(5,10) JMP
10    FORMAT (I1)
      GO TO (8,11), JMP
11    STOP
C     THIS PROGRAM WRITTEN BY JAMES VANDENDORPE. IIT/1971
      END
```

```
      SUBROUTINE STAT(N,Y,TIME,X,IDUMMY)
C          DEPENDING UPON WHERE THE ROUTINE IS ENTERED, THE ROOT-MEAN-
C     SQUARE, MEAN-SQUARE, OR MEAN IS CALCULATED.
C          THE FOLLOWING IS A DIRECTORY OF IDENTIFIERS:
C          N          --> THE NUMBER OF POINTS.
C          Y          --> AN ARRAY OF POINTS TO BE SQUARED.
C          X          --> AN ARRAY OF POINTS WHOSE MEAN IS TO BE FOUND.
C          TIME       --> THE TIME BETWEEN MEASUREMENTS OF THE POINTS.
C          INTRVL     --> THE ELAPSE TIME BETWEEN CALCULATIONS OF THE MEAN.
C          ANSWER     --> AN ARRAY OF MEAN VALUES.
C
      DIMENSION X(500), Y(500)
      LOGICAL GETRMS
      COMMON ANSWER(250), GET RMS
      EQUIVALENCE (RMS, ANSWER(1))
      REAL INTRVL
      :
C     V : : : : : : : : : : : : CALCULATE ROOT MEAN SQUARE  :
      INTRVL = TIME*N
      GET RMS = .TRUE.
      :
C     V : : : : : : : : : : : : CALCULATE MEAN SQUARE  : : :
      ENTRY MS (N,Y,TIME,X,INTRVL)
      DO 10 I=1,N
10    X(I) = Y(I) * Y(I)
      :
C     V : : : : : : : : : : : : CALCULATE MEAN  : : : : :
      ENTRY MEAN (N,X,TIME,DUMMY,INTRVL)
      INTRVL = INTRVL - .5*TIME
      SUM = 0.
      CLOCK = 0.
      CNTR =0.
      J = 1
      DO 20  I = 1,N
      SUM = SUM + X(I)
      CLOCK = CLOCK + TIME
      CNTR = CNTR + 1.
      IF (INTRVL - CLOCK) 15, 15, 20
15    ANSWER(J) = SUM / CNTR
      J=J+1
      SUM = 0.
      CLOCK = 0.
      CNTR =0.
20    CONTINUE
      ANSWER(J + 1) = 0.
      IF (GET RMS) RMS = SQRT (ANSWER(1))
      GET RMS = .FALSE.
      INTRVL = INTRVL + .5*TIME
      RETURN
      END
```

FIGURE A2-9 *Subroutine for Laboratory Exercise 7-B.*

FIGURE A2-10 *(a) Main program and (b) subroutine for Laboratory Exercise 12-B, The Digital Filter.*

```
C                    D I G I T A L    F I L T E R S
C           ARRANGE THE DATA  ONE DATUM PER CARD. DECIMAL POINTS SHOULD
C       BE INSERTED WHEN THEY ARE REQUIRED AND ONE SHOULD ALWAYS BEGIN
C       PUNCHING DATA IN COLUMN ONE. THE FIRST DATA CARD MUST BE AN
C       INTEGER, 'N', WHICH DESCRIBES THE NUMBER OF FREQUENCIES TO BE
C       INPUT TO THE FILTERS. FOLLOWING THIS VALUE ARE 'N' FREQUENCIES,
C       'W'.
C           THE LANGUAGE USED IS STANDARD ANSI FORTRAN.
C
        DIMENSION W(50)
        INTEGER LO PASS, DIFF, FILTR
        COMMON LO PASS, DIFF, FILTR
        LO PASS = 1
        DIFF = 0
        READ(5,1) N, (W(I),I=1,N)
1       FORMAT (I2/(F3.1))
       :
C                                        LOW PASS FILTER
        WRITE(6,2)
2       FORMAT('1TABLE OF INPUT FREQUENCY IN HZ. VERSUS RMS OUTPUT SIGNAL'
       A    , ' IN VOLTS FOR A LOW PASS FILTER.'//50X, 'FREQUENCY',11X,
       B    'OUTPUT')
        FILTR = LO PASS
        DO 3 I=1,N
        VRMS = FILTER(W(I))
3       WRITE(6,4) W(I), VRMS
4       FORMAT (52X, F5.2, 12X, F7.3)
       :
C                                        DIFFERENTIAL FILTER
        WRITE(6,5)
5       FORMAT ('1TABLE OF INPUT FREQUENCY IN HZ. VERSUS RMS OUTPUT ',
       A    'SIGNAL IN VOLTS FOR A DIFFERENTIATING FILTER.'//50X,
       B    'FREQUENCY', 11X, 'OUTPUT')
        FILTR = DIFF
        DO 6 I=1,N
        VRMS = FILTER(W(I))
6       WRITE(6,4) W(I), VRMS
       :
C                                        SQUARE WAVE INPUTED
        WRITE(6,7)
7       FORMAT ('1      TABLE OF TIME (IN SECONDS) VERSUS DIGITAL OUTPUT ',
       A    'SIGNAL IN VOLTS RESULTING FROM A TWO VOLT PEAK-TO-PEAK, ONE ',
       B    'HZ., SQUARE'/1X, 'WAVE INPUT SIGNAL.'//50X,'TIME',16X,
       C    'OUTPUT')
        T = 0.
        DO 8 I=1,50
        VO = SQ TEST(1., T)
        WRITE(6,4) T, VO
8       T = T + .02
        STOP
        END
```

(*a*)

```
        FUNCTION SQ WAVE(W, T)
C           THIS STEP FUNCTION GIVES THE OUTPUT AT AT TIME T OF A 2 VOLT
C       PEAK-TO-PEAK SQUARE WAVE GENERATOR SET TO FREQUENCY W.
C
        PERIOD = 1./W
        HALF = .5*PERIOD
        ACCRET = .001*HALF
C ----THE NEXT EXPRESSION IS, 'T MODULLUS PERIOD'.
        DELTA = T - IFIX(T/PERIOD)*PERIOD
        DELTA = IFIX((DELTA - HALF)/ACCRET)
        IF (DELTA .LT. -HALF/ACCRET)  DELTA = ABS(DELTA)
        IF (DELTA) 1,3,2
1       SQ WAVE = +1.0
        IF (IFIX(DELTA + HALF/ACCRET)) 4,3,4
2       SQ WAVE = -1.0
        IF (IFIX(DELTA - HALF/ACCRET)) 4,3,4
3       SQ WAVE = 0.0
4       RETURN
C       THIS PROGRAM WRITTEN BY JAMES VANDENDORPE. IIT/1971
        END
```

(*b*)

```
      FUNCTION FILTER(W)
C     SELECT    TIMES, T1 AND T2, SUCH THAT
C
C                               SIN(W*T1) = 0.
C     AND                       SIN(W*T2) = 1.
C
C         THE DIGITAL OUTPUT VO(T1) AND VO(T2) IS THEN COMPUTED FROM
C     THESE INSTANTANEOUS VALUES. FINALLY, THE RMS VOLTAGE IS GIVEN BY
C
C                   VRMS = SQRT(.5(VO(T1)**2 + VO(T2)**2)) .
C
      INTEGER LO PASS, DIFF, FILTR
      COMMON LO PASS, DIFF, FILTR
      DIMENSION V(2,10), VO(2)
C ----SAMPLING INTERVAL BETWEEN V1 AND V2 IS .01 SECONDS.
      REAL INTRVL
      INTRVL = .01
      T = 0.
      DO 1 I=1,10
      T = T - INTRVL
1     V(1,I) = SIN(W*T*6.28318)
      T = 1. / (4.*W)
      DO 2 I=1,10
      T = T - INTRVL
2     V(2,I) = SIN(W*T*6.28318)
      DO 3 I=1,2
      IF (FILTR .EQ. LO PASS)
     A   VO(I) = .12*(V(I,1) + V(I,10))
     B           -.044*(V(I,2) + V(I,9))
     C           +.127*(V(I,3) + V(I,8))
     D          -.327*(V(I,4) + V(I,7))
     E           +.90*(V(I,5) + V(I,6))
      IF (FILTR .EQ. DIFF)
     A   VO(I) = .0008*(V(I,1) - V(I,10))
     B           -.0025*(V(I,2) - V(I,9))
     C           +.0094*(V(I,3) - V(I,8))
     D           -.035*(V(I,4) - V(I,7))
     E           +.189*(V(I,5) - V(I,6))
3     CONTINUE
      FILTER = SQRT(.5*(VO(1)**2 + VO(2)**2))
      RETURN
C
C         THIS ENTRY POINT SIMULATES THE OUTPUT AT SOME TIME T OF A
C     DIFFERENTIAL FILTER. THE FILTERS INPUT IS A 2 VOLT PEAK-TO-PEAK
C     SQUARE WAVE OF FREQUENCY W.
C
      ENTRY SQ TEST(W, T)
      TAU = T
      DO 4 I=1,10
      TAU = TAU - INTRVL
4     V(1,I) = SQ WAVE(W, TAU)
      FILTER = .0008*(V(1,1) - V(1,10))
     A         -.0025*(V(1,2) - V(1,9))
     B         +.0094*(V(1,3) - V(1,8))
     C         -.035*(V(1,4) - V(1,7))
     D         +.189*(V(1,5) - V(1,6))
      RETURN
C     THIS PROGRAM WRITTEN BY JAMES VANDENDORPE. IIT/1971
      END
```

FIGURE A2-11 *Subroutine for Laboratory Exercise 12-B.*

INDEX